J. L. Lions · E. Magenes

Non-Homogeneous Boundary Value Problems and Applications

Translated from the French by
P. Kenneth

Volume I

Springer-Verlag Berlin Heidelberg New York 1972

J. L. Lions
University of Paris

E. Magenes
University of Pavia

Title of the French Original Edition:
Problèmes aux limites non homogènes et applications (tome I)
Publisher: S. A. Dunod, Paris 1968

Translator:
P. Kenneth
Paris

Geschäftsführende Herausgeber:
B. Eckmann
Eidgenössische Technische Hochschule Zürich

B. L. van der Waerden
Mathematisches Institut der Universität Zürich

AMS Subject Classifications (1970)
Primary 35J20, 35J25, 35J30, 35J35, 35J40, 35K20, 35K35, 35L20
Secondary 46E35

ISBN 3-540-05363-8 Springer-Verlag Berlin Heidelberg New York
ISBN 0-387-05363-8 Springer-Verlag New York Heidelberg Berlin

Library of Congress Catalog Card Number 71-151407

Die Grundlehren der mathematischen Wissenschaften

in Einzeldarstellungen
mit besonderer Berücksichtigung
der Anwendungsgebiete

Band 181

Preface

1. We describe, at first in a very formal manner, our essential aim. Let $\mathcal{O}$ be an open subset of $\mathbf{R}^n$, with boundary $\partial\mathcal{O}$. In $\mathcal{O}$ and on $\partial\mathcal{O}$ we introduce, respectively, linear differential operators

$$P \text{ and } Q_j, \quad 0 \leqq j \leqq \nu.$$

By "non-homogeneous boundary value problem" we mean a problem of the following type: let f and g_j, $0 \leqq j \leqq \nu$, be given in function spaces F and G_j, F being a space "on $\mathcal{O}$" and the G_j's spaces "on $\partial\mathcal{O}$"; we seek u in a function space $\mathcal{U}$ "on $\mathcal{O}$" satisfying

(1) $$Pu = f \text{ in } \mathcal{O},$$

(2) $$Q_j u = g_j \text{ on } \partial\mathcal{O}, \quad 0 \leqq j \leqq \nu^{((1))}.$$

Q_j may be identically zero on part of $\partial\mathcal{O}$, so that the number of boundary conditions may depend on the part of $\partial\mathcal{O}$ considered[2].

We take as "working hypothesis" that, for $f \in F$ and $g_j \in G_j$, the problem (1), (2) admits a unique solution $u \in \mathcal{U}$, which depends continuously on the data[3].

But for *all linear* problems, there is a large number of *choices* for the spaces $\mathcal{U}$ and $\{F; G_j\}$ (naturally linked together).

Generally speaking, our aim is to determine families of spaces $\mathcal{U}$ and $\{F; G_j\}$, associated in a "natural" way with problem (1), (2) and convenient for applications, and also *all* possible choices for $\mathcal{U}$ and $\{F; G_j\}$ in these families.

Let us make this explicit by means of two examples, chosen as the simplest possible ones, but which already demonstrate the utility of non-homogeneous problems.

((1)) The Q_j's will be called "boundary operators". Such problems are called *non-homogeneous* because if we consider the setting of *unbounded operators*, then $Pu = f$, $u \in D(P)$ (= domain of P) implies null boundary conditions; hence a certain difference between f and the boundary data g_j.

[2] This will obviously be the case for most problems of evolution.

[3] At least in general; for elliptic problems uniqueness conditions will not be satisfied, but in any case we shall deal with operators *with indices* and therefore still have uniqueness on passing to the quotient by finite-dimensional subspaces. We shall verify that this "working hypothesis" is satisfied in each particular situation.

2. Examples

2.1 For $\mathcal{O}$, we take an open subset Ω of $\mathbf{R}^n$, with boundary Γ, and for P the operator Δ, Δ = Laplacian; we take $\nu = 0$ and Q_0 = identity. Then, the problem corresponding to (1), (2) is the classical *Dirichlet problem for* Δ:

$$\Delta u = f \text{ in } \Omega, \tag{3}$$

$$u = g_0 = g \text{ on } \Gamma. \tag{4}$$

We then ask *in what spaces f and g may be chosen so that* (3), (4) *admits a unique solution* (*in an appropriate sense*).

Classical answers are furnished by potential theory and the "Dirichlet principle": for example, we may choose f, g and certain of their derivatives in Ω and Γ respectively to be square integrable and obtain u, the solution of (3), (4), as well as certain of its derivatives to be square integrable in Ω.

Therefore, a *natural* family for problem (3), (4) must be (if we limit ourselves to "L^2 theory", that is Hilbert theory) the *Hilbert family of the Sobolev spaces* $H^s(\Omega)$ and $H^s(\Gamma)$, where $H^s(\Omega)$ (resp. $H^s(\Gamma)$) is, if s is an integer $\geqq 0$, the space of u's such that u and its derivatives (in the sense of distributions) up to order s are square integrable in Ω (resp. Γ) (this definition is generalized to all *real* s by introducing the derivative of order s by Fourier transform; see Chapter 1).

Here is one of the results we shall prove (see Chapter 2):

Let s be any real, non-negative number; if $f \in H^s(\Omega)$ and $g \in H^{s+3/2}(\Gamma)$, then there exists a unique $u \in H^{s+2}(\Omega)$ which is a solution of (3), (4) (*having given an appropriate sense to* (3) *and* (4) *separately, by a natural generalization of the classical definitions*). *More precisely*: *the operator $u \to \{\Delta u, u|_\Gamma\}$ is an isomorphism of*

$$H^{s+2}(\Omega) \quad \text{onto} \quad H^s(\Omega) \times H^{s+3/2}(\Gamma).$$

It must be pointed out that the derivatives of non-integer order *necessarily* enter the problem if we want the optimal result for each s, since $s + 2$ and $s + \frac{3}{2}$ cannot be integers simultaneously!

Furthermore, it is equally natural to study the case "negative s", since many problems deal with (3), (4), where, for example, with $f = 0$, g is *very irregular*: such as g square integrable on Γ (this is the case for *optimal control theory*), or g = the Dirac-mass at the point $x_0 \in \Gamma$ (then the solution u yields the Poisson kernel of the problem) or more generally g = an arbitrary distribution on Γ.

It is still possible to solve problem (3), (4) when s is a *negative real* number, the spaces $\mathcal{U}$ and G remaining of the same type (i.e. $\mathcal{U} = H^{s+2}(\Omega)$, $G = H^{s+3/2}(\Gamma)$, but with negative s) and the space F being an appropriate subspace $\Xi^s(\Omega)$ of $H^s(\Omega)$ consisting of elements which do not grow too rapidly "in the neighborhood of Γ" (see Chapter 2, Sections 6 and 7).

It follows that (3), (4) is solvable with g an *arbitrary distribution* on Γ, since then g necessarily belongs to a space $H^s(\Gamma)$, for an appropriate s.

In fact, in volume 3 of this book, we shall see that g may belong to the space of analytic functionals on Γ (and this space is the most general for which, at least for $f = 0$, it is possible to *give meaning to problem* (3), (4)).

2.2 As a second example, we consider the *heat operator*

$$P = \frac{\partial}{\partial t} - \Delta_x$$

in

$$\mathcal{O} = \Omega \times]0, T[\subset \mathbf{R}^{n+1};$$

the part of the boundary $\partial\mathcal{O}$ on which boundary conditions are given splits up into

$$\bar{\Omega} \quad \text{and} \quad \Sigma = \Gamma \times]0, T[.$$

Then, a problem corresponding to (1), (2) is

$$\frac{\partial u}{\partial t} - \Delta_x u = f \quad \text{in} \quad \mathcal{O}, \tag{5}$$

$$u(x, 0) = u_0(x) \quad \text{in} \quad \Omega, \tag{6}$$

$$u = g \quad \text{on} \quad \Sigma. \tag{7}$$

One of our aims is to obtain the largest possible families of spaces for f, u_0 and g such that (5), (6), (7) admits a unique solution, in an appropriate sense. We shall see, in Chapter 4 of Volume 2, that a "natural" Hilbert family of spaces $\mathcal{U}$ for the solution is the family $H^{2s,s}(\mathcal{O})$, with s any real number, where $H^{2s,s}(\mathcal{O})$ is, if s is a non-negative integer, the space of u's such that u and its derivatives up to order s in t and up to order $2s$ in x are square integrable in $\mathcal{O}$. The family $H^{2s,s}(\mathcal{O})$ plays, for problem (5), (6), (7), an analogous role to the family $H^s(\Omega)$ for problem (3), (4). Also, remarks analogous to those we made for problem (3), (4) are valid for problem (5), (6), (7).

3. We now specify which are the principal systems $\{P, Q_j\}$ studied in this book.

3.1 We consider the case where P is an *elliptic* operator (denoted by A) and the Q_j's are normal boundary operators (denoted B_j), where A and B_j verify suitable ellipticity conditions (see Chapter 2).

3.2 We consider the case where

$$P = \frac{\partial}{\partial t} + A,$$

a parabolic operator with suitable boundary conditions (Chapters 3 and 4).

3.3 We also consider the cases

$$P = \frac{\partial^2}{\partial t^2} + A$$

and

$$P = \frac{\partial}{\partial t} + \mathrm{i}\, A,$$

where A is a self-adjoint elliptic operator, and still with suitable boundary conditions (Chapters 3 and 5).

4. For all these problems, we *proceed systematically as follows* (except for possibly different techniques):

(i) we study the *regularity* of problem (1), (2), i.e.: assuming the data f and g_j to be *regular* (in a sense to be specified), we study the *corresponding regularity of* u;

(ii) by *transposition* of (i) (for the "adjoint problem") we deduce therefrom (with a suitable technique and in particular the obtainment of "*trace theorems*") the solution of problem (1), (2) for data belonging to spaces of distributions;

(iii) by *interpolation* between (i) and (ii), we obtain "*intermediate*" results.

Of course, the systematic setting-up of such a program is an enormous task and many possibilities had to be put aside (we have formulated them in lists of problems in the last sections of each chapter).

In general, we consider for (i):

in volumes 1 and 2: data which are finitely often *differentiable* in the sense of L^2 (spaces such as $H^s(\Omega)$, $H^{2s,s}(\mathcal{O})$, $H^s(\Gamma)$, ...)

in volume 3: *analytic data or data belonging to suitable Gevrey classes.*

5. As we have seen, the present volume depends on regularity theorems in "*differentiable in the sense of* L^2" spaces (*Sobolev spaces*).

Therefore, the *basic tools* are:

— Sobolev spaces constructed on L^2,

— the theory of interpolation of corresponding spaces.

This is the subject of Chapter 1, where we study interpolation only *for the Hilbert cases*; the introduction of interpolation between (non-

"hilbertizable") Banach spaces and its applications to Sobolev spaces constructed on L^p, $p \neq 2$, would have complicated this work considerably.

Once in possession of these tools we need to prove regularity theorems (stage (i)) and then to implement stages (ii) and (iii).

This is done for the situations described in Section 3, above. Let us be more precise.

The elliptic case is the subject of Chapter 2. Stage (i) is studied completely by the method of J. Peetre [*2*], under the hypotheses that A is *properly elliptic* and that the B_j's *cover* A in the sense of Lopatinskii-Shapiro and Agmon-Douglis-Nirenberg.

Stages (ii) and (iii) follow our previous papers on these subjects: see Lions-Magenes [*1*], [*2*] and [*3*] (where we also study the L^p case, for $1 < p < \infty$, which we disregard here for $p \neq 2$).

"Variational" evolution operators and their applications are studied in Chapter 3.

Partial differential equations of evolution are studied in more detail in Chapters 4 and 5 of Volume 2 and applications to optimal control theory in Chapter 6 of Volume 2. Other applications will be given in Volume 3.

6. We have made an effort to make the book readable in "local" fashion; indications about the logical relations between the different subjects are given at the beginning of each chapter.

7. Each chapter ends with a section of *comments* and a section of *problems*.

The comments give bibliographical indications, which, although numerous, by no means cover the subject. This is especially the case for research work cited in the comments but not studied in this book.

The rather large number of problems to which we call attention are very unequal in difficulty. For cases where *results of type* (i) *are already available*, the execution of stages (ii) and (iii) may offer great technical difficulties if one looks for optimal results, but is certainly much more accessible if one is satisfied with results in the "neighborhood" of the optimal results. Situations for which the results of type (i) are lacking (and we indicate a number of such problems) may, of course, be much more difficult.

We wish to thank C. Baiocchi, M. S. Baouendi and G. Geymonat for reading various parts of the manuscript and giving us their comments.

Paris/Pavia, July 1967

J. L. LIONS E. MAGENES

Preface to the English Translation

The present translation follows the French edition without change, except for some corrections which were suggested to us by the remarks of M. S. Baouendi, G. Geymonat, C. Goulaouic and P. Schapira, to all of whom we express our sincerest thanks. We have added a complementary bibliography. We also wish to thank P. Kenneth for his excellent work of translation.

Paris/Pavia, October 1971

J. L. LIONS   E. MAGENES

Contents

Chapter 1

Hilbert Theory of Trace and Interpolation Spaces

Chapter 2

Elliptic Operators. Hilbert Theory

Chapter 3

Variational Evolution Equations

Contents of Volume II

Contents of Volume III

Chapter 1

Hilbert Theory of Trace and Interpolation Spaces

The aim of this chapter is to give the fundamental results of the theory of trace and interpolation spaces in the *Hilbert case*. Some indications about the possible generalizations to the non-Hilbert case are given in the Comments and in Section 14 along with the basic literature. Sections 10, 12 (except 12.1), 13–15, although used in the sequel, may be skipped on first reading.

1. Some Function Spaces

1.1 Sobolev Spaces

Let Ω be an *arbitrary* open set in $\mathbf{R}^n$; $x = \{x_1, \ldots, x_n\} \in \Omega$, $dx = dx_1, \ldots, dx_n$. We denote by $L^2(\Omega)$ the space of (classes of) functions u which are square integrable on Ω, i.e. measurable and such that

$$(1.1) \qquad \|u\|_{L^2(\Omega)} = \left(\int_\Omega |u|^2 \, dx \right)^{1/2} < \infty .$$

We shall often set

$$(1.2) \qquad L^2(\Omega) = H^0(\Omega).$$

It is a classical result that $L^2(\Omega)$ is a *Hilbert space* for the scalar product

$$(u, v)_{L^2(\Omega)} = \int_\Omega u(x)\, \overline{v(x)}\, dx$$

associated to the norm (1.1). □[1]

Let m be an integer $\geqq 1$. In short, *the Sobolev space $H^m(\Omega)$ of order m on Ω* is defined by

$$(1.3) \qquad H^m(\Omega) = \{u \mid D^\alpha u \in L^2(\Omega)\ \forall \alpha, |\alpha| \leqq m\},$$

where

$$D^\alpha = \frac{\partial^{\alpha_1 + \cdots + \alpha_n}}{\partial x_1^{\alpha_1} \ldots \partial x_n^{\alpha_n}}, \qquad \alpha = \{\alpha_1, \ldots, \alpha_n\}, \qquad |\alpha| = \alpha_1 + \cdots + \alpha_n.$$

[1] The symbol □ will be used throughout this text to indicate the end of a "logical unit".

It must be stated precisely *in what sense* $D^\alpha u$ in definition (1.3) is taken.

For this purpose, we *briefly* recall the definition of *distributions on* Ω (see Schwartz [1]).

We define

(1.4) $\mathscr{D}(\Omega) = \{\varphi \mid \varphi$ infinitely differentiable on Ω and with compact support in $\Omega\}$.

If K is a compact set in Ω, we set

$$\mathscr{D}_K(\Omega) = \{\varphi \mid \varphi \in \mathscr{D}(\Omega),\ \varphi \text{ with support in } K\}.$$

With the norms

$$p_j(\varphi) = \sup_{\substack{x \in K \\ |\alpha| \leq j}} |D^\alpha \varphi(x)|, \quad j = 0, 1, \ldots$$

$\mathscr{D}_K(\Omega)$ is a *Fréchet-space* (i.e. metrisable and complete); then if K_n is an increasing sequence of compact sets belonging to Ω and whose union is Ω, we have *algebraically*

$$\mathscr{D}(\Omega) = \bigcup_{K_n} \mathscr{D}_{K_n}(\Omega) \tag{1.5}$$

and we provide $\mathscr{D}(\Omega)$ *with the corresponding inductive limit topology* (i.e. the finest locally convex topology which makes the injections $\mathscr{D}_{K_n}(\Omega) \to \to \mathscr{D}(\Omega)$ continuous; see Schwartz [1], Horvath [1], in particular p. 165 and p. 171, where the explicit definition of a fundamental system of neighborhoods of zero is given). □

Remark 1.1. Other spaces of infinitely differentiable functions will be introduced in the following volumes. □

We *define*

(1.6) $\mathscr{D}'(\Omega)$ = dual of $\mathscr{D}(\Omega)$ = space of distributions on Ω

and provide $\mathscr{D}'(\Omega)$ with the strong dual topology.

We refer the reader to Schwartz [1] for *structure theorems* pertaining to $\mathscr{D}'(\Omega)$. □

Remark on the notation.

If $T \in \mathscr{D}'(\Omega)$ and $\varphi \in \mathscr{D}(\Omega)$, the value of T at φ will be denoted by

$$\langle T, \varphi \rangle.$$

If $\bar{\varphi}$ = complex conjugate of φ, we shall write

$$\langle T, \bar{\varphi} \rangle = (T, \varphi).$$

If $T \in L^2(\Omega)$, (T, φ) coincides with $(T, \varphi)_{L^2(\Omega)}$.

Throughout this book, $\langle \alpha, \beta \rangle$ denotes a *bilinear* couple, so that $\langle \alpha, \bar{\beta} \rangle$ is a *sesquilinear* couple (linear in α, antilinear in β), *sesquilinear couples* being generally denoted by (α, β) (with the particular notation $[\alpha, \beta]$ in Chapter 3). □

If $T \in \mathscr{D}'(\Omega)$, its derivative $\partial T/\partial x_j$ is defined by

$$\left\langle \frac{\partial T}{\partial x_j}, \varphi \right\rangle = -\left\langle T, \frac{\partial \varphi}{\partial x_j} \right\rangle \quad \forall \varphi \in \mathscr{D}(\Omega), \tag{1.7}$$

which yields a linear continuous mapping

$$T \to \frac{\partial T}{\partial x_j}$$

of $\mathscr{D}'(\Omega) \to \mathscr{D}'(\Omega)$. Of course, $D^\alpha T$ is defined by iteration.

We are now in a position to give a precise statement of definition (1.3), if we note that

$$\mathscr{D}(\Omega) \subset L^2(\Omega) \subset \mathscr{D}'(\Omega) \tag{1.8}$$

by identifying (which is permissible) every element $u \in L^2(\Omega)$ with the distribution

$$\varphi \to \langle u, \varphi \rangle.$$

Then, *the derivatives* $D^\alpha u$ *in* (1.3) *are taken in the sense of distributions on* Ω.

We provide $H^m(\Omega)$ with the norm:

$$\| u \|_{H^m(\Omega)} = \left(\sum_{|\alpha| \leqq m} \| D^\alpha u \|^2_{L^2(\Omega)} \right)^{1/2} \tag{1.9}$$

and obtain

Theorem 1.1. *With the norm* (1.9), $H^m(\Omega)$ *is a Hilbert space, the scalar product of two elements* $u, v \in H^m(\Omega)$ *being given by*

$$(u, v)_{H_m(\Omega)} = \sum_{|\alpha| \leqq m} (D^\alpha u, D^\alpha v)_{L^2(\Omega)}. \tag{1.10}$$

Proof. It is sufficient to verify that $H^m(\Omega)$ is *complete* in the norm (1.9). Let u_k be a Cauchy sequence for this norm. It follows that, for every α with $|\alpha| \leqq m$, $D^\alpha u_k$ is a Cauchy sequence *in* $L^2(\Omega)$. Since $L^2(\Omega)$ is complete, we have

$$D^\alpha u_k \to \psi_\alpha \quad \text{in} \quad L^2(\Omega) \quad \forall \alpha, \quad |\alpha| \leqq m.$$

Set $\psi_0 = u$; since $u_k \to u$ in $L^2(\Omega)$, we have, in particular, $u_k \to u$ in $\mathscr{D}'(\Omega)$ and since the derivative operator is continuous *in the sense of* $\mathscr{D}'(\Omega)$ we have

$$D^\alpha u_k \to D^\alpha u \quad \text{in} \quad \mathscr{D}'(\Omega).$$

Therefore $\psi_\alpha = D^\alpha u$ and $u \in H^m(\Omega)$. □

Remark 1.2. If $m_1 > m > 0$, we have the *strict* inclusions

$$H^{m_1}(\Omega) \subset H^m(\Omega) \subset L^2(\Omega) = H^0(\Omega). \quad □$$

1.2 The Case of the Entire Space

In the particular case $\Omega = \mathbf{R}^n$, it is possible to give – and this is important for the sequel – an equivalent definition of $H^m(\Omega)$, by making use of the *Fourier transform*. If $u \in L^2(\mathbf{R}^n)$, the Fourier transform $\hat{u}$ in $L^2(\mathbf{R}^n)$ is defined by

$$\hat{u}(y) = \frac{1}{(2\pi)^{n/2}} \int_{\mathbf{R}^n} \exp(-i x y)\, u(x)\, dx,$$

$$x y = x_1 y_1 + \cdots + x_n y_n,$$

the integral converging in the sense of L^2 and $u \to \hat{u}$ is an isomorphism of $L^2(\mathbf{R}^n)$ onto $L^2(\mathbf{R}^n)$. We set

$$\hat{u} = \mathscr{F} u$$

and

$$u = \overline{\mathscr{F}}\, \hat{u} = \frac{1}{(2\pi)^{n/2}} \int_{\mathbf{R}^n} \exp(i x y)\, \hat{u}(y)\, dy.$$

The Fourier transform extends by continuity to the space $\mathscr{S}'$ of Schwartz's tempered distributions, whose definition we now recall.

First of all, we define

$$\mathscr{S} = \{u \mid x^\alpha D^\beta u \in L^2(\mathbf{R}^n)\ \forall \alpha \text{ and } \forall \beta\}$$

$$(\text{where } x^\alpha = x_1^{\alpha_1} \ldots x_n^{\alpha_n}).$$

With the sequence of semi-norms

$$u \to \| x^\alpha D^\beta u \|_{L^2(\mathbf{R}^n)},$$

$\mathscr{S}$ is a Frechet space; of course every $u \in \mathscr{S}$ is (a.e. equal to a function) infinitely differentiable in $\mathbf{R}^n$ and every $u \in \mathscr{S}$ is *rapidly decreasing* at infinity:

$$\forall \alpha, \forall \beta, \quad |x|^\alpha D^\beta u(x) \to 0 \text{ if } |x| \to \infty$$

(*equivalent* property to the above definition).

We easily verify that

$$\begin{cases} \mathscr{F}(D^\alpha u) = (i y)^\alpha \mathscr{F} u & \forall u \in \mathscr{S}, \quad \forall \alpha \\ D^\beta \mathscr{F}(u) = \mathscr{F}((-i x)^\beta u) & \forall u \in \mathscr{S}, \quad \forall \beta, \end{cases} \tag{1.11}$$

and therefore that $\mathscr{F} \in \mathscr{L}(\mathscr{S}; \mathscr{S})$ ((1)). In the same way,

$$\overline{\mathscr{F}} \in \mathscr{L}(\mathscr{S}; \mathscr{S}) \text{ and } \mathscr{F}\, \overline{\mathscr{F}}\, u = \overline{\mathscr{F}}\, \mathscr{F} u = u, \quad \forall u \in \mathscr{S}.$$

Therefore $\mathscr{F}$ is an isomorphism of $\mathscr{S}$ onto itself, with inverse $\overline{\mathscr{F}}$.

((1)) We recall that if Φ and Ψ are two topological vector spaces, $\mathscr{L}(\Phi; \Psi)$ is defined to be the space of linear continuous mappings of $\Phi \to \Psi$.

Because of the symmetry of the kernel

$$\frac{1}{(2\pi)^{n/2}} \exp(-\mathrm{i} x y)$$

of $\mathscr{F}$, we have

$$\int_{\mathbf{R}^n} (\mathscr{F} u)\, v\, dx = \int_{\mathbf{R}^n} u(\mathscr{F} v)\, dx \qquad \forall u, \quad v \in \mathscr{S}.$$

Next, we define

$$\mathscr{S}' = \text{dual space of } \mathscr{S}, \quad \text{with the strong dual topology}$$

and we define

$$\mathscr{F}, \overline{\mathscr{F}} \in \mathscr{L}(\mathscr{S}'; \mathscr{S}') \quad \textit{by transposition.}$$

Thus, $\forall u \in \mathscr{S}'$, we have

$$\langle \mathscr{F} u, \varphi \rangle = \langle u, \mathscr{F} \varphi \rangle \qquad \forall \varphi \in \mathscr{S}$$

where the brackets denote the duality between $\mathscr{S}'$ and $\mathscr{S}$.

The formulas (1.11) are still valid $\forall u \in \mathscr{S}'$.

Theorem 1.2. *If* $\Omega = \mathbf{R}^n$, $H^m(\mathbf{R}^n)$ *may be defined by* (1.3) *or by*

$$H^m(\mathbf{R}^n) = \{u \mid u \in \mathscr{S}', (1 + |y|^2)^{m/2}\, \hat{u} \in L^2(\mathbf{R}^n)\} \tag{1.12}$$

$$(\text{where } |y|^2 = y_1^2 + \cdots + y_n^2),$$

the norm

$$|\!|\!| u |\!|\!|_{H^m(\mathbf{R}^n)} = \|(1 + |y|^2)^{m/2}\, \hat{u}\|_{L^2(\mathbf{R}^n)} \tag{1.13}$$

being equivalent to the norm (1.9).

Proof. From (1.11) and Plancherel's theorem

$$\|D^\alpha u\|_{L^2(\mathbf{R}^n)} = \|y^\alpha\, \hat{u}\|_{L^2(\mathbf{R}^n)},$$

so that (1.9) yields (for $\Omega = \mathbf{R}^n$)

$$\|u\|^2_{H^m(\mathbf{R}^n)} = \int_{\mathbf{R}^n} \Big(\sum_{|\alpha| \leqq m} y^{2\alpha} \Big) |\hat{u}(y)|^2\, dy. \tag{1.14}$$

But for a suitable constant C:

$$(1 + |y|^2)^m \leqq \sum_{|\alpha| \leqq m} y^{2\alpha} \leqq C(1 + |y|^2)^m,$$

which, together with (1.13), (1.14) yields

$$|\!|\!| u |\!|\!|_{H^m(\mathbf{R}^n)} \leqq \|u\|_{H^m(\mathbf{R}^n)} \leqq C^{1/2} |\!|\!| u |\!|\!|_{H^m(\mathbf{R}^n)}. \quad \square$$

Remark 1.3. $\mathscr{D}(\mathbf{R}^n)$ is *dense* in $H^m(\mathbf{R}^n)$ (this is easily seen by regularization and truncation), but this is not true in general: on the contrary $\mathscr{D}(\Omega)$ is not generally dense in $H^m(\Omega)$. We shall return to this point later. $\square$

1.3 The Half-Space Case

In case Ω is the half-space $\{x \mid x_n > 0\}$, we introduce another equivalent definition of $H^m(\Omega)$. □

The space $L^2(a, b; X)$.

Let X be a Hilbert space. $L^2(a, b; X)$ denotes the space of (classes of) function f, strongly measurable on $[a, b]$ with range in X (for the Lebesgue measure dt on $[a, b]$) and such that

$$(1.15) \qquad \left(\int_a^b \|f(t)\|_X^2 \, dt \right)^{1/2} = \|f\|_{L^2(a,b;X)} < +\infty,$$

where $\| \ \|_X$ is the Hilbert norm of X.

With the norm (1.15), $L^2(a, b; X)$ is a *Hilbert space* (Bourbaki [1]). □

Distributions on $]a, b[$ *with range in* X.

We recall that, if Φ and Ψ are two topological vector spaces, we have set

$$(1.16) \qquad \mathscr{L}(\Phi; \Psi) = \text{space of linear continuous mappings of } \Phi \text{ into } \Psi$$

(the "adequate" topology on this space being defined in each particular case).

We call (following Schwartz [6]) *space of distributions on* $]a, b[$ *with range in* X, *denoted by* $\mathscr{D}'(]a, b[; X)$, *the space*

$$(1.17) \qquad \mathscr{D}'(]a, b[; X) = \mathscr{L}(\mathscr{D}(]a, b[); X),$$

provided with the topology of uniform convergence on the bounded sets of $\mathscr{D}(]a, b[)$.

Therefore, if $f \in \mathscr{D}'(]a, b[; X)$, then $\forall \varphi \in \mathscr{D}(]a, b[)$, $\langle f, \varphi \rangle$ (value of f at φ) is in X and $\varphi \to \langle f, \varphi \rangle$ is a continuous mapping of $\mathscr{D}(]a, b[)$ into X.

The *derivative* $\dfrac{df}{dt}$ of $f \in \mathscr{D}'(]a, b[; X)$ is defined as the unique element of this space which satisfies

$$(1.18) \qquad \left\langle \frac{df}{dt}, \varphi \right\rangle = -\left\langle f, \frac{d\varphi}{dt} \right\rangle \quad \forall \varphi \in \mathscr{D}(]a, b[),$$

(the equality (1.18) takes place *in the space* X).

The mapping

$$f \to \frac{df}{dt}$$

is a continuous mapping of $\mathscr{D}'(]a, b[; X)$ into itself. □

Remark 1.4. The fact that X is a *Hilbert space* plays *no* role in the preceding notions on distributions taking their values in X; later on (in volume 3), we shall work with much more general notions. But for the purposes of this chapter, the hypothesis that X is a Hilbert space is sufficient. □

Now, if $f \in L^2(a, b; X)$, we define

$$\tilde{f} \in \mathscr{D}'(]a, b[; X)$$

by

$$\langle \tilde{f}, \varphi \rangle = \int_a^b f(t)\, \varphi(t)\, dt \quad (\in X) \quad \forall \varphi \in \mathscr{D}(]a, b[), \tag{1.19}$$

so that we have a (continuous linear) mapping $f \to \tilde{f}$ of $L^2(a, b; X) \to$ $\to \mathscr{D}'(]a, b[; X)$. This mapping is *one-to-one*; and so we identify $\tilde{f}$ with f and obtain

$$L^2(a, b; X) \subset \mathscr{D}'(]a, b[; X). \tag{1.20}$$

Consequently:

Proposition 1.1. *For $f \in L^2(a, b; X)$, $\dfrac{df}{dt}, \dfrac{d^2 f}{dt^2}, \ldots$ may be defined as distributions on $]a, b[$ with range in X.* □

Remark 1.5. The following may be verified as an exercise (as in Theorem 1.1, but now using the continuity of the derivative mapping for vector distributions). Define:

$$H^m(a, b; X) = \left\{ f \mid f,\ f^{(1)} = \frac{df}{dt}, \ldots,\ f^{(m)} = \frac{d^m f}{dt^m} \in L^2(a, b; X) \right\}$$

with the scalar product

$$\sum_{j=0}^{m} \int_a^b (f^{(j)}(t), g^{(j)}(t))_X\, dt;$$

$H^m(a, b; X)$ is a Hilbert space. □

We are now ready to prove

Theorem 1.3. *When $\Omega = \{x \mid x_n > 0\}$, $H^m(\Omega)$ may be defined by (1.3) or by*

$$H^m(\Omega) = \Big\{ u \mid u \in L^2(0, \infty; H^m(\mathbf{R}^{n-1}_{x'})), \ldots, \tag{1.21}$$

$$\frac{\partial^j u}{\partial x_n^j} \in L^2(0, \infty; H^{m-j}(\mathbf{R}^{n-1}_{x'})), \ldots, \frac{\partial^m u}{\partial x_n^m} \in L^2(0, \infty; H^0(\mathbf{R}^{n-1}_{x'})) \Big\}$$

and

$$\| u \|^2_{H^m(\Omega)} = \sum_{j=0}^{m} \left\| \frac{\partial^j u}{\partial x_n^j} \right\|^2_{L^2(0, \infty; H^{m-j}(\mathbf{R}^{n-1}_{x'}))} \qquad (x' = \{x_1, \ldots, x_{n-1}\}).$$

Proof. 1) If $u \in L^2(0, \infty; H^m(\mathbf{R}^{n-1}_{x'}))$, then $\dfrac{\partial^j u}{\partial x_n^j}$ is defined in the sense of

$$\mathscr{D}'(]0, \infty[; H^m(\mathbf{R}^{n-1}_{x'}))$$

and therefore the conditions "$\dfrac{\partial^j u}{\partial x_j^n} \in L^2(0, \infty; H^{m-j}(\mathbf{R}^{n-1}_{x'}))$" *have meaning.*

2) If $u \in L^2(0, \infty; H^m(\mathbf{R}^{n-1}_{x'}))$, then since, for all $|\alpha| \leqq m$, $D^\alpha_{x'}$ is a continuous linear mapping of $H^m(\mathbf{R}^{n-1}_{x'})$ into $H^0(\mathbf{R}^{n-1}_{x'})$, we have

$$D^\alpha_{x'} u \in L^2(0, \infty; H^0(\mathbf{R}^{n-1}_{x'})) \qquad \forall |\alpha| \leqq m \tag{1.22}$$

and since by Fubini's theorem

$$L^2(0, \infty; H^0(\mathbf{R}^{n-1}_{x'})) = L^2(\Omega), \tag{1.23}$$

we have

$$D^\alpha_{x'} u \in L^2(\Omega) \quad \forall \alpha, \quad |\alpha| \leqq m. \tag{1.24}$$

Conversely, if u satisfies (1.24), then u satisfies (1.22); therefore, for almost all x_n, $u(\cdot, x_n) \in H^m(\mathbf{R}^{n-1}_{x'})$ and

$$\int_0^\infty \| u(., x_n) \|^2_{H^m(\mathbf{R}^{n-1}_{x'})} \, dx_n = \sum_{|\alpha| \leqq m} \int_\Omega |(D^\alpha_{x'} u)|^2 \, dx < \infty,$$

so that $u \in L^2(0, \infty; H^m(\mathbf{R}^{n-1}_{x'}))$ (since u is measurable and takes its values in $H^m(\mathbf{R}^{n-1}_{x'})$).

Therefore, the fact that $u \in L^2(0, \infty; H^m(\mathbf{R}^{n-1}_{x'}))$ is equivalent to (1.24).

3) Consequently

$$\frac{\partial^j u}{\partial x_n^j} \in L^2(0, \infty; H^{m-j}(\mathbf{R}^{n-1}_{x'})) \quad \forall j \Leftrightarrow D^\alpha_{x'} \frac{\partial^j u}{\partial x_n^j} \in L^2(\Omega) \quad \forall |\alpha| \leqq m - j,$$

$$\forall j \Leftrightarrow D^\alpha u \in L^2(\Omega) \quad \forall |\alpha| \leqq m,$$

from which the theorem follows. □

1.4 Orientation

Property (1.21) justifies the introduction of the notions of the next section, these notions playing an essential role in the following chapters. This will bring us to:

- ○ trace theorems and "intermediate spaces",
- ○ interpolation theory.

2. Intermediate Derivatives Theorem

2.1 Intermediate Spaces

Let X and Y be two Hilbert spaces which, in order to slightly simplify our account (and also because this will suffice for our purposes), we suppose to be *separable* with

(2.1) $X \subset Y$, X dense in Y with continuous injection. □

Remark 2.1. The space $H^m(\Omega)$, introduced in (1.3), is separable. Indeed, via the mapping

$$u \to \{D^\alpha u \mid |\alpha| \leqq m\}$$

it can be identified with a closed vector subspace of a product of $L^2(\Omega)$, this product space being separable because $L^2(\Omega)$ is. □

Let $(\ ,\ )_X$ and $(\ ,\ )_Y$ be the scalar products in X and Y respectively.

The operator Λ.

The space X may be defined as the domain of an operator Λ, which is self adjoint, positive and unbounded in Y (in fact, Λ is not unique!), the norm in X being *equivalent* to the *norm of the graph*

(2.2) $$(\|u\|_Y^2 + \|\Lambda u\|_Y^2)^{1/2}, \quad u \in D(\Lambda) = X.$$

The result is classical (see Riesz-Nagy [1]). Let us briefly recall a procedure (which, as a matter of fact, is linked to the variational formulation of elliptic boundary value problems, Chapter 2, Section 9).

We denote by $D(S)$ the set of u's such that the antilinear form

(2.3) $$v \to (u, v)_X, \quad v \in X$$

is continuous *in the topology induced by* Y. Then

(2.4) $$(u, v)_X = (S u, v)_Y,$$

which defines S as an unbounded operator in Y, with domain $D(S)$.

It is easy to see that:

$D(S)$ *is dense in* Y,

S *is a self-adjoint operator*, i.e. S coincides with S^* = adjoint of S (which also means that their domains coincide) and S is *strictly positive*; indeed,

$$(S v, v)_Y = \|v\|_X^2 \geqq (\text{constant}) \|v\|_Y^2.$$

(For the definition and the principal notions relative to self-adjoint operators we refer the reader to Stone [1], Riesz-Nagy [1], Yosida [2], among others). Using the *spectral decomposition* of self-adjoint operators, the *powers* S^θ of S, $\theta \in \mathbf{R}$ (or even $\theta \in \mathbf{C}$), may be defined (see the texts just cited and the reminders given in Section 2.3, below).

In particular, we shall use

$$\Lambda = S^{1/2}. \tag{2.5}$$

The operator Λ is self-adjoint and positive in Y, with domain X. From (2.4), (2.5) we deduce

$$(u, v)_X = (\Lambda u, \Lambda v)_Y \quad \forall u, v \in X. \quad \square \tag{2.6}$$

Remark 2.2. The operator S *depends* on the *choice* of the scalar products on X and Y (of course, without changing the topology of X and Y) and therefore Λ also depends on these scalar products; thus, it is not intrinsically linked to the *spaces* X and Y. □

We give the following definition of the *intermediate spaces* $[X, Y]_\theta$:

Definition 2.1. *Under hypothesis* (2.1) *and with* Λ *defined by* (2.5), *we set*

$$[X, Y]_\theta = D(\Lambda^{1-\theta}), \quad (\text{domain of } \Lambda^{1-\theta}), \quad 0 \leqq \theta \leqq 1, \tag{2.7}$$

with

(2.8) *norm on* $[X, Y]_\theta$ = *norm of the graph of* $\Lambda^{1-\theta}$, *i.e.*

$$(\|u\|_Y^2 + \|\Lambda^{1-\theta} u\|_Y^2)^{1/2}. \quad \square$$

From the properties of the spectral decomposition we immediately have that X *is dense in* $[X, Y]_\theta$. □

Remark 2.3. According to Remark 2.2, it is not obvious that the space $[X, Y]_\theta$ is *intrinsically* linked to X and Y; this will be shown in theorems 3.2 and 4.2, below. We shall obtain:

if Λ_1 and Λ_2 are two positive, self-adjoint operators in Y, with domain X, then

$$D(\Lambda_1^{1-\theta}) = D(\Lambda_2^{1-\theta}), \text{ with equivalent norms. } \square$$

Remark 2.4. We have:

$$[X, Y]_0 = X$$

$$[X, Y]_1 = Y. \quad \square$$

2.2 Density and Extension Theorems

The space $W(a, b)$.

Let a and b be two real numbers, finite or not, $a < b$.

Let X, Y be Hilbert spaces as in Section 2.1, and m an integer $\geqq 1$. We set

$$W(a, b) = \left\{ u \mid u \in L^2(a, b; X), \frac{d^m u}{dt^m} = u^{(m)} \in L^2(a, b; Y) \right\}, \tag{2.9}$$

where $u^{(m)}$ is taken in the sense of distributions in $\mathscr{D}'(]a, b[; X)$. Provided with the norm

$$\| u \|_{W(a,b)} = [\| u \|^2_{L^2(a,b;X)} + \| u^{(m)} \|^2_{L^2(a,b;Y)}]^{1/2} \tag{2.10}$$

$W(a, b)$ is a *Hilbert space* (it is *complete* in the norm (2.10) because of the continuity of differentiation in t in the sense of $\mathscr{D}'(]a, b[; X)$).

The space $\mathscr{D}([a, b]; X)$.

We denote by $\mathscr{D}([a, b]; X)$ the space of functions which are infinitely differentiable for $a \leqq t \leqq b$, with range in X and of compact support.

There are three cases:

1) $a = -\infty$, $b = +\infty$; then, the space coincides with $\mathscr{D}(\mathbf{R}; X)$;

2) a finite, $b = +\infty$; then, the functions of $\mathscr{D}([a, b]; X)$ vanish for sufficiently large t (the case $a = -\infty$, b finite, follows by symmetry);

3) a and b finite.

Density theorem:

Theorem 2.1. *The space* $\mathscr{D}([a, b]; X)$ *is dense in* $W(a,b)$.

Proof. First case: $a = -\infty$, $b = +\infty$.

Let $\{\varrho_n\}$ be a *regularizing sequence*:

$$\begin{cases} \varrho_n \in \mathscr{D}(\mathbf{R}), \quad \int_{+\infty}^{-\infty} \varrho_n(t)\, dt = 1, \quad \varrho_n(t) \geqq 0, \\ \varrho_n \text{ with support in } [\alpha_n, \beta_n], \ \alpha_n, \beta_n \to 0. \end{cases}$$

If $u \in W(-\infty, +\infty) = W(\mathbf{R})$, then

$$u * \varrho_n(t) = \int_{+\infty}^{-\infty} \varrho_n(t - \sigma)\, u(\sigma)\, d\sigma \to u \quad \text{in} \quad L^2(-\infty, +\infty; X),$$

$$(u * \varrho_n)^{(m)} = u^{(m)} * \varrho_n \to u^{(m)} \qquad \text{in} \quad L^2(-\infty, +\infty; Y).$$

Therefore, it is sufficient to approximate (in the sense of $W(a, b)$) the functions v of the form

$$v = u * \varphi, \qquad \varphi \in \mathscr{D}(\mathbf{R}). \tag{2.11}$$

by elements of $\mathscr{D}([a, b]; X)$.

We already have:

$$v^{(k)} = u * \varphi^{(k)} \quad \forall k,$$

therefore

$$v^{(k)} \in L^2(-\infty, +\infty; X) \quad \forall k, \tag{2.12}$$

so that, as is easily verified:

(2.13) v is an infinitely differentiable function of $\mathbf{R} \to X$.

It is now sufficient to *truncate*. For instance, let $\psi \in \mathcal{D}(\mathbf{R})$, $\psi(t) = 1$ for $|t| \leqq 1$, $\psi(t) = 0$ for $|t| \geqq 2$. Let ψ_N be defined by

$$\psi_N(t) = \psi\left(\frac{t}{N}\right). \tag{2.14}$$

Then, because of (2.13):

$$\psi_N v \in \mathcal{D}([a, b]; X). \tag{2.15}$$

We have:

$$\psi_N v \to v \quad \text{in} \quad L^2(-\infty, +\infty; X)$$

and therefore, we shall obtain the desired result if

$$(\psi_N v)^{(m)} \to v^{(m)} \quad \text{in} \quad L^2(-\infty, +\infty; X) \tag{2.16}$$

(therefore, in particular in $L^2(-\infty, +\infty; Y)$).

But $\psi_N v^{(m)} \to v^{(m)}$ in $L^2(-\infty, +\infty; X)$ and (2.16) follows from

$$\psi_N^{(k)} v^{(m-k)} \to 0 \quad \text{in} \quad L^2(-\infty, +\infty; X),\ N \to \infty,\ k \geqq 1;$$

but this last statement is an immediate consequence of the definition (2.14) of ψ_N. □

Second case: a finite, $b = +\infty$.

Of course, the situation is invariant by translation and we may assume $a = 0$.

Let $u \in W(0, \infty)$; for $h > 0$, we define

$$u_h(t) = u(t + h), \quad t > 0 \text{ (which has meaning e.a.).} \tag{2.17}$$

Since

$$u_h^{(m)}(t) = u^{(m)}(t + h) \quad \text{a.e. for } t > 0,$$

we see that

$$u_h \to u \quad \text{in} \quad W(0, \infty) \text{ as } h \to 0. \tag{2.18}$$

Therefore, it is sufficient to approach u_h (in the sense of $W(0, \infty)$) with elements of $\mathcal{D}([0, \infty]; X)$, with h *fixed*.

Consider a scalar function Φ, intinitely differentiable on $\mathbf{R}$, $\Phi(t) = 1$ if $t \geqq -h/2$, $\Phi(t) = 0$ if $t \leqq -h$, and let

$$v(t) = \begin{cases} \Phi(t)\, u(t + h) & \text{if} \quad t \geqq -h \\ 0 & \text{if} \quad t \leqq -h. \end{cases}$$

Then

$v = u_h$ a.e. on $t > 0$ (we did what was necessary for this)

and

$$v \in W(-\infty, +\infty).$$

From the first case, there exist $f_n \in \mathscr{D}(\mathbf{R}; X)$, with

$$f_n \to v \quad \text{in} \quad W(-\infty, +\infty).$$

If g_n = restriction of f_n to $[0, +\infty[$, we have:

$$g_n \to (\text{restriction of } v) = u_h \quad \text{in} \quad W(0, +\infty),$$
$$g_n \in \mathscr{D}([a, b]; X);$$

hence the desired result. □

Third case: a and b finite.

Let α and β be two scalar functions, with the properties

$$(2.19) \qquad \begin{cases} \alpha, \beta \in \mathscr{D}([a, b)], \ \alpha(t) + \beta(t) = 1 \quad \text{for} \quad a \leqq t \leqq b, \\ \alpha \,(\text{resp. } \beta) \text{ vanishing in the neighbourhood of } b \text{ (resp. } a). \end{cases}$$

Then every $u \in W(a, b)$ may be written

$$u = \alpha u + \beta u$$

and if we define

$$\widetilde{\alpha u} = \begin{cases} \alpha u & \text{for} \quad a \leqq t \leqq b \\ 0 & \text{for} \quad t > b \end{cases}, \qquad \widetilde{\beta u} = \begin{cases} \beta u & \text{for} \quad a \leqq t \leqq b, \\ 0 & \text{for} \quad t < a, \end{cases}$$

we have, thanks to (2.19):

$$\widetilde{\alpha u} \in W(a, +\infty), \qquad \widetilde{\beta u} \in W(-\infty, b).$$

Therefore, following the second case, there exist sequences

$$f_n (\text{resp.}\, g_n) \in \mathscr{D}([a, +\infty]; X) \ (\text{resp.}\, \mathscr{D}([-\infty, b]; X))$$

such that

$$f_n \to \widetilde{\alpha u} \,(\text{resp.}\, g_n \to \widetilde{\beta u}) \quad \text{in} \quad W(a, +\infty) \ (\text{resp.}\, W(-\infty, b))$$

and by restriction to $[a, b]$ (as at the end of the second case) the desired result follows. □

Remark 2.5. It is possible to reason directly on the third case by "extending" the definition of the functions by *homothetic mappings* (instead of the translations used in the second case).

Extension theorem:

Theorem 2.2. *It is assumed that at least one of a or b is finite. There exists a continuous linear operator $u \to p(u)$ from*

$$W(a, b) \to W(-\infty, +\infty)$$

such that

$$(2.20) \qquad p(u) = u \quad \text{a.e.} \quad \textit{on }]a, b[.$$

Proof. By the method of the "third case" in the proof of Theorem 2.1, we are brought back to the case $[a, +\infty]$ (the case $[-\infty, b]$ of course being analogous by a change of t) and we may assume $a = 0$. Therefore, let us define $p(u)$ for the case $[0, +\infty]$. We use the method of "*extension by reflection*".

For $u \in \mathscr{D}([0, \infty]; X)$, we define $p(u)$ by

$$(2.21) \qquad p(u)(t) = \begin{cases} u(t) & \text{if } t > 0, \\ \sum_{k=1}^{m} \alpha_k u(-kt) & \text{if } t < 0, \end{cases}$$

where the numbers α_k are defined by the conditions

$$(2.22) \qquad \frac{d^j}{dt^j} p(u)(0) = u^{(j)}(0), \quad 0 \leqq j \leqq m-1, \quad \forall u \in \mathscr{D}([0, \infty]; X),$$

i.e.

$$(2.23) \qquad \sum_{k=1}^{m} (-1)^j k^j \alpha_k = 1, \quad 0 \leqq j \leqq m-1$$

(the α_k's are well-defined by this system).

The function $p(u)$ is in $W(-\infty, +\infty)$, thanks to (2.22), and

$$p(u)^{(m)}(t) = \begin{cases} u^{(m)}(t) & t > 0, \\ \sum_{k=1}^{m} (-k)^m \alpha_k u^{(m)}(-kt) & t < 0, \end{cases}$$

so that

$$(2.24) \qquad \|p(u)\|_{W(-\infty, +\infty)} \leqq c \|u\|_{W(0, \infty)} \qquad (c = \text{constant})$$

is easy to verify, and therefore (from Theorem 2.1) the mapping $u \to p(u)$ extends by continuity to a linear mapping, still denoted $u \to p(u)$, of

$$W(0, \infty) \to W(-\infty, +\infty).$$

The property: $p(u) = u$ on $t > 0$ (which is satisfied for u regular) yields $p(u) = u$ a.e. for $t > 0$ by passage to the limit (and for $u \in W(0, \infty)$, $p(u)$ is again defined by (2.21), this time a.e. in t). □

2.3 Intermediate Derivatives Theorem

First, we recall the concept of *measurable hilbertian sums*.

On $[\lambda_0, +\infty[$, let there be given a Radon measure $d\mu(\lambda) \geqq 0$. For each $\lambda \in [\lambda_0, +\infty[$, let $\mathfrak{h}(\lambda)$ be a Hilbert space on $\mathbf{C}$ (for which the scalar product and the norm are denoted $(\, , \,)_{\mathfrak{h}(\lambda)}$ and $\| \; \|_{\mathfrak{h}(\lambda)}$, respectively).

The spaces $\mathfrak{h}(\lambda)$ are said to form a *μ-measurable field* if the $\mathfrak{h}(\lambda)$ "depend μ-measurably on λ", which means: define a family $\mathcal{M}$ of functions:

$$\lambda \to f(\lambda) \quad \text{of} \quad [\lambda_0, +\infty[\to \mathfrak{h}(\lambda)$$

having the following properties:

(i) $\forall f \in \mathcal{M}$, the function $\lambda \to \| f(\lambda) \|_{\mathfrak{h}(\lambda)}$ is μ-measurable;

(ii) $\begin{cases} \text{if } g \text{ is a function with values in } \mathfrak{h}(\lambda) \text{ such that, } \forall f \in \mathcal{M}, \\ \lambda \to (f(\lambda), g(\lambda))_{\mathfrak{h}(\lambda)} \text{ is } \mu\text{-measurable, then } g \in \mathcal{M}; \end{cases}$

(iii) $\begin{cases} \text{there exists a sequence } f_1 \ldots f_n \ldots \text{ of elements of } \mathcal{M} \text{ such that,} \\ \forall \lambda \in [\lambda_0, +\infty[, \text{ the sequence } f_1(\lambda), \ldots, f_n(\lambda), \ldots \text{ generates } \mathfrak{h}(\lambda). \end{cases}$

The elements of $\mathcal{M}$ are the measurable functions taking their values in $\mathfrak{h}(\lambda)$.

Then the space:

$$\mathfrak{h} = \int^{\oplus} \mathfrak{h}(\lambda)\, d\mu(\lambda)$$

is defined as follows. A measurable function $\lambda \to f(\lambda)$ is in $\mathfrak{h}$ if and only if

$$\| f \|_{\mathfrak{h}}^2 = \int_{\lambda_0}^{+\infty} \| f(\lambda) \|_{\mathfrak{h}(\lambda)}^2 \, d\mu(\lambda) < \infty .$$

For $f, g \in \mathfrak{h}$, their *scalar product* is defined by

$$(f, g)_{\mathfrak{h}} = \int_{\lambda_0}^{+\infty} (f(\lambda), g(\lambda))_{\mathfrak{h}(\lambda)} \, d\mu(\lambda) .$$

It can be shown (Dixmier [1], p. 146; see also Gelfand-Vilenkin [1], p. 114 of the English edition) *that the $\mathfrak{h}$ so defined is a Hilbert space.* The space $\mathfrak{h}$ is called the *measurable hilbertian sum* (or the direct hilbertian integral) of the spaces $\mathfrak{h}(\lambda)$.

We shall make use of this notion in the proof of the next result; a result upon which we shall frequently call in this book (under the name of "*intermediate derivatives theorem*").

Theorem 2.3. *Let X, Y be two Hilbert spaces with the properties* (2.1) *and let $[X, Y]_\theta$ be defined by* (2.7). *If $u \in W(a, b)$ (defined by* (2.9)), *then*

$$u^{(j)} \in L^2(a, b; [X, Y]_{j/m}), \quad 1 \leqq j \leqq m - 1 \tag{2.25}$$

(for $j = 0$ and m, (2.25) still holds and reduces, according to Remark 2.4, to the definition of $W(a, b)$). *Furthermore, $u \to u^{(j)}$ is a continuous linear mapping of*

$$W(a, b) \to L^2(a, b; [X, Y]_{j/m}) .$$

Proof. 1) Thanks to the extension operator p of Theorem 2.2, it is sufficient to prove the theorem for $a = -\infty$, $b = +\infty$. We may use the *Fourier transform in t*:

$$u \to \hat{u} = \mathscr{F} u, \quad \hat{u}(\tau) = \frac{1}{\sqrt{2\pi}} \int_{-\infty}^{+\infty} \exp(-\mathrm{i}t\tau)\, u(t)\, dt$$

which is a unitary isomorphism of $L^2(\mathbf{R}_t; E) \to L^2(\mathbf{R}_\tau; E)$ whenever E is a Hilbert space.

Following Schwartz [6], we define the space $\mathscr{S}'(E)$ of tempered distributions with values in E by

$$\mathscr{S}'(E) = \mathscr{S}'(\mathbf{R}_t; E) = \mathscr{L}(\mathscr{S}; E);$$

every $u \in \mathscr{S}'(E)$ may be written, in non-unique fashion, as a finite sum of derivatives in t of functions of the form $(1 + |t|)^k f_k$, $f_k \in L^2(E)$, where the derivatives are given by

$$\frac{d^m u}{dt^m}(\varphi) = (-1)^m u\left(\frac{d^m \varphi}{dt^m}\right), \qquad \forall \varphi \in \mathscr{S}.$$

For $u \in \mathscr{S}'(E)$, we define $\mathscr{F} u \in \mathscr{S}'(E)$ by

$$\mathscr{F} u(\varphi) = u(\mathscr{F} \varphi) \qquad \forall \varphi \in \mathscr{S},$$

which again defines an isomorphism of $\mathscr{S}'(E)$ onto itself. We still have the "usual" rules, in particular

$$\mathscr{F}\left(\frac{d^j u}{dt^j}\right) = (\mathrm{i}\tau)^j \mathscr{F} u \qquad \forall u \in \mathscr{S}'(E). \tag{2.26}$$

Consequently:

$$u \in W(-\infty, +\infty) \Leftrightarrow \begin{cases} \hat{u} \in L^2(\mathbf{R}_\tau; X) \\ \tau^m \hat{u} \in L^2(\mathbf{R}_\tau; Y) \end{cases} \tag{2.27}$$

and the properties (2.25) to be demonstrated are equivalent to

$$\tau^j \hat{u} \in L^2(\mathbf{R}_\tau; [X, Y]_{j/m}). \tag{2.28}$$

2) We now use the *diagonalization* of Λ. According to the theory of spectral decomposition (see Dixmier [1], and Gelfand-Vilenkin [1]), there exist:

(i) a measurable hilbertian sum $\mathfrak{h} = \int^{\oplus} \mathfrak{h}(\lambda)\, d\mu(\lambda)$, $0 < \lambda_0 \leqq \lambda < \infty$,

$d\mu(\lambda) =$ Radon measure $\geqq 0$ on $[\lambda_0, +\infty[$,

(ii) a unitary operator $\mathscr{U}$ of Y onto $\mathfrak{h}$ which maps X onto $\mathfrak{h}_1$, where

$$\mathfrak{h}_1 = \{v \mid v \in \mathfrak{h}, \lambda v \in \mathfrak{h}\}, \tag{2.29}$$

with

$$\| v \|_{\mathfrak{h}_1} = \left(\int_{\lambda_0}^{+\infty} \lambda^2 \, \| v(\lambda) \|^2_{\mathfrak{h}(\lambda)} \, d\mu(\lambda) \right)^{1/2}$$

and such that

(2.30) $\mathscr{U}(\Lambda u) = \lambda(\mathscr{U} u) \qquad \forall u \in D(\Lambda) = X. \quad \square$

We verify right away that

(2.31) *$\mathscr{U}$ is an isomorphism of $[X, Y]_\theta$ onto $\mathfrak{h}_{1-\theta}$,*

where

(2.32) $\mathfrak{h}_{1-\theta} = \{ v \mid v \in \mathfrak{h}, \lambda^{1-\theta} v \in \mathfrak{h} \} \qquad 0 \leqq \theta \leqq 1,$

with

$$\| v \|_{\mathfrak{h}_{1-\theta}} = \left(\int_{\lambda_0}^{+\infty} \lambda^{2(1-\theta)} \, \| v(\lambda) \|^2_{\mathfrak{h}(\lambda)} \, d\mu(\lambda) \right)^{1/2}.$$

We note that $v \in \mathfrak{h}_{1-\theta}$ if and only if $\lambda^{1-\theta} v \in \mathfrak{h}$ (since $\lambda_0 > 0$ and $1 - \theta \geqq 0$), and the norm on $\mathfrak{h}_{1-\theta}$ is equivalent to the norm of the graph.

3) For $f \in L^2(\mathbf{R}; Y)$, we define $\mathscr{U} f$ by

$$(\mathscr{U} f)(t) = \mathscr{U}(f(t)) \quad \text{a.e.}$$

and similarly for $f \in L^2(\mathbf{R}; [X; Y]_\theta)$. Then

(2.33) *$\mathscr{U}$ is an isomorphism of $L^2(\mathbf{R}; [X, Y]_\theta)$ onto $L^2(\mathbf{R}; \mathfrak{h}_{1-\theta})$.*

With the notations of (2.27), we also have:

(2.34) $$u \in W(-\infty, +\infty) \Leftrightarrow \begin{cases} \mathscr{U} \hat{u} \in L^2(\mathbf{R}_\tau; \mathfrak{h}_1) \\ \tau^m \, \mathscr{U} \hat{u} \in L^2(\mathbf{R}_\tau; \mathfrak{h}). \end{cases}$$

We set $v = \mathscr{U} \hat{u} (= v(\lambda, \tau))$. Then conditions (2.34) may be written

(2.35) $$\int_{-\infty}^{+\infty} \int_{\lambda_0}^{+\infty} (\lambda^2 + \tau^{2m}) \, \| v(\lambda, \tau) \|^2_{\mathfrak{h}(\lambda)} \, d\mu(\lambda) \, d\tau < +\infty,$$

which is equivalent to

(2.36) $(\lambda + |\tau|^m) \, v(\lambda, \tau) \in L^2(\mathbf{R}_\tau; \mathfrak{h}).$

Still with the same notation and using (2.28) and (2.33), we see that the property to be demonstrated is *equivalent* to

(2.37) $\lambda^{1-j/m} \, |\tau|^j \, v(\lambda, \tau) \in L^2(\mathbf{R}_\tau; \mathfrak{h}),$

with

(2.38) $\| \lambda^{1-j/m} \, |\tau|^j \, v(\lambda, \tau) \|_{L^2(\mathbf{R}_\tau; \mathfrak{h})} \leqq C_1 \, \| (\lambda + |\tau|^m) \, v(\lambda, \tau) \|_{L^2(\mathbf{R}_\tau; \mathfrak{h})}.$

But these last properties result from the inequality:

$$\lambda^{1-j/m}\,|\tau|^j \leqq C_2(\lambda + |\tau|^m)$$

(apply the inequality

$$\lambda^{1-j/m}\,|\tau|^j \leqq \frac{1}{p}\lambda^{(1-j/m)p} + \frac{1}{p'}|\tau|^{jp'},$$

$$\frac{1}{p} + \frac{1}{p'} = 1, \quad \text{with} \quad \left(1 - \frac{j}{m}\right)p = 1, \quad jp' = m\right). \quad \square$$

Remark 2.6. From (2.31) *and* (2.32), *we have that X is dense in* $[X; Y]_\theta$ $\forall\theta$. □

2.4 A Simple Example

We introduce a situation which we shall often meet in later chapters.

Let V and H be two Hilbert spaces, $V \subset H$, V dense in H with continuous injection.

Identifying H with its *anti-dual* and V' denoting the anti-dual of V (by abuse of language, we adopt the *same* notation for the dual and anti-dual, in order to prevent meaningless distinctions in the practical examples), we have:

(2.39) $$V \subset H \subset V',$$

each space being dense in the following one.

If (u, v) (resp. $((u, v))$) is a scalar product on H (resp. V), we define the operator A as S in Section 2.1 by taking $V = X$ and $H = Y$. Then

(2.40) $$D(A) \subset V \subset H \subset V', \qquad V = D(A^{1/2}).$$

We diagonalize the operator A (see the proof of Theorem 2.3) by a measurable sum $\mathfrak{h}$ and a unitary operator $\mathscr{U}$ of H onto $\mathfrak{h}$ and $D(A)$ onto

$$\mathfrak{h}_1 = \{v \mid v \in \mathfrak{h}, \lambda v \in \mathfrak{h}\} = \{v \mid \lambda v \in \mathfrak{h}\}.$$

Then $\mathscr{U}$ is an isomorphism of V onto $\mathfrak{h}_{1/2}$ and of V' onto $\mathfrak{h}_{-1/2}$, where

$$\mathfrak{h}_\alpha = \{v \mid, \lambda^\alpha v \in \mathfrak{h}\}, \qquad \alpha \in \mathbf{R}.$$

We immediately obtain:

Proposition 2.1. *With the notations* (2.39), (2.40), *we have*

(2.41) $$[V, V']_{1/2} = H,$$

(2.42) $$[D(A), H]_{1/2} = V.$$

In particular, the intermediate derivatives theorem yields:

Proposition 2.2. *If* $u \in L^2(0, \infty; V)$ *and* $u'' \in L^2(0, \infty; V')$, *then*

$$u' \in L^2(0, \infty; H). \quad \square$$

2.5 Interpolation Inequality

The following inequality is an immediate consequence of the definition of the spaces $[X, Y]_\theta$ and the fact that

$$\| \Lambda^{1-\theta} u \|_Y \leqq \| \Lambda u \|_Y^{1-\theta} \| u \|_Y^\theta:$$

Proposition 2.3. *For every* $u \in X$,

(2.43) $$\| u \|_{[X, Y]_\theta} \leqq C \| u \|_X^{1-\theta} \| u \|_Y^\theta. \quad \square$$

3. Trace Theorem

3.1 Continuity Properties of Elements of $W(a, b)$

Generally, if E is a Hilbert space, we set

(3.1) $$\mathscr{B}(a, b; E) = \begin{cases} C^0([a, b]; E) = \text{continuous functions of } [a, b] \to E \\ \quad \text{if } a \text{ and } b \text{ are finite;} \\ \text{Continuous bounded functions of } t \geqq a \to E \text{ if } a \\ \quad \text{is finite and } b = +\infty; \\ \text{continuous bounded functions of } \mathbf{R} \to E \text{ if} \\ \quad a = -\infty,\ b = +\infty. \end{cases}$$

We provide $\mathscr{B}(a, b; E)$ with the norm

$$\| \varphi \|_{\mathscr{B}(a, b; E)} = \sup_{t \in [a, b]} \| \varphi(t) \|_E.$$

Theorem 3.1. *With the notation* (3.1), *for* $u \in W(a, b)$ *we have:*

(3.2) $$u^{(j)} \in \mathscr{B}(a, b; [X, Y]_{(j+1/2)/m}), \quad 0 \leqq j \leqq m - 1,$$

$u \to u^{(j)}$ *being a continuous and linear mapping of* $W(a, b)$

$$\to \mathscr{B}(a, b; [X, Y]_{(j+1/2)/m}).$$

Remark 3.1. We have:

$$[X, Y]_{j/m} \subset [X, Y]_{(j+1/2)/m}.$$

Therefore, if $u \in W(a, b)$, then $u^{(j)}$ is *square integrable* with values in $[X, Y]_{j/m}$ and *continuous* with values in a *larger space.* $\square$

Remark 3.2. We must make (3.2) somewhat more precise: the function $u^{(j)}$, which is known to belong to $L^2(a, b; [X, Y]_{j/m})$ (intermediate derivatives theorem) is, *after a possible modification on a set of measure zero*, a continuous mapping of

$$[a, b] \to [X, Y]_{(j+1/2)/m}$$

(with the modifications appearing in (3.1) if a and b are infinite).

In this form, the last part of the theorem is slightly ambiguous (the set on which $u^{(j)}$ is modified could depend on u); a more precise statement of the theorem is (see Theorem 2.1):

$$\text{(3.3)}\quad \begin{cases} \textit{the mapping } u \to u^{(j)} \textit{ of } \mathscr{D}([a,b];X) \to \mathscr{D}([a,b];X) \textit{ extends by} \\ \textit{continuity to a mapping of} \\ \qquad W(a,b) \to \mathscr{B}(a,b;[X,Y]_{(j+1/2)/m}) \end{cases}$$

Proof. 1) As in the proof of Theorem 2.3, we are led back (with the help of Theorem 2.2) to the case $a = -\infty$, $b = +\infty$. Again we use the Fourier transform in t and the diagonalization $\mathscr{U}$. Therefore, for $u \in \mathscr{D}(\mathbf{R}; X)$, let

$$v = \mathscr{U} \hat{u}.$$

We show that

$$\text{(3.4)}\qquad \| u^{(j)} \|_{\mathscr{B}(a,b;[X\,Y]_{(j+1/2)/m})} \leqq c \| u \|_{W(a,b)},$$

which proves the theorem.

Now

$$\mathscr{U} u^{(j)}(t_0) = \int_{-\infty}^{+\infty} \exp(2\pi i t_0 \tau)(2\pi i \tau)^j v(\tau)\, d\tau, \qquad t_0 \in \mathbf{R},$$

and (3.4) will follow from

$$\text{(3.5)}\quad \| \mathscr{U} u^{(j)}(t_0) \|_{\mathfrak{h}_{1-(j+1/2)/m}} \leqq C \| w \|_{L^2(\mathbf{R}_\tau;\, \mathfrak{h})}, \qquad w = (\lambda + |\tau|^m)\, v.$$

2) We have

$$\| \mathscr{U} u^{(j)}(t_0) \|^2_{\mathfrak{h}_{1-(j+1/2)/m}} = \int_{\lambda_0}^{\infty} \lambda^{2-\frac{2j+1}{m}} \| \mathscr{U} u^{(j)}(\lambda, t_0) \|^2_{\mathfrak{h}(\lambda)}\, d\mu(\lambda) \leqq$$

$$\leqq c_1 \int_{\lambda_0}^{\infty} \lambda^{2-\frac{2j+1}{m}} d\mu(\lambda) \left(\int_{-\infty}^{+\infty} \frac{|\tau|^j}{\lambda + |\tau|^m} \| w(\lambda,\tau) \|_{\mathfrak{h}(\lambda)}\, d\tau \right)^2 \leqq$$

$$\leqq c_1 \int_{0}^{\infty} \lambda^{2-\frac{2j+1}{m}} d\mu(\lambda) \left(\int_{-\infty}^{+\infty} \frac{|\tau|^{2j}\, d\tau}{(\lambda + |\tau|^m)^2} \right) \left(\int_{-\infty}^{+\infty} \| w(\lambda,\tau) \|^2_{\mathfrak{h}(\lambda)}\, d\tau \right)$$

$$= (\text{set } \tau = \lambda^{1/m} \sigma) = c_1 \int_{-\infty}^{+\infty} \left(\frac{|\sigma|^j}{1 + |\sigma|^m} \right)^2 d\sigma \int_{\lambda_0}^{+\infty} d\mu(\lambda) \int_{-\infty}^{+\infty} \| w(\lambda,\tau) \|^2_{\mathfrak{h}(\lambda)}\, d\tau$$

$$= c_2 \int_{\lambda_0}^{\infty} \int_{-\infty}^{+\infty} \| w \|^2_{\mathfrak{h}(\lambda)}\, d\tau\, d\mu(\lambda),$$

whence (3.5).

3.2 Trace Theorem

We state the result for $a = 0$, $b = +\infty$ (as will be the case for most of the applications and with no loss of generality!):

Theorem 3.2. *Let* $u \in W(0, \infty)$. *According to Theorem 3.1*:

$$u^{(j)}(0) \in [X, Y]_{(j+1/2)/m}, \quad 0 \leqq j \leqq m-1.$$

The mapping

$$(3.6)\quad u \to \{u^{(j)}(0) \mid 0 \leqq j \leqq m-1\} \quad \text{of} \quad W(0, \infty) \to \prod_{j=0}^{m-1} [X, Y]_{(j+1/2)/m}$$

is surjective.

Proof. 1) We start with a very general remark which is independent of the Hilbert structure of the spaces:

$$(3.7)\quad \begin{cases} \text{to show the surjectivity of (3.6) it is sufficient to verify the} \\ \text{surjectivity of the mapping } u \to u^j(0) \text{ of } W(0, \infty) \to \\ \to [X, Y]_{(j+1/2)/m}, \text{ for } \textit{arbitrary fixed } j. \end{cases}$$

Indeed, assume the surjectivity for arbitrary fixed j.

Let $\{a_j\} \in \prod_{j=0}^{m-1} [X, Y]_{(j+1/2)/m}$. Then, there exists $u_j \in W(0, \infty)$ with

$$u_j^{(j)}(0) = a_j.$$

We construct $U_j \in W(0, \infty)$, with

$$(3.8)\qquad U_j^{(k)}(0) = \begin{cases} 0 & \text{if } k \neq j, \quad 0 \leqq k \leqq m-1 \\ a_j & \text{if } k = j. \end{cases}$$

Then

$$U = \sum_{j=0}^{m-1} U_j \in W(0, \infty) \quad \text{and} \quad U^j(0) = a_j \quad \forall j, \; 0 \leqq j \leqq m-1;$$

whence the desired surjectivity – after the construction of U_j.

For this purpose, we define

$$(3.9)\qquad U_j(t) = \sum_{r=1}^{m} c_{rj}\, u_j(r t),$$

with

$$(3.10)\qquad \sum_{r=1}^{m} r^k c_{rj} = \begin{cases} 0 & \text{if } k \neq j, \quad 0 \leqq k \leqq m-1 \\ 1 & \text{if } k = j \end{cases}$$

(which properly defines the c_{rj}'s); of course, we have $U_j \in W(0, \infty)$ with (3.8), since $u_j \to U_j$ is a continuous mapping of $W(0, \infty)$ into itself.

2) Therefore, there remains to show the surjectivity with *fixed* j. Using the diagonalization (as in Theorem 2.3), we consider

$$a(\lambda) = a \in \mathfrak{h}_{1-(j+1/2)/m}.$$

We shall construct w to satisfy

$$(3.11) \qquad \begin{cases} w \in L^2(0, \infty; \mathfrak{h}_1), \\ w^{(m)} \in L^2(0, \infty; \mathfrak{h}), \\ w^{(j)}(0) = a, \end{cases}$$

$$(3.12) \qquad \| w \|_{L^2(0,\infty;\mathfrak{h}_1)} + \| w^{(m)} \|_{L^2(0,\infty;\mathfrak{h})} \leqq C \| a \|_{\mathfrak{h}_{1-(j+1/2)/m}}.$$

(Then $u(t) = \mathscr{U}^{-1}(w(t))$ will yield $u \in W(0, \infty)$ with $u^{(j)}(0) = a$ and the desired result will follow).

For the construction of w, we consider a function $\varphi \in \mathscr{D}([0, +\infty[)$, with

$$\varphi^{(j)}(0) = 1$$

and we set

$$(3.13) \qquad w(\lambda, t) = \lambda^{-1/m} a(\lambda) \varphi(\lambda^{1/m} t).$$

Then, the properties (3.11) and (3.12) may be verified by an elementary calculation. □

Remark 3.3. The proof of the theorem shows that there exists a continuous linear *right inverse* $\mathscr{R}$

$$(3.14) \qquad \{a_j\}_{j=0}^{m-1} \overset{\mathscr{R}}{\rightarrow} u \quad \text{of} \quad \prod_{j=0}^{m-1} [X, Y]_{(j+1/2)/m} \rightarrow W(0, \infty)$$

such that

$$(3.15) \qquad u^{(j)}(0) = a_j, \quad 0 \leqq j \leqq m-1. \quad \square$$

Remark 3.4. The "intermediate spaces" $[X, Y]_{(j+1/2)/m}$ appear as the spaces described by the "traces" $u^{(j)}(0)$; for this reason, the spaces $[X, Y]_{(j+1/2)/m}$ are often called "*trace spaces*". □

Remark 3.5. Since $u \rightarrow u^{(j)}(0)$ is a continuous mapping of $W(0, \infty) \rightarrow$ $\rightarrow [X, Y]_{(j+1/2)/m}$, we have

$$(3.16) \qquad \| u^{(j)}(0) \|_{[X,Y]_{(j+1/2)/m}} \leqq c_j (\| u \|_{L^2(0,\infty;X)} + \| u^{(m)} \|_{L^2(0,\infty;Y)}),$$

for a suitable constant c_j.

But, with q denoting a positive parameter, set

$$u_q(t) = q^{-j} u(q t),$$

and apply (3.16) to u_q; since $u_q^{(j)}(0) = u^{(j)}(0)$, we have

$$(3.17) \qquad \| u^{(j)}(0) \|_{[X,Y]_{(j+1/2)/m}} \leqq c_j (q^{-j-1/2} \| u \|_{L^2(0,\infty;X)} + q^{m-j-1/2} \| u^{(m)} \|_{L^2(0,\infty;Y)}),$$

for arbitrary q. Choose q such that (u being fixed)

$$q^{-j-1/2} \| u \|_{L^2(0,\infty;X)} = q^{m-j-1/2} \| u^{(m)} \|_{L^2(0,\infty;Y)}.$$

Then (3.17) yields

$$(3.18) \qquad \| u^{(j)}(0) \|_{[X,Y]_{(j+1/2)/m}} \leqq 2 c_j \| u \|_{L^2(0,\infty;X)}^{1-(j+1/2)/m} \| u^{(m)} \|_{L^2(0,\infty;Y)}^{(j+1/2)/m}. \quad \square$$

Remark 3.6. Since the mapping $u \to u^{(j)}(0)$ is a *surjection* of $W(0,\infty) \to \to [X, Y]_{(j+1/2)/m}$, it is natural to introduce the "*quotient-norm*"

$$(3.19) \qquad ||| a |||_{[X,Y]_{(j+1/2)/m}} = \inf_{\substack{u \in W(0,\infty) \\ u^{(j)}(0) = a}} \| u \|_{W(0,\infty)}.$$

on $[X, Y]_{(j+1/2)/m}$.

Then, the norm (3.19) *is equivalent to the norm* (2.8). Indeed, provided with the norm (3.19), $[X, Y]_{(j+1/2)/m}$ is the quotient of the Hilbert space $W(0, \infty)$ by the closed subspace of functions φ such that $\varphi^{(j)}(0) = 0$ and is, therefore, also a *Hilbert space*. Since $u \to u^{(j)}(0)$ is a continuous mapping of $W(0, \infty) \to [X, Y]_{(j+1/2)/m}$, we have (3.16) and therefore

$$\| a \|_{[X,Y]_{(j+1/2)/m}} \leqq c_j \inf_{u^{(j)}(0) = a} \| u \|_{W(0,\infty)} = c_j ||| a |||_{[X,Y]_{(j+1/2)/m}},$$

so that the norms (2.8) and (3.19) are equivalent (closed graph theorem; we could also use the right inverse (3.14)). $\square$

Remark 3.7. Remark 3.5 shows that the *space* $[X, Y]_{(j+1/2)/m}$, *with its topology*, is *intrinsically* linked to X and Y (see Remark 2.3). For $[X, Y]_\theta$, with arbitrary $\theta \subset]0, 1[$, the analogous question will be answered in Sections 4 and 10, below. $\square$

4. Trace Spaces and Non-Integer Order Derivatives

4.1 Orientation. Definitions

An analysis of the proofs of Theorems 2.3 and 3.1 easily shows that, especially when $a = -\infty$, $b = +\infty$, the hypotheses "integer m" and "integer j" do not intervene in an essential manner.

For instance, (2.27) leads to the definition (we replace m by s to avoid any confusion):

$$(4.1) \quad W(-\infty, +\infty; s; X, Y) = \{u \mid \hat{u} \in L^2(\mathbf{R}_\tau; X), |\tau|^s \hat{u} \in L^2(\mathbf{R}_\tau; Y)\}$$

for arbitrary $s > 0$, *integer or not*. Provided with the norm

$$(4.2)\qquad (\| \hat{u} \|^2_{L^2(\mathbf{R}_\tau; X)} + \| |\tau|^s \hat{u} \|^2_{L^2(\mathbf{R}_\tau; Y)})^{1/2} = \| u \|_{W(-\infty, +\infty; s; X, Y)},$$

$W(-\infty, +\infty; s; X, Y)$ is a Hilbert space.

Of course, $W(-\infty, +\infty; m; X, Y) = W(-\infty, +\infty)$, with the notation of Section 2. □

Remark 4.1. We shall often say that *the derivative of order s of u* is in $L^2(\mathbf{R}_\tau; Y)$. □

The subspace of functions u such that $D_t^k u \in L^2(\mathbf{R}; X)$ $\forall k$ is *dense* in $W(-\infty, +\infty; s; X, Y)$ (immediate, by regularization). □

We shall now extend Theorems 2.3 and 3.2 to this setting.

4.2 "Intermediate Derivatives" and Trace Theorems

Theorem 4.1 ("intermediate derivatives"). *For every*

$$u \in W(-\infty, +\infty; s; X, Y),$$

we have

$$(4.3)\qquad |\tau|^r \hat{u} \in L^2(\mathbf{R}_\tau; [X, Y]_{r/s}), \quad 0 \leqq r \leqq s,$$

and

$$(4.4)\qquad \| |\tau|^r \hat{u} \|_{L^2(\mathbf{R}_\tau; [X, Y]_{r/s})} \leqq c \, \| u \|_{W(-\infty, +\infty; s; X, Y)}.$$

Proof. In the same way as for the proof of Theorem 2.3, we are led to verify (compare with (2.38)) that

$$\lambda^{1-r/s} |\tau|^r \leqq \text{constant } (\lambda + |\tau|^s),$$

which is immediate (in fact, simply let $j = r$, $m = s$ in the proof of Theorem 2.3!). □

Theorem 4.2 (traces). *For every*

$$u \in W(-\infty, +\infty; s; X, Y)$$

we have

$$(4.5)\qquad u^{(j)}(0) \in [X, Y]_{(j+1/2)/s}, \quad 0 \leqq j < s - \tfrac{1}{2}.$$

Furthermore, the mapping

$$(4.6)\qquad u \to \{u^{(j)}(0)\}_{0 \leqq j < s - 1/2}$$

of

$$W(-\infty, +\infty; s; X, Y) \to \prod_{0 \leqq j < s - 1/2} [X, Y]_{(j+1/2)/s}$$

is surjective.

Proof. 1) In fact, we have (which makes (4.5) more precise):

$$(4.7)\qquad u^{(j)} \in \mathscr{B}(-\infty, +\infty; [X, Y]_{(j+1/2)/s}).$$

By the method of the proof of Theorem 3.1, this brings us to the necessity of verifying that

$$\int_{-\infty}^{+\infty} \left(\frac{|\tau|^j \lambda^{1-(j+1/2)/s}}{|\tau|^s + \lambda} \right)^2 d\tau \leqq \text{constant}.$$

But setting $\tau = \lambda^{1/s} \sigma$, this integral becomes

$$\int_{-\infty}^{+\infty} \left(\frac{|\sigma|^j}{|\sigma|^s + 1} \right)^2 d\sigma,$$

which is finite if (and only if) $j < s - \frac{1}{2}$.

2) The same procedure as in the proof of Theorem 3.2 (valid on $(-\infty, +\infty)$ as well as on $(0, +\infty)$) shows that it is sufficient to prove that $u \to u^j(0)$ is a surjection for *fixed* j (note that "the homothetic mapping": $u \to u(r\,.)$ is a continuous mapping of $W(-\infty, +\infty; s; X, Y)$ into itself).

As right inverse, we take (compare with (3.13))

$$w(\lambda, t) = \lambda^{-j/s} a(\lambda)\, \varphi(\lambda^{1/s} t). \tag{4.8}$$

We verify that $w \in W(-\infty, +\infty; s; X, Y)$ and in particular that

$$|\tau|^s \hat{w}(\lambda, \tau) \in L^2(\mathbf{R}_\tau; \mathfrak{h}). \tag{4.9}$$

But

$$\hat{w}(\lambda, \tau) = \frac{1}{\sqrt{2\pi}} \int_{-\infty}^{+\infty} \exp(-i\tau t)\, w(\lambda, t)\, dt$$

$$= \lambda^{-(j+1)/s} a(\lambda)\, \hat{\varphi}(\lambda^{-1/s} \tau)$$

and therefore

$$\int_{-\infty}^{+\infty} \int_{\lambda_0}^{+\infty} |\tau|^{2s} \| \hat{w}(\lambda, \tau) \|^2_{\mathfrak{h}(\lambda)}\, d\mu(\lambda)\, d\tau$$

$$= \int_{\lambda_0}^{+\infty} \lambda^{-2[(j+1)/s]} \| a(\lambda) \|^2_{\mathfrak{h}(\lambda)}\, d\mu(\lambda) \int_{-\infty}^{+\infty} |\tau|^{2s} |\hat{\varphi}(\lambda^{-1/s}\tau)|^2\, d\tau$$

$$= \int_{\lambda_0}^{+\infty} \lambda^{-2[(j+1)/s]} \| a(\lambda) \|^2_{\mathfrak{h}(\lambda)}\, d\mu(\lambda)\, \lambda^{2+1/s} \int_{-\infty}^{+\infty} |\hat{\varphi}(\xi)|^2\, d\xi$$

$$= (\text{constant})\, \| \lambda^{1-(j+1/2)/s} a \|^2_{\mathfrak{h}} < \infty,$$

whence (4.9). □

Now, an obvious question is the following: consider again the mapping

$$u \to u^{(j)}(0) \tag{4.10}$$

of $\mathscr{D}(\mathbf{R}; X) \to X$. If $j \geqq s - \frac{1}{2}$, is it possible to find a topology on X such that (4.10) is *continuous* when $\mathscr{D}(\mathbf{R}; X)$ is provided with the topology induced by $W(-\infty, +\infty; s; X, Y)$? The answer is *no*:

Theorem 4.3. *For any fixed $\xi \neq 0$ in X, the mapping*

$$u \to (u^{(j)}(0), \xi)_X \tag{4.11}$$

of $\mathscr{D}(\mathbf{R}; X) \to \mathbf{C}$ is not continuous when $\mathscr{D}(\mathbf{R}; X)$ is provided with the topology induced by $W(-\infty, +\infty; s; X, Y)$, if

$$j \geqq s - \tfrac{1}{2}. \tag{4.12}$$

Proof. By Fourier transform in t and if $v = \mathscr{U}\hat{u}$ and $\eta = \eta(\lambda) = \mathscr{U}\xi$, we have

$$\begin{aligned}(u^{(j)}(0), \xi)_X &= \int_{\lambda_0}^{+\infty} \int_{-\infty}^{+\infty} (\mathrm{i}\,\tau)^j (v(\lambda, \tau), \eta(\lambda))_{\mathfrak{h}(\lambda)} \lambda\, d\mu(\lambda)\, d\tau \\ &= ((|\tau|^s + \lambda)\, v(\lambda, \tau), g(\lambda, \tau))_{L^2(\mathbf{R}_\tau; \mathfrak{h})},\end{aligned}$$

where

$$g(\lambda, \tau) = (\mathrm{i}\tau)^j \frac{\lambda}{|\tau|^s + \lambda} \eta(\lambda).$$

Then, since $\| u \|_{W(-\infty, +\infty; s; X, Y)}$ is equivalent to

$$\| (|\tau|^s + \lambda)\, v(\lambda, \tau) \|_{L^2(\mathbf{R}_\tau; \mathfrak{h})},$$

if (4.11) was continuous for the topology induced by

$$W(-\infty, +\infty; s; X, Y),$$

we would have

$$g(\lambda, \tau) \in L^2(\mathbf{R}_\tau; \mathfrak{h}),$$

which is false if (4.12) holds. □

Remark 4.2. It can be seen, as in Remarks 3.5 and 3.7, that $[X, Y]_{(j+1/2)/s}$, $0 \leqq j < s - \frac{1}{2}$, may be defined as the *trace space* described by $u^{(j)}(0)$ as u describes $W(-\infty, +\infty; s; X, Y)$, with the quotient norm:

$$\| a \|_{[X, Y]_{(j+1/2)/s}} = \inf_{u^{(j)}(0) = a} \| u \|_{W(-\infty, +\infty; s; X, Y)}. \tag{4.13}$$ □

Since we are free to choose the parameter s, we see that $[X, Y]_\theta$ and its topology are intrinsically linked to the couple $\{X, Y\}$. Another demonstration of this property follows from Theorem 10.1, below. □

5. Interpolation Theorem

5.1 Main Theorem

Let $\{\mathscr{X}, \mathscr{Y}\}$ be a second couple of Hilbert spaces having properties analogous to the couple $\{X, Y\}$.

Let π be a continuous linear operator of Y into $\mathscr{Y}$ and of X into $\mathscr{X}$; in short

(5.1) $$\pi \in \mathscr{L}(X; \mathscr{X}) \cap \mathscr{L}(Y; \mathscr{Y}).$$

Then, we have

Theorem 5.1. *If* π *satisfies* (5.1), *then*

(5.2) $$\pi \in \mathscr{L}([X, Y]_\theta; [\mathscr{X}, \mathscr{Y}]_\theta) \quad \forall \theta, \quad 0 < \theta < 1.$$

Proof. Let θ be fixed. Set $s = 1/(2\theta)$; then $[X, Y]_\theta$ is the space described by $u(0)$ as u describes $W(-\infty, +\infty; s; X, Y) = W$. Therefore, if $a \in [X, Y]_\theta$, there exists a $u \in W$ such that $u(0) = a$. Define v by

$$v(t) = \pi(u(t)) \quad \text{a.e.}$$

Since $\pi \in \mathscr{L}(X; \mathscr{X})$ (resp. $\mathscr{L}(Y; \mathscr{Y})$), we see that

$$v \in L^2(\mathbf{R}_t; \mathscr{X}) \text{ (resp. } |\tau|^s \hat{v}(\tau) = \pi(|\tau|^s \hat{u}(\tau)) \in L^2(\mathbf{R}_\tau; \mathscr{Y}))$$

and that, with α (resp. β) denoting the norm of π in $\mathscr{L}(X, \mathscr{X})$ (resp. $\mathscr{L}(Y; \mathscr{Y})$):

$$\| v \|_{L^2(\mathbf{R}_t; \mathscr{X})} \leqq \alpha \| u \|_{L^2(\mathbf{R}_t; X)},$$

$$\| |\tau|^s v \|_{L^2(\mathbf{R}_\tau; \mathscr{Y})} \leqq \beta \| |\tau|^s \hat{u} \|_{L^2(\mathbf{R}_\tau; Y)}.$$

Therefore $v \in W(-\infty, +\infty; s; \mathscr{X}, \mathscr{Y}) = \mathscr{W}$ and

(5.3) $$\| v \|_{\mathscr{W}} \leqq \max(\alpha, \beta) \| u \|_W.$$

Then $v(0) \in [\mathscr{X}, \mathscr{Y}]_\theta$; since $v \to v(0)$ is a continuous mapping of $\mathscr{W} \to$ $\to [\mathscr{X}, \mathscr{Y}]_\theta$ and since $v(0) = \pi a$, we have

(5.4) $$\| \pi a \|_{[\mathscr{X}, \mathscr{Y}]_\theta} \leqq c \| v \|_{\mathscr{W}},$$

which, together with (5.3), yields

$$\| \pi a \|_{[\mathscr{X}, \mathscr{Y}]_\theta} \leqq c \max(\alpha, \beta) \inf \| u \|_W, \quad u(0) = a,$$

from which (5.2) follows. □

5.2 Interpolation of a Family of Operators

Theorem 5.2. *Let* π *satisfy* (5.1). *Let* π_ϱ *be a family of operators* $(0 \leqq \varrho \leqq \varrho_0)$ *satisfying*

(5.5) $$\pi_\varrho \in \mathscr{L}(X; \mathscr{X}) \cap \mathscr{L}(Y; \mathscr{Y})$$

(and therefore also satisfying (5.2)). *Assume that*

(5.6) $\pi_\varrho a \to \pi a$ *in* $\mathscr{X}$ (*resp.* $\mathscr{Y}$) $\forall a \in X$ (resp. Y), *as* $\varrho \to 0$.

Then

(5.7) $\pi_\varrho a \to \pi a$ *in* $[\mathscr{X}, \mathscr{Y}]_\theta \, \forall a \in [X, Y]_\theta$, *as* $\varrho \to 0$.

Proof. We may assume $\pi = 0$.

With the notation of the proof of Theorem 5.1, we have:

(5.8) $\|\pi_\varrho a\|_{[\mathscr{X},\mathscr{Y}]_\theta} \leqq (\text{constant}) \left(\|\pi_\varrho u\|_{L^2(\mathbf{R}_t; X)} + \|\pi_\varrho(|\tau|^s \hat{u})\|_{L^2(\mathbf{R}_\tau; Y)}\right)$.

We shall verify that

(5.9) $$\|\pi_\varrho u\|_{L^2(\mathbf{R}_t; X)} \to 0.$$

In the same manner, we would verify that $\|\pi_\varrho(|\tau|^s \hat{u})\|_{L^2(\mathbf{R}_\tau; Y)} \to 0$ and therefore (5.8) yields the desired result.

But (5.9) is a consequence of Lebesgue's Theorem; indeed, according to (5.6) (with $\pi = 0$), $\pi_\varrho u(t) \to 0$ a.e. in $\mathscr{X}$, and still according to (5.6):

$$\|\pi_\varrho\|_{\mathscr{L}(X;\mathscr{X})} \leqq \text{constant}, \quad \text{therefore } \|\pi_\varrho u(t)\|_{\mathscr{X}} \leqq c\,\|u(t)\|_X. \quad \square$$

Remark 5.1. Because of the preceding properties, the spaces $[X, Y]_\theta$ are also called *interpolation spaces between X and Y* (but there are *other* intermediate and interpolation spaces – in particular *non*-Hilbert spaces). □

6. Reiteration Properties and Duality of the Spaces $[X, Y]_\theta$

6.1 Reiteration

Let θ_0 and θ_1 be fixed in $]0, 1[$, with

(6.1) $$\theta_0 < \theta_1.$$

Then

(6.2) $$[X, Y]_{\theta_0} \subset [X, Y]_{\theta_1}.$$

Proposition 6.1. *The space* $[X, Y]_{\theta_0}$ *is dense in* $[X, Y]_{\theta_1}$.

Proof. By diagonalization (see Proof of Theorem 2.3), we see that $\mathscr{U}$ is an isomorphism of $[X, Y]_\theta$ onto $\mathfrak{h}_{1-\theta} = \{v \mid v \in \mathfrak{h}, \lambda^{1-\theta} v \in \mathfrak{h}\}$ and $\mathfrak{h}_{1-\theta_0}$ is dense in $\mathfrak{h}_{1-\theta_1}$. □

Therefore, we may apply the theory of intermediate – or interpolation – spaces to the couple $\{[X, Y]_{\theta_0}, [X, Y]_{\theta_1}\}$.

Theorem 6.1. *For all* $\theta \in]0, 1[$, *we have*

(6.3) $$[[X, Y]_{\theta_0}, [X, Y]_{\theta_1}]_\theta = [X, Y]_{(1-\theta)\theta_0 + \theta\theta_1},$$

with equivalent norms.

Proof. We use the diagonalization $\mathscr{U}$. Then, the property to be demonstrated is equivalent to:

$$[\mathfrak{h}_{1-\theta_0}, \mathfrak{h}_{1-\theta_1}]_\theta = \mathfrak{h}_{1-((1-\theta)\theta_0+\theta\theta_1)}. \tag{6.4}$$

But $\mathfrak{h}_{1-\theta_0}$ is the domain *in the space* $\mathfrak{h}_{1-\theta_1}$ of the operator $v \to \lambda^{\theta_1-\theta_0} v$, therefore (by definition of the spaces $[X, Y]_\theta$) the space $[\mathfrak{h}_{1-\theta_0}, \mathfrak{h}_{1-\theta_1}]_\theta$ is the domain in the space $\mathfrak{h}_{1-\theta_1}$ of the operator $v \to \lambda^{(1-\theta)(\theta_1-\theta_0)} v$ and therefore coincides with the v's such that

$$\lambda^{1-\theta_1}(\lambda^{(1-\theta)(\theta_1-\theta_0)} v) \in \mathfrak{h}, \quad \text{whence (6.4).} \quad \square$$

Property (6.3) is called *reiteration property* or *stability property*: by successive applications of the "operation" $\{X, Y\} \to [X, Y]_\theta$, for various values of θ, we recover a space of the same type (for different values of the "interpolation parameter" θ). $\square$

6.2 Duality

Since $X \subset [X, Y]_\theta \subset Y$, each space being dense in the following one, we have, by duality (without any identification between the space and its dual):

$$Y' \subset [X, Y]'_\theta \subset X',$$

each space being dense in the following ones. We have the following *duality theorem*:

Theorem 6.2. *For all* $\theta \in]0, 1[$,

$$[X, Y]'_\theta = [Y', X']_{1-\theta}, \tag{6.5}$$

with equivalent norms.

Proof. Set (still in the notation of the diagonalization of the proof of Theorem 2.3):

$$\mathfrak{h}_s = \{v \mid \lambda^s v \in \mathfrak{h}\}, \quad s \in \mathbf{R}$$

of arbitrary sign (and the norm of the graph $\| \lambda^s v \|_{\mathfrak{h}}$).

For the demonstration, we identify — which is permissible — Y' with Y, therefore $\mathfrak{h}'$ with $\mathfrak{h} = \mathfrak{h}_0$. Since $\mathscr{U}$ is an isomorphism of X' onto $\mathfrak{h}_{-1}$ (dual of $\mathfrak{h}_1$) and of $[X, Y]'_\theta$ onto $\mathfrak{h}_{\theta-1}$ (dual of $\mathfrak{h}_{1-\theta}$), the property (6.5) to be demonstrated is *equivalent* to

$$\mathfrak{h}_{\theta-1} = [\mathfrak{h}, \mathfrak{h}_{-1}]_{1-\theta},$$

which is a consequence of the definition of the spaces $[X, Y]_\theta$, since $\mathfrak{h}$ may be defined as the domain of $v \to \lambda v$ in $\mathfrak{h}_{-1}$. $\square$

7. The Spaces $H^s(\mathbf{R}^n)$ and $H^s(\Gamma)$

7.1 $H^s(\mathbf{R}^n)$-Spaces

First, we apply the concepts of Section 2.1, with

$$X = H^m(\mathbf{R}^n), \quad \text{integer } m > 0, \quad Y = H^0(\mathbf{R}^n) = L^2(\mathbf{R}^n).$$

According to Theorem 1.2, $\mathscr{F}$ defines an isomorphism of X onto

$$\hat{X} = \hat{H}^m(\mathbf{R}^m) = \{v \mid (1 + |y|^2)^{m/2}\, v \in L^2(\mathbf{R}^n_y)\}$$

and of Y onto $L^2(\mathbf{R}^n_y) = \hat{Y}$. Then, if $\hat{\Lambda}$ is the operator associated to the couple $\{\hat{X}, \hat{Y}\}$ as Λ is associated to the couple $\{X, Y\}$ (see Section 2.1), we have:

$$\hat{\Lambda}\, v = (1 + |y|^2)^{m/2}\, v,$$

from which we immediately obtain

$$(\hat{\Lambda})^\theta\, v = (1 + |y|^2)^{\theta m/2}\, v.$$

We define the space ($s \in \mathbf{R}$)

$$H^s(\mathbf{R}^n) = \{v \mid v \in \mathscr{S}'(\mathbf{R}^n),\ (1 + |y|^2)^{s/2}\, \hat{v} \in L^2(\mathbf{R}^n_y)\}^{((1))} \tag{7.1}$$

and provide it with the norm

$$\|v\|_{H^s(\mathbf{R}^n)} = \|(1 + |y|^2)^{s/2}\, \hat{v}\|_{L^2(\mathbf{R}^n_y)}, \tag{7.2}$$

which makes it *a Hilbert space.*

Remark 7.1. We insist on the fact that s may be negative in definition (7.1). We have:

$$H^s(\mathbf{R}^n) \subset H^0(\mathbf{R}^n) \subset H^\sigma(\mathbf{R}^n) \quad \text{if} \quad \sigma < 0 < s. \tag{7.3}$$

The spaces $H^s(\mathbf{R}^n)$ are often called "fractional order Sobolev spaces". □

We may now state

Theorem 7.1. *In the notation* (2.7) *of Definition* 2.1, *we have*

$$[H^m(\mathbf{R}^n), H^0(\mathbf{R}^n)]_\theta = H^{(1-\theta)m}(\mathbf{R}^n), \tag{7.4}$$

with equivalent norms.

Remark 7.2. In general, let $L^2(d\mu)$ be the space of (classes of) square integrable functions on a locally compact space for the measure $d\mu$;

((1)) Equivalently, $H^s(\mathbf{R}^n)$ may be defined as the space of Fourier antitransforms of the measurable functions $\hat{v}$ such that $(1 + |y|^2)^{s/2}\, \hat{v} \in L^2(\mathbf{R}^n_y)$.

if M is a – say, continuous – positive function, let

$$L^2_M(d\mu) = \{f \mid f \text{ measurable}, \ M f \in L^2(d\mu)\}.$$

Then the *dual space* $(L^2_M(d\mu))'$ *is identified with* $L^2_{(1/M)}(d\mu)$. Consequently, (from Definition 7.2) we have:

Theorem 7.2. *The space* $H^0(\mathbf{R}^n)$ *being identified with its dual* (or antidual), *we have, for all* $s > 0$:

$$(H^s(\mathbf{R}^n))' = H^{-s}(\mathbf{R}^n). \quad \square \tag{7.5}$$

Remark 7.3. This is also a consequence of the duality Theorem 6.2. □

Local properties of $H^s(\mathbf{R}^n)$-spaces.

Theorem 7.3. *For all* $s \in \mathbf{R}$ *and all* $\varphi \in \mathscr{D}(\mathbf{R}^n)$,

$$u \to \varphi u \tag{7.6}$$

is a continuous linear mapping of $H^s(\mathbf{R}^n)$ *into itself.*

Proof. 1) If the property holds for $H^s(\mathbf{R}^n)$, it holds (by transposition and with (7.5)) for $H^{-s}(\mathbf{R}^n)$ as well. Therefore, it is sufficient to prove the result for $s > 0$.

2) Choose integer $m \geqq s$. It is easy to verify that (7.6) is a continuous linear mapping of $H^m(\mathbf{R}^n)$ into itself and of $H^0(\mathbf{R}^n)$ into itself. Therefore, by interpolation (Theorem 5.1) it is a continuous mapping of $[H^m(\mathbf{R}^n), H^0(\mathbf{R}^n)]_\theta$ into itself. It is permissible to choose θ such that $(1 - \theta)\, m = s$. If we do so and use (7.4), we obtain the desired result. □

Remark 7.4. *The space* $\mathscr{D}(\mathbf{R}^n)$ *is dense in* $H^s(\mathbf{R}^n)$, $\forall s$.

Indeed, X is dense in $[X, Y]_\theta$ and therefore if $X_0 \subset X$, X_0 dense in X, we have: X_0 dense in $[X, Y]_\theta$. It is sufficient, for $s \geqq 0$, to apply this with $X = H^m(\mathbf{R}^n)$ (integer $m \geqq s$), $X_0 = \mathscr{D}(\mathbf{R}^n)$, $Y = L^2(\mathbf{R}^n)$; for $s < 0$, we may reason by duality or take $X = L^2(\mathbf{R}^n)$ and $Y = H^{-m}(\mathbf{R}^n)$. It is also possible to verify that the usual approximation by regularization and truncation is valid in $H^s(\mathbf{R}^n)$. □

Theorem 7.3 can be supplemented with

Lemma 7.1. *Let* s *be real and let* φ *be given in* $\mathscr{D}(\mathbf{R}^n)$. *Then there exists a constant* c *and a constant* $c_{s,\varphi}$ *which depends on* s *and* φ *such that,* $\forall u \in H^s(\mathbf{R}^n)$:

$$\|\varphi u\|_{H^s(\mathbf{R}^n)} \leqq c(\sup|\varphi|)\, \|u\|_{H^s(\mathbf{R}^n)} + c_{s,\varphi} \|u\|_{H^{s-1}(\mathbf{R}^n)}. \quad \square$$

We shall give two proofs of this lemma, one now and another in Remark 10.6.

Proof. Thanks to Remark 7.4 we may assume that $u \in \mathscr{D}(\mathbf{R}^n)$.

Since $\widehat{\varphi u} = \hat{\varphi} * \hat{u}$ and since the norm of v in $H^s(\mathbf{R}^n)$ is equivalent to $\|(1+|\xi|)^s \hat{v}\|_{L^2(\mathbf{R}^n)}$, we have

$$\|\varphi u\|_{H^s(\mathbf{R}^n)} = c \left\| (1+|\xi|)^s \int_{\mathbf{R}^n} \hat{\varphi}(\eta)\, \hat{u}(\xi-\eta)\, d\eta \right\|_{L^2(\mathbf{R}^s)} \leqq$$

$$\leqq c \left\| \int_{\mathbf{R}^n} \hat{\varphi}(\eta)\, (1+|\xi-\eta|)^s\, \hat{u}(\xi-\eta)\, d\eta \right\|_{L^2(\mathbf{R}^n)} +$$

$$+ c \left\| \int_{\mathbf{R}^n} \hat{\varphi}(\eta)\, ((1+|\xi|)^s - (1+|\xi-\eta|)^s)\, \hat{u}(\xi-\eta)\, d\eta \right\|_{L^2(\mathbf{R}^n)} .$$

But

$$\left\| \int_{\mathbf{R}^n} \hat{\varphi}(\eta)\, (1+|\xi-\eta|)^s\, \hat{u}(\xi-\eta)\, d\eta \right\|_{L^2(\mathbf{R}^n)}$$

$$= \left\| \int_{\mathbf{R}^n} \hat{\varphi}(\xi-t)\, g(t)\, dt \right\|_{L^2(\mathbf{R}^n)} \quad (\text{where } g(t) = (1+|t|)^s\, \hat{u}(t))$$

$$= \|\hat{\varphi} * g\|_{L^2(\mathbf{R}^n)} = (\text{using Parseval's equality})$$

$$= \|\varphi \hat{g}\|_{L^2(\mathbf{R}^n)} \leqq (\sup|\varphi|)\, \|\hat{g}\|_{L^2(\mathbf{R}^n)} \leqq$$

$$\leqq c(\sup|\varphi|)\, \|u\|_{H^s(\mathbf{R}^n)} .$$

The inequality of the lemma is therefore a consequence of the following inequality:

$$\left\| \int_{\mathbf{R}^n} \hat{\varphi}(\eta)\, ((1+|\xi|)^s - (1+|\xi-\eta|)^s)\, \hat{u}(\xi-\eta)\, d\eta \right\|_{L^2(\mathbf{R}^n)} \leqq$$

$$\leqq c\,|s| \left(\int_{\mathbf{R}^n} |\eta|\, (1+|\eta|)^{|s-1|}\, |\hat{\varphi}(\eta)|\, d\eta \right) \|u\|_{H^{s-1}(\mathbf{R}^n)},$$

which in turn follows from the elementary inequality:

$$|(1+|\xi|)^s - (1+|\xi-\eta|)^s| \leqq$$

$$\leqq |s|\, |\eta|\, (1+|\xi-\eta|)^{s-1}\, (1+(1+|\eta|)^{|s-1|}). \quad \square$$

We also note that from Theorem 7.1 and the definition of $H^s(\mathbf{R}^n)$ it follows that:

Lemma 7.2. *Let φ be given in $\mathscr{D}(\mathbf{R}^n)$ and let D^α be a derivative of arbitrary order $|\alpha|$. Then*

$$u \to \varphi\, D^\alpha u$$

is a continuous linear mapping of $H^s(\mathbf{R}^n) \to H^{s-|\alpha|}(\mathbf{R}^n)$, for all real s.

Remark 7.5. According to the reiteration theorem, if s_1 and $s_2 \in \mathbf{R}$,

$$[H^{s_1}(\mathbf{R}^n), H^{s_2}(\mathbf{R}^n)]_\theta = H^{(1-\theta)s_1+\theta s_2}(\mathbf{R}^n).$$

7.2 Traces on the Boundary of a Half-Space

Consider now $H^m(\Omega)$ for $\Omega = \{x \mid x_n > 0\}$. We note that:

(7.7)
$$\begin{cases} \text{if} & u \in L^2(0, \infty; H^m(\mathbf{R}_{x'}^{n-1})) \\ \text{and} & \dfrac{\partial^m u}{\partial x_n^m} \in L^2(0, \infty; H^0(\mathbf{R}_{x'}^{n-1})), \\ \text{then} & \dfrac{\partial^j u}{\partial x_n^j} \in L^2(0, \infty; H^{m-j}(\mathbf{R}_{x'}^{n-1})). \end{cases}$$

Indeed, if we apply the intermediate derivatives theorem (Theorem 2.3), with $a = 0$, $b = +\infty$, $x_n = t$, $X = H^m(\mathbf{R}_{x'}^{n-1})$, $Y = H^0(\mathbf{R}_{x'}^{n-1})$, then

$$u^{(j)} = \frac{\partial^j u}{\partial x_n^j} \in L^2(0, \infty; [H^m(\mathbf{R}_{x'}^{n-1}), H^0(\mathbf{R}_{x'}^{n-1})]_{j/m}),$$

from which (7.7) follows, since according to (7.4) (replacing n by $(n-1)$):

$$[H^m(\mathbf{R}_{x'}^{n-1}), H^0(\mathbf{R}_{x'}^{n-1})]_{j/m} = H^{(1-j/m)m}(\mathbf{R}_{x'}^{n-1}).$$

Thus, from Theorem 1.3, we obtain

Theorem 7.4. *For $\Omega = \{x \mid x_n > 0\}$, the space $H^m(\Omega)$ defined by* (1.3) *or by* (1.21) *may also be defined by*

(7.8)
$$H^m(\Omega) = \left\{u \mid u \in L^2(0, \infty; H^m(\mathbf{R}_{x'}^{n-1})), \quad \frac{\partial^m u}{\partial x_n^m} \in L^2(0, \infty; H^0(\mathbf{R}_{x'}^{n-1}))\right\}$$

(with equivalent norms). □

Therefore, still with $X = H^m(\mathbf{R}_{x'}^{n-1})$, $Y = H^0(\mathbf{R}_{x'}^{n-1})$, the space $H^m(\Omega)$ may be identified with $W(0, \infty)$ (see Definitions (2.9), (2.10)). Theorem 3.2 joined with Theorem 2.1 then yields the important consequence:

Theorem 7.5. *Let $\Omega = \{x \mid x_n > 0\}$. The space $\mathcal{D}([0, \infty[; H^m(\mathbf{R}_{x'}^{n-1}))$ is dense in $H^m(\Omega)$. The mapping*

(7.9)
$$u \to \left\{\frac{\partial^j u}{\partial x_n^j}(x', 0)\right\}_{j=0}^{m-1}$$

of $\mathcal{D}([0, \infty[; H^m(\mathbf{R}_{x'}^{n-1})) \to (H^m(\mathbf{R}_{x'}^{n-1}))^m$ extends by continuity to a continuous linear mapping of $H^m(\Omega) \to \prod_{j=0}^{m-1} H^{m-j-1/2}(\mathbf{R}_{x'}^{n-1})$, which is surjective.

Proof. It is sufficient to write out

$$[X, Y]_{(j+1/2)/m} = [H^m(\mathbf{R}_{x'}^{n-1}), H^0(\mathbf{R}_{x'}^{n-1})]_{(j+1/2)/m}$$

explicitly, using (7.4). □

Our immediate aim is to extend, in appropriate fashion, the results of this theorem to spaces $H^m(\Omega)$ with Ω an *arbitrary open* set in $\mathbf{R}^n$. In fact, we shall impose *very strong* hypotheses on Ω (and which are amenable to numerous generalizations).

7.3 $H^s(\Gamma)$-Spaces

For the remainder of this book, Ω shall denote an open set in $\mathbf{R}^n$, with

(7.10) $\left\{\begin{array}{l}\text{the boundary } \Gamma \text{ of } \Omega \text{ is a } (n-1) \text{ dimensional infinitely} \\ \text{differentiable variety, } \Omega \text{ being locally on one side of } \Gamma \text{ (i.e.} \\ \text{we consides } \bar{\Omega} \text{ a variety with boundary of class } C^\infty\text{, the boundary} \\ \text{being } \Gamma).\end{array}\right.$

We shall denote by $d\Gamma$ or $d\tau$ the surface measure on Γ, induced by dx.

In general, we shall assume that

(7.11) $$\Omega \text{ is bounded}$$

(except in particular cases such as $\Omega = \mathbf{R}^n$ or $\Omega =$ half-space).

It is hopeless to try to extend Theorem 7.5 to the case of an open set Ω with (7.10) and (7.11) unless $H^s(\Gamma)$ is properly defined. □

Definition of $H^s(\Gamma)$. Let $\mathcal{O}_j$, $j = 1, \ldots, \nu$ be a family of open bounded sets in $\mathbf{R}^n$, covering Γ, such that, for each j, there exists an infinitely differentiable mapping

$$x \to \varphi_j(x) = y$$

of $\mathcal{O}_j \to \mathcal{Q} = \{y \mid y = \{y', y_n\}, |y'| < 1, -1 < y_n < 1\}$ such that φ_j has an inverse,

$$y \to \varphi_j^{-1}(y) = x$$

which is also an infinitely differentiable mapping of $\mathcal{Q} \to \mathcal{O}_j$, φ_j mapping

$$\mathcal{O}_j \cap \Omega \to \mathcal{Q}_+ = \{y \mid y \in \mathcal{Q}, y_n > 0\}$$

(resp. $\mathcal{O}_j \cap \complement\bar{\Omega} \to \mathcal{Q}_- = \{y \mid y \in \mathcal{Q},\ y_n < 0\}$, resp. $\mathcal{O}_j \cap \Gamma \to \mathcal{Q} \cap \{y_n = 0\}$).

Furthermore, let the following compatibility conditions hold: if $\mathcal{O}_j \cap \mathcal{O}_i \neq \emptyset$, there exists an infinitely differentiable homeomorphism J_{ij} of $\varphi_i(\mathcal{O}_i \cap \mathcal{O}_j)$ onto $\varphi_j(\mathcal{O}_i \cap \mathcal{O}_j)$, with positive jacobian, such that

$$\varphi_j(x) = J_{ij}(\varphi_i(x)) \qquad \forall x \in \mathcal{O}_i \cap \mathcal{O}_j.$$

Let $\{\alpha_j\}$ be a partition of unity on Γ having the properties:

$$(7.12)\quad \begin{cases} \alpha_j \in \mathscr{D}(\Gamma) = \text{space of infinitely differentiable functions on } \Gamma, \\ \alpha_j \text{ with compact support in } \mathscr{O}_j \cap \Gamma,\ \sum_{j=1}^{\nu} \alpha_j = 1 \text{ on } \Gamma. \end{cases}$$

If u is a function on Γ (for instance integrable), we decompose

$$(7.13)\qquad u = \sum_{j=1}^{\nu} (\alpha_j u),$$

and define

$$(7.14)\quad \varphi_j^*(\alpha_j u)(y', 0) = (\alpha_j u)(\varphi_j^{-1}(y', 0)), \quad y' \in \mathscr{Q} \cap \{y_n = 0\}.$$

Since α_j has compact support in $\Gamma \cap \mathscr{O}_j$, the function $\varphi_j^*(\alpha_j u)$ has compact support in $\mathscr{Q} \cap \{y_n = 0\}$ and therefore we may also consider $\varphi_j^*(\alpha_j u)$ to be defined in $\mathbf{R}_{y'}^{n-1}$ by extending it by zero out of $\mathscr{Q} \cap \{y_n = 0\}$.

$$u \to \varphi_j^*(\alpha_j u)$$

is a continuous linear mapping of $L^1(\Gamma) \to L^1(\mathbf{R}_{y'}^{n-1})$, of $\mathscr{D}(\Gamma) \to \mathscr{D}(\mathbf{R}_{y'}^{n-1})$ and extends by continuity to a continuous linear mapping of $\mathscr{D}'(\Gamma) \to \to \mathscr{D}'(\mathbf{R}_{y'}^{n-1})$ (easily verified by duality).

Now, we define

$$(7.15)\qquad H^s(\Gamma) = \{u \mid \varphi_j^*(\alpha_j u) \in H^s(\mathbf{R}_{y'}^{n-1})\ j = 1, \ldots, \nu\},$$

which is valid for any real s.

Thanks to the local character of $H^s(\mathbf{R}_{y'}^{n-1})$ (see Theorem 7.3) we see that *the* (algebraic) *definition* (7.15) *is independent of the choice of the system of local maps* $\{\mathscr{O}_j, \varphi_j\}$ *and of the partition of unity* $\{\alpha_j\}$. □

Next, we may take the norm

$$(7.16)\qquad \| u \|_{H^s(\Gamma)} = \left(\sum_{j=1}^{\nu} \| \varphi_j^*(\alpha_j u) \|^2_{H^s(\mathbf{R}_{y'}^{n-1})} \right)^{1/2},$$

which, of course, *depends* on the system $\{\mathscr{O}_j, \varphi_j, \alpha_j\}$. We easily verify that, with (7.16), $H^s(\Gamma)$ is a *Hilbert space* and that the different norms (7.16) are equivalent. We note that

$$(7.17)\qquad \mathscr{D}(\Gamma) \textit{ is dense in } H^s(\Gamma), \quad s \geq 0.$$

Indeed, $\varphi_j^*(\alpha_j u) = \lim_{n \to \infty} \psi_{j,n}$, $\psi_{jn} \in \mathscr{D}(\mathbf{R}_{y'}^{n-1})$ and we may suppose that ψ_{jn} has compact support in $|y'| < 1$, the limit taking place in $H^s(\mathbf{R}_{y'}^{n-1})$ (see Remark 7.4; we arrive at compact supports in $|y'| < 1$ by infinitely differentiable truncation), from which results that $\alpha_j u = \text{limit in } H^s(\Gamma)$ of $(\varphi_j^*)^{-1}(\psi_{jn}) \in \mathscr{D}(\Gamma)$. □

Theorem 7.6. *Identifying $H^0(\Gamma)$ with its dual, we have*

$$(H^s(\Gamma))' = H^{-s}(\Gamma). \tag{7.18}$$

Proof. It is sufficient to consider $s > 0$.

According to definitions (7.15), (7.16) and the Hahn-Banach theorem, every continuous linear form $u \to L(u)$ on $H^s(\Gamma)$ may be written, in non-unique fashion,

$$\begin{cases} L(u) = \sum_{j=1}^{\nu} \langle f_j, \varphi_j^*(\alpha_j u)\rangle, \\ f_j \in (H^s(\mathbf{R}_{y'}^{n-1}))' = H^{-s}(\mathbf{R}_{y'}^{n-1}). \end{cases} \tag{7.19}$$

If $\psi_j \in \mathscr{D}(\mathbf{R}_{y'}^{n-1})$, has compact support in $|y'| < 1$ and is equal to one in a neighborhood of the support of $\varphi_j^*(\alpha_j u)$, we may write (7.19) in the form

$$L(u) = \sum_{j=1}^{\nu} \langle \psi_j f_j, \varphi_j^*(\alpha_j u)\rangle. \tag{7.20}$$

Therefore,

$$L(u) = \left\langle \sum_{j=1}^{\nu} \alpha_j \, {}^t\varphi_j^*(\psi_j f_j), u \right\rangle \tag{7.21}$$

and L may be represented by

$$\sum_{j=1}^{\nu} \alpha_j \, {}^t\varphi_j^*(\psi_j f_j) = f.$$

But then $\varphi_k^* \alpha_k f \in H^{-s}(\mathbf{R}_{y'}^{n-1})$, so that $f \in H^{-s}(\Gamma)$ by definition.

Reciprocally, if $f \in H^{-s}(\Gamma)$, we define $\langle f, u\rangle$ for $u \in \mathscr{D}(\Gamma)$ (then for $u \in H^s(\Gamma)$) by $\sum_{j,k} \langle \alpha_j f, \alpha_k u\rangle$ (for $\alpha_j \cdot \alpha_k \neq 0$) and we define $\langle \alpha_j f, \alpha_k u\rangle$ by returning, via local maps, to the duality between

$$H^{-s}(\mathbf{R}_{y'}^{n-1}) \quad \text{and} \quad H^s(\mathbf{R}_{y'}^{n-1}). \quad \square$$

We also have

Theorem 7.7. *For all $s_1, s_2 \in \mathbf{R}$, $s_1 > s_2$:*

$$[H^{s_1}(\Gamma), H^{s_2}(\Gamma)]_\theta = H^{(1-\theta)s_1 + \theta s_2}(\Gamma), \tag{7.22}$$

with equivalent norms.

Proof. 1) Let $a_1 > a_2$, $\xi \in]0, 1[$. We show that

$$[H^{a_1}(\Gamma), H^{a_2}(\Gamma)]_\xi \subset H^{(1-\xi)a_1 + a_2\xi}(\Gamma) \tag{7.23}$$

(which yields the inclusion of the first space in the second in (7.22)). Indeed, $\forall j$, $u \to \varphi_j^*(\alpha_j u)$ is a continuous linear mapping of

$$H^{a_i}(\Gamma) \to H^{a_i}(\mathbf{R}_{y'}^{n-1})$$

(by definition), therefore, according to the interpolation Theorem 5.1 and Remark 7.5, of

$$[H^{a_1}(\Gamma), H^{a_2}(\Gamma)]_\xi \to [H^{a_1}(\mathbf{R}_{y'}^{n-1}), H^{a_2}(\mathbf{R}_{y'}^{n-1})]_\xi = H^{(1-\xi)a_1+\xi a_2}(\mathbf{R}_{y'}^{n-1}).$$

Therefore, if

$$u \in [H^{a_1}(\Gamma), H^{a_2}(\Gamma)]_\xi,$$

we have

$$\varphi_j^*(\alpha_j u) \in H^{(1-\xi)a_1+\xi a_2}(\mathbf{R}_{y'}^{n-1})$$

and thus

$$u \in H^{(1-\xi)a_1+\xi a_2}(\Gamma), \quad \text{whence (7.23)}.$$

2) Since $\mathscr{D}(\Gamma) \subset [H^{a_1}(\Gamma), H^{a_2}(\Gamma)]_\xi$, this last space is dense in $H^{(1-\xi)a_1+\xi a_2}(\Gamma)$, which together with Theorem 7.6 yields

$$H^{-(1-\xi)a_1-\xi a_2}(\Gamma) \subset [H^{a_1}(\Gamma), H^{a_2}(\Gamma)]_\xi'.$$

But according to the duality Theorem 7.2, this means

$$H^{-(1-\xi)a_1-\xi a_2}(\Gamma) \subset [H^{-a_2}(\Gamma), H^{-a_1}(\Gamma)]_{1-\xi}. \tag{7.24}$$

Choose $-a_1 = s_2$, $-a_2 = s_1$, $1-\xi = \theta$; then (7.24) yields the inclusion from right to left in (7.22). □

Remark 7.6. Use of the Laplace-Beltrami operator.

It is possible to give a more intrinsic definition of the spaces $H^s(\Gamma)$ by using the Laplace-Beltrami operator Δ_Γ on Γ (see de Rham [1]). Indeed, it will follow from Chapter 2, Section 3 that

$$H^{2m}(\Gamma) = \{u \mid u \in L^2(\Gamma), \Delta_\Gamma^m u \in L^2(\Gamma)\}, \tag{7.25}$$

any one of the norms defined by (7.16) (with $s = 2m$) being equivalent to $\|u\|_{L^2(\Gamma)} + \|\Delta^m u\|_{L^2(\Gamma)}$. We may then define $H^s(\Gamma)$ *by interpolation*.

Therefore, we also have (using (7.22) and (7.23)):

$$H^s(\Gamma) = \text{domain of } (-\Delta_\Gamma)^s. \tag{7.26}$$

We use the *eigenfunctions* w_j of $-\Delta_\Gamma$ on Γ (see for example Berezanksi [4], Minakshisundaran-Plejel [1]):

$$-\Delta_\Gamma w_j = \lambda_j w_j, \quad j = 1, \ldots, \tag{7.27}$$

where the w_j's are orthonormalized in $H^0(\Gamma)$.

If u is a distribution on Γ ($u \in \mathscr{D}(\Gamma)$), let $\langle u, \bar{w}_j\rangle$ be *its Fourier coefficients* (relative to $\{w_j\}$). Then, we obtain ($s \in \mathbf{R}$):

$$H^s(\Gamma) = \left\{u \mid u \in \mathscr{D}'(\Gamma), \left(\sum_{j=1}^{\infty} \lambda_j^{2s} |\langle u, \bar{w}_j\rangle|^2\right)^{1/2} < \infty\right\}. \tag{7.28}$$ □

8. Trace Theorem in $H^m(\Omega)$

8.1 Extension and Density Theorems

Theorem 8.1. *Let Ω satisfy* (7.10) *and* (7.11). *Then, for any integer $m > 0$, there exists an operator p with the properties*

(8.1) $$p \in \mathscr{L}(H^k(\Omega), H^k(\mathbf{R}^n)) \quad \forall k, \quad 0 \leqq k \leqq m,$$

(8.2) $$p\, u = u \quad \text{a.e. on } \Omega \quad \forall u \in H^0(\Omega).$$

Proof. 1) Let $\beta_0, \beta_1, \ldots, \beta_\nu$ be a partition of unity in Ω with:

$$\beta_0 \in \mathscr{D}(\Omega), \quad \beta_j \in \mathscr{D}(\bar{\Omega}), \quad 1 \leqq j \leqq \nu, \quad \beta_j \text{ with support in } \mathscr{O}_j \cap \Omega,$$

$$\sum_{j=0}^{\nu} \beta_j = 1 \quad \text{in} \quad \Omega.$$

Then for arbitrary $u \in H^0(\Omega)$ we have:

$$u = \beta_0 u + \sum_{j=1}^{\nu} \beta_j u.$$

We shall define $p(\beta_j u)$, from which we will obtain that p defined by

(8.3) $$p\, u = \beta_0 u + \sum_{j=1}^{\nu} p(\beta_j u) \qquad (\beta_0 u \text{ extended by } 0)$$

has properties (8.1) and (8.2).

2) Let $u \in H^k(\Omega)$. The function $\varphi_j^*(\beta_j u)$ (see Definition of $H^s(\Gamma)$ in Section 7, for the introduction of φ_j and φ_j^*) is in $H^k(\mathbf{R}_+^n)$ ($\mathbf{R}_+^n = \{y \mid y_n > 0\}$), with support in $\mathscr{Q}_+$ and vanishes in the neighborhood of the boundary of $\mathscr{Q}_+$ distinct from $\{y_n = 0\}$. We *extend* $\varphi_j^*(\beta_j u)$ with the help of the *extension by reflection* introduced in the proof of Theorem 2.2 *for* $H^m(\mathbf{R}_+^n)$.

Here we set:

(8.4) $$p * v = \begin{cases} v(x) & \text{if} \quad x_n > 0 \\ \sum_{k=1}^{m} \alpha_k\, v(x', -k\, x_n) & \text{if} \quad x_n < 0 \end{cases}$$

(where α_k is defined by (2.23)).

Then $p * \varphi_j^*(\beta_j u) \in H^k(\mathbf{R}^n)$ if $u \in H^k(\Omega)$ *for* $k \leqq m$ and has compact support in $\mathscr{Q}$. We may set:

(8.5) $$p(\beta_j u) = (\varphi_j^*)^{-1}(p * \varphi_j^*(\beta_j u)),$$

from which the theorem follows. □

Remark 8.1. The *restriction* to Ω

(8.6) $$u \to r_\Omega u = \text{restriction of } u \text{ to } \Omega,$$

is a continuous linear mapping of $H^k(\mathbf{R}^n) \to H^k(\Omega)$, for every integer $k \geqq 0$. □

Theorem 8.2. *If* Ω *satisfies* (7.10) *and* (7.11), *the space* $\mathscr{D}(\overline{\Omega})$ *is dense in* $H^m(\Omega)$.

Proof. There are two methods:

(i) approach $\varphi_j^*(\beta_j u)$ (notation of the proof of Theorem 8.1) with functions of $\mathscr{D}(\overline{\mathscr{Q}}_+)$ vanishing in the neighborhood of the boundary of $\mathscr{Q}_+$ distinct from $\{y_n = 0\}$ (which is permissible according to Theorem 2.1);

(ii) approach $p u$ with $\Phi_\varrho \in \mathscr{D}(\mathbf{R}^n)$ in the sense of $H^m(\mathbf{R}^n)$ (Remark 1.3), then

$$u = r_\Omega(p\, u) = \lim r_\Omega\, \Phi_\varrho \quad \text{in} \quad H^m(\Omega) \quad \text{(Remark 8.1)}$$

and since $r_\Omega\, \Phi_\varrho \in \mathscr{D}(\overline{\Omega})$, the result follows. □

8.2 Trace Theorem

The following result plays a fundamental role in the rest of the book:

Theorem 8.3. *Assume that* Ω *satisfies* (7.10) *and* (7.11). *The mapping*

$$u \to \left\{ \frac{\partial^j u}{\partial \nu^j} \,\middle|\, j = 0, 1, \ldots, m-1 \right\} \tag{8.7}$$

$\left(\text{where } \frac{\partial^j u}{\partial \nu^j} = \text{normal } j\text{-order derivative on } \Gamma, \text{ oriented toward the interior of } \Omega, \text{ to fix the ideas}^1\right)$ *of* $\mathscr{D}(\overline{\Omega}) \to (\mathscr{D}(\Gamma))^m$ *extends by continuity to a continuous linear mapping, still denoted*

$$u \to \left\{ \frac{\partial^j u}{\partial \nu^j} \,\middle|\, j = 0, \ldots, m-1 \right\},$$

of

$$H^m(\Omega) \to \prod_{j=0}^{m-1} H^{m-j-1/2}(\Gamma).$$

This mapping is surjective and there exists a continuous linear right inverse

$$\vec{g} = \{g_j\} \to \mathscr{R}\vec{g} \quad \text{of} \quad \prod_{j=0}^{m-1} H^{m-j-1/2}(\Gamma) \to H^m(\Omega),$$

such that

$$\frac{\partial^j}{\partial \nu^j} \mathscr{R}\vec{g} = g_j, \quad 0 \leqq j \leqq m-1. \tag{8.8}$$

---

[1] In the following chapters, we shall often use the notation

$$\frac{\partial^j u}{\partial \nu^j} = \gamma_j u.$$

Proof. 1) Let $u \in \mathscr{D}(\bar{\Omega})$. With the help of local maps, choosing, as it is permissible to do, the local coordinates so that the image of $\frac{\partial}{\partial \nu}(\alpha_j u)$ under φ_j be

$$\frac{\partial}{\partial y_n} \varphi_j^* (\alpha_j u) |_{y_n=0},$$

we see that according to Theorem 7.5 and the definition of $H^s(\Gamma)$ the mapping (8.7) is continuous for the topologies indicated in the statement of the theorem. According to Theorem 8.2, we may extend to $H^m(\Omega)$.

2) The surjectivity follows from the surjectivity in Theorem 7.5. The existence of the right inverse follows from Remark 3.3 (which gives an *explicit* right inverse) or from the following general remark (but particular to Hilbert spaces, which is not the case with the first method): if $\mathscr{H}$ and $\mathscr{K}$ are two Hilbert spaces and if $\mathscr{T} \in \mathscr{L}(\mathscr{H}; \mathscr{K})$, where $\mathscr{T}$ is surjective, then $\mathscr{T}$ is an isomorphism of $\mathscr{H}/_{\mathrm{Ker}\,\mathscr{T}}$ onto $\mathscr{K}$ ($\mathrm{Ker}\,\mathscr{T}$ = kernel of $\mathscr{T}$); but $\mathscr{H}$ being a Hilbert space, $\mathscr{H}/_{\mathrm{Ker}\,\mathscr{T}}$ is identified to a subspace of $\mathscr{H}$, say $\mathscr{H}_0$, and it is sufficient to "right invert" $\mathscr{T}$ by the inverse of the isomorphism of $\mathscr{H}_0$ onto $\mathscr{K}$, defined by $\mathscr{T}$. □

9. The Spaces $H^s(\Omega)$, Real $s \geqq 0$

9.1 Definition by Interpolation

By definition, we set

(9.1)
$$H^s(\Omega) = [H^m(\Omega), H^0(\Omega)]_\theta, \quad (1-\theta)\, m = s, \quad m \text{ integer}, \quad 0 < \theta < 1.$$

Clearly, there are several items to be verified:

(i) up to an equivalence in norms, definition (9.1) depends only on s and not on the *choice of* m (as long as $(1-\theta)\, m = s$);

(ii) *when* s *is an integer,* definition (9.1) gives back definition (1.1) (still up to an equivalence in norms).

Such is the case *when* Ω *has a sufficiently regular boundary,* as the following theorems show.

Theorem 9.1. *Assume that* Ω *satisfies* (7.10) *and* (7.11). *Then* $H^s(\Omega)$ *coincides* (algebraically) *with the space of restrictions to* Ω *of the elements of* $H^s(\mathbf{R}^n)$.

Proof. 1) Consider the operator p of Theorem 8.1. We have:

$$p \in \mathscr{L}(H^0(\Omega); H^0(\mathbf{R}^n)) \cap \mathscr{L}(H^m(\Omega); H^m(\mathbf{R}^n)),$$

therefore, by the interpolation Theorem 5.1:

$$p \in \mathscr{L}([H^m(\Omega), H^0(\Omega)]_\theta; [H^m(\mathbf{R}^n), H^0(\mathbf{R}^n)]_\theta).$$

But according to definition (9.1) and Theorem 7.1, we have:

$$p \in \mathscr{L}(H^s(\Omega); H^s(\mathbf{R}^n)).$$

2) In the same way, using the "restriction to Ω" operator r_Ω, we have:

$$r_\Omega \in \mathscr{L}(H^s(\mathbf{R}^n); H^s(\Omega)).$$

3) Then every $u \in [H^m(\Omega), H^0(\Omega)]_\theta = H^s(\Omega)$ is of the form:

$$u = r_\Omega\, p\, u,$$

i.e. is a restriction to Ω of $p\, u \in H^s(\mathbf{R}^n)$ and every $u = r_\Omega\, v$, $v \in H^s(\mathbf{R}^n)$, belongs to $H^s(\Omega)$. □

Theorem 9.2. *The norm of $H^s(\Omega)$ defined by (9.1) is equivalent to the norm*

(9.2) $$|\!|\!| u |\!|\!|_{H^s(\Omega)} = \operatorname{Inf} \| U \|_{H^s(\mathbf{R}^n)}, \qquad U = u \quad \text{a.e. on } \Omega.$$

Proof. The norm (9.2) amounts to providing $H^s(\Omega)$ with the norm of the quotient of $H^s(\mathbf{R}^n)$ by the *closed* subspace of functions vanishing a.e. outside of Ω. Then, $H^s(\Omega)$ is a Hilbert space for the norm $\| u \|_{H^s(\Omega)}$ associated to (9.1) and for (9.2); and since

$$|\!|\!| u |\!|\!|_{H^s(\Omega)} \leqq \| p\, u \|_{H^s(\mathbf{R}^n)} \leqq C \, \| u \|_{H^s(\Omega)},$$

we have the equivalence of norms. □

Also, note

Theorem 9.3. *Assume that Ω satisfies* (7.10) *and* (7.11). *Then $\mathscr{D}(\bar{\Omega})$ is dense in $H^s(\Omega)$, $s \geqq 0$.*

Proof. According to Remark 2.6, $H^m(\Omega)$ is dense in $H^s(\Omega)$, $((1-\theta)m = s)$. Hence, the theorem follows as a consequence of Theorem 8.2. □

Furthermore, we deduce from the preceding results that

$$H^s(\Omega) = [H^{k+1}(\Omega), H^k(\Omega)]_{1-\theta} \quad \text{if} \quad s = k + \theta, \quad 0 < \theta < 1,$$
$$\text{integer } k \geqq 0,$$

which allows us to define $H^s(\Omega)$ by interpolation "between two *consecutive* integers". □

9.2 Trace Theorem in $H^s(\Omega)$

Theorem 9.4. *Assume that Ω satisfies* (7.10) *and* (7.11). *Then, the mapping*

(9.3) $$u \to \left\{ \frac{\partial^j u}{\partial \nu^j} \;\middle|\; j = 0, 1, \ldots, \mu \right\}$$

of $\mathcal{D}(\bar{\Omega}) \to (\mathcal{D}(\Gamma))^m$ *extends by continuity to a continuous linear mapping*

$$u \to \left\{ \frac{\partial^j u}{\partial \nu^j} \,\middle|\, j = 0, \dots, \mu \right\} \quad \text{of} \quad H^s(\Omega) \to \prod_{j=0}^{\mu} H^{s-j-1/2}(\Gamma),$$

where μ *is the greatest integer such that*

$$\mu < s - \tfrac{1}{2}. \tag{9.4}$$

The mapping $u \to \left\{ \frac{\partial^j u}{\partial \nu^j} \,\middle|\, j = 0, \dots, \mu \right\}$ *is surjective and there exists a continuous right inverse mapping of*

$$\prod_{j=0}^{\mu} H^{s-j-1/2}(\Gamma) \to H^s(\Omega).$$

Proof. 1) Let $u \in H^s(\Omega)$. The space $H^s(\Omega)$ has a *local* character (as in Theorem 7.3), i.e., if $\beta_0, \beta_1, \dots, \beta_\nu$ is the partition of unity introduced in part 1) of the proof of Theorem 8.1, then

$$\beta_j u \in H^s(\Omega), \quad 0 \leqq j \leqq \nu.$$

We may reason on $U \in H^s(\mathbf{R}^n)$, $U = u$ a.e. on Ω, as well (according to Theorems 9.1 and 9.2). We may assume that $\beta_j \in \mathcal{D}(\mathbf{R}^n)$, with support in $\mathcal{O}_j$, $1 \leqq j \leqq \nu$, $\sum_{j=0}^{\nu} \beta_j = 1$ in a neighbourhood of $\bar{\Omega}$. Then $\beta_j U \in H^s(\mathbf{R}^n)$, $\beta_j U$ has support in $\mathcal{O}_j$, $1 \leqq j \leqq \nu$ ($\beta_0 U$ with compact support in Ω).

$$U \to \varphi_j^*(\beta_j U)$$

defined by φ_j (local map), is a continuous linear mapping of

$$H^m(\mathbf{R}_x^n) \quad (\text{resp. } H^0(\mathbf{R}_x^n)) \to H^m(\mathbf{R}_y^n) \quad (\text{resp. } H^0(\mathbf{R}_y^n))$$

and therefore, by interpolation, of $H^s(\mathbf{R}_x^n) \to H^s(\mathbf{R}_y^n)$. So that:

$$\varphi_j^*(\beta_j U) \in H^s(\mathbf{R}_y^n) \ \forall j. \tag{9.5}$$

2) But it is easily verified that we may "separate the variables" in the definition of $H^s(\mathbf{R}^n)$ (compare with (7.2));

$$H^s(\mathbf{R}^n) = \{ v \mid v \in L^2(\mathbf{R}_{x_n}; H^s(\mathbf{R}_{x'}^{n-1})), |y_n|^s \tilde{v} \in L^2(\mathbf{R}_{y_n}; H^0(\mathbf{R}_{x'}^{n-1})) \} \tag{9.6}$$

where

$$\begin{cases} \tilde{v}(x', y_n) = \text{Fourier transform in } x_n \text{ of } v(x', x_n), \text{ therefore equal to} \\ \dfrac{1}{\sqrt{2\pi}} \displaystyle\int_{-\infty}^{+\infty} \exp(-i x_n y_n)\, v(x', x_n)\, d x_n. \end{cases} \tag{9.7}$$

Indeed, applying the Fourier transform *in* x' to definition (9.6), we obtain

$$(1 + |y'|^2)^{s/2}\, \hat{v} \in L^2(\mathbf{R}^n_y),$$

$$|y_n|^s\, \hat{v} \in L^2(\mathbf{R}^n_y),$$

which is equivalent to $(1 + |y|^2)^{s/2}\, \hat{v} \in L^2(\mathbf{R}^n_y)$.

3) We apply Theorem 4.2 with $t = x_n$ ($\tau = y_n$) and

$$X = H^s(\mathbf{R}^{n-1}_{x'}), \qquad Y = H^0(\mathbf{R}^{n-1}_{x'}).$$

We obtain that

$$\frac{\partial^k}{\partial y_n^k}(\varphi_j^*(\beta_j U))|_{y_n=0} \in H^{s-k-1/2}(\mathbf{R}^{n-1}_{x'}), \qquad s - k - \tfrac{1}{2} > 0,$$

from which by "getting onto Γ again" (and choosing the local coordinates so that $\dfrac{\partial}{\partial \nu}$ becomes $\left.\dfrac{\partial}{\partial y_n}\right|_{y_n=0}$):

$$\frac{\partial^k}{\partial \nu^k}(\beta_j U)\left(\text{or } \frac{\partial^k}{\partial \nu^k}(\beta_j u)\right) \in H^{s-k-1/2}(\Gamma).$$

4) The surjectivity results, by the same procedure, from the surjectivity in Theorem 4.2. The continuity of the right inverse results, either from an explicit construction starting with (4.8) or from the general remark at the end of the proof of Theorem 8.3. □

Condition (9.4) cannot be weakened:

Theorem 9.5. *Assume that* Ω *satisfies* (7.10) *and* (7.11). *Suppose*

(9.8) $$\mu \geqq s - \tfrac{1}{2}.$$

Then, for any $\varphi \in \mathcal{D}(\Gamma)$, $\varphi \neq 0$, *the linear form*

(9.9) $$u \to \int_\Gamma \frac{\partial^\mu u}{\partial \nu^\mu} \varphi \, d\Gamma, \qquad u \in \mathcal{D}(\bar{\Omega})$$

is not continuous for the topology induced by $H^s(\Omega)$.

Proof. As above, we are brought back, via local maps, to the application of Theorem 4.3. □

9.3 Interpolation of $H^s(\Omega)$-Spaces

Theorem 9.6. *Assume that* Ω *satisfies* (7.10) *and* (7.11). *Then*

(9.10) $$[H^{s_1}(\Omega), H^{s_2}(\Omega)]_\theta = H^{(1-\theta)s_1+\theta s_2}(\Omega)$$

for all $s_i > 0$, $s_2 < s_1$, $0 < \theta < 1$ (with equivalent norms).

Proof. Let integer $m \geqq \max(s_1, s_2)$. Then

$$H^{s_i}(\Omega) = [H^m(\Omega), H^0(\Omega)]_{\theta_i}, \quad (1 - \theta_i)\, m = s_i, \quad i = 1, 2,$$

so that

$$[H^{s_1}(\Omega), H^{s_2}(\Omega)]_\theta = [[H^m(\Omega), H^0(\Omega)]_{\theta_1}, [H^m(\Omega), H^0(\Omega)]_{\theta_2}]_\theta$$
$$= [H^m(\Omega), H^0(\Omega)]_{(1-\theta)\theta_1 + \theta\theta_2}$$

(the last step follows from Theorem 6.1), from which (9.10) follows. □

Remark 9.1. From (9.10) and (2.43), we deduce the *interpolation inequality*

$$\| u \|_{H^{(1-\theta)s_1 + \theta s_2}(\Omega)} \leqq c(\theta, s_1, s_2)\, \| u \|_{H^{s_1}(\Omega)}^{1-\theta}\, \| u \|_{H^{s_2}(\Omega)}^{\theta}. \quad □$$

Theorem 9.7. *Assume that Ω satisfies* (7.10) *and* (7.11). *Then, the two conditions*

(9.11) $$u \in H^s(\Omega),$$

(9.12) *if m is an integer* $\leqq s$, $u \in H^{s-m}(\Omega)$, $D^p u \in H^{s-m}(\Omega)$, $\forall |p| = m$,

are equivalent.

Furthermore, the norms

$$\| u \|_{H^s(\Omega)} \quad \text{and} \quad \Big(\| u \|^2_{H^{s-m}(\Omega)} + \sum_{|p| = m} \| D^p u \|^2_{H^{s-m}(\Omega)} \Big)^{1/2}$$

are equivalent.

Proof. 1) It is sufficient to show the result for $m = 1$ and to apply it iteratively.

2) Let m_1 and m_2 be integers, $m_1 > s$, $m_2 \leqq s$, $m_2 \geqq 1$.

Since $\partial/\partial x_i$ is a continuous linear operator of $H^{m_j}(\Omega) \to H^{m_j - 1}(\Omega)$, $j = 1, 2$, it is, by interpolation, a continuous operator of

$$[H^{m_1}(\Omega), H^{m_2}(\Omega)]_\theta \to [H^{m_1 - 1}(\Omega), H^{m_2 - 1}(\Omega)]_\theta.$$

Therefore, taking $(1 - \theta)\, m_1 + \theta\, m_2 = s$, we see that

$$\frac{\partial}{\partial x_i} \in \mathscr{L}((H^s(\Omega); H^{s-1}(\Omega)),$$

which shows that (9.11) implies (9.12).

3) Conversely, let

$$u \in H^{s-1}(\Omega) \quad \text{with} \quad \frac{\partial u}{\partial x_i} \in H^{s-1}(\Omega), \quad i = 1, \ldots, n.$$

According to the proof of Theorem 9.1 and the construction of p (see also Theorem 8.1), we can find an operator p having the properties

(9.13) $$p \in \mathscr{L}(H^\sigma(\Omega); H^\sigma(\mathbf{R}^n)), \quad 0 \leqq \sigma \leqq s,$$

(9.14) $$p\, u = u \quad \text{a.e. on } \Omega, \quad \forall u \in H^0(\Omega),$$

(9.15) $$\begin{cases} \dfrac{\partial}{\partial x_i}(p\, u) = q\left(\dfrac{\partial u}{\partial x_i}\right). \\ \text{where } q \in \mathscr{L}(H^\sigma(\mathbf{R}^n); H^\sigma(\mathbf{R}^n)), \ 0 \leqq \sigma \leqq s-1 \\ q\, u = u \quad \text{a.e. on } \Omega \quad \forall u \in H^0(\Omega). \end{cases}$$

Then:

$$p\, u \in H^{s-1}(\mathbf{R}^n),$$

$$\frac{\partial}{\partial x_i}(p\, u) = q\left(\frac{\partial u}{\partial x_i}\right) \in H^{s-1}(\mathbf{R}^n),$$

which implies $p\, u \in H^s(\mathbf{R}^n)$ (immediate verification by the use of the Fourier transform), from which the theorem follows. □

9.4 Regularity Properties of $H^s(\Omega)$-Functions

It is natural to *compare* $H^s(\Omega)$-spaces to *other* spaces, the simplest being $L^p(\Omega)$-spaces, $p \neq 2$, and the usual spaces of differentiable functions on $\bar{\Omega}$.

The comparison of $H^s(\Omega)$ with $L^p(\Omega)$ (or with Sobolev-spaces constructed on $L^p(\Omega)$) results, for *integer* s, from the *Sobolev inequalities*, for which we refer the reader to Sobolev [2] and Gagliardo [1].

For non-integer s, there are, in this connection, the "fractional Sobolev inequalities"; we refer the reader to J. Peetre [12] and to the bibliography of this work.

Here, we give a very simple result pertaining to the comparison between $H^s(\Omega)$ and

(9.16) $$C^0(\bar{\Omega}) = \{v \mid v \text{ continuous on } \bar{\Omega}\},$$

which is a Banach space for the norm $\max\limits_{x \in \Omega} |v(x)|$.

Theorem 9.8. *Assume that* Ω *satisfies* (7.10) *and* (7.11). *Then, if*

(9.17) $$s > \frac{n}{2} \quad (\Omega \subset \mathbf{R}^n),$$

we have

(9.18) $$H^s(\Omega) \subset C^0(\bar{\Omega})$$

with continuous injection.

Proof. Let $u \in H^s(\Omega)$. We introduce (see proof of Theorem 9.1)

$$(9.19)\quad \begin{cases} v = p\, u \in H^s(\mathbf{R}^n) \\ v = u \quad \text{a.e. on } \Omega. \end{cases}$$

Let $\hat{v}$ be the Fourier transform of v. Then

$$\hat{v} = (1 + |\xi|)^{-s}(1 + |\xi|)^s\, \hat{v} \in L^1(\mathbf{R}^n_\xi)$$

and

$$\|\hat{v}\|_{L^1(\mathbf{R}^n_\xi)} \leqq c\, \|v\|_{H^s(\mathbf{R}^n)}.$$

Therefore $v =$ (inverse) Fourier transform of $\hat{v} \in L^1(\mathbf{R}^n_\xi)$ is (a.e. equal to) a *continuous* function (vanishing at infinity), so that by restriction to Ω we obtain the desired result. □

Remark 9.2. The result of Theorem 9.8 cannot be improved. For example, if $n = 2$, $s = 1$, $H^1(\Omega)$ is not contained in $C^0(\bar{\Omega})$; indeed, the function

$$x \to \log|\log|x||,$$

defined on the disk Ω: $|x| < \frac{1}{2}$, belongs to $H^1(\Omega)$ but not to $C^0(\bar{\Omega})$. □

Remark 9.3. The injection (9.18) is to be understood in the sense: if $u \in H^s(\Omega)$, it is a.e. equal to an element of $C^0(\bar{\Omega})$, identified to u. □

Remark 9.4. We have implicitly shown that

(9.20) if (9.17) holds, then $H^s(\mathbf{R}^n) \subset$ space of continuous functions vanishing at infinity (and even, more precisely, $H^s(\mathbf{R}^n) \subset \mathscr{F} L^1$ = image of $L^1(\mathbf{R}^n)$ under Fourier transformation). □

From Theorem 9.8, we immediately deduce the following corollaries:

Corollary 9.1. *Assume that* Ω *satisfies* (7.10) *and* (7.11). *Then, if*

$$(9.21)\quad s > \frac{n}{2} + m, \quad \text{integer } m > 0,$$

we have

$$(9.22)\quad H^s(\Omega) \subset C^m(\bar{\Omega}) = \text{space of } m\text{-times continuously differentiable functions on } \bar{\Omega}.$$

$\left(\text{with continuous injection, } C^m(\bar{\Omega}) \text{ having the norm } \sum_{|\alpha| \leqq m} \max_{x = \bar{\Omega}} |D^\alpha v(x)|\right).$

Corollary 9.2. *Assume that* Ω *satisfies* (7.10) *and* (7.11). *Then*

$$(9.23)\quad \bigcap_{k=0}^{\infty} H^k(\Omega) = \mathscr{D}(\bar{\Omega}).$$

Remark 9.5. If X is a Hilbert space, we define (integer $m \geqq 0$)

$$H^m(\Omega; X) = \{v \mid D^p v \in L^2(\Omega; X)\ \forall\, |p| \leqq m\},$$

where $L^2(\Omega; X)$ = space of (classes of) square integrable functions on Ω, taking their values in X; provided with the norm

$$\left(\sum_{|p| \leqq m} \int_{\Omega} \| D^p v(x) \|_X^2 dx \right)^{1/2},$$

it is a Hilbert space. Of course

$$H^0(\Omega; X) = L^2(\Omega; X).$$

Then, for real $s > 0$, we define

$$H^s(\Omega; X) = [H^m(\Omega; X), H^0(\Omega; X]_\theta, \quad (1-\theta)\, m = s;$$

it can be shown that this space depends only on s (and not on m), up to equivalence of norms.

Further, it can be shown (by the same type of methods as in the preceding proofs) that, if Y is a second Hilbert space (as in Section 2), we have

(9.24) $$[H^{s_1}(\Omega; X), H^{s_2}(\Omega; Y)]_\theta = H^{(1-\theta)s_1+\theta s_2}(\Omega; [X, Y]_\theta).$$

In the same way, the *scalar* setting considered in the following sections extends to the *vector* case, as in (9.24). □

10. Some Further Properties of the Spaces $[X, Y]_\theta$

10.1 Domains of Semi-Groups

Let X, Y be two Hilbert spaces satisfying (2.1). Let $G(t)$ be *a continuous semi-group* on Y, that is:

(10.1) $$\begin{cases} G(t) \in \mathscr{L}(Y; Y) \quad \forall t \geqq 0, \quad G(0) = \text{identity} = I, \\ \forall y \in Y, \quad t \to G(t)y \text{ is a continuous mapping of } t \geqq 0 \to Y, \\ G(t)\,G(s) = G(t+s) \quad \forall t, s \geqq 0. \end{cases}$$

Let A be the *infinitesimal generator of* $G(t)$, with domain $D(A)$ (see Hille-Phillips [1], Yosida [2]), a Hilbert space for the norm of the graph $(\|y\|_Y^2 + \|A\, y\|_Y^2)^{1/2}$.

We assume that:

(10.2) $$D(A) = X \text{ (with equivalent norms).} \quad □$$

Remark 10.1. There exists an infinity of semi-groups having property (10.2); for example, Λ being defined by (2.5) and (2.6) (and there is an infinity of such Λ's), we can take

(10.3) $$G(t) = \exp(-t\Lambda)$$

(or $G(t) = \exp(i\, t\, \Lambda)$, unitary group). □

Remark 10.2. At the risk of having to change $G(t)$ to $e^{-\omega t} G(t)$, for a suitable ω (which changes A, but not $D(A)$), we may always assume that $\| G(t) \|_{\mathscr{L}(Y;Y)}$ is bounded. □

Theorem 10.1. *Let X, Y satisfy* (2.1). *The three following statements are equivalent* $(0 < \theta < 1)$:

$$a \in [X, Y]_\theta, \tag{10.4}$$

$$\begin{cases} a = u(0), & t^\alpha u \in L^2(0, \infty; X), \\ t^\alpha \dfrac{du}{dt} \in L^2(0, \infty; Y), & \theta = \dfrac{1}{2} + \alpha, \end{cases} \tag{10.5}$$

$$t^{\alpha-1}(G(t)\, a - a) \in L^2(0, \infty; Y). \tag{10.6}$$

Furthermore, the norms

$$\| a \|_{[X,Y]_\theta} \quad \textit{and} \quad \left(\| a \|_Y^2 + \int_0^\infty t^{2(\alpha-1)} \| G(t)\, a - a \|_Y^2\, dt \right)^{1/2}$$

are equivalent.

Proof. 1) (10.5) ⇒ (10.6).
Set

$$-A\, u + \frac{du}{dt} = f \tag{10.7}$$

and note that $t^\alpha f \in L^2(0, \infty; Y)$.
Then (solution of Cauchy's problem; see for example Yosida [2])

$$u(t) = G(t)\, a + \int_0^t G(t - \sigma)\, f(\sigma)\, d\sigma$$

and therefore

$$G(t)\, a - a = u(t) - u(0) - \int_0^t G(t - \sigma)\, f(\sigma)\, d\sigma,$$

so that

$$t^{-1}(G(t)\, a - a) = \frac{1}{t} \int_0^t u'(\sigma)\, d\sigma - \frac{1}{t} \int_0^t G(t - \sigma)\, f(\sigma)\, d\sigma. \tag{10.8}$$

Therefore

$$\| t^{-1}(G(t)\, a - a) \|_Y \leq \frac{1}{t} \int_0^t \| u'(\sigma) \|_Y\, d\sigma + \frac{c}{t} \int_0^t \| f(\sigma) \|_Y\, d\sigma$$

and (10.6) will follow as a consequence of

Lemma 10.1. *For* $\alpha < \frac{1}{2}$:

$$\left\| t^{\alpha} \left(\frac{1}{t} \int_0^t g(\sigma)\, d\sigma \right) \right\|_{L^2(0,\infty)} \leqq (\text{constant}) \, \| t^{\alpha} g(t) \|_{L^2(0,\infty)} . \tag{10.9}$$

Proof. Setting $t = e^x$, $\sigma = e^y$, $e^{(\alpha - 1/2)x} g(e^x) = \tilde{g}(x)$, (10.9) is equivalent to

$$\left\| \int_{-\infty}^{x} e^{(\alpha - 1/2)(x-y)} \tilde{g}(y)\, dy \right\|_{L^2(\mathbf{R})} \leqq (\text{constant}) \, \| \tilde{g} \|_{L^2(\mathbf{R})} .$$

But

$$\int_{-\infty}^{x} e^{(\alpha - 1/2)(x-y)} \tilde{g}(y)\, dy = E * \tilde{g}(x),$$

where

$$E(x) = \begin{cases} \exp(\alpha - \frac{1}{2})\, x, & x > 0, \\ 0, & x \leqq 0. \end{cases}$$

Since $\theta < 1$, we have $\alpha < \frac{1}{2}$ and therefore

$$\int_{-\infty}^{+\infty} E(x)\, dx = \frac{1}{\frac{1}{2} - \alpha},$$

whence (10.9) (and since $E(x) \geqq 0$, *the best constant* in (10.9) is $1/(\frac{1}{2} - \alpha)$). □

2) (10.6) ⇒ (10.5).

Let a satisfy (10.6); construct u as follows:

$$\begin{cases} u(t) = q(t) \dfrac{1}{t} \displaystyle\int_0^t G(\sigma)\, a\, d\sigma, \\ q \in C^1([0, +\infty[) \text{ and has compact support}, \\ q(0) = 1 \end{cases} \tag{10.10}$$

We have: $u(0) = a$. We have to show:

$$t^{\alpha} A u \in L^2(0, \infty; Y), \tag{10.11}$$

$$t^{\alpha} u' \in L^2(0, \infty; Y). \tag{10.12}$$

But in general (see Yosida [2])

$$A\left(\int_0^t G(\sigma)\, a\, d\sigma\right) = G(t)\, a - a,$$

so that

$$t^\alpha A\, u = q(t)\, t^{\alpha-1}(G(t)\, a - a)$$

and (10.11) follows from (10.6). Next

$$u' = q'\left(\frac{1}{t}\int_0^t G(\sigma)\, a\, d\sigma\right) + q\, v' \quad \text{if} \quad v(t) = \frac{1}{t}\int_0^t G(\sigma)\, a\, d\sigma$$

and (10.12) will obviously follow from

(10.13) $$t^\alpha v' \in L^2(0, \infty; Y).$$

But

$$v'(t) = \frac{1}{t} G(t)\, a - \frac{1}{t^2}\int_0^t G(\sigma)\, a\, d\sigma$$

$$= \frac{1}{t}(G(t)\, a - a) - \frac{1}{t^2}\int_0^t (G(\sigma)\, a - a)\, d\sigma$$

and therefore (10.13) will follow from

$$w(t) = t^{\alpha-2}\int_0^t (G(\sigma)\, a - a)\, d\sigma \in L^2(0, \infty; Y).$$

Now $G(\sigma)\, a - a = \sigma^{1-\alpha}\varphi(\sigma)$, $\varphi \in L^2(0, \infty; Y)$, therefore

$$w(t) = t^{\alpha-2}\int_0^t \sigma^{1-\alpha}\varphi(\sigma)\, d\sigma,$$

and (since $\sigma \leqq t$)

$$\| w(t) \|_Y \leqq t^\alpha \frac{1}{t}\int_0^t \sigma^{-\alpha} \| \varphi(\sigma) \|_Y\, d\sigma,$$

from which the result follows, by (10.9). □

3) (10.4) ⇒ (10.6).

Since (10.5) ⇔ (10.6) and (10.5) is *independent* of the semi-group $G(t)$ (as long as $D(A) = X$), it is sufficient to show the equivalence of (10.4) and (10.6) for *one* particular semi-group $G(t)$; we consider (10.3). Using spectral decomposition, (10.6) is equivalent to (setting

$b(\lambda) = \mathcal{U}\, a(\lambda))$

$$\int_0^{+\infty}\int_{\lambda_0}^{+\infty} t^{(\alpha-1)}(1 - e^{-t\lambda})^2\, \|b(\lambda)\|^2_{\mathfrak{h}(\lambda)}\, dt\, d\mu(\lambda) < \infty .$$

Since $\int_0^{+\infty} t^{2(\alpha-1)}(1 - e^{-t\lambda})^2\, dt = c\,\lambda^{1-2\alpha}$, the above integral becomes

$$c\int_{\lambda_0}^{+\infty} \lambda^{1-2\alpha}\, \|b(\lambda)\|^2_{\mathfrak{h}(\lambda)}\, d\mu(\lambda) = c\int_{\lambda_0}^{+\infty} \lambda^{2(1-\theta)}\, \|b(\lambda)\|^2_{\mathfrak{h}(\lambda)}\, d\mu(\lambda)$$
$$= c\, \|a\|_{[X,Y]_\theta},$$

which gives the desired result and the equivalence of norms. □

Remark 10.3. The equivalence of (10.5) and (10.6) extends to the case of Banach spaces and with L^2 replaced by L^p. □

Remark 10.4. Let $G_1, G_2, \ldots, G_n$ be n continuous semi-groups on Y, which are *commutative*, i.e.

(10.14) $$G_i(s)\, G_j(t) = G_j(t)\, G_i(s) \quad \forall i, j, s, t.$$

Let A_i be the infinitesimal generator of G_i with domain $D(A_i)$. Suppose

(10.15) $$\begin{cases} X = \bigcap_{i=1}^{n} D(A_i) \\ \|u\|_X \text{ is equivalent to } \left(\|u\|_Y^2 + \sum_{i=1}^{n} \|A_i u\|_Y^2\right)^{1/2}. \end{cases}$$

Then, if we assume $\|G_i(t)\|_{\mathcal{L}(Y;Y)} \leq$ (constant) $\forall i$, we have:

(10.16) $$\begin{cases} \textit{condition (10.5) is equivalent to} \\ t^{\alpha-1}(G_i(t)\, a - a) \in L^2(0, \infty; Y) \quad \forall i = 1, \ldots, n. \end{cases}$$

Indeed, (10.5) ⇒ (10.16) is obvious since (10.5) ⇒ (10.6). To show that (10.16) ⇒ (10.5), we "right invert" a by the formula (compare with (10.10)):

(10.17) $$u(t) = q(t)\frac{1}{t^n}\int_0^t G_1(\sigma)\, d\sigma \int_0^t G_2(\sigma)\, d\sigma \ldots \int_0^t G_n(\sigma)\, d\sigma \cdot a .$$

We finish the proof as above. □

10.2 Application to $H^s(\mathbf{R}^n)$

We apply Theorem 10.1 and Remark 10.4 to $H^s(\mathbf{R}^n)$, for $0 < s < 1$. We have:

(10.18) $$H^s(\mathbf{R}^n) = [H^1(\mathbf{R}^n), H^0(\mathbf{R}^n)]_\theta, \quad 1 - \theta = s.$$

We consider the *translation semi-groups* (in fact, groups):

$$G_i(t) f(x) = f(x_1, \ldots, x_{i-1}, x_i + t, x_{i+1}, \ldots, x_n) \tag{10.19}$$

in

$$Y = L^2(\mathbf{R}^n) = H^0(\mathbf{R}^n). \tag{10.20}$$

The theory applies. Therefore, according to (10.16), we have:

$$u \in H^s(\mathbf{R}^n) \Leftrightarrow t^{\alpha-1}(G_i(t)\, u(x) - u(x)) \in L^2(0, \infty; Y)$$

i.e.

$$\int_0^\infty t^{(\alpha-1)}\, dt \int_{\mathbf{R}^n} |u(x_1, \ldots, x_i + t, \ldots, x_n) - u(x)|^2\, dx < \infty.$$

Therefore, since $2(\alpha - 1) = -(2s + 1)$, we have:

Theorem 10.2. *For $0 < s < 1$, the following conditions are equivalent:*

$$u \in H^s(\mathbf{R}^n), \tag{10.21}$$

$$\begin{cases} u \in L^2(\mathbf{R}^n) \text{ and} \\ \displaystyle\int_0^\infty t^{-(2s+1)}\, dt \int_{\mathbf{R}^n} |u(x_1, \ldots, x_i + t, \ldots, x_n) - u(x)|^2\, dx < \infty, \\ i = 1, \ldots, n. \end{cases} \tag{10.22}$$

Furthermore, the norms $\| u \|_{H^s(\mathbf{R}^n)}$ and

$$\left(\| u \|_{L^2(\mathbf{R}^n)} + \sum_{i=1}^n \int_0^\infty t^{-(2s+1)}\, dt \int_{\mathbf{R}^n} |u(x_1, \ldots, x_i + t, \ldots, x_n) - u(x)|^2\, dx \right)^{1/2}$$

are equivalent. □

Remark 10.5. Conditions (10.22) may be replaced by the *equivalent* conditions:

$$\begin{cases} u \in L^2(\mathbf{R}^n), \\ \displaystyle\iint_{\mathbf{R}^n \times \mathbf{R}^n} \frac{1}{|x - y|^{n+2s}} |u(x) - u(y)|^2\, dx\, dy < \infty \end{cases}$$

(the norm in Theorem 10.2 being equivalent to

$$\left(\| u \|^2_{L_2(\mathbf{R}^n)} + \iint_{\mathbf{R}^n \times \mathbf{R}^n} \frac{1}{|x - y|^{n+2s}} |u(x) - u(y)|^2\, dx\, dy \right)^{1/2} \Bigg).$$

This equivalence may be verified by Fourier transformation, for example. □

Remark 10.6. We now give another (neater) proof of Lemma 7.1.

Let us define

$$A_0 = H^{m+1} \subset A_1 = H^m = B_0 \subset B_1 = H^{m-1}, \tag{10.23}$$

where m is a positive or negative integer.

We assume that

$$s = m + (1 - \theta). \tag{10.24}$$

Let $G(t)$ be the semi-group defined in *all* these spaces by

$$\widehat{G(t)\, v}(\xi) = \exp(-t(1 + |\xi|))\, \hat{v}(\xi).$$

Then A_0 (resp. B_0) is the domain of the infinitesimal generator of G considered in A_1 (resp. B_1).

Consider two Hilbert spaces $C_0 \subset C_1$ and let π be given to satisfy

$$\pi \in \mathscr{L}(A_0; C_0) \cap \mathscr{L}(A_1; C_1) \tag{10.25}$$

and

$$\|\pi u\|_{C_i} \leqq \alpha \|u\|_{A_i} + \beta \|u\|_{B_i}, \quad \forall u \in A_i, \quad i = 0, 1. \tag{10.26}$$

Example. We let $C_0 = A_0$, $C_1 = A_1$, $\pi =$ the operator $u \to \varphi u$, φ given in $\mathscr{D}(\Omega)$. With the choices (10.23), we have (10.26) with

$$\alpha = \sup_x |\varphi(x)|. \quad \square \tag{10.27}$$

The hypotheses (10.25) and the interpolation theorem imply that

$$\pi \in \mathscr{L}(A_\theta; C_\theta), \quad \text{where } A_0 = [A_0, A_1]_\theta, \quad \text{etc.} \ldots$$

In the example, $A_\theta = H^s(\mathbf{R}^n)$ (due to (10.24)) and $C_\theta = A_\theta$, $B_\theta = H^{s-1}(\mathbf{R}^n)$.

The general result is that we can "interpolate" (10.26), namely that there exists a constant γ such that

$$\|\pi u\|_{C_\theta} \leqq \alpha\gamma \|u\|_{A_\theta} + \beta\gamma \|u\|_{B_\theta} \quad \forall u \in A_\theta. \tag{10.28}$$

Lemma 7.1 follows from (10.28) applied to the example.

Proof of (10.28). Define (see (10.10))

$$v(t) = \mathscr{R} u(t) = \left(\frac{1}{t} \int_0^t G(\sigma)\, u\, d\sigma\right) q(t). \tag{10.29}$$

Then, if we define W_A (analogous definitions for W_B and W_C) as the space of functions u such that

$$t^\alpha u \in L^2(0, \infty, A_0), \quad t^\alpha u' = t^\alpha \frac{du}{dt} \in L^2(0, \infty, A_1), \quad \theta = \frac{1}{2} + \alpha,$$

provided with the norm

$$\| u \|_{W_A} = (\| t^\alpha u \|^2_{L_2(0,\infty,A_0)} + \| t^\alpha u' \|^2_{L_2(0,\infty,A_1)})^{1/2},$$

we have

$$\mathcal{R} \in \mathcal{L}(A_\theta; W_A) \quad \textit{and} \quad \mathcal{R} \in \mathcal{L}(B_\theta; W_B) \tag{10.30}$$

(since A_0 and B_0 are defined in A_1 and B_1 through the *same* semi group G). If we define

$$w(t) = \pi v(t),$$

then, using (10.26):

$$\begin{aligned} \| w \|_{W_C} &\leqq c_1 \alpha \| v \|_{W_A} + c_1 \beta \| v \|_{W_B} \\ &\leqq c_2 \alpha \| u \|_{A_\theta} + c_2 \beta \| u \|_{B_\theta} \end{aligned}$$

and since

$$\| \pi u \|_{C_\theta} \leqq c \| w \|_{W_C}$$

the result (10.28) follows. □

10.3 Application to $H^s(0, \infty)$

We now apply Section 10.1 to

$$H^s(0,\infty) = [H^1(0,\infty), L^2(0,\infty)]_\theta, \quad 1 - \theta = s.$$

We consider the semi-group $G(t)$ defined in $L^2(0,\infty)$ by

$$G(t) f(x) = f(x+t) \quad \text{a.e.,} \quad x > 0,$$

with infinitesimal generator $A = \dfrac{d}{dx}$, having domain $H^1(0,\infty)$.

We obtain

$$\left\{ \begin{array}{l} \text{for } 0 < s < 1, \text{ the condition “} u \in H^s(0,\infty) \text{” is equivalent} \\ \text{to } u \in L^2(0,\infty) \text{ and} \\ \displaystyle\int_0^\infty t^{-(2s+1)} \, dt \int_0^\infty |u(x+t) - u(x)|^2 \, dx < \infty . \quad \square \end{array} \right. \tag{10.31}$$

11. Subspaces of $H^s(\Omega)$. The Spaces $H_0^s(\Omega)$

11.1 $H_0^s(\Omega)$-Spaces

Since the mapping $u \to \left\{ \dfrac{\partial^j u}{\partial \nu^j} \middle| 0 \leqq j < s - \dfrac{1}{2} \right\}$ (see Theorem 9.4) *vanishes* on $\mathcal{D}(\Omega)$ and is a *surjection* of

$$H^s(\Omega) \to \prod_{0 \leqq j < s - 1/2} H^{s-j-1/2}(\Gamma),$$

there results that, if $s > \frac{1}{2}$, *the space* $\mathscr{D}(\Omega)$ *is not dense in* $H^s(\Omega)$. In general, we shall set:

(11.1) $$H_0^s(\Omega) = \text{closure of } \mathscr{D}(\Omega) \text{ in } H^s(\Omega).$$

We have

Theorem 11.1. *Assume that* Ω *satisfies* (7.10) *and* (7.11).

The space $\mathscr{D}(\Omega)$ *is dense in* $H^s(\Omega)$ *if and only if*

(11.2) $$s \leqq \tfrac{1}{2}$$

(*then* $H_0^s(\Omega) = H^s(\Omega)$). *If* $s > \frac{1}{2}$, *we have*: $H_0^s(\Omega)$ *is strictly contained in* $H^s(\Omega)$.

Proof. 1) There only remains to show that $\mathscr{D}(\Omega)$ is dense in $H^s(\Omega)$ if (11.2) is satisfied. Since (according to the general properties of the spaces $[X, Y]_\theta$):

$$H^{1/2}(\Omega) \subset H^s(\Omega), \quad H^{1/2}(\Omega) \textit{ dense } \text{in } H^s(\Omega), \text{ if } s < \tfrac{1}{2},$$

it is sufficient to show that

(11.3) $$\mathscr{D}(\Omega) \textit{ is dense in } H^{1/2}(\Omega).$$

2) For the time being, we admit

Lemma 11.1. *Let* X *and* Y *be two Hilbert spaces satisfying* (2.1). *Consider the space* $\mathscr{V}$ *of* v*'s such that*

(11.4) $$v \in L^2(\mathbf{R}_t; X), \quad |\tau|^{1/2}\hat{v} \in L^2(\mathbf{R}_\tau; Y)$$

$\left(\textit{where } \hat{v}(\tau) = \dfrac{1}{\sqrt{2\pi}}\displaystyle\int_{-\infty}^{+\infty} \exp(-i\,t\,\tau)\, v(t)\, dt\right)$, *with the (hilbertian) norm*

(11.5) $$(\|v\|^2_{L^2(\mathbf{R}_t; X)} + \||\tau|^{1/2}\hat{v}\|^2_{L^2(\mathbf{R}_\tau; Y)})^{1/2}.$$

The subspace of $\varphi \in \mathscr{D}(\mathbf{R}_t; X)$, *with* $\varphi = 0$ *in a (variable) neighborhood of* $t = 0$, *is dense in the space* $\mathscr{V}$. □

Now, we show (11.3). Via local maps, we are brought back to the following: let

$$v \in H^{1/2}(\mathbf{R}^n_+),$$

with compact support in $\overline{\mathbf{R}}^n_+$; we need to show that $v = \lim_{j\to\infty} \varphi_j$, $\varphi_j \in \mathscr{D}(\mathbf{R}^n_+)$, limit in $H^{1/2}(\mathbf{R}^n_+)$.

Using an extension operator of $H^s(\mathbf{R}^n_+) \to H^s(\mathbf{R}^n)$, which exists for all s, we are led to:

(11.6) $$\left\{\begin{array}{l}\text{let } v \in H^{1/2}(\mathbf{R}^n); \text{ show that } v = \lim\limits_{j\to\infty} \Phi_j,\ \Phi_j \in \mathscr{D}(\mathbf{R}^n),\ \Phi_j \equiv 0 \\ \text{in the neighbourhood of } \{x_n = 0\}, \text{ limit in } H^{1/2}(\mathbf{R}^n).\end{array}\right.$$

But applying Lemma 11.1, with $X = H^{1/2}(\mathbf{R}_{x'}^{n-1})$, $Y = H^0(\mathbf{R}_{x'}^{n-1})$, we already know that

$$(11.7)\qquad \begin{cases} v = \lim_{j\to\infty} \psi_j \quad \text{in} \quad H^{1/2}(\mathbf{R}^n), \\ \psi_j \in \mathscr{D}(\mathbf{R}_{x_n}; H^{1/2}(\mathbf{R}_{x'}^{n-1})), \\ \psi_j \equiv 0 \quad \text{in the neighborhood of } \{x_n = 0\}. \end{cases}$$

The assertion (11.6) follows by truncation and regularization of ψ_j *in* x' (which is permissible in $H^s(\mathbf{R}_{x'}^{n-1})\ \forall s$). Therefore, the proof of the theorem will be complete as soon as we have shown Lemma 11.1 to hold.

Proof of Lemma 11.1. 1) Let $v \to N(v)$ be an antilinear continuous form on the space $\mathscr{V}$ defined by (11.4) and (11.5). Then, if we use *diagonalization* and set

$$w(\lambda, \tau) = (\mathscr{U}\, \hat{v}(\tau))(\lambda),$$

we may represent $N(v)$ (according to the Hahn-Banach theorem) in non-unique fashion by

$$(11.8)\qquad N(v) = \int_{-\infty}^{+\infty} \int_{\lambda_0}^{\infty} (\lambda f(\lambda, \tau) + |\tau|^{1/2} g(\lambda, \tau), w(\lambda, \tau))_{\mathfrak{h}(\lambda)}\, d\mu(\lambda)\, d\tau,$$

where

$$(11.9)\qquad f \in L^2(\mathbf{R}_\tau; \mathfrak{h}), \quad g \in L^2(\mathbf{R}_\tau; \mathfrak{h}).$$

If

$$(11.10)\qquad h = \overline{\mathscr{F}}_t\, \mathscr{U}^{-1}(\lambda f(\lambda, \tau) + |\tau|^{1/2} g(\lambda, \tau)),$$

we have, *in particular*

$$(11.11)\qquad h \in \mathscr{S}'(\mathbf{R}_t; X')\ (X \subset Y \subset X')$$

($\mathscr{S}'(\mathbf{R}_t; X')$ = space of tempered distributions taking their values in X') and

$$(11.12)\qquad N(\varphi) = (h, \varphi) \quad \forall \varphi \in \mathscr{S}(\mathbf{R}_t; X)$$

($\mathscr{S}(\mathbf{R}_t; X)$ = space of rapidly decreasing, infinitely differentiable functions taking their values in X, the parentheses in (11.12) denoting the antiduality between $\mathscr{S}'(\mathbf{R}_t; X')$ and $\mathscr{S}(\mathbf{R}_t; X)$; see L. Schwartz [6]).

2) Suppose that $N(\varphi) = 0$ for $\varphi \in \mathscr{D}(\mathbf{R}_t; X)$, $\varphi = 0$ in the neighborhood of 0. Then:

$$h \text{ has support in } t \text{ concentrated at } \{t = 0\}$$

and therefore (L. Schwartz [1], [6])

$$(11.13)\qquad \begin{cases} h = \sum_{\text{finite}} \delta^{(j)}(t) \otimes \xi_j', \\ \xi_j' \in X', \quad \delta(t) = \text{unit mass at the origin.} \end{cases}$$

But then computing $\mathscr{F}_t\, \mathscr{U}(h)$ and setting

(11.14) $$\mathscr{U}\,\xi'_j = \eta_j(\lambda)$$

and comparing with (11.10), we obtain

(11.15) $$\lambda f(\lambda, \tau) + |\tau|^{1/2} g(\lambda, \tau) = \sum_{\text{finite}} (\mathrm{i}\,\tau)^j\, \eta_j(\lambda).$$

But, according to (11.9), we have in particular

$$\frac{1}{1 + |\tau|^{1/2}} (\lambda f + |\tau|^{1/2} g) \in L^2(\mathbf{R}_\tau; \mathfrak{h}_{-1})$$

and therefore, for all $\chi \in \mathfrak{h}_1$, we have:

$$\begin{cases} \sum_{\text{finite}} \dfrac{(\mathrm{i}\,\tau)^j}{1 + |\tau|^{1/2}} (\eta_j, \chi) \text{ (where } (\eta_j, \chi) \text{ denotes the antidual between} \\ \mathfrak{h}_{-1} \text{ and } \mathfrak{h}_1) \text{ belongs to } L^2(\mathbf{R}_\tau). \end{cases}$$

This is possible only if, *for all* χ, $(\eta_j, \chi) = 0$, therefore $\eta_j = 0$, therefore $h = 0$, therefore $\lambda f + |\tau|^{1/2} g = 0$ and then (11.8) shows that $N(v) = 0 \ \forall v$. □

Remark 11.1. The proof of Lemma 11.1 essentially uses the Fourier transform and the fact that we are dealing with the spaces L^2 with hilbertian range, but points out the rôle of the *exceptional parameter* $\frac{1}{2}$. For an analogous result in the case of the spaces L^p, $p \neq 2$, consult Lions-Magenes [1] (IV), Theorem 1.1, p. 313. □

11.2 A Property of $H^s(\Omega)$, $0 \leqq s < \frac{1}{2}$

We introduce a function ϱ by

(11.16) $$\begin{cases} \varrho \text{ is infinitely differentiable on } \bar{\Omega}, \text{ positive on } \Omega, \text{ vanishing} \\ \text{on } \Gamma \text{ of the order of } d(x, \Gamma) \ (= \text{distance from } x \text{ to } \Gamma), \text{ i.e.} \\ \text{such that} \\ \qquad \lim_{x \to x_0} \dfrac{\varrho(x)}{d(x, \Gamma)} = d \neq 0 \quad \text{if} \quad x_0 \in \Gamma. \end{cases}$$

Such functions do exist, since Γ is an infinitely differentiable variety.

Theorem 11.2. *Assume that* Ω *satisfies* (7.10) *and* (7.11). *Let* ϱ *be defined by* (11.16). *Suppose that*

(11.17) $$0 \leqq s < \tfrac{1}{2}.$$

Then

(11.18) $$u \to \varrho^{-s} u$$

is a continuous linear mapping of $H^s(\Omega) \to H^0(\Omega) = L^2(\Omega)$.

If $\Omega = \mathbf{R}_+^n = \{x \mid x_n > 0\}$, *the theorem holds with* $\varrho(x) = x_n$.

Proof. 1) According to the density Theorem 11.1, it is sufficient to show (the c_i's denoting suitable constants)

$$\|\varrho^{-s}\varphi\|_{H^0(\Omega)} \leqq c_1 \|\varphi\|_{H^s(\Omega)} \quad \forall \varphi \in \mathscr{D}(\Omega). \tag{11.19}$$

With the help of local maps, to show (11.19) amounts to verifying:

$$\|x_n^{-s}\varphi\|_{H^0(\mathbf{R}_+^n)} \leqq c_2 \|\varphi\|_{H^s(\mathbf{R}_+^n)} \quad \forall \varphi \in \mathscr{D}(\mathbf{R}_+^n). \tag{11.20}$$

2) In general, Y being a Hilbert space, we define (see Remark 9.5)

$$H^s(0, \infty; Y) = [H^1(0, \infty; Y), L^2(0, \infty; Y)]_\theta, \quad 1 - \theta = s, \tag{11.21}$$

where

$$H^1(0, \infty; Y) = \{v \mid v \in L^2(0, \infty; Y), \quad v' \in L^2(0, \infty; Y)\},$$

$$\|v\|_{H^1(0,\infty;Y)} = \left(\int_0^\infty (\|v(t)\|_Y^2 + \|v'(t)\|_Y^2)\, dt\right)^{1/2}.$$

The properties we have seen to hold for the scalar case extend, without difficulty, to (11.21).

Therefore, in particular (Theorem 9.1), every $v \in H^s(0, \infty; Y)$ is a restriction of $w \in H^s(\mathbf{R}; Y)$ to $]0, \infty[$, i.e. satisfying

$$(1 + |\tau|^s)\, \hat{w} \in L^2(\mathbf{R}_\tau; Y), \quad (\hat{w} = \text{Fourier transform of } w).$$

Therefore

$$\left\{\begin{array}{l} H^s(\mathbf{R}_+^n) \subset H^s(0, \infty; Y), \quad Y = L^2(\mathbf{R}_{x'}^{n-1}), \\ \text{with continuous injection.} \end{array}\right. \tag{11.22}$$

(In fact, we have precisely (compare with (9.6))

$$H^s(\mathbf{R}_+^n) = L^2(0, \infty; H^s(\mathbf{R}_{x'}^{n-1})) \cap H^s(0, \infty; Y)).$$

3) Consequently, (11.20) follows from the stronger inequality

$$\|t^{-s}\varphi\|_{L^2(0,\infty;Y)} \leqq c_3 \|\varphi\|_{H^s(0,\infty;Y)}, \quad \forall \varphi \in \mathscr{D}(]0, \infty[; Y). \tag{11.23}$$

We shall use the identity

$$\varphi(x) = v(x) - w(x), \quad x > 0, \quad \varphi \in \mathscr{D}(]0, +\infty[; Y), \tag{11.24}$$

where

$$v(x) = \frac{1}{x}\int_0^x (\varphi(x) - \varphi(\xi))\, d\xi, \tag{11.25}$$

$$w(x) = \int_x^\infty \frac{1}{\xi} v(\xi)\, d\xi. \tag{11.26}$$

Indeed, if $\varphi \in \mathscr{D}(]0, +\infty[;\ Y)$, $v(x)$ and $w(x) \to 0$ as $x \to +\infty$, from which (11.24) follows, if we verify $\varphi' = v' - w'$, which is immediate.

4) Then (11.23) will, according to (11.24), follow from the two inequalities:

(11.27) $$\|x^{-s} v\|_{L^2(0,\infty;Y)} \leqq c_5 \|\varphi\|_{H^s(0,\infty;Y)},$$

(11.28) $$\|x^{-s} w\|_{L^2(0,\infty;Y)} \leqq c_6 \|\varphi\|_{H^s(0,\infty;Y)}.$$

Proof of (11.27). According to the definition (11.25) of v, we have:

$$\|v(x)\|_Y^2 \leqq \frac{1}{x}\int_0^x \|\varphi(x) - \varphi(\xi)\|_Y^2\, d\xi,$$

therefore

$$\int_0^\infty x^{-2s}\|v(x)\|_Y^2\, dx \leqq \int_0^\infty x^{-2s-1}\, dx \int_0^x \|\varphi(x) - \varphi(\xi)\|_Y^2\, d\xi$$

$$= \int_0^\infty d\xi \int_\xi^\infty x^{-2s-1}\|\varphi(x) - \varphi(\xi)\|_Y^2\, dx$$

$$= \int_0^\infty d\xi \int_0^\infty (\xi + t)^{-2s-1}\|\varphi(\xi + t) - \varphi(\xi)\|_Y^2\, dt \leqq$$

$$\leqq \int_0^\infty \xi^{-(2s+1)}\, d\xi \int_0^\infty \|\varphi(\xi + t) - \varphi(\xi)\|_Y^2\, dt,$$

from which we obtain the desired result, according to (10.23).

Proof of (11.28). The inequality (11.28) is a consequence of (11.27) and of

(11.29) $$\|x^{-s} w\|_{L^2(0,\infty;Y)} \leqq c_7 \|x^{-s} v\|_{L^2(0,\infty;Y)} \quad (0 < s < \tfrac{1}{2}).$$

This last inequality may be verified as in Lemma 10.1 — and also can be deduced from it by noting that the *transpose* of the mapping

$$g \to \frac{1}{x}\int_0^x g(y)\, dy \quad \text{is} \quad g \to \int_x^\infty \frac{1}{\xi} g(\xi)\, d\xi;$$

therefore (11.29) follows from (10.9), taking $\alpha = -s$. □

Theorem 11.2 is completed by

Theorem 11.3. *Assume that* (7.10), (7.11) *and* (11.16) *hold and that* $\frac{1}{2} < s \leqq 1$; *then* (11.18) *is a continuous mapping of* $H_0^s(\Omega) \to L^2(\Omega)$ (and the same is true with $\varrho(x) = x_n$, when $\Omega = \{x \mid x_n > 0\}$).

Proof. Since by definition of $H_0^s(\Omega)$, $\mathscr{D}(\Omega)$ is dense in $H_0^s(\Omega)$, we have to prove (11.23) – this time, with $\frac{1}{2} < s \leqq 1$.

Instead of formulas (11.24), (11.25), (11.26), we use:

(11 25a) $$\varphi(x) = v(x) + w_1(x)$$

(11.26a) $$w_1(x) = \int_0^x \frac{1}{\xi} v(\xi)\, d\xi$$

(which follows from (11.24), (11.25), (11.26) and $\int_0^\infty \frac{1}{\xi} v(\xi)\, d\xi = 0$ if $\varphi \in \mathscr{D}(]0, \infty[; Y))$.

We still have (11.27) if $s > \frac{1}{2}$. Finally (11.28) is replaced by

(11.28a) $$\|x^{-s} w_1\|_{L^2(0,\infty;Y)} \leqq c_8 \|\varphi\|_{H^s(0,\infty;Y)},$$

which follows from

(11.29a) $$\|x^{-s} w\|_{L^2(0,\infty;Y)} \leqq c_9 \|x^{-s} v\|_{L^2(0,\infty;Y)}.$$

Setting $\frac{1}{x} v = f$, this inequality is equivalent to

$$\left\| x^{1-s} \left(\frac{1}{x} \int_0^x f(\xi)\, d\xi \right) \right\|_{L^2(0,\infty;Y)} \leqq c_9 \|x^{1-s} f\|_{L^2(0,\infty;Y)},$$

which is the same as (10.9) (setting $1 - s = \alpha < \frac{1}{2}$). □

11.3 The Extension by 0 outside Ω

Theorem 11.4. *Assume that Ω satisfies* (7.10) *and* (7.11).

(11.30) $$u \to \tilde{u} = \textit{extension of } u \textit{ by } 0 \textit{ outside } \Omega,$$

is a continuous mapping of $H^s(\Omega) \to H^s(\mathbf{R}^n)$ if and only if $0 \leqq s < \frac{1}{2}$. (11.30) *is a continuous mapping of $H_0^s(\Omega) \to H^s(\mathbf{R}^n)$ for $s > \frac{1}{2}$ if and only if $s \neq$ integer $+ \frac{1}{2}$.*

Proof. 1) According to Theorem 9.4, if $s > \frac{1}{2}$, we may define $u|_\Gamma$ *and* $\tilde{u}|_\Gamma$. But since $\tilde{u} = 0$ outside Ω, we must have $\tilde{u}|_\Gamma = 0$, therefore $u|_\Gamma = 0$, which is not the case (since $u \to u|_\Gamma$ is a surjective mapping of $H^s(\Omega) \to H^{s-1/2}(\Gamma)$).

Therefore, the mapping (11.30) may *eventually be* continuous only if

$$0 \leqq s \leqq \tfrac{1}{2}.$$

2) We show that the mapping is continuous for $0 \leqq s < \frac{1}{2}$. Since (Theorem 11.1) the space $\mathscr{D}(\Omega)$ is dense in $H^s(\Omega)$, it is sufficient to show that

(11.31) $$\|\tilde{\varphi}\|_{H^s(\mathbf{R}^n)} \leqq c_1 \|\varphi\|_{H^s(\Omega)} \quad \forall \varphi \in \mathscr{D}(\Omega) \quad (0 \leqq s < \tfrac{1}{2}).$$

(The case $s = 0$ is obvious; we have equality with $c_1 = 1$.)

Via local maps, it is sufficient to show (11.31) with $\Omega = \mathbf{R}^n_+$. Using Sections 10.2 and 10.3, we see that (11.31) will follow from the inequalities:

(11.32) $$\begin{cases} \displaystyle\int_0^\infty t^{-(2s+1)}\, dt \int_{\mathbf{R}^n} |\tilde{\varphi}(x_1, \dots, x_i + t, \dots, x_n) - \tilde{\varphi}(x)|^2\, dx \\ \displaystyle\leqq c_2 \int_0^\infty t^{-(2s+1)}\, dt \int_{\mathbf{R}^n_+} |\varphi(x_1, \dots, x_i + t, \dots, x_n) - \varphi(x)|^2\, dx \\ \text{for } 1 \leqq i \leqq n. \end{cases}$$

The inequalities are obvious for $1 \leqq i \leqq n - 1$. To prove the case "$i = n$", it is sufficient to show that

(11.33) $$\begin{cases} \displaystyle\int_0^\infty t^{-(2s+1)}\, dt \int_{-\infty}^{+\infty} |\tilde{\varphi}(x + t) - \tilde{\varphi}(x)|^2\, dx \leqq \\ \displaystyle\leqq c_3 \int_0^\infty t^{-(2s+1)}\, dt \int_0^\infty |\varphi(x + t) - \varphi(x)|^2\, dx \\ = c_3 N(\varphi) \quad \forall \varphi \in \mathscr{D}(]0, \infty[). \end{cases}$$

(Apply (11.33) to $x_n \to \varphi(x', x_n)$ and integrate with respect to x'.) This, in turn, reduces to showing that

$$\int_0^\infty t^{-(2s+1)}\, dt \int_{-\infty}^0 |\varphi(x + t)|^2\, dx \leqq c_4 N(\varphi) \quad \text{(notation of 11.33)}$$

or that

$$\int_0^\infty t^{-(2s+1)}\, dt \int_0^t |\varphi(y)|^2\, dy \leqq c_4 N(\varphi),$$

i.e. that

(11.34) $$\frac{1}{2s} \int_0^\infty y^{-2s} |\varphi(y)|^2\, dy \leqq c_4 N(\varphi),$$

which is true, according to Theorem 11.2. □

3) We shall now prove that (11.30) *is not* continuous when $s = \frac{1}{2}$. Indeed, (11.34) is necessary for continuity; if this inequality would hold

with $s = \frac{1}{2}$, it would be true for all $\varphi \in H^{1/2}(0, \infty)$; but if φ is once continuously differentiable, with compact support and $\varphi(0) = 1$, (11.34) is not true for $s = \frac{1}{2}$. □

4) Finally, the continuity of (11.30) as a mapping of $H_0^s(\Omega) \to H^s(\mathbf{R}^n)$ for $s > \frac{1}{2}$ if $s \neq$ integer $+ \frac{1}{2}$, follows from the following considerations:

(i) when $\frac{1}{2} < s < 1$, the continuity may be verified as in 2), using Theorem 11.3 instead of Theorem 11.2;

(ii) for *integer* $s \geqq 1$, the result follows immediately, since

$$\|\tilde{\varphi}\|_{H^m(\mathbf{R}^n)} = \|\varphi\|_{H^m(\Omega)} \quad \forall \text{ integer } m \geqq 1;$$

(iii) for $m < s < m + 1$, we use Theorem 9.7; indeed, if $\varphi \in \mathscr{D}(\Omega)$, we have, according to this theorem, the equivalence of

$$\|\varphi\|_{H^s(\Omega)} \quad \text{with} \quad \left(\sum_{|p| \leqq m} \|D^p \varphi\|^2_{H^{s-m}(\Omega)} \right)^{1/2}.$$

As we have seen that $\varphi \to \tilde{\varphi}$ is a continuous mapping of $H_0^\sigma(\Omega) \to H^\sigma(\mathbf{R}^n)$ for $\sigma \neq \frac{1}{2}$, $0 \leqq \sigma < 1$ (if $\sigma < \frac{1}{2}$, $H_0^\sigma(\Omega) = H^\sigma(\Omega)$), we see that

$$\left(\sum_{|p| \leqq m} \|D^p \tilde{\varphi}\|^2_{H^{s-m}(\mathbf{R}^n)} \right)^{1/2} \leqq c_1 \left(\sum_{|p| \leqq m} \|D^p \varphi\|^2_{H^{s-m}(\Omega)} \right)^{1/2}$$

and since, again according to Theorem 9.7, the first term is equivalent to $\|\tilde{\varphi}\|_{H^s(\mathbf{R}^n)}$, we finally have

$$\|\tilde{\varphi}\|_{H^s(\mathbf{R}^n)} \leqq c_2 \|\varphi\|_{H^s(\Omega)} \quad \forall \varphi \in \mathscr{D}(\Omega),$$

from which the desired result follows. □

11.4 Characterization of $H_0^s(\Omega)$-Spaces

Theorem 11.5. *Assume that Ω satisfies* (7.10) *and* (7.11).
Let $s > \frac{1}{2}$. Then the following two conditions are equivalent:

(11.35) $$u \in H_0^s(\Omega),$$

(11.36) $$\begin{cases} u \in H^s(\Omega) \\ \dfrac{\partial^j u}{\partial \nu^j} = 0, \quad 0 \leqq j < s - \frac{1}{2}. \end{cases}$$

Proof. We have already seen that (11.35) ⇒ (11.36). There remains to be shown that the converse holds. Via local maps, this amounts to showing that if $u \in H^s(\mathbf{R}_+^n)$, with

(11.37) $$\frac{\partial^j u}{\partial x_n^j}(x', 0) = 0, \quad 0 \leqq j < s - \tfrac{1}{2},$$

then $u \in H_0^s(\mathbf{R}_+^n)$.

For the moment, we denote by E the space of $u \in H^s(\mathbf{R}^n_+)$ satisfying (11.37); it is a *closed* subspace of $H^s(\mathbf{R}^n_+)$ (according to Theorem 9.4). We need to show that $\mathscr{D}(\mathbf{R}^n_+)$ is dense in E. But, in the same way as for Theorem 11.1, Lemma 11.1, we note that everything follows from

Lemma 11.2. *Let X and Y be two Hilbert spaces satisfying* (2.1). *Consider the space of v's such that*

$$\begin{cases} v \in L^2(\mathbf{R}_t; X), & |\tau|^s \hat{v} \in L^2(\mathbf{R}_\tau; Y) \\ v^{(j)}(0) = 0, & 0 \leqq j < s - \frac{1}{2}, \end{cases} \tag{11.38}$$

with the norm

$$(\| v \|^2_{L^2(\mathbf{R}_t; X)} + \| |\tau|^s \hat{v} \|^2_{L^2(\mathbf{R}_\tau; Y)})^{1/2}. \tag{11.39}$$

The subspace of $\varphi \in \mathscr{D}(\mathbf{R}_t; X)$, with $\varphi = 0$ in a (variable) neighborhood of $t = 0$, is dense in the space of v's.

Proof of Lemma 11.2. As for Lemma 11.1, if $v \to N(v)$ is continuous antilinear on the space defined by (11.38) and (11.39), we have (compare with (11.8)):

$$N(v) = \int_{-\infty}^{+\infty} \int_{\lambda_0}^{\infty} (\lambda f + |\tau|^s g, v)_{\mathfrak{h}(\lambda)} \, d\mu(\lambda) \, d\tau,$$

where

$$f, g \in L^2(\mathbf{R}_\tau; \mathfrak{h}).$$

We introduce h (compare with (11.10)) by

$$h = \overline{\mathscr{F}}_\tau \, \mathscr{U}^{-1} (\lambda f + |\tau|^s g).$$

Then, if $N(\varphi) = 0$ for all $\varphi \in \mathscr{D}(\mathbf{R}_t; X)$ vanishing in the neighborhood of 0, we have (compare with (11.13)):

$$h = \sum_{\text{finite}} \delta^{(j)}(t) \otimes \xi'_j, \quad \xi'_j \in X'$$

and since

$$\mathscr{F} \mathscr{U} h = \sum_{\text{finite}} (\mathrm{i}\,\tau)^j \eta_j(\lambda) \quad (\mathscr{U} \xi_j = \eta_j)$$

and

$$(\mathscr{F} \mathscr{U} h) \frac{1}{1 + |\tau|^s} \in L^2(\mathbf{R}_\tau; \mathfrak{h}_{-1}),$$

it follows that $0 \leqq j < s - \frac{1}{2}$. Then

$$h = \sum_{0 \leqq j < s - 1/2} \delta^{(j)}(t) \otimes \xi'_j.$$

$$\frac{\mathscr{F} \mathscr{U} h}{|\lambda| + |\tau|^s} \in L^2(\mathbf{R}_\tau; \mathfrak{h})$$

if and only if

$$\int_{\lambda_0}^{\infty} \|\eta_j(\lambda)\|^2_{\mathfrak{h}(\lambda)} \int_{-\infty}^{\infty} \frac{\tau^{2j}}{(\lambda + |\tau|^s)^2} d\tau < \infty,$$

i.e.

$$\int_{\lambda_0}^{\infty} \|\eta_j(\lambda)\|^2_{\mathfrak{h}(\lambda)} \lambda^{2(j+1/2)/s-2} d\mu(\lambda) < \infty \quad \text{i.e.}$$

$$\xi_j' \in ([X, Y]_{(j+1/2)/s})'. \tag{11.40}$$

Then, if $v \in L^2(\mathbf{R}_t; X)$, $|\tau|^s \hat{v} \in L^2(\mathbf{R}_\tau; Y)$:

$$N(v) = \sum_{0 \leq j < s - 1/2} (-1)^j (v^{(j)}(0), \xi_j'),$$

where the parenthesis corresponding to the index "j" denotes the antiduality between $[X, Y]_{(j+1/2)/s}$ and its antidual.

But then, if $v^j(0) = 0$, $0 \leqq j < s - \frac{1}{2}$, we have $N(v) = 0$, whence the desired result. □

11.5 Interpolation of $H_0^s(\Omega)$-Spaces

Theorem 11.6. *Assume that Ω satisfies* (7.10) *and* (7.11). *Let*

$$s_1 > s_2 \geqq 0,$$

s_1 *and* $s_2 \neq$ *integer* $+ \frac{1}{2}$. *If*

$$(1 - \theta) s_1 + \theta s_2 \neq \textit{integer} + \tfrac{1}{2}, \tag{11.41}$$

then

$$[H_0^{s_1}(\Omega), H_0^{s_2}(\Omega)]_\theta = H_0^{(1-\theta)s_1 + \theta s_2}(\Omega) \tag{11.42}$$

(*with equivalent norms*).

Remark 11.2. Hypothesis (11.41) is essential; as we shall see, the statement of the theorem is false if, for example, $(1 - \theta) s_1 + \theta s_2 = \frac{1}{2}$. □

Proof. 1) Let us assume for the moment that

$$\begin{cases} [H_0^m(\Omega), H^0(\Omega)]_\theta = H_0^{(1-\theta)m}(\Omega) \\ \text{if } m \text{ is an integer, } (1 - \theta) m \neq \text{integer} + \frac{1}{2} \end{cases} \tag{11.43}$$

Then (11.42) follows. Indeed, choose integer $m \geqq s_i$, $i = 1, 2$. Then

$$H_0^{s_i}(\Omega) = [H_0^m(\Omega), H^0(\Omega)]_{\theta_i}, \quad (1 - \theta_i) m = s_i.$$

Therefore

$$[H_0^{s_1}(\Omega), H_0^{s_2}(\Omega)]_\theta = [[H_0^m(\Omega), H^0(\Omega)]_{\theta_1}, [H_0^m(\Omega), H^0(\Omega)]_{\theta_2}]_\theta$$

and according to the *reiteration Theorem* 6.1, this last space is

$$[H_0^m(\Omega), H^0(\Omega)]_{(1-\theta)\theta_1+\theta\theta_2},$$

from which we obtain the desired result by applying (11.43) once more. We still have to show (11.43).

2) First, we show that

$$[H_0^m(\Omega), H^0(\Omega)]_\theta \subset H_0^{(1-\theta)m}(\Omega). \tag{11.44}$$

We consider the mapping (11.30), which is a continuous mapping of

$$H_0^m(\Omega) \to H_0^m(\mathbf{R}^n) \quad \text{and of } H^0(\Omega) \to H^0(\mathbf{R}^n),$$

therefore, by interpolation, of

$$[H_0^m(\Omega), H^0(\Omega)]_\theta \to [H^m(\mathbf{R}^n), H^0(\mathbf{R}^n)]_\theta = H^{(1-\theta)m}(\mathbf{R}^n).$$

Then, if $u \in [H_0^m(\Omega), H^0(\Omega)]_\theta$, we have:

$$u = (\tilde{u})_\Omega, \quad \tilde{u} \in H^{(1-\theta)m}(\mathbf{R}^n)$$

and since $\tilde{u} = 0$ outside of Ω, we have

$$\frac{\partial^j \tilde{u}}{\partial \nu^j} = 0 \quad \text{on } \Gamma, \quad 0 \leqq j < s - \tfrac{1}{2},$$

therefore

$$\frac{\partial^j u}{\partial \nu^j} = 0 \quad \text{on } \Gamma, \quad 0 \leqq j < s - \tfrac{1}{2},$$

and therefore, according to Theorem 11.5,

$$u \in H_0^{(1-\theta)m}(\Omega), \quad \text{whence} \quad (11.44).$$

3) Assume, for the moment, the truth of

Lemma 11.3. *If Ω satisfies* (7.10) *and* (7.11), *then for every integer m there exists an operator*

$$u \to R\,u, \tag{11.45}$$

having the properties:

(11.46) $R \in \mathscr{L}(H^k(\mathbf{R}^n); H_0^k(\Omega)) \quad \forall k, \quad 0 \leqq k \leqq m, \quad k = \textit{integer},$

$$R(\tilde{u}) = u \quad \forall u \in H_0^k(\Omega). \quad \square \tag{11.47}$$

Then, by interpolation

$$R \in \mathscr{L}(H^{(1-\theta)m}(\mathbf{R}^n); [H_0^m(\Omega), H^0(\Omega)]_\theta).$$

So that, if $u \in H_0^{(1-\theta)m}(\Omega)$ *and if* $(1-\theta)\,m \neq$ integer $+ \frac{1}{2}$, the function $\tilde{u}$ is, according to Theorem 11.4, in $H^{(1-\theta)m}(\mathbf{R}^n)$ and therefore

$$u = R\,\tilde{u} \in [H_0^m(\Omega), H^0(\Omega)]_\theta,$$

rom which we have the inverse inclusion of (11.44) and the theorem. $\square$

Proof of Lemma 11.3. Via local maps, we are brought back to the case "$\Omega = \mathbf{R}_+^n$". Set (compare with (2.21)):

$$R\,u(x) = u(x) - \sum_{k=1}^{m} \alpha_k\, u(x' - k\, x_n), \tag{11.48}$$

where the α_k's are defined by

$$\sum_{k=1}^{m} (-k)^j \alpha_k = 1, \quad 0 \leqq j \leqq m-1. \ \square \tag{11.49}$$

The case $(1-\theta)\, s_1 + \theta\, s_2 =$ integer $+ \frac{1}{2}$ in Theorem 11.6 is *in fact* singular, as the following theorem shows.

Theorem 11.7. *Assume that Ω satisfies* (7.10) *and* (7.11). *Let*

$$s_1 > s_2 \geqq 0, \quad s_1 \text{ and } s_2 \neq \text{integer} + \tfrac{1}{2}.$$

Assume

$$(1-\theta)\, s_1 + \theta\, s_2 = \mu + \tfrac{1}{2}, \quad \mu \text{ an integer} \geqq 0. \tag{11.50}$$

Then

$$\left\{ \begin{array}{l} [H_0^{s_1}(\Omega), H_0^{s_2}(\Omega)]_{\theta_1} \text{ is a space independent of } s_i \text{ and } \theta \\ \text{with (11.50).} \end{array} \right. \tag{11.51}$$

Set: $[H_0^{s_1}(\Omega), H_0^{s_2}(\Omega)]_\theta = H_{00}^{\mu+1/2}(\Omega)$. *Then*

$$\left\{ \begin{array}{l} H_{00}^{\mu+1/2}(\Omega) = \{u \mid u \in H_0^{\mu+1/2}(\Omega),\ \varrho^{-1/2} D^p u \in L^2(\Omega) \\ \forall p \text{ with } |p| = \mu \ (\varrho \text{ defined by (11.16)})\}, \end{array} \right. \tag{11.52}$$

the interpolation norm being equivalent to

$$|\!|\!| u |\!|\!|_{H_{00}^{\mu+1/2}(\Omega)} = \left(\| u \|^2_{H^{\mu+1/2}(\Omega)} + \sum_{|p|=\mu} \| \varrho^{-1/2} D^p u \|^2_{L^2(\Omega)} \right)^{1/2}. \tag{11.53}$$

The space $H_{00}^{\mu+1/2}(\Omega)$ is strictly contained in $H_0^{\mu+1/2}(\Omega)$, with a strictly finer topology.

Proof. 1) Choose an integer $m \geqq \max(s_1, s_2)$, then θ_i from $(1-\theta_i)\, m = s_i$, $i = 1, 2$. According to Theorem 11.6, we have (since $s_i \neq$ integer $+ \frac{1}{2}$):

$$H_0^{s_i}(\Omega) = [H_0^m(\Omega), H^0(\Omega)]_{\theta_i}$$

and, according to the reiteration Theorem 6.1, we have (with equivalent norms):

$$[H_0^{s_1}(\Omega), H_0^{s_2}(\Omega)]_\theta = [H_0^m(\Omega), H^0(\Omega)]_{(1-\theta)\theta_1 + \theta\theta_2}.$$

2) But still according to the reiteration theorem:

$$[H_0^m(\Omega), H^0(\Omega)]_{\theta_3} = [[H_0^m(\Omega), H^0(\Omega)]_{\alpha_1}, [H_0^m(\Omega), H^0(\Omega)]_{\alpha_2}]_\alpha$$

if $(1-\alpha)\, \alpha_1 + \alpha\, \alpha_2 = \theta_3$. Choose α_1 and α_2 according to

$$(1-\alpha_1)\, m = \mu + 1, \quad (1-\alpha_2)\, m = \mu.$$

Then $(1-\alpha)\,\alpha_1 + \alpha\,\alpha_2 = \theta_3$ implies $\alpha = \frac{1}{2}$ and therefore (using Theorem 11.6), we have:

$$[H_0^{s_1}(\Omega), H_0^{s_2}(\Omega)]_\theta = [H_0^{\mu+1}(\Omega), H_0^{\mu}(\Omega)]_{1/2}.$$

Therefore we have (11.51).

3) To obtain the characterization (11.52), we shall use the fact that

$$[H_0^{\mu+1}(\Omega), H_0^{\mu}(\Omega)]_{1/2} = H_{00}^{\mu+1/2}(\Omega).$$

We come back to the case

$$\Omega = \{x \mid x_n > 0\}, \qquad \varrho = x_n.$$

We shall apply Theorem 10.1 and Remark 10.4, as well as the following fact: define *in* $H_0^{\mu}(\Omega)$,

$G_i(t)$ = group of translations of t on the variable x_i, $1 \leqq i \leqq n-1$,

$G_n(t)$ = semi-group, defined by

$$G_n(t)\,f(x) = \begin{cases} 0 & \text{if } x_n < t \\ f(x', x_n - t) & \text{if } x > t. \end{cases}$$

If $-\Lambda_i$ is the infinitesimal generator of G_i, $1 \leqq i \leqq n$, we have

$$H_0^{\mu+1}(\Omega) = \bigcap_{i=1}^{n} D(\Lambda_i)$$

and therefore, according to (10.16):

$$(11.54)\qquad u \in H_{00}^{\mu+1/2}(\Omega) \Leftrightarrow \begin{cases} u \in H_0^{\mu+1/2}(\Omega) \\ \displaystyle\int_0^\infty t^{-2}\, \| G_n(t)\, u - u \|^2_{H_0^{\mu}(\Omega)}\, dt < \infty. \end{cases}$$

The condition on the integral is equivalent to

$$\int_0^\infty t^{-2} \sum_{|p|=\mu} \left(\int_\Omega |G_n(t)\, D_x^p u - D_x^p u|^2\, dx \right) dt +$$
$$+ \int_0^\infty t^{-2} \sum_{|p|<\mu} \left(\int_\Omega |G_n(t)\, D_x^p u - D_x^p u|^2\, dx \right) dt < \infty.$$

But if $u \in H_0^{\mu}(\Omega)$, the second integrals are finite and therefore (11.54) yields:

$$u \in H_{00}^{\mu+1/2}(\Omega) \Leftrightarrow \begin{cases} u \in H_0^{\mu+1/2}(\Omega) \\ \displaystyle\int_0^\infty t^{-2} \int_\Omega |G_n(t)\, D_x^p u - D_x^p u|^2\, dx\, dt < \infty \\ \forall p, \quad |p| = \mu. \end{cases}$$

We rewrite the integral condition:

$$\int_0^\infty t^{-2}\,dt \int_\Omega |D^p u(x', x_n + t) - D^p u(x)|^2\,dx_n + \\ + \int_0^\infty t^{-2}\,dt \int_{\mathbf{R}_{x'}^{n-1}} dx' \int_0^t |D^p u(x)|^2\,dx_n < \infty .$$

The first integral is finite for all $u \in H_0^{\mu+1/2}(\Omega)$ and we see that:

$$u \in H_{00}^{\mu+1/2}(\Omega) \Leftrightarrow \begin{cases} u \in H_0^{\mu+1/2}(\Omega) \\ \int_0^\infty t^{-2}\,dt \int_{\mathbf{R}_{x'}^{n-1}} dx' \int_0^t |D^p u(x)|^2\,dx_n < \infty \\ \forall p \text{ with } |p| = \mu . \end{cases}$$

But this last integral amounts to

$$\int_\Omega \frac{1}{x_n} |D^p u(x)|^2\,dx,$$

whence the theorem. □

Remark 11.3. The same proof, for the case s_i = integer + $\frac{1}{2}$, yields

$$[H_{00}^{s_1}(\Omega), H_{00}^{s_2}(\Omega)]_\theta \begin{cases} = H_0^{(1-\theta)s_1-\theta s_2}(\Omega) \text{ if } (1-\theta)s_1 + \theta s_2 \neq \text{integer} + \frac{1}{2} \\ = H_{00}^{(1-\theta)s_1+\theta s_2}(\Omega) \text{ if } (1-\theta)s_1 + \theta s_2 = \text{integer} + \frac{1}{2} \end{cases}$$

and the analogous result if *only one* of the s_i is in the form "integer + $\frac{1}{2}$". □

Remark 11.4. Each $H_0^{s_i}(\Omega)$ is *closed* in $H^{s_i}(\Omega)$; nevertheless, $[H_0^{s_1}(\Omega), H_0^{s_2}(\Omega)]_\theta$ is (unfortunately) not always closed in $[H^{s_1}(\Omega), H^{s_2}(\Omega)]_\theta$; more precisely, it is closed *except* when θ satisfies (11.50).

Remark 11.3 yields the *same* conclusion starting from two spaces $(H_{00}^{s_i}(\Omega))$ which *are not closed* in $H^{s_i}(\Omega)$ ... □

Remark 11.5. Let $\Omega =]0, T[$ and

$${}_0H^1(\Omega) = \{u \mid u \in H^1(\Omega), u(0) = 0\}.$$

Then (same proof as above)

$$(11.55) \qquad [{}_0H^1(\Omega), H^0(\Omega)]_{1/2} = {}_0H_0^{1/2}(\Omega) \\ = \{u \mid u \in H^{1/2}(\Omega), t^{-1/2} u \in L^2(\Omega)\}.$$

This can be extended to the case of functions taking their values in a Hilbert space and will be used in Chapter 4, Volume 2. □

Remark 11.6. We shall not go into a general study – somewhat teratological – of spaces $[H_0^{s_1}(\Omega), H_0^{s_2}(\Omega)]_\theta$ *when* s_1 *or* s_2 *is equal to* $k + \frac{1}{2}$, *with* k *an integer.* □

Remark 11.7. The problem to which we called attention in the preceding remark is probably also tied to a new characterization of $H_0^s(\Omega)$-spaces, which we shall only give here for integer s (we shall make use of this characterization in Chapter 2, Section 7):

Theorem 11.8. *Let integer $s > 0$. Then $u \in H_0^s(\Omega)$ if and only if $u \in \mathscr{D}'(\Omega)$ and*

$$\varrho^{-s+|\alpha|} D^\alpha u \in L^2(\Omega) \quad \forall \alpha \text{ with } |\alpha| \leqq s \tag{11.56}$$

(ϱ defined by (11.16)).

Proof. 1) Let $u \in H_0^s(\Omega)$; we prove (11.56). Via "local maps" and "partition of unity", we are led to the following situation: we have a function $v(x', x_n)$ belonging to $H_0^s(\mathbf{R}_+^n)$ and with compact support in $\mathbf{R}_+^n$. We must show that

$$x_n^{-s+j+|\gamma|} D_{x_n}^j D_{x'}^\gamma v(x', x_n) \in L^2(\mathbf{R}_+^n), \quad 0 \leqq j + |\gamma| \leqq s. \tag{11.57}$$

But since $v \in H_0^s(\mathbf{R}_+^n)$, we have that $D_{x'}^\gamma v \in H_0^{s-|\gamma|}(\mathbf{R}_+^n)$. Thus, we may apply Lemma 10.1 to the (vector-valued) function:

$$x_n \to w(x_n) = D_{x'}^\gamma v(x', x_n) \in L^2(\mathbf{R}^{n-1}) \quad \text{and to its derivatives;}$$

we obtain

$$\int_0^\infty x_n^{-2(s-|\gamma|)} \| D_{x'}^\gamma v(x', x_n) \|_{L^2(\mathbf{R}^{n-1})}^2 \, dx_n \leqq$$

$$\leqq C \int_0^\infty x_n^{-2(s-|\gamma|)} \| x_n D_{x_n} D_{x'}^\gamma v(x', x_n) \|_{L^2(\mathbf{R}^{n-1})}^2 \, dx_n$$

$$= C \int_0^\infty x_n^{-2(s-|\gamma|)+2} \| D_{x_n} D_{x'}^\gamma v(x', x_n) \|_{L^2(\mathbf{R}^{n-1})}^2 \, dx_n \leqq$$

$$\leqq C^2 \int_0^\infty x_n^{-2(s-|\gamma|)+2} \| x_n D_{x}^2 D_{x'}^\gamma v(x', x_n) \|_{L^2(\mathbf{R}^{n-1})}^2 \, dx_n$$

$$= C^2 \int_0^\infty x_n^{-2(s-|\gamma|)+4} \| D_{x_n}^2 D_{x'}^\gamma v(x', x_n) \|_{L^2(\mathbf{R}^{n-1})}^2 \, dx_n \leqq$$

$$\leqq C^{s-|\gamma|} \int_0^\infty x_n^{-2} \| x_n D_{x_n}^{s-|\gamma|} D_{x'}^\gamma v(x', x_n) \|_{L^2(\mathbf{R}^{n-1})}^2 \, dx_n$$

$$= C^{s-|\gamma|} \| D_{x_n}^{s-|\gamma|} D_{x'}^\gamma v \|_{L^2(\mathbf{R}_+^n)}^2 < +\infty,$$

and therefore (11.57) is verified.

2) Conversely, suppose $u \in \mathscr{D}'(\Omega)$ and (11.56) holds. Then, we obtain $u \in H^s(\Omega)$ immediately from (11.56). We show that u may be approached in $H^s(\Omega)$ by functions of $\mathscr{D}(\Omega)$ and therefore that $u \in H_0^s(\Omega)$. Indeed, let $\delta_\nu(x)$ be a sequence of functions of $\mathscr{D}(\Omega)$ such that

$$\delta_\nu(x) = 1 \quad \text{if} \quad d(x, \Gamma) \geqq \frac{2}{\nu}$$

and

$$\delta_\nu(x) = 0 \quad \text{if} \quad d(x, \Gamma) \leqq \frac{1}{\nu}$$

and

$$|D^\alpha \delta_\nu(x)| \leqq \frac{C_\alpha}{d(x, \Gamma)^{|\alpha|}} \quad (\text{where } C_\alpha \text{ depends on } \alpha);$$

such a sequence exists.

Then $\delta_\nu u \in H^s(\Omega)$ has compact support in Ω. We verify that $\delta_\nu u \to u$ in $H^s(\Omega)$, as $\nu \to +\infty$, i.e. $D^\alpha(\delta_\nu u) \to D^\alpha u$ in $L^2(\Omega)$ for $|\alpha| \leqq s$.

But $\delta_\nu D^\alpha u \to D^\alpha u$ in $L^2(\Omega)$; therefore it is sufficient to show that

$$\text{(11.58)} \quad D^\beta \delta_\nu D^\gamma u \to 0 \quad \text{in} \quad L^2(\Omega) \quad \text{if} \quad |\beta| \geqq 1, \quad |\gamma| + |\beta| = |\alpha| \leqq s.$$

Now, $D^\beta \delta_\nu D^\gamma u$ vanishes if $d(x, \Gamma) \geqq 2/\nu$ since $|\beta| \geqq 1$, so that, according to Lebesgue's Theorem, we have (11.58) if we note that, thanks to (11.56), we have

$$|D^\beta \delta_\nu D^\gamma u| \leqq \frac{C_\beta}{\varrho^{|\beta|}} |D^\gamma u| \leqq c_\beta\, \varrho^{-s+|\gamma|} |D^\gamma u| \in L^2(\Omega).$$

Thus it is sufficient to approach $\delta_\nu u$ (ν fixed) in $H^s(\Omega)$ with functions of $\mathscr{D}(\Omega)$; and this is possible by the usual method of regularization, since the support of $\delta_\nu u$ is in a compact subset of Ω. □

Remark 11.8. Denote by $L^2_{\varrho^{-\sigma}}(\Omega)$ (real $\sigma \geqq 0$) the space of functions $u \in L^2(\Omega)$ such that $\varrho^{-\sigma} u \in L^2(\Omega)$, with the norm of the graph. Then

$$[L^2_{\varrho^{-s}}(\Omega), L^2(\Omega)]_\theta = L^2_{\varrho^{-s(1-\theta)}}(\Omega)$$

by definition of $[X, Y]_\theta$-spaces, since $L^2_{\varrho^{-s}}(\Omega)$ is the domain in $L^2(\Omega)$ of the self-adjoint operator of multiplication by ϱ^{-s}. Since Theorem 11.8 implies that $H_0^s(\Omega) \subset L^2_{\varrho^{-s}}(\Omega)$, we have

$$\text{(11.59)} \quad [H_0^s(\Omega), L^2(\Omega)]_\theta \subset L^2_{\varrho^{-s(1-\theta)}}(\Omega), \quad \text{integer } s > 0,\ 0 < \theta < 1.$$

12. The Spaces $H^{-s}(\Omega)$, $s > 0$

12.1 Definition. First Properties

By definition, we set

$$\text{(12.1)} \qquad H^{-s}(\Omega) = (H_0^s(\Omega))', \quad s > 0.$$

Since $\mathscr{D}(\Omega)$ iw dense in $H_0^s(\Omega)$ (by definition), we have:

$$\text{(12.2)} \qquad H^{-s}(\Omega) \subset \mathscr{D}'(\Omega). \quad □$$

Example 12.1. Definition (12.1) coincides with (7.5) when $\Omega = \mathbf{R}^n$ and if we take definition (7.1). □

Example 12.2. When s is an integer, we have the *structure theorem*:

Theorem 12.1. *Let m be a positive integer. Then every $f \in H^{-m}(\Omega)$ may be represented, in non-unique fashion, by*

$$f = \sum_{|p| \leqq m} D^p f_p, \quad f_p \in L^2(\Omega). \tag{12.3}$$

Proof. 1) Through the mapping

$$u \to \{D^p u \mid |p| \leqq m\},$$

the space $H^m(\Omega)$ is identified to a closed subspace of a product of $L^2(\Omega)$ and therefore, according to the Hahn-Banach theorem, every continuous linear form $u \to L(u)$ on $H^m(\Omega)$ may be represented (non-uniquely) by

$$L(u) = \sum_{|p| \leqq m} \int_\Omega g_p D^p u \, dx, \quad g_p \in L^2(\Omega). \tag{12.4}$$

2) Now if $L \in (H_0^m(\Omega))'$, it may again be represented by (12.4) and this time L is uniquely defined by $L(\varphi)$ $\forall \varphi \in \mathscr{D}(\Omega)$; but $L(\varphi) = \langle f, \varphi \rangle$, where f is defined by (12.3) with $f_p = (-1)^{|p|} g_p$, the brackets denoting the duality between $\mathscr{D}'(\Omega)$ and $\mathscr{D}(\Omega)$. Thus, the theorem is proved. □

Remark 12.1. We must be careful to distinguish between the spaces

$$H^{-1/2}(\Omega) = \text{dual of } (H^{1/2}(\Omega) = H_0^{1/2}(\Omega))$$

and

$$(H_{00}^{1/2}(\Omega))' = \text{dual of } H_{00}^{1/2}(\Omega)$$

which, according to (11.52), we may also represent by

$$f \in (H_{00}^{1/2}(\Omega))' \Leftrightarrow f = f_0 + f_1, \quad f_0 \in H^{-1/2}(\Omega), \quad \varrho^{1/2} f_1 \in L^2(\Omega). \tag{12.5}$$

Thus, the function

$$\frac{1}{\varrho^{1-\varepsilon}} \in (H_{00}^{1/2}(\Omega))' \quad \forall \varepsilon > 0. \; □$$

12.2 Interpolation between the Spaces $H^{-s}(\Omega)$, $s > 0$

According to the duality Theorem 6.2 and with Theorems 11.6 and 11.7, we have

Theorem 12.2. *Assume that Ω satisfies* (7.10) *and* (7.11). *Let $s_2 > s_1 \geqq 0$, s_1 and $s_2 \neq$ integer $+ \frac{1}{2}$. Then, if*

$$(1 - \theta) s_1 + \theta s_2 \neq \text{integer} + \tfrac{1}{2}, \tag{12.6}$$

we have

$$[H^{-s_1}(\Omega), H^{-s_2}(\Omega)]_\theta = H^{-(1-\theta)s_1 - \theta s_2}(\Omega). \tag{12.7}$$

If

(12.8) $$(1-\theta)\, s_1 + \theta\, s_2 = \tfrac{1}{2} + \mu \quad (\textit{integer } \mu),$$

then

(12.9) $$[H^{-s_1}(\Omega), H^{-s_2}(\Omega)] = (H_{00}^{\mu+1/2}(\Omega))'. \quad \square$$

It seems natural now to "interpolate" between $H_0^{s_1}(\Omega)$ (resp. $H^{s_1}(\Omega)$) and $H^{-s_2}(\Omega)$. We shall successively examine these points. □

Remark 12.2. The case s_i = integer + $\frac{1}{2}$ will not be discussed (see also Remark 11.6). □

12.3 Interpolation between $H_0^{s_1}(\Omega)$ and $H^{-s_2}(\Omega)$, $s_i > 0$

Theorem 12.3. *Assume that* Ω *satisfies* (7.10) *and* (7.11). *Let* s_1 *and* $s_2 \geqq 0$ *and* $\neq$ *integer* $+ \frac{1}{2}$. *Assume*

(12.10) $$(1-\theta)\, s_1 + \theta\, s_2 \neq \mu + \tfrac{1}{2} \quad \textit{and} \quad \neq -\mu - \tfrac{1}{2} \quad (\textit{integer } \mu \geqq 0).$$

Then

(12.11) $$[H_0^{s_1}(\Omega), H^{-s_2}(\Omega)]_\theta = \begin{cases} H_0^{(1-\theta)s_1-\theta s_2}(\Omega) & \textit{if } (1-\theta)\, s_1 - \theta\, s_2 \geqq 0 \\ H^{(1-\theta)s_1-\theta s_2}(\Omega) & \textit{if } (1-\theta)\, s_1 - \theta\, s_2 \leqq 0. \end{cases}$$

Furthermore, still if $s_i \neq$ *integer* $+ \frac{1}{2}$, *we have*:

(12.12) $$[H_0^{s_1}(\Omega), H^{-s_2}(\Omega)]_\theta = H_{00}^{\mu+1/2}(\Omega) \quad \textit{if } (1-\theta)\, s_1 - \theta\, s_2 = \mu + \tfrac{1}{2}$$

and

(12.13) $$[H_0^{s_1}(\Omega), H^{-s_2}(\Omega)]_\theta = (H_{00}^{\mu+1/2}(\Omega))' \quad \textit{if } (1-\theta)\, s_1 - \theta\, s_2 = -\mu - \tfrac{1}{2}.$$

Proof. Theorem 12.3 follows from the preceding results and the reiteration theorem. First of all, from (2.41), with $V = H_0^m(\Omega)$, $H = H^0(\Omega)$, we deduce:

$$[H_0^m(\Omega), H^{-m}(\Omega)]_{1/2} = H^0(\Omega).$$

Applying the reiteration theorem, we have

(i) if $\alpha < \frac{1}{2}$,

$$[H_0^m(\Omega), H^{-m}(\Omega)]_\alpha = [[H_0^m(\Omega), H^{-m}(\Omega)]_0, [H_0^m(\Omega), H^{-m}(\Omega)]_{1/2}]_{2\alpha}$$

$$= [H_0^m(\Omega), H^0(\Omega)]_{2\alpha} = H_0^{(1-2\alpha)m}(\Omega) \quad \text{(by Theorem 11.6)},$$

if $(1-2\alpha)\, m \neq \mu + \frac{1}{2}$, and $= H_{00}^{\mu+1/2}(\Omega)$ (Theorem 11.7),

if $(1-2\alpha)\, m = \mu + \frac{1}{2}$;

(ii) if $\alpha > \frac{1}{2}$,

$$[H_0^m(\Omega), H^{-m}(\Omega)]_\alpha = [[H_0^m(\Omega), H^{-m}(\Omega)]_{1/2}\, [H_0^m(\Omega), H^{-m}(\Omega)]_1]_{2\alpha-1}$$
$$= [H^0(\Omega), H^{-m}(\Omega)]_{2\alpha-1} = H^{(1-2\alpha)m}(\Omega) \quad \text{(by Theorem 12.2)},$$

if $(1-2\alpha)\, m \neq -\mu - \frac{1}{2}$, and $= (H_{00}^{\mu+1/2}(\Omega))'$

if $(1-2\alpha)\, m = -\mu - \frac{1}{2}$.

Next, having *chosen* integer $m \geqq \max(s_1, s_2)$, we note that, according to (i):

$$H_0^{s_1}(\Omega) = [H_0^m(\Omega), H^{-m}(\Omega)]_{\alpha_1}, \qquad (1-2\alpha_1)\, m = s_1$$

and according to (ii):

$$H^{-s_2}(\Omega) = [H_0^m(\Omega), H^{-m}(\Omega)]_{\alpha_2}, \qquad (1-2\alpha_2)\, m = s_2.$$

Then,

$$[H_0^{s_1}(\Omega), H^{-s_2}(\Omega)]_\theta = [[H_0^m(\Omega), H^{-m}(\Omega)]_{\alpha_1}, [H_0^m(\Omega), H^{-m}(\Omega)]_{\alpha_2}]_\theta$$
$$= [H_0^m(\Omega), H^{-m}(\Omega)]_{(1-\theta)\alpha_1+\theta\alpha_2}$$

and the theorem follows by applying (i) and (ii). □

12.4 Interpolation between $H^{s_1}(\Omega)$ and $H^{-s_2}(\Omega)$, $s_i > 0$

The interpolation between $H^{s_1}(\Omega)$ and $H^{-s_2}(\Omega)$ is somewhat more delicate than between $H_0^{s_i}(\Omega)$ and $H^{-s_2}(\Omega)$.

Theorem 12.4. *Assume that* Ω *satisfies* (7.10) *and* (7.11). *Let* s_1 *and* $s_2 \geqq 0$, *with* $s_2 \neq \mu + \frac{1}{2}$ *(integer* $\mu \geqq 0$*). We have:*

(12.14) $$[H^{s_1}(\Omega), H^{-s_2}(\Omega)]_\theta = H^{(1-\theta)s_1-\theta s_2}(\Omega)$$

if $(1-\theta)\, s_1 - \theta\, s_2 \neq -\frac{1}{2} - \nu$ *(integer* $\nu \geqq 0$*)*,

(12.15) $$[H^{s_1}(\Omega), H^{-s_2}(\Omega)]_\theta = (H_{00}^{\nu+1/2}(\Omega))'$$

if $(1-\theta)\, s_1 - \theta\, s_2 = -\frac{1}{2} - \nu$.

Proof. For the time being, we assume

Lemma 12.1. *If* Ω *satisfies* (7.10) *and* (7.11), *then for all integer* $m \geqq 1$:

(12.16) $$[H^m(\Omega), H^{-m}(\Omega)]_{1/2} = H^0(\Omega).$$

Then, *choosing* $m \geqq \max(s_1, s_2)$, we have, if $(1-\theta_1)\, m = s_1$:

$$H^{s_1}(\Omega) = [H^m(\Omega), H^0(\Omega)]_{\theta_1} = [H^m(\Omega), [H^m(\Omega), H^{-m}(\Omega)]_{1/2}]_{\theta_1}$$
$$= [H^m(\Omega), H^{-m}(\Omega)]_{\theta_1/2}.$$

If $(1-\theta_2)\,m = s_2 \neq \mu + \frac{1}{2}$, it follows from Theorem 11.6 and the duality theorem that

$$H^{-s_2}(\Omega) = [H^0(\Omega), H^{-m}(\Omega)]_{1-\theta_2} = [[H^m(\Omega), H^{-m}(\Omega)]_{1/2}, H^{-m}(\Omega)]_{1-\theta}$$
$$= [H^m(\Omega), H^{-m}(\Omega)]_{1-\theta_2/2}.$$

Therefore

$$[H^{s_1}(\Omega), H^{-s_2}(\Omega)]_\theta = [[H^m(\Omega), H^{-m}(\Omega)]_{\theta_1/2}, [H^m(\Omega), H^{-m}(\Omega)]_{1-\theta_2/2}]_\theta$$
$$= [H^m(\Omega), H^{-m}(\Omega)]_\alpha,$$

where

$$\alpha = (1-\theta)\frac{\theta_1}{2} + \theta\left(1 - \frac{\theta_2}{2}\right),$$

so that $(1-2\alpha)\,m = (1-\theta)\,s_1 - \theta\,s_2$.

If $\alpha \leqq \frac{1}{2}$, we have:

$$[H^m(\Omega), H^{-m}(\Omega)]_\alpha = [[H^m(\Omega), H^{-m}(\Omega)]_0, [H^m(\Omega), H^{-m}(\Omega)]_{1/2}]_{2\alpha}$$
$$= [H^m(\Omega), H^0(\Omega)]_{2\alpha} = H^{(1-2\alpha)m}(\Omega).$$

If $\alpha > \frac{1}{2}$, we have:

$$[H^m(\Omega), H^{-m}(\Omega)]_\alpha = [[H^m(\Omega), H^{-m}(\Omega)_{1/2}, [H^m(\Omega), H^{-m}(\Omega)]_1]_{2\alpha-1}$$
$$= [H^0(\Omega), H^{-m}(\Omega)]_{2\alpha-1}$$
$$= ([H_0^m(\Omega), H^0(\Omega)]_{1-(2\alpha-1)})',$$

which is equal to (Theorem 11.6)

$$(H_0^{(2\alpha-1)m}(\Omega))' = H^{(1-2\alpha)m}(\Omega)$$

if $(2\alpha-1)\,m \neq \frac{1}{2} + \nu$, integer $\nu \geqq 0$, i.e. if $(1-\theta)\,s_1 - \theta\,s_2 \neq$ integer $-\frac{1}{2}$, and is equal to (Theorem 11.7)

$$(H_{00}^{1/2+\nu}(\Omega))' \quad \text{if} \quad (1-\theta)\,s_1 - \theta\,s_2 = -\tfrac{1}{2} - \nu. \quad \sqcup$$

Proof of Lemma 12.1. 1) Consider the mapping r_Ω = restriction to Ω. This is a continuous linear mapping of $H^m(\mathbf{R}^n)$ (resp. $H^{-m}(\mathbf{R}^n)$) $\to$ $\to H^m(\Omega)$ (resp. $H^{-m}(\Omega)$), therefore by interpolation, it is a continuous linear mapping of

$$[H^m(\mathbf{R}^n), H^{-m}(\mathbf{R}^n)]_{1/2} = H^0(\mathbf{R}^n) \to [H^m(\Omega), H^{-m}(\Omega)]_{1/2}.$$

Then, if $u \in H^0(\Omega)$, its extension $\tilde{u}$ by 0 outside Ω is in $H^0(\mathbf{R}^n)$, therefore

$$r_\Omega\,\tilde{u} = u \in [H^m(\Omega), H^{-m}(\Omega)]_{1/2},$$

therefore

$$H^0(\Omega) \subset [H^m(\Omega), H^{-m}(\Omega)]_{1/2}. \tag{12.17}$$

(Note that this inclusion is valid without any regularity hypothesis on the boundary Γ of Ω.)

Variant: $[H_0^m(\Omega), H^{-m}(\Omega)]_{1/2} = H^0(\Omega) \subset [H^m(\Omega), H^{-m}(\Omega)]_{1/2}$.

2) The inverse inclusion of (12.17) follows from the following lemma:

Lemma 12.2. *If Ω satisfies* (7.10) *and* (7.11), *there exists, for every integer m, an extension operator P having the following properties*:

(12.18) $$P \in \mathscr{L}(H^m(\Omega); H^m(\mathbf{R}^n))$$

(12.19) $$P \in \mathscr{L}(H^{-m}(\Omega); H^{-m}(\mathbf{R}^n))$$

(12.20) $$r_\Omega P u = u \qquad \forall u \in H^{-m}(\Omega).$$

Then, by interpolation:

$$P \in \mathscr{L}([H^m(\Omega), H^{-m}(\Omega)]_{1/2}; [H^m(\mathbf{R}^n), H^{-m}(\mathbf{R}^n)]_{1/2} = H^0(\mathbf{R}^n)),$$

therefore, if $u \in [H^m(\Omega), H^{-m}(\Omega)]_{1/2}$, we have: $P u \in H^0(\mathbf{R}^n)$, therefore $r_\Omega P u (= u) \in H^0(\Omega)$, from which the inverse inclusion of (12.17) follows. □

Proof of Lemma 12.2. Via local maps, it is sufficient to construct P with properties (12.18), (12.19), (12.20) when $\Omega = \mathbf{R}_+^n$. We define P by

(12.21) $$P u(x) = \begin{cases} u(x) & \text{if } x_n > 0, \\ \sum_{k=1}^{2m} \alpha_k u(x', a_k x_n) = u_*(x) & \text{if } x_n < 0 \quad (a_k < 0), \end{cases}$$

where the α_k's and a_k's are chosen to satisfy the relations

(12.22) $$\sum_{k=1}^{2m} \alpha_k a_k^j = 1, \qquad -m \leq j \leq m-1,$$

which is possible: first, choose the a_k's such that

(12.23) $$\det. a_k^j \neq 0,$$

then the α_k's as the solutions of (12.22).

Conditions (12.22) for $0 \leqq j \leqq m-1$ imply (12.18), since they interpret the fact that

$$\frac{\partial^j}{\partial x_n^j} u(x', 0) = \frac{\partial^j u_*}{\partial x_n^j}(x', 0), \qquad 0 \leqq j \leqq m-1.$$

Conditions (12.22) for $-m \leqq j \leqq -1$ imply (12.19); indeed, (12.19) is equivalent to (${}^tP$ denoting the transpose of P considered as an element of $\mathscr{L}(\mathscr{D}(\Omega); \mathscr{D}(\mathbf{R}^n))$)

(12.24) $${}^tP \in \mathscr{L}(H^m(\mathbf{R}^n); H_0^m(\Omega)).$$

But, as can easily be verified, if $f \in \mathscr{D}(\mathbf{R}^n)$, then

(12.25) $${}^tP f = f(x) - g(x), \qquad x_n > 0,$$

where

$$g(x) = \sum_{k=1}^{2m} a_k^{-1} \alpha_k f(x', a_k^{-1} x_n) \tag{12.26}$$

and then conditions (12.22) for $-m \leqq j \leqq -1$ interpret the fact that

$$\frac{\partial^j f}{\partial x_n^j}(x', 0) - \frac{\partial^j g}{\partial x_n^j}(x', 0) = 0, \quad 0 \leqq j \leqq m-1,$$

and therefore we obtain (12.24) (applying Theorem 11.5 with $s = m$).
Property (12.20) being obvious, the lemma is proved. □

12.5 Interpolation between $H^{s_1}(\Omega)$ and $(H^{s_2}(\Omega))'$

Let s_1 and s_2 be *fixed* and *positive*.

The space $(H^{s_2}(\Omega))'$, dual (or anti-dual) of $H^{s_2}(\Omega)$, is not necessarily identified with an "ordinary" function space but is an "abstract" space. We may *identify* $H^{s_1}(\Omega)$ *with a dense subspace of* $(H^{s_2}(\Omega))'$, in the following manner. Let $u \in H^{s_1}(\Omega)$. Then

$$v \to \int_\Omega u \bar{v} \, dx$$

is a continuous antilinear form on $H^{s_2}(\Omega)$, say u_*. Thus, we have a linear mapping $u \to u_*$ of $H^{s_1}(\Omega) \to (H^{s_2}(\Omega))'$, which is continuous and *one-to-one*: if $u_* = 0$, then

$$\int_\Omega u \bar{v} \, dx = 0 \quad \forall v \in H^{s_2}(\Omega),$$

therefore $\forall v \in \mathscr{D}(\Omega)$, therefore $u = 0$. Identifying u_* and u, we have: $H^{s_1}(\Omega) \subset (H^{s_2}(\Omega))'$ and $H^{s_1}(\Omega)$ is *dense*. Indeed if $w \in ((H^{s_2}(\Omega))')' = H^{s_2}(\Omega)$ and $w = 0$ on $H^{s_1}(\Omega)$, we have:

$$\int_\Omega w \bar{\varphi} \, dx = 0 \quad \forall \varphi \in H^{s_1}(\Omega), \text{ therefore } w = 0.$$

Furthermore, note that this identification is independent of s_1 and s_2. □

Theorem 12.5. *Assume that* Ω *satisfies* (7.10) *and* (7.11). *Let* s_1 *and* $s_2 \geqq 0$. *We have*

$$[H^{s_1}(\Omega), (H^{s_2}(\Omega))']_\theta = \begin{cases} H^{(1-\theta)s_1 - \theta s_2}(\Omega), & \text{if } (1-\theta)s_1 - \theta s_2 \geqq 0; \\ (H^{-((1-\theta)s_1 - \theta s_2)}(\Omega))' & \text{if } (1-\theta)s_1 - \theta s_2 \leqq 0. \end{cases} \tag{12.27}$$

(This time, there is no exceptional parameter such as in the preceding theorems.)

Proof. We use the same type of reasoning as for Theorems 12.3 and 12.4, for example. First, note that, according to (2.41):

$$[H^m(\Omega), (H^m(\Omega))']_{1/2} = H^0(\Omega).$$

Choose integer $m \geqq \max(s_1, s_2)$. We have (Theorem 9.6):

$$\begin{aligned} H^{s_i}(\Omega) &= [H^m(\Omega), H^0(\Omega)]_{\theta_i}, \quad (1-\theta_i)\, m = s_i \\ &= [H^m(\Omega), [H^m(\Omega), (H^m(\Omega))']_{1/2}]_{\theta_i} \\ &= \text{(according to the reiteration theorem)} \\ &\quad [H^m(\Omega), (H^m(\Omega))']_{\theta_i/2}. \end{aligned}$$

By the duality Theorem 6.2,

$$(H^{s_2}(\Omega))' = [H^m(\Omega), (H^m(\Omega))']_{1-\theta_2/2}$$

and therefore (by the reiteration theorem)

$$[H^{s_1}(\Omega), (H^{s_2}(\Omega))']_\theta = [H^m(\Omega), (H^m(\Omega))']_\alpha,$$

$$\alpha = (1-\theta)\frac{\theta_1}{2} + \theta\left(1 - \frac{\theta_2}{2}\right).$$

If $\alpha \leqq \frac{1}{2}$, $\quad [H^m(\Omega), (H^m(\Omega))']_\alpha = [H^m(\Omega), [H^m(\Omega), (H^m(\Omega))']_{1/2}]_{2\alpha}$

$$= H^{(1-2\alpha)m}(\Omega),$$

whence the first equality in (12.27).

If $\alpha > \frac{1}{2}$,

$$\begin{aligned} [H^m(\Omega), (H^m(\Omega))']_\alpha &= [[H^m(\Omega), (H^m(\Omega))']_{1/2}, (H^m(\Omega))']_{2\alpha-1} \\ &= [H^0(\Omega), (H^m(\Omega))']_{2\alpha-1} \\ &= ([H^m(\Omega), H^0(\Omega)]_{2-2\alpha})' = (H^{(2\alpha-1)m}(\Omega))', \end{aligned}$$

whence the second equality in (12.27). □

12.6 Interpolation between $H_0^{s_1}(\Omega)$ and $(H^{s_2}(\Omega))'$

Theorem 12.6. *Assume that Ω satisfies* (7.10) *and* (7.11). *Suppose $s_i \geqq 0$ and $s_1 \neq$ integer $+ \frac{1}{2}$. Then*

(12.28)
$$[H_0^{s_1}(\Omega), (H^{s_2}(\Omega))']_\theta = \begin{cases} H_0^{(1-\theta)s_1-\theta s_2}(\Omega) \\ \text{if } (1-\theta)\, s_1 - \theta\, s_2 \geqq 0 \quad \text{and} \neq \text{integer} + \frac{1}{2}; \\ (H^{-((1-\theta)s_1-\theta s_2)}(\Omega))' \\ \text{if } (1-\theta)\, s_1 - \theta\, s_2 \leqq 0; \\ H_{00}^{\mu+1/2}(\Omega) \quad \text{if } (1-\theta)\, s_1 - \theta\, s_2 = \frac{1}{2} + \mu, \\ \qquad \text{integer } \mu \geqq 0. \end{cases}$$

Remark 12.3. Theorems 12.5 and 12.6 may be compared as follows: *except* for certain exceptional values of the parameter, the interpolated space between $H_0^{s_1}(\Omega)$ and $(H^{s_2}(\Omega))'$ is a closed *subspace* of the interpolated space between $H^{s_1}(\Omega)$ and $(H^{s_2}(\Omega))'$ (for the same parameter). □

Proof of Theorem 12.6. By duality we deduce from (12.16):

$$[H_0^m(\Omega), (H^m(\Omega))']_{1/2} = H^0(\Omega). \tag{12.29}$$

Since $s_1 \neq \text{integer} + \frac{1}{2}$, we have:

$$H_0^{s_1}(\Omega) = [H_0^m(\Omega), H^0(\Omega)]_{\theta_1}, \quad (1 - \theta_1)\, m = s_1,$$

and applying (12.29):

$$\begin{aligned} H_0^{s_1}(\Omega) &= \big[H_0^m(\Omega), [H_0^m(\Omega), (H^m(\Omega))']_{1/2}\big]_{\theta_1} \\ &= [H_0^m(\Omega), (H^m(\Omega))']_{\theta_1/2}. \end{aligned} \tag{12.30}$$

Next

$$H^{s_2}(\Omega) = [H^m(\Omega), H^0(\Omega)]_{\theta_2}, \quad (1 - \theta_2)\, m = s_2,$$

therefore, by duality,

$$\begin{aligned} (H^{s_2}(\Omega))' &= [H^0(\Omega), (H^m(\Omega))']_{1-\theta_2} \\ &= \big[[H_0^m(\Omega), (H^m(\Omega))']_{1/2}, (H^m(\Omega))'\big]_{1-\theta_2} \\ &= [H_0^m(\Omega), (H^m(\Omega))']_{\theta_2/2+1-\theta_2}, \end{aligned}$$

which, together with (12.30) (and the reiteration Theorem 6.1), yields

$$[H_0^{s_1}(\Omega), (H^{s_2}(\Omega))']_\theta = [H_0^m(\Omega), (H^m(\Omega))']_\alpha,$$

with

$$\alpha = (1 - \theta)\frac{\theta_1}{2} + \theta\left(1 - \frac{\theta_2}{2}\right).$$

If $\alpha \leqq \frac{1}{2}$, we have:

$$\begin{aligned} [H_0^m(\Omega), (H^m(\Omega))']_\alpha &= \big[H_0^m(\Omega), [H_0^m(\Omega), (H^m(\Omega))']_{1/2}\big]_{2\alpha} \\ &= [H_0^m(\Omega), H^0(\Omega)]_{2\alpha} \\ &= \begin{cases} H^{(1-2\alpha)m}(\Omega) & \text{if } (1-2\alpha)\, m \neq \text{integer} + \frac{1}{2} \\ H_{00}^{\mu+1/2}(\Omega) & \text{if } (1-2\alpha)\, m = \frac{1}{2} + \mu \\ & \qquad (\text{integer } \mu \geqq 0). \end{cases} \end{aligned}$$

If $\alpha > \frac{1}{2}$, we have:

$$[H_0^m(\Omega), (H^m(\Omega))']_\alpha = ([H^m(\Omega), H^0(\Omega)]_{2-2\alpha})' = (H^{(2\alpha-1)m}(\Omega))'. \quad \square$$

Remark 12.4. Let $s > 0$. r_Ω (restriction to Ω) maps $H^s(\mathbf{R}^n) \to H^s(\Omega)$ and according to Theorem 9.1 it is *surjective*. Therefore, if

$$\operatorname{Ker}_s(r_\Omega) = \{u \mid u \in H^s(\mathbf{R}^n), r_\Omega u = 0\}, \tag{12.31}$$

we see that r_Ω is, by passage to the quotient, an isomorphism of $H^s(\mathbf{R}^n)/\mathrm{Ker}_s(r_\Omega)$ onto $H^s(\Omega)$ and therefore, by transposing:

(12.32) $\quad {}^t(r_\Omega)$ *is an isomorphism of* $(H^s(\Omega))'$ *onto* $(H^s(\mathbf{R}^n)/\mathrm{Ker}_s(r_\Omega))'$.

But

$$(H^s(\mathbf{R}^n)/\mathrm{Ker}_s(r_\Omega))' =$$
$$= \{f \mid f \in H^{-s}(\mathbf{R}^n), \langle f, \varphi\rangle = 0 \quad \forall \varphi \in \mathscr{D}(\mathbf{R}^n), \quad \text{with } \varphi = 0 \text{ on } \complement\,\Omega\}.$$

Therefore set

(12.33) $\quad H_{\bar\Omega}^{-s}(\mathbf{R}^n) = \{f \mid f \in H^{-s}(\mathbf{R}^n),\ f \text{ with support in } \bar\Omega\}$.

Then

(12.34) $\quad {}^t(r_\Omega)$ *is an isomorphism of* $(H^s(\Omega))'$ *onto* $H_{\bar\Omega}^{-s}(\mathbf{R}^n)$. □

Remark 12.5. Since $H_0^s(\Omega)$ is a closed vector subspace of $H^s(\Omega)$ $(s > 0)$, we may identify (through the isomorphism (12.34)) the dual $H^{-s}(\Omega)$ of $H_0^s(\Omega)$ with the quotient of ${}^t(r_\Omega)\,(H^s(\Omega))' = H_{\bar\Omega}^{-s}(\mathbf{R}^n)$ by the subspace of $H_{\bar\Omega}^{-s}(\mathbf{R}^n)$ which is orthogonal to $H_0^s(\Omega)$, i.e. by the distributions $f \in H_{\bar\Omega}^{-s}(\mathbf{R}^n)$ such that

$$\langle f, r_\Omega\,\varphi\rangle = 0 \qquad \forall \varphi \in \mathscr{D}(\mathbf{R}^n), \qquad r_\Omega\,\varphi \in H_0^s(\Omega).$$

Therefore $f = 0$ in Ω and

$$f \in H_\Gamma^{-s}(\mathbf{R}^n) = \text{distributions of } H^{-s}(\mathbf{R}^n) \textit{ with support in } \Gamma.$$

Therefore, in short:

$$H^{-s}(\Omega) \approx H_{\bar\Omega}^{-s}(\mathbf{R}^n)/H_\Gamma^{-s}(\mathbf{R}^n).$$

Also, in this connection, note that the completion of $\widetilde{\mathscr{D}(\Omega)}$ = (space of $\mathscr{D}(\Omega)$-functions extended by 0 outside Ω) for the topology of $H^{-s}(\mathbf{R}^n)$ may be identified with $H_{\bar\Omega}^{-s}(\mathbf{R}^n)$ (and is therefore essentially distinct from $H^{-s}(\Omega)$). □

Remark 12.6. If $s_1 = \text{integer} + \frac{1}{2}$, *result* (12.29) *is still valid if we replace* $H_0^{s_1}(\Omega)$ *with* $H_{00}^{s_1}(\Omega)$, as the same proof shows; indeed, if we use

$$H_{00}^{s_1}(\Omega) = [H_0^m(\Omega), H^0(\Omega)]_{\theta_1}, \qquad (1 - \theta_1)\, m = s_1,$$

the proof of Theorem 12.6 applies immediately. Therefore, *in particular*

$$[H_{00}^{s_1}(\Omega), (H^{s_2}(\Omega))']_\theta = H_0^{(1-\theta)s_1 - \theta s_2}(\Omega)$$

if $(1 - \theta)\, s_1 - \theta\, s_2 \geqq 0$, $\neq \text{integer} + \frac{1}{2}$. □

12.7. A Lemma

We shall give below (Lemma 12.3) a technical result which will be useful for the study of the regularity of solutions of elliptic boundary value problems (Chapter 2). In this section, we set:

$$\Omega = \{x \mid x_n > 0\} \tag{12.35}$$

$$K(\Omega) = \left\{u \mid u,\ \frac{\partial u}{\partial x_i} \in H^{r-1}(\Omega),\quad 1 \leqq i \leqq n-1,\quad \frac{\partial^\mu u}{\partial x_n^\mu} \in H^{r-\mu}(\Omega)\right\}, \tag{12.36}$$

where r and μ are two integers $\geqq 1$, $r - \mu$ of arbitrary sign. We provide $K(\Omega)$ with the norm

$$\left(\| u \|^2_{H^{r-1}(\Omega)} + \sum_{i=1}^{n-1} \left\| \frac{\partial u}{\partial x_i} \right\|^2_{H^{r-1}(\Omega)} + \left\| \frac{\partial^\mu u}{\partial x_n^\mu} \right\|^2_{H^{r-\mu}(\Omega)}\right)^{1/2}$$

which makes it a Hilbert space. We define $K(\mathbf{R}^n)$ in the same way.

Lemma 12.3. *With* $K(\Omega)$ *defined by* (12.36), *we have*

$$K(\Omega) = H^r(\Omega). \tag{12.37}$$

Proof. The case "$r \geqq \mu$" is immediate. Thus, the interesting case is "$r < \mu$".

For the time being, we assume

Lemma 12.4. *The space* $\mathscr{D}(\overline{\Omega})$ *is dense in* $K(\Omega)$.

and we show that Lemma 12.3 results from it:

for $u \in \mathscr{D}(\overline{\Omega})$, we define the function $\underline{P u}$ in $\mathbf{R}^n$ by

$$\underline{P u}(x) = \begin{cases} u(x) & \text{if } x_n > 0 \\ \sum_{j=1}^{\mu} \lambda_j u(x', -j x_n) & \text{if } x_n < 0, \end{cases} \tag{12.38}$$

($x' = \{x_1, \ldots, x_{n-1}\}$), where the λ_j's are chosen so that

$$\sum_{j=1}^{\mu} (-j)^k \lambda_j = 1, \quad 0 \leqq k \leqq \mu - 1. \tag{12.39}$$

Then, we verify that:

$$\begin{cases} (u \to \underline{P u} \text{ is a continuous linear mapping of } \mathscr{D}(\overline{\Omega}) \\ \text{(provided with the topology induced by } K(\Omega)) \text{ into } K(\mathbf{R}^n). \end{cases} \tag{12.40}$$

First of all, thanks to (12.39), we have

$$D^\alpha \underline{P u}(x) = \begin{cases} D^\alpha u(x) & \text{if } x_n > 0 \\ \sum_{j=1}^{\mu} \lambda_j D^\alpha(u(x', -j x_n)) & \text{if } x_n < 0 \end{cases}$$

for all α such that $|\alpha| \leqq \mu$, therefore in particular $\forall \alpha$ such that $|\alpha| \leqq r$ and therefore

$$\| \underline{P u} \|_{H^{r-1}(\mathbf{R}^n)} + \sum_{i=1}^{n-1} \left\| \frac{\partial}{\partial x_i} (\underline{P u}) \right\|_{H^{r-1}(\mathbf{R}^n)} \leqq \tag{12.41}$$

$$\leqq c \left(\| u \|_{H^{r-1}(\Omega)} + \sum_{i=1}^{n-1} \left\| \frac{\partial u}{\partial x_i} \right\|_{H^{r-1}(\Omega)} \right).$$

Next

$$\frac{\partial^\mu}{\partial x_n^\mu} \underline{P u} = \begin{cases} \dfrac{\partial^\mu u}{\partial x_n^\mu}(x) & \text{if } x_n > 0, \\ \displaystyle\sum_{j=1}^{\mu} (-j)^\mu \lambda_j \frac{\partial^\mu}{\partial x_n^\mu} u(x', -j x_n) & \text{if } x_n < 0 \end{cases} \tag{12.42}$$

and we shall obtain (12.40) if we can show that

(12.43) $\begin{cases} v \to Q v \text{ is a continuous linear mapping of } \mathscr{D}(\bar{\Omega}) \text{ (provided} \\ \text{with the topology induced by } H^{r-\mu}(\Omega)) \text{ into } H^{r-\mu}(\mathbf{R}^n), \end{cases}$

where

$$Q v = \begin{cases} v & \text{if } x_n > 0 \\ \displaystyle\sum_{j=1}^{\mu} (-j)^\mu \lambda_j v(x', -j x_n) & \text{if } x_n < 0. \end{cases} \tag{12.44}$$

For this purpose, it is equivalent to show that Q^* (adjoint of Q) is a continuous linear mapping of $H^{\mu-r}(\mathbf{R}^n) \to H_0^{\mu-r}(\Omega)$. But if $\varphi \in H^{\mu-r}(\mathbf{R}^n)$, we have

$$Q^* \varphi = \varphi(x) - \sum_{j=1}^{\mu} (-j)^{\mu-1} \lambda_j \varphi\left(x', -\frac{1}{j} x_n\right) \quad (x_n > 0) \tag{12.45}$$

and

$$\frac{\partial^k Q^* \varphi}{\partial x_n^k}(x', 0) = \frac{\partial^k \varphi}{\partial x_n^k}(x', 0) \left[1 - \sum_{j=1}^{\mu} (-j)^{\mu-1} \left(-\frac{1}{j} \right)^k \lambda_j \right].$$

Therefore, thanks to (12.39), $\dfrac{\partial^k}{\partial x_n^k} Q^* \varphi(x', 0) = 0$ for $0 \leqq k \leqq \mu - 1$, therefore for $0 \leqq k \leqq \mu - r$, whence the result (12.43) and therefore (12.40).

Because of Lemma 12.4, we may extend $u \to \underline{P u}$ by continuity to a continuous linear mapping of $K(\Omega) \to K(\mathbf{R}^n)$, which we still denote by $u \to \underline{P u}$.

Now, we verify that

$$K(\mathbf{R}^n) = H^r(\mathbf{R}^n). \tag{12.46}$$

We use the Fourier transform; let $\hat{v}$ be the transform of $v \in K(\mathbf{R}^n)$, $\xi = \{\xi', \xi_n\}$ the dual variable of $x = \{x', x_n\}$; we have:

$$(12.47) \qquad (1 + |\xi'|)\,(1 + |\xi|)^{r-1}\,\hat{v} \in L^2(\mathbf{R}^n)$$

and

$$(12.48) \qquad |\xi|^\mu \frac{1}{(1 + |\xi|)^{\mu - r}}\,\hat{v} \in L^2(\mathbf{R}^n).$$

We have to show that $(1 + |\xi|)^r\,\hat{v} \in L^2(\mathbf{R}^n)$; since

$$(1 + |\xi|)^r \leqq c_1[(1 + |\xi'|)\,(1 + |\xi|)^{r-1} + |\xi_n|^r]$$

it is sufficient to show that $|\xi_n|^r\,\hat{v} \in L^2(\mathbf{R}^n)$. But

$$|\xi_n|^r\,(1 + |\xi|)^{\mu - r} \leqq c_2[|\xi_n|^\mu + |\xi_n|^r\,(1 + |\xi'|)^{\mu - r}] \leqq$$
$$\leqq c_2[|\xi_n|^\mu + (1 + |\xi'|)\,(1 + |\xi|)^{\mu - 1}],$$

from which we obtain

$$|\xi_n|^r \leqq c_2\left[\frac{|\xi_n|^\mu}{(1 + |\xi|)^{\mu - r}} + (1 + |\xi'|)\,(1 + |\xi|)^{r-1}\right],$$

from which the result follows according to (12.47) and (12.48).

Thus, if $u \in K(\Omega)$, we have: $v = \underline{P\,u} \in K(\mathbf{R}^n)$, therefore according to (12.46) $v \in H^r(\mathbf{R}^n)$ and the restriction of v to Ω (that is u) is in $H^r(\Omega)$. □

We still have to prove Lemma 12.4.

We shall make use of a new lemma, which is of interest in itself:

Lemma 12.5. *Let X be a Hilbert space; for s a positive integer, we define:*

$$H^{-s}(0, \infty; X) = (H_0^s(0, \infty; X))'.$$

Let v satisfy

$$(12.49) \qquad v \in H^{-s}(0, \infty; X), \quad v^{(k)} \in H^{-s}(0, \infty; X), \quad \left(v^{(k)} = \frac{d^k v}{d t^k}\right).$$

Then

$$(12.50) \qquad v^{(j)} \in H^{-s}(0, \infty; X), \quad 0 \leqq j \leqq k.$$

Proof. 1) The space $\mathscr{D}([0, \infty[; X)$ is dense in the space of v's satisfying (12.49) (with the natural norm). Indeed if L is a continuous anti-linear form on this space, we may write it

$$(12.51) \qquad L(v) = (w_0, v) + (w_1, v^{(k)}), \quad w_0, w_1 \in H_0^s(0, \infty; X).$$

We assume that

$$(12.52) \qquad L(\varphi) = 0 \quad \forall \varphi \in \mathscr{D}([0, \infty[; X)$$

and we want to show that $L = 0$.

If $\tilde{w}_i$ = extension of w_i by 0 for $t < 0$, we deduce from (12.52) that

$$\tilde{w}_0 + (-1)^k \frac{d^k}{dt^k} \tilde{w}_1 = 0. \tag{12.53}$$

But $\tilde{w}_0 \in H^s(\mathbf{R}; X)$, therefore (12.53) implies

$$\tilde{w}_1 \in H^{s+k}(\mathbf{R}; X) \quad \text{and} \quad w_1 \in H_0^{s+k}(0, \infty; X).$$

Then:

$$L(v) = (w_0 + (-1)^k w_1^{(k)}, v) = 0.$$

2) For $v \in \mathscr{D}([0, \infty[; X)$, we define

$$\pi v(t) = \begin{cases} v(t) & \text{if} \quad t > 0 \\ \sum_{j=1}^{s+k+1} c_j v(-jt) & \text{if} \quad t < 0, \end{cases} \tag{12.54}$$

with

$$\sum_{j=1}^{s+k+1} (-j)^p c_j = 1, \qquad -s \leqq p \leqq k. \tag{12.55}$$

Then $v \to \pi v$ is a continuous mapping of $\mathscr{D}([0, \infty[; X)$, provided with the topology induced by $H^{-s}(0, \infty; X)$, $\to H^{-s}(\mathbf{R}, X)$; next

$$(\pi v)^{(k)} = \left\{ \begin{matrix} v^{(k)}, & \text{if} \quad t > 0 \\ \sum_{j=1}^{s+k+1} (-j)^k c_j v^{(k)}(-jt), & \text{if} \quad t \leqq 0 \end{matrix} \right\} = \tilde{\pi} v^{(k)}$$

and the mapping $v^{(k)} \to \tilde{\pi} v^{(k)}$ is continuous, for the topology induced by $H^{-s}(0, \infty; X)$, $\to H^{-s}(\mathbf{R}; X)$.

Therefore, through extension by continuity, using 1), we see that every v satisfying (12.49) is a restriction of πv to $]0, \infty[$ with

$$\pi v \in H^{-s}(\mathbf{R}; X), \qquad (\pi v)^{(k)} \in H^{-s}(\mathbf{R}; X). \tag{12.56}$$

But by Fourier transform in t, we immediately verify that (12.56) implies

$$(\pi v)^{(j)} \in H^{-s}(\mathbf{R}; X), \qquad 1 \leqq j \leqq k - 1,$$

and (12.50) follows. □

Proof of Lemma 12.4. 1) Let $u \in K(\Omega)$. Via regularization in x', we immediately see that u is the limit in $K(\Omega)$ of functions v satisfying, for example

$$\begin{cases} v \in H^0(0, \infty; H^k_{x'}), \qquad v^{(r-1)} \in H^0(0, \infty; H^k_{x'}), \\ v^{(\mu)} \in H^{r-\mu}(0, \infty; H^k_{x'}), \end{cases} \tag{12.57}$$

where k is chosen *arbitrarily* and $H^k_{x'} = H^k(\mathbf{R}^{n-1}_{x'})$.

2) According to Lemma 12.5, we obtain in particular from (12.57) that

(12.58) $$v^{(j)} \in H^{r-\mu}(0, \infty; H^k_{x'}), \quad r \leqq j \leqq \mu.$$

3) We define

(12.59) $$\begin{cases} w_h(x) = v(x', x_n + h), & x_n > -h \quad (h > 0) \\ v_h = \text{restriction of } w_h \text{ to } \Omega \end{cases}$$

and we verify that

(12.60) $$v_h \to v \text{ in } K(\Omega) \text{ weakly as } h \to 0.$$

In fact, the only thing to verify is that $v_h^{(\mu)} \in H^{r-\mu}(0, \infty; H^k_{x'})$ and that $v_h^{(\mu)} \to v^{(\mu)}$ in $H^{r-\mu}(0, \infty; H^h_{x'})$ as $h \to 0$. But, for $\varphi \in \mathscr{D}(\Omega)$:

$$(v_h^{(\mu)}, \varphi) = (-1)^\mu \int_\Omega v \frac{\partial^\mu}{\partial x_n^\mu} \varphi(x', x_n - h)\, dx$$

(where $\varphi(x', x_n - h)$ is extended by 0 for $x_n < h$), from which the result easily follows.

4) To show the lemma, it is therefore sufficient to approach, in the sense of $K(\Omega)$, with a sequence of functions of $\mathscr{D}(\bar{\Omega})$, a function v satisfying (12.57) and (12.58) and which is a restriction to Ω of w having the same properties as v but on $\Omega_h = \{x \mid x_n > -h\}$. Therefore, in particular

(12.61) $$w^{(j)} \in H^{r-\mu}(-h, \infty; H^k_{x'}), \quad r \leqq j \leqq \mu.$$

We consider a function $\theta = \theta(x_n)$, C^∞ on $\mathbf{R}$, and such that $\theta(x_n) = 1$ if $x_n \geqq -h/3$, $\theta(x_n) = 0$ if $x_n \leqq -2h/3$. Thanks to (12.61), θw has the same properties as v, this time on $\mathbf{R}^n$, and therefore in particular

$$v = \text{restriction to } \Omega \text{ of } \Phi \in K(\mathbf{R}^n).$$

But, through regularization and truncation, $\Phi = \lim_j \Phi_j$ in $K(\mathbf{R}^n)$, $\Phi_j \in \mathscr{D}(\mathbf{R}^n)$, and by restriction to Ω: $v = \lim_j \varphi_j$ in $K(\Omega)$, φ_j = restriction of Φ_j to Ω, $\in \mathscr{D}(\bar{\Omega})$. □

Remark 12.7. With the same type of procedure, we could show the following (a variant of the intermediate derivatives theorem of Section 2.3):

if $u \in L^2(0, \infty; X)$, $u^{(m)} \in H^{-k}(0, \infty; Y)$, *then*

$$u^{(j)} \in H^{-kj/m}(0, \infty; [X, Y]_{j/m}).$$ □

12.8 Differential Operators on $H^s(\Omega)$

Let Ω satisfy (7.10), (7.11) and let Λ be a differential operator of order N with infinitely differentiable coefficients in $\bar{\Omega}$.

We want to investigate how Λ operates on $H^s(\Omega)$, with for example $s > 0$. This is a simple exercise starting with the preceding results on interpolation and the fact that

(12.62) $\Lambda \in \mathscr{L}(H^m(\Omega); H^{m-N}(\Omega))$ if m is an integer ($\geqq 0$ or < 0).

(Moreover, we shall obtain, in the same fashion, the properties of Λ on $H_0^s(\Omega)$, $H_{00}^s(\Omega)$, $H^{-s}(\Omega)$, etc.).

Proposition 12.1. *We have*

(12.63) $\Lambda \in \mathscr{L}(H^s(\Omega); H^{s-N}(\Omega))$ *if* $s - N \neq -\nu - \frac{1}{2}$, *integer* $\nu \geqq 0$

and

(12.64) $\Lambda \in \mathscr{L}(H^s(\Omega); (H_{00}^{N-s}(\Omega))')$ *if* $s - N = -\nu - \frac{1}{2}$, *integer* $\nu \geqq 0$.

Proof. Let μ be an integer, $\mu \geqq \max(N, s)$. We apply (12.62) with $m = \mu$ and $m = 0$. By interpolation, we obtain

(12.65) $$\Lambda \in \mathscr{L}([H^\mu(\Omega), H^0(\Omega)]_\theta; [(H^{\mu-N}(\Omega), H^{-N}(\Omega)]_\theta).$$

We choose θ from $(1 - \theta) = s$. The first space in (12.65) is $H^s(\Omega)$ (definition (9.1)). Through an application of Theorem 12.4, in order to interpret

$$[H^{\mu-N}(\Omega), H^{-N}(\Omega)]_\theta,$$

we obtain the desired results. □

Remark 12.8. We note that (particular case of the preceding Proposition):

(12.66) $$\begin{cases} \dfrac{\partial}{\partial x_i} \in \mathscr{L}(H^s(\Omega); H^{s-1}(\Omega)) \quad \text{for} \quad s \geqq 0, \quad s \neq \frac{1}{2}, \\ \dfrac{\partial}{\partial x_i} \in \mathscr{L}(H^{1/2}(\Omega); (H_{00}^{1/2}(\Omega))'). \end{cases}$$

By transposition (for example), we verify that

$$\frac{\partial}{\partial x_i} \in \mathscr{L}(H^s(\Omega); H^{s-1}(\Omega)) \qquad \forall s < 0.$$

12.9 Invariance by Diffeomorphism of $H^s(\Omega)$-Spaces

Let Ω' be another open set in $\mathbf{R}^n$ having properties analogous to those of Ω, and let Θ be an infinitely differentiable diffeomorphism of $\bar{\Omega}$ onto $\bar{\Omega}'$. Then Θ induces, in a natural manner, an isomorphism of $H^s(\Omega)$ onto $H^s(\Omega')$ (resp. $H_0^s(\Omega)$ onto $H_0^s(\Omega')$, resp. $H_{00}^s(\Omega)$ onto $H_{00}^s(\Omega')$ if $s =$ integer $+ \frac{1}{2}$); this can be shown directly, first for integer $s \geqq 0$, then, by interpolation for real $s > 0$; finally, by duality we pass to the case $s < 0$.

13. Intersection Interpolation

13.1 A General Result

Theorem 13.1. *Let $\mathscr{H}$ be a separable Hilbert space and Λ_0, Λ_1 two positive, self-adjoint, commutative operators in $\mathscr{H}$. Let $D(\Lambda_i)$ be the domain of Λ_i, provided with the norm of the graph.*

We have, $\forall \theta \in]0, 1[$:

$$[D(\Lambda_0) \cap D(\Lambda_1), \mathscr{H}]_\theta = [D(\Lambda_0), \mathscr{H}]_\theta \cap [D(\Lambda_1), \mathscr{H}]_\theta \tag{13.1}$$

(with equivalent norms).

Proof. Because Λ_0 and Λ_1 are *commutative*, there exists a simultaneous "diagonalization" of the Λ_i's (see Dixmier [1], p. 217).

More precisely, there exists a measurable sum:

$$\mathfrak{h} = \int^{\oplus} \mathfrak{h}(\lambda)\, d\mu(\lambda), \qquad \lambda = \{\lambda_0, \lambda_1\},$$

where $d\mu(\lambda)$ is a measure $\geqq 0$ on $\lambda_0 \geqq \lambda_{00} > 0$, $\lambda_1 \geqq \lambda_{10} > 0$, and a unitary operator $\mathscr{U}$ of $\mathscr{H}$ onto $\mathfrak{h}$ such that $\mathscr{U}(D(\Lambda_i)) = \{f \mid f \in \mathfrak{h}, \lambda_i f \in \mathfrak{h}\} =$ domain of $\{\lambda_i\}$, the operator of multiplication by λ_i, and

$$\mathscr{U}(\Lambda_i u) = \lambda_i(\mathscr{U} u) \qquad \forall u \in D(\Lambda_i).$$

Then

$$\mathscr{U}(D(\Lambda_0) \cap D(\Lambda_1)) = \{f \mid f \in \mathfrak{h}, \lambda_0 f \in \mathfrak{h}, \lambda_1 f \in \mathfrak{h}\},$$

a condition *equivalent to* $(\lambda_0 + \lambda_1) f \in \mathfrak{h}$.

Therefore, $\{\lambda_0 + \lambda_1\}$ denoting the operator of multiplication by $(\lambda_0 + \lambda_1)$, we have

$$\mathscr{U}(D(\Lambda_0) \cap D(\Lambda_1)) = D(\{\lambda_0 + \lambda_1\})$$

and property (13.1) is equivalent to

$$[D(\{\lambda_0 + \lambda_1\}), \mathscr{H}]_\theta = [D(\{\lambda_0\}), \mathscr{H}]_\theta \cap [D(\{\lambda_1\}), \mathscr{H}]_\theta. \tag{13.2}$$

But according to the definition of the spaces $[X, Y]_\theta$, the first space is the domain of $\{\lambda_0 + \lambda_1\}^{1-\theta}$, that is the space of functions $f \in \mathfrak{h}$ such that

$$(\lambda_0 + \lambda_1)^{1-\theta} f \in \mathfrak{h},$$

which is a condition *equivalent to*

$$\lambda_0^{1-\theta} f \in \mathfrak{h}, \qquad \lambda_1^{1-\theta} f \in \mathfrak{h},$$

whence the theorem. □

13.2 Example of Application (I)

We shall make frequent use of Theorem 13.1 in the following chapters: the theorem is useful in situations where we have to work with function spaces defined on *product spaces of the type* $\mathbf{R}_x^n \times \mathbf{R}_y^{n'}$ (or $\Omega_x \times \Omega_y$, Ω_x (resp. Ω_y) an open set in $\mathbf{R}_x^n$ (resp. $\mathbf{R}_y^{n'}$)), with *different properties according to the variables.*

This is the case for evolution equations, quasi-elliptic equations etc., in fact for *all* applications to *non-elliptic* differential operators. Here, we give a very simple example. □

On the space $\mathbf{R}_x^n \times \mathbf{R}_y^{n'}$, we define

$$(13.3)\quad H_{x,y}^{s,\sigma} = \{u \mid [(1 + |\xi|^2)^{s/2} + (1 + |\eta|^2)^{\sigma/2}]\, \hat{u} \in L^2(\mathbf{R}_{\xi,\eta}^{n+n'})\},$$

where

$$\hat{u}(\xi, \eta) = \frac{1}{(2\pi)^{(n+n')/2}} \int\limits_{\mathbf{R}_x^n \times \mathbf{R}_y^{n'}} \exp(-\mathrm{i}(x\,\xi + y\,\eta))\, u(x, y)\, dx\, dy.$$

Note that

$$H_{x,y}^{0,0} = H_{x,y}^0 = L_{x,y}^2(\mathbf{R}_{x,y}^{n+n'}).$$

When s_i and $\sigma_i \geqq 0$, we immediately deduce from Theorem 13.1 that

$$(13.4)\quad [H_{x,y}^{s_1,\sigma_1}, H_{x,y}^{s_2,\sigma_2}]_\theta = H_{x,y}^{(1-\theta)s_1+\theta s_2, (1-\theta)\sigma_1+\theta\sigma_2}. \quad \square$$

13.3 Example of Application (II)

We now introduce spaces which are to play a fundamental rôle in Chapters 4, 5, 6 of Volume 2.

Let Ω satisfy (7.10) and (7.11), and let Γ be its boundary. In the space $\mathbf{R}_x^n \times \mathbf{R}_t$, we consider the cylinder

$$Q = \Omega \times]0, T[,$$

with lateral boundary:

$$\Sigma = \Gamma \times]0, T[.$$

The spaces $H^{\alpha,\beta}(\Sigma)$, $\alpha, \beta \geqq 0$.

As for the spaces $H^s(\Omega)$ we *may* first *define* the spaces $H^{\alpha,\beta}(\Gamma \times \mathbf{R}_t)$ and then $H^{\alpha,\beta}(\Sigma)$ *by restriction to* Σ.

By analogy with (13.3), we define

$$(13.5)\quad H^{\alpha,\beta}(\Gamma \times \mathbf{R}) = \{u \mid u \in L^2(\mathbf{R}_t; H^\alpha(\Gamma)),\ |\tau|^\beta\, \hat{u} \in L^2(\mathbf{R}_\tau; H^0(\Gamma))\},$$

where

$$\hat{u}(\tau) = \frac{1}{\sqrt{2\pi}} \int\limits_{-\infty}^{+\infty} \mathrm{e}^{-\mathrm{i}t\tau}\, u(t)\, dt.$$

It is a Hilbert space for the norm

$$(\| u \|^2_{L^2(\mathbf{R}_t; H^\alpha(\Gamma))} + \| |\tau|^\beta u \|^2_{L^2(\mathbf{R}_\tau; H^0(\Gamma))})^{1/2}.$$

(We shall see in Chapter 4 of Volume 2, *why* these spaces are indispensable for the study of evolution equations, and the *properties of traces* of these spaces.) □

We define $H^{\alpha,\beta}(\Sigma)$ as the image of $H^{\alpha,\beta}(\Gamma \times \mathbf{R}_t)$ under the mapping $r_\Sigma =$ "restriction to Σ", provided with the corresponding quotient norm. Note that $H^{0,0}(\Sigma) = H^0(\Sigma) = L^2(\Sigma)$. □

Here is an equivalent definition.

Consider the space $H^{1,0}(\Sigma) = L^2(0, T; H^1(\Gamma))$ as a subspace of $Y = L^2(\Sigma)$ for the measure $d\Sigma = d\Gamma\, dt$; considering $H^{1,0}(\Sigma)$ as the space X in Section 2, we see that

(13.6) $H^{1,0}(\Sigma) =$ domain of Λ_0, positive self-adjoint operator in $L^2(\Sigma)$.

In fact, Δ_Γ being the Laplace-Beltrami operator on Γ, we have:

$$\Lambda_0 = (-\Delta_\Gamma)^{1/2} \tag{13.7}$$

and more precisely:

(13.7a) $\Lambda_0 u(t) = (-\Delta_\Gamma)^{1/2} (u(t))$ for almost all $t \in]0, T[$.

Next, we consider the space

$$H^{0,1}(\Sigma) = \left\{ u \mid u \in L^2(\Sigma), \frac{\partial u}{\partial t} \in L^2(\Sigma) \right\};$$

then, as above:

(13.8) $H^{0,1}(\Sigma) =$ domain of Λ_1, positive self-adjoint operator in $L^2(\Sigma)$

and in fact it can be verified that

(13.9) $\begin{cases} \Lambda_1 = \left(-\dfrac{d^2}{dt^2} + 1 \right)^{1/2}, \text{ with, as boundary conditions for} \\ -\dfrac{d^2}{dt^2} + 1, \text{ the conditions } \dfrac{\partial u(0)}{\partial t} = \dfrac{\partial u(T)}{\partial t} = 0. \end{cases}$

Then, we have:

(13.10) $H^{\alpha,\beta}(\Sigma) = D(\Lambda_0^\alpha) \cap D(\Lambda_1^\beta), \quad 0 \leqq \alpha, \beta \leqq 1.$ □

For $H^{\alpha,\beta}(\Gamma \times \mathbf{R})$, we shall take

(13.11) $\Lambda_1 = \left(-\dfrac{d^2}{dt^2} + 1 \right)^{1/2}$, defined by Fourier transform,

and then we shall have the result analogous to (13.10) *for all* $\alpha, \beta \geqq 0$. □

As for Theorem 13.1, we can verify that, $\forall \alpha_i, \beta_i \geqq 0$:

(13.12) $$[H^{\alpha_1,\beta_1}(\Gamma \times \mathbf{R}_t), H^{\alpha_2,\beta_2}(\Gamma \times \mathbf{R}_t)]_\theta = H^{(1-\theta)\alpha_1+\theta\alpha_2,\,(1-\theta)\beta_1+\theta\beta_2}(\Gamma \times \mathbf{R}_t).$$

We extend this result to the case wehre Σ replaces $\Gamma \times \mathbf{R}$ (by using the *extension* operators of $H^{\alpha,\beta}(\Sigma) \to H^{\alpha,\beta}(\Gamma \times \mathbf{R}_t)$). □

Now, we introduce (because we shall *need* spaces of this type in Chapter 4):

(13.13) $$\begin{cases} {}_0H^{\alpha,\beta}(\Sigma) = \{u \mid u \in H^{\alpha,\beta}(\Sigma), \quad u(.,0) = 0\} \\ \text{fixed } \beta > \frac{1}{2}. \end{cases}$$

This definition has meaning since according to Theorem 4.2, if $\beta > \frac{1}{2}$, $t \to u(\cdot, t)$ is a continuous function of

$$[0, T] \to [H^\alpha(\Gamma), H^0(\Gamma)]_{1/2\beta} = H^{\alpha(1-1/2\beta)}(\Gamma),$$

so that (13.13) has meaning and defines a *closed* subspace of $H^{\alpha,\beta}(\Sigma)$. We have

Proposition 13.1. *Let* $\alpha, \beta > 0$, $\beta > \frac{1}{2}$. *Then*

(13.14) $$[{}_0H^{\alpha,\beta}(\Sigma), H^0(\Sigma)]_\theta = H^{(1-\theta)\alpha,(1-\theta)\beta}(\Sigma), \quad \textit{if} \quad (1-\theta)\beta < \tfrac{1}{2}.$$

Proof. Let

$$H_0^{\alpha,\beta}(\Sigma) = \text{closure of } D(\Sigma) \text{ in } H^{\alpha,\beta}(\Sigma)$$

$$= \left\{u \mid u \in H^{\alpha,\beta}(\Sigma), \frac{\partial^j u}{\partial t^j}(.,0) = \frac{\partial^j u}{\partial t^j}(.,T) = 0, \quad 0 \leqq j < \beta - \frac{1}{2}\right\}.$$

We have

$$H_0^{\alpha,\beta}(\Sigma) \subset {}_0H^{\alpha,\beta}(\Sigma) \subset H^{\alpha,\beta}(\Sigma)$$

and therefore

(13.15) $$[H_0^{\alpha,\beta}(\Sigma), H^0(\Sigma)]_\theta \subset [{}_0H^{\alpha,\beta}(\Sigma), H^0(\Sigma)]_\theta \subset [H^{\alpha,\beta}(\Sigma), H^0(\Sigma)]_\theta.$$

In (13.15), the last space equals $H^{(1-\theta)\alpha,(1-\theta)\beta}(\Sigma)$. The first (proof analogous to those of Theorems 11.6 and 11.7) is equal to

$$H_0^{(1-\theta)\alpha,(1-\theta)\beta}(\Sigma) \quad \text{if} \quad (1-\theta)\beta \neq \text{integer} + \tfrac{1}{2}$$

(and to the strict subspace $H_{00}^{(1-\theta)\alpha,1/2}(\Sigma)$ of $H_0^{(1-\theta)\alpha,1/2}(\Sigma)$

$$= H_0^{(1-\theta)\alpha,1/2}(\Sigma), \quad \text{if} \quad (1-\theta)\beta = \tfrac{1}{2}),$$

whence (13.14), since

$$H_0^{(1-\theta)\alpha,(1-\theta)\beta}(\Sigma) = H^{(1-\theta)\alpha,(1-\theta)\beta}(\Sigma) \quad \text{if} \quad (1-\theta)\beta < \tfrac{1}{2}. \quad \square$$

13.4 Interpolation of Quotient Spaces

Let X and Y be two Hilbert spaces as in Section 2 and

$$N = \text{closed vector subspace of } X \text{ and of } Y. \tag{13.16}$$

Let π be the canonical mapping of $X \to X^{\bullet} = X/N$ (and of $Y \to Y^{\bullet} = Y/N$); π is a continuous linear mapping of $X \to X^{\bullet}$, $Y \to Y^{\bullet}$, therefore, by interpolation

$$\pi \in \mathscr{L}([X, Y]_\theta; [X^{\bullet}, Y^{\bullet}]_\theta). \tag{13.17}$$

Since π is obviously a continuous linear surjection of

$$[X, Y]_\theta \to ([X, Y]_\theta)/N,$$

we have

Proposition 13.2. *Let X and Y be two Hilbert spaces satisfying* (2.1) *and let N be defined by* (13.16). *Then*

$$([X, Y]_\theta)/N \subset [X/N, Y/N]_\theta, \quad 0 < \theta < 1, \tag{13.18}$$

(*with continuous injection*). □

The following theorem gives a very restrictive sufficient condition for *equality* in (13.18) (but which will be useful in Chapter 2).

Theorem 13.2. *Let X and Y be two Hilbert spaces satisfying* (2.1) *and let N be defined by* (13.16). *Then, if N is finite-dimensional, we have:*

$$([X, Y]_\theta)/N = [X/N, Y/N]_\theta, \quad 0 < \theta < 1. \tag{13.19}$$

Proof. Let $z_1, \ldots, z_\nu$ form a basis for N, chosen to be orthonormal in Y. For all $u \in Y$, we define

$$R u = u - \sum_{i=1}^{\nu} (u, z_i)_Y z_i. \tag{13.20}$$

We have: $R \in \mathscr{L}(X; X) \cap \mathscr{L}(Y; Y)$ and $R = 0$ on N, therefore, $R^{\bullet}$ denoting the quotient mapping:

$$R^{\bullet} \in \mathscr{L}(X^{\bullet}; X) \cap \mathscr{L}(Y^{\bullet}; Y),$$

therefore, by interpolation

$$R^{\bullet} \in \mathscr{L}([X^{\bullet}, Y^{\bullet}]_\theta; [X, Y]_\theta).$$

Thus if $u^{\bullet} \in [X^{\bullet}, Y^{\bullet}]_\theta$, we have

$$R^{\bullet} u^{\bullet} \in [X, Y]_\theta \quad \text{and} \quad \pi R^{\bullet} u^{\bullet} = u^{\bullet} \in ([X, Y]_\theta)/N,$$

whence the inverse inclusion in (13.18) and the theorem. □

Remark 13.1. The preceding arguments are evidently general, as long as *N is finite-dimensional*; neither the hilbertian structure, nor the particular construction of the interpolation spaces came into play. □

Remark 13.2. For Chapter 2, we shall also need the following "dual" viewpoint: let

$$(13.21)\quad \begin{cases} N_* = \text{closed vector subspace of } Y' \ (X \text{ and } Y \text{ are not identified} \\ \text{with their duals; therefore } Y' \subset X'). \end{cases}$$

We define

$$(13.22)\quad \begin{cases} \{X; N_*\} = \{u \mid u \in X, \langle u, z_*\rangle = 0 \quad \forall z_* \in N_*\} \\ \{Y; N_*\} = \{u \mid u \in Y, \langle u, z_*\rangle = 0 \quad \forall z_* \in N_*\}. \end{cases}$$

They are closed vector subspaces of X and Y.

Theorem 13.3. *Let X and Y be two Hilbert spaces satisfying* (2.1) *and let N_* be defined by* (13.21) *and finite-dimensional. Then* (with notations analogous to (13.22))

$$(13.23)\quad \{[X, Y]_\theta; N_*\} = [\{X; N_*\}, \{Y; N_*\}]_\theta, \quad 0 < \theta < 1.$$

Proof. Let $z'_1, \ldots, z'_\nu$ form a basis for N_*, orthonormal in Y', and let $z_1, \ldots, z_\nu$, be the elements of Y defined by

$$\langle z_i, z'_j\rangle = \delta_i^j,$$

the brackets denoting the duality between Y and Y'. If N denotes the space generated by $z_1, \ldots, z_\nu$, we have the decompositions:

$$X = N \dotplus \{X; N_*\}, \qquad Y = N \dotplus \{Y; N_*\},$$

which brings (13.23) back to Theorem 13.2. □

14. Holomorphic Interpolation

14.1 General Result

Let us again take up the setting of Section 2, with the two Hilbert spaces X and Y.

The space $\mathscr{H}(X, Y)$.

We denote by $\mathscr{H}(X, Y)$ the space of functions

$$z \to f(z), \quad z = \xi + i\eta, \quad 0 \leqq \xi \leqq 1, \eta \in \mathbf{R}, f(z) \in Y$$

having the following properties:

$$(14.1)\quad \begin{cases} f \text{ is a continuous and bounded function of the strip} \\ \{0 \leqq \xi \leqq 1\} \to Y \\ \text{and a holomorphic function of the open strip } \{0 < \xi < 1\} \to Y, \end{cases}$$

(the holomorphic property is equivalent to the *scalar* holomorphic property:

$$z \to (f(z), y)_Y \text{ is holomorphic in } 0 < \xi < 1, \ \forall y \in Y),$$

(14.2) $$\begin{cases} f(\mathrm{i}\,\eta) \in X \text{ and } \eta \to f(\mathrm{i}\,\eta) \text{ is a bounded continuous function} \\ \text{of } \mathbf{R}_\eta \to X \text{ (therefore } f(\mathrm{i}\,\eta) \in \mathscr{B}(\mathbf{R}_\eta; X)), \end{cases}$$

(14.3) $$f(1 + \mathrm{i}\,\eta) \in \mathscr{B}(\mathbf{R}_\eta; Y).$$

In general, we set

$$\| g \|_{\mathscr{B}(\mathbf{R};Z)} = \sup_{\eta \in \mathbf{R}_\eta} \| g(\eta) \|_Z$$

and provide $\mathscr{H}(X, Y)$ with the norm

(14.4) $$\| f \|_{\mathscr{H}(X,Y)} = \max \left(\| f(\mathrm{i}\,\eta) \|_{\mathscr{B}(\mathbf{R}_\eta;X)}, \| f(1 + \mathrm{i}\,\eta) \|_{\mathscr{B}(\mathbf{R}_\eta;Y)} \right).$$

Thanks to the classical *three-line theorem*, we see that $\mathscr{H}(X, Y)$, provided with the norm (14.4), *is a Banach space.*

Theorem 14.1. *Let* X *and* Y *be two Hilbert spaces satisfying* (2.1). *For* $0 < \theta < 1$, *the mapping*

(14.5) $$f \to f(\theta)$$

of $\mathscr{H}(X, Y) \to Y$ *is in fact a continuous linear surjection of* $\mathscr{H}(X, Y) \to$ $\to [X, Y]_\theta$.

The norm on $[X, Y]_\theta$ *is equivalent to the norm*

(14.6) $$|\!|\!| a |\!|\!|_{[X,Y]_\theta} = \inf \| f \|_{\mathscr{H}(X,Y)}, \quad f(\theta) = a.$$

Remark 14.1. We may therefore also use (14.5) to *define* $[X, Y]_\theta$ as the image of $\mathscr{H}(X, Y)$ under this mapping: this *definition* is valid for *Banach spaces and even for locally convex spaces*; this point will be made more precise and applied in Section 14.2. □

Remark 14.2. We may also use traces to *define* interpolation spaces (for example, by applying the notions introduced in Section 10). We would obtain *different* spaces in general, but which all coincide in the hilbertian case (and only in this case).

Proof[1]. We use spectral decomposition (see Riesz-Nagy [1], J. Dixmier [1]), which allows us to define Λ^z, $z \in \mathbf{C}$, if Λ is positive self-adjoint in Y. We take Λ to be an operator such that $D(\Lambda) = X$ (see Section 2).

1) Let $f \in \mathscr{H}(X, Y)$. Consider

(14.7) $$g(z) = \Lambda^{-z} f(z), \quad 0 \leqq \xi \leqq 1.$$

---

[1] This proof, which predates the general introduction of $[X, Y]_\theta$-spaces, was communicated to us by N. Aronszajn in 1958.

This function is holomorphic in $0 < \xi < 1$; we have:

$$g(\mathrm{i}\,\eta) = \Lambda^{-\mathrm{i}\eta} f(\mathrm{i}\,\eta) \in \mathscr{B}(\mathbf{R}_\eta; X), \tag{14.8}$$

$$\| g(\mathrm{i}\,\eta) \|_{\mathscr{B}(\mathbf{R}_\eta; X)} \leqq \| f(\mathrm{i}\,\eta) \|_{\mathscr{B}(\mathbf{R}_\eta; X)}. \tag{14.9}$$

Then

$$g(1 + \mathrm{i}\,\eta) = \Lambda^{-1-\mathrm{i}\eta} f(1 + \mathrm{i}\,\eta) \in \mathscr{B}(\mathbf{R}_\eta; X),$$

since $\Lambda^{-1-\mathrm{i}\eta} \in \mathscr{L}(Y; X)$ and since $\| \Lambda^{-1-\mathrm{i}\eta} \|_{\mathscr{L}(Y;X)} \leqq \text{constant} = c_1$, we have:

$$g(1 + \mathrm{i}\,\eta) \in \mathscr{B}(\mathbf{R}_\eta; X) \tag{14.10}$$

and

$$\| g(1 + \mathrm{i}\,\eta) \|_{\mathscr{B}(\mathbf{R}_\eta; X)} \leqq c_1 \| f(1 + \mathrm{i}\,\eta) \|_{\mathscr{B}(\mathbf{R}_\eta; X)}. \tag{14.11}$$

According to (14.8), (14.10), the function g takes its *values in* X (since g is a continuous bounded function of $0 \leqq \xi \leqq 1 \to Y$, a holomorphic function of $0 < \xi < 1 \to Y$ and continuous and bounded, taking its *values in* X, for $\xi = 0$ *and* $\xi = 1$). Therefore

$$g(\theta) \in X,$$

therefore

$$f(\theta) = \Lambda^\theta g(\theta) \in D(\Lambda^{1-\theta})$$

and

$$\| g(\theta) \|_X \leqq c_2 \max(\| g(\mathrm{i}\,\eta) \|_{\mathscr{B}(\mathbf{R}_\eta; X)}, \| g(1 + \mathrm{i}\,\eta) \|_{\mathscr{B}(\mathbf{R}_\eta; Y)});$$

from which, according to (14.9), (14.11), we obtain:

$$\| f(\theta) \|_{D(\Lambda^{1-\theta})} \leqq c_3 \| f \|_{\mathscr{H}(X, Y)}. \tag{14.12}$$

2) We still have to show the *surjectivity* of (14.5). Let $a \in [X, Y]_\theta = D(\Lambda^{1-\theta})$. Define

$$f(z) = \Lambda^{z-\theta} a. \tag{14.13}$$

This function is in $\mathscr{H}(X, Y)$ and

$$\| f \|_{\mathscr{H}(X, Y)} \leqq c_4 \| a \|_{D(\Lambda^{1-\theta})}.$$

Since $f(\theta) = a$, we obtain the surjectivity – and, in addition, we have constructed a continuous linear right inverse $a \to f$ of $[X, Y]_\theta \to \to \mathscr{H}(X, Y)$ – which completes the proof of the theorem. □

Remark 14.3. In the preceding definition of $[X, Y]_\theta$-spaces, the hypothesis that the functions $\eta \to f(\mathrm{i}\,\eta)$ (resp. $f(\xi + \mathrm{i}\,\eta)$) be bounded in X (resp. Y) is *not essential*. For example (and what is to follow is *also not* the maximum degree of generality) consider the space

$$\mathscr{H}_{\exp}(X, Y)$$

of functions $z \to f(z)$ having the same properties as $f \in \mathscr{H}(X, Y)$, but with

$$\|f(\mathrm{i}\,\eta)\|_X \leqq C\, \mathrm{e}^{K|\eta|},$$

$$\|f(\xi + \mathrm{i}\,\eta)\|_Y \leqq C\, \mathrm{e}^{K|\eta|}, \quad \text{fixed arbitrary } K > 0,\ 0 \leqq \xi \leqq 1.$$

We define a natural norm on $\mathscr{H}_{\exp}(X, Y)$ by

$$\|f\|_{\mathscr{H}(X,Y)} = \max\left[\sup_\eta \mathrm{e}^{-K|\eta|}\, \|f(\mathrm{i}\,\eta)\|_X,\ \sup_\eta \mathrm{e}^{-K|\eta|}\, \|f(1 + \mathrm{i}\,\eta)\|_Y\right]$$

and we define

$[X, Y]_{\theta, \exp}$ = space described by $f(\theta)$ as f describes $\mathscr{H}_{\exp}(X, Y)$, with the norm of the quotient by the kernel of $f \to f(\theta)$.

Then:

Proposition 14.1. *We have*

$$[X, Y]_{\theta, \exp} = [X, Y]_\theta,$$

(where, as usual, the equality is understood with norm equivalence).

Proof. Clearly $\mathscr{H}(X, Y) \subset \mathscr{H}_{\exp}(X, Y)$ and therefore $[X, Y]_\theta \subset$ $\subset [X, Y]_{\theta, \exp}$. Conversely, let $a \in [X, Y]_{\theta, \exp}$; therefore, there exists $f \in \mathscr{H}_{\exp}(X, Y)$ such that $f(\theta) = a$. But we introduce

$$g(z) = \exp(z - \theta)^2\, f(z).$$

We easily verify that $g \in \mathscr{H}(X, Y)$ and since $g(\theta) = f(\theta) = a$, we see that $a \in [X, Y]_\theta$ and therefore $[X, Y]_{\theta, \exp} \subset [X, Y]_\theta$.

Corollary 14.1. *We still obtain the same spaces $[X, Y]_\theta$ by starting with functions $z \to f(z)$ of polynomial growth in X for $z = \mathrm{i}\,\eta$ and in Y for $z = \zeta + \mathrm{i}\,\eta$.* □

14.2 Interpolation of Spaces of Continuous Functions with Hilbert Range

We first state Remark 14.1 in precise terms.

Let $\tilde{X}$, $\tilde{Y}$ be two *Banach* spaces such that

$$\left\{ \begin{array}{l} \tilde{X} \subset \Phi,\ \tilde{Y} \subset \Phi, \text{ with continuous injection, where } \Phi \text{ is a locally} \\ \text{convex topological vector space.} \end{array} \right.$$

(We do not assume that $\tilde{X} \subset \tilde{Y}$.)

Let $\tilde{X} + \tilde{Y}$ be the space of elements $x + y$, $x \in \tilde{X}$, $y \in \tilde{Y}$, provided with the norm

$$\|a\|_{\tilde{X}+\tilde{Y}} = \inf_{x+y=a} (\|x\|_{\tilde{X}} + \|y\|_{\tilde{Y}}),$$

which makes it a Banach space.

We define $\mathscr{H}(\tilde{X}, \tilde{Y})$ = space of functions $z \to f(z)$ such that (compare with (14.1), (14.2), (14.3)) $z \to f(z)$ is a continuous and bounded function of $0 \leq \xi \leq 1 \to \tilde{X} + \tilde{Y}$ and a holomorphic function of $0 < \xi < 1 \to \to \tilde{X} + \tilde{Y}$, with

$$f(i\eta) \in \mathscr{B}(\mathbf{R}_\eta; \tilde{X}), \quad f(1 + i\eta) \in \mathscr{B}(\mathbf{R}_\eta; \tilde{Y})$$

(the definition of $\mathscr{B}(\mathbf{R}_\eta; X)$, X a Hilbert space, extends immediately to the case $\mathscr{B}(\mathbf{R}_\eta; X)$, X a Banach space); we provide it with the norm

$$\max\left(\sup_\eta \|f(i\eta)\|_{\tilde{X}}, \sup_\eta \|f(1 + i\eta)\|_{\tilde{Y}}\right) = \|f\|_{\mathscr{H}(\tilde{X}, \tilde{Y})},$$

which makes it a Banach space.

We again *define* $[\tilde{X}, \tilde{Y}]_\theta$ as the space described by $f(\theta)$ as f describes $\mathscr{H}(\tilde{X}, \tilde{Y})$ and provide it with the (quotient) norm

$$\|a\|_{[\tilde{X}, \tilde{Y}]_\theta} = \inf_{f(\theta)=a} \|f\|_{\mathscr{H}(\tilde{X}, \tilde{Y})},$$

which makes it a Banach space.

We verify, *exactly as for the Hilbert case,* that these spaces have the *interpolation property*; let $\tilde{X}_1$, $\tilde{Y}_1$ be a second couple of Banach spaces (with properties analogous to $\tilde{X}$, $\tilde{Y}$) and $\pi \in \mathscr{L}(\tilde{X}; \tilde{X}_1) \cap \cap \mathscr{L}(\tilde{Y}; \tilde{Y}_1)$, then

$$\pi \in \mathscr{L}([\tilde{X}, \tilde{Y}]_\theta; [\tilde{X}_1, \tilde{Y}_1]_\theta). \quad \square$$

We consider again the *Hilbert couple* X, Y, and introduce

(14.14) $\tilde{X} = C^0([0, T]; X)$ = space of continuous functions of $[0, T] \to X$, with the norm $\sup_{t \in [0,T]} \|f(t)\|_X$ (which makes it a *Banach space*)

and in the same way

(14.15)
$$\tilde{Y} = C^0([0, T]; Y).$$

Theorem 14.2. *Let* X, Y *satisfy* (2.1) *and let* $\tilde{X}$, $\tilde{Y}$ *be defined by* (14.14) *and* (14.15). *Then:*

(14.16)
$$[\tilde{X}, \tilde{Y}]_\theta = C^0([0, T]; [X, Y]_\theta),$$

with equivalent norms.

Proof. The proof follows the same line of arguments as the proof of Theorem 14.1. Let $f \in \mathscr{H}(\tilde{X}, \tilde{Y})$. Then, as may be easily verified,

$$\Lambda^{-z} f(z) = g(z) \in \mathscr{H}(\tilde{X}; \tilde{Y})$$

(indeed, setting $(\Lambda^{-z}\varphi)(t) = \Lambda^{-z}(\varphi(t))$, we see that

$$\Lambda^{-z} \in \mathscr{L}(\tilde{X}; \tilde{Y}), \qquad \Lambda^{-1-i\eta} \in \mathscr{L}(\tilde{Y}; \tilde{X})).$$

Then $g(\theta) = \Lambda^{-\theta} f(\theta) \in \tilde{X} = C^0([0, T]; X)$ and therefore

$$f(\theta) = \Lambda^{\theta} g(\theta) \in C^0([0, T]; D(\Lambda^{1-\theta})) = C^0([0, T]; [X, Y]_\theta).$$

Conversely, if $a \in C^0([0, T]; [X, Y]_\theta)$, then $\Lambda^{z-\theta} a = f(z)$ is in $\mathscr{H}(\tilde{X}, \tilde{Y})$ and $f(\theta) = a$, from which the theorem follows. □

Remark 14.4. In general, for a given Banach space E, let $L^p(0, T; E)$ be the space of (classes of) strongly measurable functions f taking their values in E, and such that

$$\left(\int_0^T \| f(t) \|_E^p \, dt\right)^{1/p} = \| f \|_{L^p(0,T;E)}. \tag{14.17}$$

$L^p(0, T; E)$ is a Banach space for the norm (14.17) (see Bourbaki [1]). Then, with the *same* proof as above, we obtain

$$[L^p(0, T; X), L^p(0, T; Y)]_\theta = L^p(0, T; [X, Y]_\theta).$$

14.3 A Result Pertaining to Interpolation of Subspaces

Let X and Y be two *Banach spaces* such that

$$\begin{cases} X \subset \Phi, \ Y \subset \Phi, \text{ with continuous injection, where } \Phi \text{ is a} \\ \text{locally convex topological vector space.} \end{cases} \tag{14.18}$$

We do not assume that $X \subset Y$.

We shall now define subspaces of X and Y in a way which occurs frequently in the applications, as we shall see in Chapter 2, in particular.

For this purpose, let Ψ be a locally convex topological vector space; let

$$\partial \in \mathscr{L}(\Phi; \Psi), \tag{14.19}$$

(14.20) $\mathscr{X}$ (resp. $\mathscr{Y}$) be a Banach space, with $\mathscr{X} \subset \Psi$, $\mathscr{Y} \subset \Psi$,

(*in particular,* we may take $\mathscr{X} = \mathscr{Y} = \{0\}$). Define

$$(X)_{\partial,\mathscr{X}} = \{u \mid u \in X, \partial u \in \mathscr{X}\}, \tag{14.21}$$

$$(Y)_{\partial,\mathscr{Y}} = \{u \mid u \in Y, \partial u \in \mathscr{Y}\}. \tag{14.22}$$

Each of these spaces is provided with the *norm of the graph* (for example

$$\| u \|_{(X)_{\partial,\mathscr{X}}} = \| u \|_X + \| \partial u \|_{\mathscr{X}}),$$

which makes it a Banach space.

Our problem is to compare the spaces

$$[(X)_{\partial,\mathscr{X}}, (Y)_{\partial,\mathscr{Y}}]_\theta \quad \text{and} \quad ([X, Y]_\theta)_{\partial,[\mathscr{X},\mathscr{Y}]_\theta},$$

the spaces $[\ ,\ ]_\theta$ being understood in the sense of the definition in Section 14.2.

We shall make the following hypotheses:

(14.23) $\begin{cases} \text{there exist Banach spaces } \tilde{\mathscr{X}}, \tilde{\mathscr{Y}} \text{ such that:} \\ \text{i) } \mathscr{X} \subset \tilde{\mathscr{X}} \subset \Psi, \quad \mathscr{Y} \subset \tilde{\mathscr{Y}} \subset \Psi, \\ \text{ii) } \partial \in \mathscr{L}(X; \tilde{\mathscr{X}}) \cap \mathscr{L}(Y; \tilde{\mathscr{Y}}), \\ \text{iii) there exist } \mathscr{G} \in \mathscr{L}(\tilde{\mathscr{X}}; X) \cap \mathscr{L}(\tilde{\mathscr{Y}}; Y) \text{ and} \\ \qquad r \in \mathscr{L}(\tilde{\mathscr{X}}; \mathscr{X}) \cap \mathscr{L}(\tilde{\mathscr{Y}}; \mathscr{Y}) \\ \text{such that} \\ \qquad \partial \mathscr{G} \chi = \chi + r\chi \quad \forall \chi \in \tilde{\mathscr{X}} + \tilde{\mathscr{Y}}. \end{cases}$

Theorem 14.3. *Assume that* (14.18), (14.19), (14.20) *and* (14.23) *are satisfied. Then*

(14.24) $$[(X)_{\partial,\mathscr{X}}, (Y)_{\partial,\mathscr{Y}}]_\theta = ([X, Y]_\theta)_{\partial,[\mathscr{X},\mathscr{Y}]_\theta}, \quad 0 < \theta < 1$$

(with equivalent norms).

Proof. 1) In (14.24) we *always* (i.e. without (14.23)) have the inclusion of the first space in the second. Indeed, clearly

$$[(X)_{\partial,\mathscr{X}}, (Y)_{\partial,\mathscr{Y}}]_\theta \subset [X, Y]_\theta.$$

On the other hand (we did what was necessary for that):

$$\partial \in \mathscr{L}((X)_{\partial,\mathscr{X}}; \mathscr{X}) \cap \mathscr{L}((Y)_{\partial,\mathscr{Y}}; \mathscr{Y}),$$

therefore, by interpolation

$$\partial \in \mathscr{L}([(X)_{\partial,\mathscr{X}}, (Y)_{\partial,\mathscr{Y}}]_\theta; [\mathscr{X}, \mathscr{Y}]_\theta),$$

from which our assertion follows.

2) Conversely, let $a \in ([X, Y]_\theta)_{\partial,[\mathscr{X},\mathscr{Y}]_\theta}$.

This means that there exist $f \in \mathscr{H}(X, Y)$, $g \in \mathscr{H}(\mathscr{X}, \mathscr{Y})$ such that

(14.25) $$a = f(\theta), \quad \partial a = g(\theta).$$

We *define* h by

(14.26) $$h(z) = f(z) - \mathscr{G}\, \partial f(z) + \mathscr{G}\, g(z).$$

We shall verify that

(14.27) $$h \in \mathscr{H}((X)_{\partial,\mathscr{X}}\, (Y)_{\partial,\mathscr{Y}}),$$

in a moment.

Then

$$h(\theta) = f(\theta) - \mathscr{G}\, \partial f(\theta) + \mathscr{G}\, g(\theta) = a - \mathscr{G}\, \partial a + \mathscr{G}\, \partial a = a,$$

therefore $a \in [(X)_{\partial,\mathscr{X}}, (Y)_{\partial,\mathscr{Y}}]_\theta$, from which the inverse of the inclusion in 1) and the result follow.

There remains to show (14.27). But $h(z) \in X + Y$, is holomorphic in z with values in $X + Y$. Furthermore

$$\begin{aligned}\partial h(z) &= \partial f(z) - \partial \mathcal{G}\, \partial f(z) + \partial \mathcal{G}\, g(z) = \text{(using 14.23), iii)}\\ &= g(z) + r(g(z) - \partial f(z));\end{aligned}$$

so that $\partial h(i\eta) = g(i\eta) + r(g(i\eta) - \partial f(i\eta))$; but $g(i\eta) \in \mathcal{X}$, therefore $r\, g(i\eta)$ also belongs to $\mathcal{X}$ and $\partial f(i\eta) \in \tilde{\mathcal{X}}$ (since $\partial \in \mathcal{L}(X; \tilde{\mathcal{X}})$), therefore $r(\partial f(i\eta)) \in \mathcal{X}$ and

$$\partial h(i\eta) \in \mathcal{B}(\mathcal{X}).$$

In the same way $\partial h(1 + i\eta) \in \mathcal{B}(\mathcal{Y})$, whence (14.27). □

Examples of applications of the preceding result are given in Chapter 2, Section 7.

15. Another Intrinsic Definition of the Spaces $[X, Y]_\theta$

In general, if X and Y are two Banach spaces satisfying (14.18), we set, for every $a \in X + Y$ and for all $t > 0$:

$$K(t, a; X, Y) = \inf_{a_0 + a_1 = a} (\| a_0 \|_X^2 + t^2 \| a_1 \|_Y^2)^{1/2}, \quad a_0 \in X, \quad a_1 \in Y. \tag{15.1}$$

It is then possible to define *new interpolation spaces* by considering the set of a's for which the function $t \to K(t, a; X, Y)$ has various properties (belongs to $L^p(0, \infty; t^\alpha\, dt)$, etc.).

In this way, we have

Theorem 15.1. *Let X and Y be two Hilbert spaces satisfying* (2.1). *Then*

$$[X, Y]_\theta = \{a \mid a \in Y, t^{-(\theta + 1/2)} K(t, a; X, Y) \in L^2(0, \infty)\}. \tag{15.2}$$

Furthermore, the norms $\| a \|_{[X,Y]_\theta}$ and

$$\left(\| a \|_Y^2 + \int_0^\infty t^{-(2\theta+1)} K(t, a; X, Y)^2\, dt \right)^{1/2}$$

are equivalent.

Proof. We start by showing the formula

$$K(t, a; X, Y)^2 = t^2 (\Lambda^2 (\Lambda^2 + t^2)^{-1} a, a)_Y \tag{15.3}$$

if $X = D(\Lambda)$ (see Section 2), with $\| u \|_X = \| \Lambda u \|_Y$.

Indeed

$$K(t, a; X, Y)^2 = \inf_{a_0 \in D(\Lambda)} (\| \Lambda a_0 \|_Y^2 + t^2 \| a - a_0 \|_Y^2).$$

The Euler equation for this minimization problem is

(15.4) $(\Lambda^2 + t^2)\, a_0 = t^2\, a$, a_0 being the element realizing the optimum.

Then a_0 being given by (15.4), we have:

$$K(t, a; X, Y)^2 = \|\Lambda\, a_0\|_Y^2 + t^2\, \|a_0\|_Y^2 + t^2\, \|a\|_Y^2 - 2t^2\, \mathrm{Re}\,(a, a_0)_Y$$

and since according to (15.4), $\|\Lambda\, a_0\|_Y^2 + t^2\, \|a_0\|_Y^2 = t^2\,(a, a_0)_Y$, we have:

$$K(t, a; X, Y)^2 = t^2\,(a, a - a_0)_Y = (a, \Lambda^2\, a_0)_Y$$
$$= (a, \Lambda^2\, t^2\,(\Lambda^2 + t^2)^{-1}\, a)_Y, \quad \text{whence (15.3)}.$$

Therefore

$$\int_0^\infty t^{-(2\theta+1)}\, K(t, a, X, Y)^2\, dt = \int_0^\infty (t^{(1-2\theta)}\, \Lambda^2\,(\Lambda^2 + t^2)^{-1}\, a, a)_Y\, dt.$$

But (see Riesz-Nagy [1], Dunford-Schwartz [1], or use the spectral decomposition as in Section 2, assuming Y to be separable):

$$\int_0^\infty t^{(1-2\theta)}\, \Lambda^2\,(\Lambda^2 + t^2)^{-1}\, dt = c\,\Lambda^{2(1-\theta)} \left(c = \int_0^\infty \frac{s^{1-2\theta}}{1 + s^2}\, ds\right), \tag{15.5}$$

so that

$$\int_0^\infty t^{-(2\theta+1)}\, K(t, a, X, Y)^2\, dt = c\,\|\Lambda^{1-\theta}\, a\|_Y^2. \tag{15.6}$$

In fact, we have to be somewhat more precise: we have (15.5) for example in $D(\Lambda^2)$, therefore (15.6) for $a \in D(\Lambda^2)$ for example. Then we pass to the completions.

The theorem follows from 15.6. □

16. Compactness Properties

The following result will be used in a fundamental way in Chapter 2.

Theorem 16.1. *Assume that Ω satisfies* (7.10) *and* (7.11). *Let $s \in \mathbf{R}$. Then, for every $\varepsilon > 0$, the injection*

$$H^s(\Omega) \to H^{s-\varepsilon}(\Omega)$$

is compact.

Proof. 1) It is sufficient to show the theorem for $s > 0$. Since then, by passage to closed subspaces, the injection of $H_0^s(\Omega) \to H_0^{s-\varepsilon}(\Omega)$ is also compact and therefore, by transposition $H^{-s+\varepsilon}(\Omega) \to H^{-s}(\Omega)$ is compact, from which we obtain the desired result.

2) Therefore, fix $s > 0$. There exists (see proof of Theorem 9.1) an operator p:

$$u \to p\,u,$$

a continuous linear mapping of $H^s(\Omega) \to H^s(\mathbf{R}^n)$ such that

$$p\,u = u \quad \text{a.e. on } \Omega, \tag{16.1}$$

(16.2) $p\,u$ has support in a fixed compact set K, $\Omega \subset K$.

(For (16.2) it is sufficient to *truncate* after having extended by reflection.)

Let $u_n \in H^s(\Omega)$ and $u_n \to 0$ in $H^s(\Omega)$ weakly. We must show that $u_n \to 0$ in $H^{s-\varepsilon}(\Omega)$ *strongly*. However,

$$v_n = p\,u_n \to 0 \quad \text{in} \quad H^s(\mathbf{R}^n) \text{ weakly} \tag{16.3}$$

and since the operation "restriction to Ω" is a *continuous* mapping of $H^{s-\varepsilon}(\mathbf{R}^n) \to H^{s-\varepsilon}(\Omega)$, it is sufficient to show that (16.3) implies:

$$v_n \to 0 \quad \text{in} \quad H^{s-\varepsilon}(\mathbf{R}^n) \text{ strongly}, \tag{16.4}$$

or, $\hat{v}_n$ denoting the Fourier transform of v_n, that

$$X_n = \int_{R^n_\xi} (1 + |\xi|)^{2(s-\varepsilon)} |\hat{v}_n(\xi)|^2 \, d\xi \leqq \eta \quad \text{for} \quad n \geqq n(\eta). \tag{16.4a}$$

But

$$X_n = \int_{|\xi| \geqq M} (1 + |\xi|)^{-2\varepsilon} (1 + |\xi|)^{2s} |\hat{v}(\xi)|^2 \, d\xi + \\ + \int_{|\xi| \geqq M} (1 + |\xi|^{2(s-\varepsilon)} |\hat{v}_n(\xi)|^2 \, d\xi,$$

therefore

$$X_n \leqq (1 + M)^{-2\varepsilon} \int_{R^n_\xi} (1 + |\xi|)^{2s} |\hat{v}_n(\xi)|^2 \, d\xi + \\ + (1 + M)^{2(s-\varepsilon)} \int_{|\xi| \leqq M} |\hat{v}_n(\xi)|^2 \, d\xi. \tag{16.5}$$

According to (16.3), v_n remains in a bounded set of $H^s(\mathbf{R}^n)$, therefore (16.5) implies:

$$X_n \leqq c_1 (1 + M)^{-2\varepsilon} + (1 + M)^{2(s-\varepsilon)} \int_{|\xi| \leqq M} |\hat{v}_n(\xi)|^2 d\xi. \tag{16.6}$$

Choose M so that $c_1(1+M)^{-2\varepsilon} \leqq \eta/2$ and (16.4a) follows if we can show that

$$\int_{|\xi|\leqq M} |\hat{v}_n(\xi)|^2\, d\xi \leqq \frac{\eta}{2(1+M)^{2(s-\varepsilon)}} \quad \text{for} \quad n \geqq n(\eta). \tag{16.7}$$

But, thanks to (16.2), v_n has support in K; therefore if θ denotes a function of $\mathscr{D}(\mathbf{R}^n)$, we have:

$$\hat{v}_n(\xi) = \langle v_n, \theta \exp(-2\pi \mathrm{i}\, x\, \xi)\rangle, \tag{16.8}$$

where the brackets denote the duality between $H^s(\mathbf{R}^n)$ and $H^{-s}(\mathbf{R}^n)$. But when $|\xi| \leqq M$, $\theta \exp(-2\pi \mathrm{i}\, x\, \xi)$ belongs to a relatively compact set of $H^{-s}(\mathbf{R}^n)$ and therefore $\langle v_n, \theta \exp(-2\pi \mathrm{i}\, x\, \xi)\rangle \to 0$ *uniformly* for $|\xi| \leqq M$.

So that, by (16.8), $\hat{v}_n(\xi) \to 0$ uniformly for $|\xi| \leqq M$, therefore

$$\int_{|\xi|\leqq M} |\hat{v}_n(\xi)|^2\, d\xi \to 0 \quad \text{and (16.7) holds.} \quad \square$$

Remark 16.1. The preceding proof does not economize the hypotheses on Ω. On the one hand, the result still holds if the boundary Γ is Lipschitzian (see Adams-Aronszajn-Smith [1]).

On the other hand, the result may still be true if Ω is not bounded but "sufficiently small" at infinity. (Note that the result is *false* for $\Omega = \mathbf{R}^n$).

We may introduce, in general for $s > 0$, the open sets Ω having property (C_s): they are the open sets such that the property of Theorem 16.1 holds. This notion indeed depends on s. $\square$

Remark 16.2. We again consider the general setting of $[X, Y]_\theta$-spaces.

Theorem 16.2. *If the injection* $X \to Y$ *is compact, then the injection*

$$[X, Y]_{\theta_1} \to [X, Y]_{\theta_2}, \quad \theta_1 < \theta_2 \quad (0 < \theta_i < 1) \tag{16.9}$$

is compact.

Proof. We consider (see Section 1) an operator Λ, positive and self-adjoint in Y, such that

$$X = D(\Lambda).$$

Since $X \to Y$ is compact, the operator Λ^{-1} is compact in Y and the spectral decomposition of Λ is given by the eigenfunctions w_j:

$$\begin{cases} \Lambda w_j = \lambda_j w_j, & j = 1, 2, \dots \\ \lambda_j > 0, & \lambda_j \to +\infty; \end{cases} \tag{16.10}$$

choosing the functions w_j so that

$$(w_j, w_k)_Y = \delta_j^k,$$

the w_j form a complete orthonormal system in Y. Then

(16.11) $$X = D(\Lambda) = \left\{ u \mid u = \sum_{j=1}^{\infty} x_j w_j, \sum_{j=1}^{\infty} \lambda_j^2 |x_j|^2 < +\infty \right\}$$

and

(16.12) $$[X, Y]_\theta = \left\{ u \mid u = \sum_{j=1}^{\infty} x_j w_j, \sum_{j=1}^{\infty} \lambda_j^{2(1-\theta)} |x_j|_2 < +\infty \right\}.$$

The property to be demonstrated follows without difficulty. □

Theorem 16.2 shows that in order to verify Theorem 16.1 it is *sufficient* to consider the case "integer s". However, this does not simplify the situation appreciably. □

Remark 16.3. We also have the following result, which is very useful for obtaining a priori estimates (see Chapter 2):

Theorem 16.3. *Assume that Ω satisfies* (7.10) *and* (7.11). *Let s_1, s_2 and s_3 be given in* **R** *and*

(16.13) $$s_1 > s_2 > s_3 .$$

Then, for all $\eta > 0$, there exists a constant $c(\eta)$ such that

(16.14) $$\| u \|_{H^{s_2}(\Omega)} \leqq \eta \, \| u \|_{H^{s_1}(\Omega)} + c(\eta) \, \| u \|_{H^{s_3}(\Omega)}, \qquad \forall u \in H^{s_1}(\Omega).$$

Because of Theorem 16.1, this last theorem is a consequence of the following general result:

Theorem 16.4. *Let X, Y, Z be three Banach spaces such that*

(16.15) $$\begin{cases} X \subset Y \subset Z \\ \text{the injection } X \to Y \text{ is compact.} \end{cases}$$

Then, $\forall \eta > 0$, there exists a constant $c(\eta)$ such that

(16.16) $$\| u \|_Y \leqq \eta \, \| u \|_X + c(\eta) \, \| u \|_Z \qquad \forall u \in X .$$

Proof. Suppose (16.16) does not hold. Then for given $\eta > 0$, there exist

$$u_n \in X \quad \text{and} \quad c_n \to +\infty$$

with

$$\| u_n \|_Y \geqq \eta \, \| u_n \|_X + c_n \, \| u_n \|_Z .$$

Introducing $v_n = u_n / \| u_n \|_X$, we have

(16.17) $$\| v_n \|_Y \geqq \eta + c_n \, \| v_n \|_Z .$$

But since $\| v_n \|_Y \leqq$ (constant) $\| v_n \|_X =$ constant, there results from (16.17) that

$$\| v_n \|_Z \to 0. \tag{16.18}$$

Now, since $\| v_n \|_X = 1$ and $X \to Y$ is compact, we may extract a subsequence v_j, which is strongly convergent in Y and (according to (16.18)) necessarily tends to 0; thus $\| v_j \|_Y \to 0$, which contradicts (16.17) and the theorem is proved. □

17. Comments

The spaces $H^s(\Omega)$ (integer $s \geqq 0$) were introduced by Sobolev [1]. The spaces $H^s(\Omega)$, non-integer $s > 0$, were introduced by numerous authors and via numerous methods. We have seen that all reasonable definitions coincide (at least when Ω has a sufficiently regular boundary). The spaces $H^s(\Omega)$, for integer $s < 0$, were introduced by Schwartz [4] in connection with the works of Gårding [1] and Vishik [1] on the Dirichlet problem.

Analogous spaces may be constructed by replacing $L^2(\Omega)$ with $L^p(\Omega)$, $p \neq 2$ (and even Orlicz spaces). The corresponding spaces, also introduced by Sobolev, *loc. cit.*, are generally denoted $W_p^s(\Omega)$ or $W^{s,p}(\Omega)$. Theorems of the *type* discussed here are valid in the case of the spaces $W^{s,p}(\Omega)$, $p \neq 2$, but with serious additional difficulties; for example it is known (J. P. Kahane, E. M. Stein) that for the case $p \neq 2$, the method of holomorphic interpolation gives a *different* result than the method of interpolation by "traces". For $p \neq 2$, we therefore have different families of spaces (Sobolev spaces $W^{s,p}(\Omega)$, Besov spaces $B^{s,p}(\Omega)$, Lebesgue spaces or Bessel potentials $H^{s,p}(\Omega)$, for arbitrary real s; for the definitions, see, for example, Magenes [3]) which coincide for $p = 2$; we refer the reader to Aronszajn-Smith [2], Aronszajn-Mulla-Szeptycki [1], Aronszajn [4], Baiocchi [1], Besov [3], Berezanski [3], Besov [1, 2], Calderon [1, 2], Gagliardo [1, 2], Krein-Petunin [1], Lions-Magenes [1] (III), (IV), (V), Lions-Peetre [1], Magenes [3], Nikolskii [1, 4], Nikolski-Lions-Lizorkin [1], Peetre [12], Stein [1], Shamir [2], Slobodetski [1], Taibleson [1, 2], Uspenski [1] and to the bibliographies of these works. We shall limit ourselves to the "L^2" situation in this volume and in Volume 3 to situations which may be deduced from it by "passage to the inductive or projective limit" and where it is unnecessary to consider the L^p-theory for $p \neq 2$.

The $H^s(\Omega)$-spaces are our "basic tools" for the study of "elliptic" boundary value problems which we take up in Chapter 2; for boundary value problems, it is clearly fundamental to define the *values of the functions on the boundary*, and this is why it is so important to study *traces* in $H^s(\Omega)$. This is our essential aim in Sections 1–5 and 8, where the theory of traces in $H^s(\Omega)$ appears as a particular case of a more general theory (which, by the way, is no more difficult to present than the particular case); we follow the account of Lions [9].

This leads in a natural way to the *theory of interpolation of linear operators* (Sections 5 and 6), which plays a fundamental role in the sequel. The properties in Section 6 are simple particular cases (we limit ourselves to the *Hilbert* case) of the results of Lions-Peetre [1] pertaining to a generalization of trace spaces in the form of "averaged spaces" (see also Lions [14, 19]). They are *also* simple particular cases of the results of Calderon [2, 3] pertaining to holomorphic interpolation, introduced by Calderon, *loc. cit.*, S. Krein [3] and Lions [15], which we briefly present in Section 14. (Our aim in this chapter has not been in any way to present a complete theory of interpolation. Aside from the articles just cited, the reader can consult the works of Gagliardo and of Peetre mentioned in the bibliography, as well as Aronszajn [3], Aronszajn-Gagliardo [1], Krein-Petunin [1], Krein-Petunin-Semenov [1], Semenov [1, 2]).

The trace theorems in $H^s(\Omega)$ are due to Aronszajn [2], Prodi [1, 2], Slobodetski [1] and other authors.

The extension method given in Section 3.2 is due to Hestenes [2] and Lichtenstein [1] (see also Babitch [1]). Another method (using singular integrals) is given in Calderon [1] under weaker hypotheses on Ω (but the extension depends in an essential way on the order s of the Sobolev space). An extension valid for all positive s, when Ω has a C^∞-boundary, was constructed by Seeley [3]; the differentiability hypotheses on Ω have been considerably weakened by Adams-Aronszajn-Smith [1].

The density result (Theorem 8.1) is still valid under much more general hypotheses on Ω (see Gagliardo [1]).

Section 10 follows Lions [11], in which further results can be found; applied to L^p-spaces, the results of this section reproduce a result of Gagliardo [2].

Lemma 10.1 is known as the inequality of Hardy-Littlewood-Polya [1] (Section 3.30).

The interpolation results of Sections 11 and 12 follow Lions-Magenes [1] (II), (III), (IV); in particular, in (IV), the somewhat ... dangerous question of the distinction between $[H_0^1(\Omega), H^0(\Omega)]_{1/2}$ and $[H^1(\Omega), H^0(\Omega)]_{1/2}$ was resolved, the first space *not being* a closed vector subspace of the second. The presentation given here is simpler than the one followed in Lions-Magenes, *loc. cit.*; on the whole, we adopt the presentation of Grisvard [4] (who systematically uses formula (11.25), already used, for different purposes, but still in Sobolev spaces, by V. P. Il'in [1]). Another simple presentation of this result is due to S. Jones (personal communication). For Theorem 11.8, often used by numerous authors, see for example Kadlec-Kufner [1].

Concerning Lemma 12.2, see also Baiocchi [4]; for an extension valid for $H^s(\Omega)$, $s \geqq 0$ or $s < 0$ (which extends the result of Seeley [3]) see Geymonat [4].

The result of Section 14.3 is due to Baiocchi [5]; a generalization is given in Baouendi-Goulaouic [1].

The function $K(t, a)$ was introduced and used systematically by Peetre [8]; see also a presentation of the theory in Goulaouic [1], Chapter 1. Another point of view, different, but connected with $K(t, a)$, is developed in Golovkin [1].

The determination of "all" interpolation spaces between two Hilbert spaces is given in Foias-Lions [1], to be completed with Peetre [10] and Goulaouic [1].

Also useful for the applications to partial differential equations, are the Sobolev spaces which bring into play *derivatives of different order according to the direction* (*non-isotropic* Sobolev spaces); particular cases of this situation will be met in Chapters 4 and 5, Volume 2 (where the time and space variables play different roles). We have not taken up the systematic study of this question, for which we refer the reader to Baiocchi [1], Besov [4], Besov-Kadlec-Kufner [1], Besov-Il'in-Lizorkin [1], Cattabriga [2–4], Cavallucci [2, 3], Gårding-Malgrange [1], Grisvard [4], Hörmander [6], Itano [1], Jones [1], Krée [1], . . ., [4], Lions-Magenes [1] (I), Lizorkin [1], Malgrange [1], Nikolski [2, 3], Pagni [2, 4], Pini [10, 12], Solonnikov [2], Ramazanov [1], Uspenski [2], Volevich-Panejach [1], etc.

Similarly, it is useful to introduce *weighted Sobolev spaces* in the applications (we introduced the weights t^α in (10.5), but they appeared only as a "tool"); we shall meet such spaces (the spaces $\Xi(\Omega)$, among others) in Chapter 2 and we shall establish interpolation properties for these spaces (Chapter 2, Section 7).

A systematic study of weighted Sobolev spaces is not attempted; the reader may consult Besov-Kadlec-Kufner [1], Grisvard [1], Geymonat-Grisvard [2], Lizorkin-Nikolski [1], Morel [1], Necas [2], Nikolski [1], etc.

We also call attention to the results of Baouendi [1], where the weights are used in an essential way.

For a study of Sobolev spaces, isotropic or not, and their variants, systematically using the theory of approximation by entire functions of exponential type, see Nikolski [5]. One may also (see Yoshikawa [1]) use the fractional powers of operators (see Komatsu [1, 2]).

Using the Fourier transform, we may also define the spaces $H^{s(x)}$ of order varying with x; these spaces come up in a natural manner in the theory of pseudo-differential operators; see Vishik-Eskin [2], Unterberger-Bokobza [1].

In Volume 3, we shall meet situations for which it would be interesting to interpolate "between" spaces *without norms*; we do not study the corresponding theory here, for it might well fill . . . a volume. The reader can consult, aside from the last chapter of Lions-Peetre [1], the works of Deutsch [1] and Goulaouic [1]; see also Girardeau [1].

18. Problems[1]

18.1 As we pointed out in Remark 11.6, we have not studied the spaces $[H_0^{s_1}(\Omega), H_0^{s_2}(\Omega)]_\theta$ when at least one of the s_i's is of the form integer $+ \frac{1}{2}$. The same is true for the case mentioned in Remark 12.2.

18.2 A study as complete as the one presented in Sections 11 and 12, for the spaces $W^{s,p}(\Omega)$, $B^{s,p}(\Omega)$, $H^{s,p}(\Omega)$ $(p \neq 1, 2, \infty)$ remains to be done, especially concerning the exceptional parameters.

We call attention to the following question: $B^{\theta,p}(\mathbf{R}^n)$ being defined as a trace space (or an averaged space) between $W^{1,p}(\mathbf{R}^n)$ and $W^{0,p}(\mathbf{R}^n) = L^p(\mathbf{R}^n)$, and setting

$$B^{-\theta,p}(\mathbf{R}^n) = (B^{\theta,p'}(\mathbf{R}^n))',$$

is the space $L^p(\mathbf{R}^n)$ an interpolation space between $B^{\theta_1,p}(\mathbf{R}^n)$ and $B^{-\theta_2,p}(\mathbf{R}^n)$ for $0 < \theta_i < 1$, or not? Does interpolation between *triplets* (instead of pairs) of spaces help to solve problems of this type?

18.3 If $W_\gamma^{s,p}(\Omega) = \{v \mid v \in W^{s,p}(\Omega), \gamma_j v = 0, 0 \leqq j < s - 1/p\}$, does this space coincide with the closure of $\mathcal{D}(\Omega)$ in $W^{s,p}(\Omega)$, *for all values of* s? In this text, we prove that the answer is yes for $p = 2$, and in Lions-Magenes [1] (V) for $p \neq 1, \infty$ and $s \neq$ integer $+ 1/p$.

18.4 In Chapter 2, Section 6, we shall see how $\Xi^s(\Omega)$-spaces, of the same type as $H^s(\Omega)$, but *with weights on the boundary*, are introduced. The interpolation between these spaces is studied in Chapter 2, Section 7. Analogous constructions and results pertaining to $\Xi^{s,p}(\Omega)$, which are defined like the spaces $\Xi^s(\Omega)$ but replacing L^2 by L^p, $p \neq 1, 2, \infty$, would be of interest.

---

[1] Other problems pertaining to the theory of interpolation will be met in the following chapters. We call attention only to problems directly tied to the applications of interpolation to partial differential equations, and not to "general questions" connected with interpolation (for example, of the type: when does there exist a *Hilbert space* or a *hilbertizable space* of interpolation between two Banach spaces?).

Likewise for the compactness properties of the injection of $\Xi^{s,p}(\Omega)$ into $\Xi^{s-\varepsilon,p}(\Omega)$, $\varepsilon > 0$.

Also, interpolation between spaces of the type "weighted $W^{s,p}$" is not completely clear yet.

18.5 Interpolation of Subspaces. Remark 11.4 emphasized one of the main difficulties of the use of interpolation: *the interpolated space between closed subspaces is not necessarily a closed subspace in the interpolated space.*

It would be of great interest to obtain criteria allowing to affirm a priori that, except for certain values of the parameters, the interpolated space is *closed.*

From the point of view of applications, here is one situation in which we meet this problem: let $H^m_B(\Omega)$ be the subspace of $H^m(\Omega)$ made up of the elements $u \in H^m(\Omega)$ such that

$$B_j u = 0 \quad \text{on } \Gamma, \quad 0 \leqq j \leqq \nu,$$

where the B_j's are differential operators of order $m_j < m$. Do we have:

$$[H^m_B(\Omega), H^0(\Omega)]_\theta$$

$$= \{v \mid v \in H^{(1-\theta)m}(\Omega), \quad B_j v = 0 \quad \text{on } \Gamma, \quad \text{if } m_j < (1-\theta)m - \tfrac{1}{2}\}$$

for $(1-\theta)m \neq \text{integer} + \frac{1}{2}$?

The answer, is yes if the B_j's are *normal* operators (Grisvard [8], who also investigated the case of the exceptional parameters and studied the analogous problem for $W^{m,p}(\Omega)$); see Chapter 4, Volume 2, Section 14.5.

Does the result still hold if the B_j's are not necessarily normal? (see also Fujiwara [2]).

Of course, problems of this type exist for Sobolov spaces which bring in a *number of derivatives different according to the directions* (which is useful for evolution equations – see Chapters 4 and 5 in Volume 2 – and for quasi-elliptic equations).

Likewise, it would be very interesting to dispose of other criteria than the one in Section 14.3 (due to Baiocchi [5], and which is very useful).

18.6 Similarly, it would be useful to extend the criteria given in Section 13 for the *interpolation of intersections.*

18.7 In a Banach space E, let Λ be an operator which is the infinitesimal generator of a semi-group $G(t)$.

Consider the space of functions u_i $(i = 1, 2)$ such that

$$t^\alpha u_i \in L^p(0, \infty, D(\Lambda)),$$

$$t^\alpha \frac{du_i}{dt} \in L^p(0, \infty; E)$$

and

$$\Lambda u_1 + \frac{du_2}{dt} = 0.$$

What space does $\{u_1(0), u_2(0)\}$ describe?

This problem is solved in Lions [16] for the *Hilbert* case.

The preceding problem is tied to trace problems of the following type: let $u \in (H^m(\Omega))^N$ be such that $Du = 0$, where D is a differential system: what space is described by $\{\gamma_j u_k\}$, $0 \leqq j \leqq m - 1$, $1 \leqq k \leqq N$?

Chapter 2

Elliptic Operators, Hilbert Theory

For Sections 1–5, with the exception of 5.4, only the definition of $H^s(\Omega)$-spaces, for s of arbitrary sign, and the trace theorem of Section 8.2 are required from Chapter 1. Section 8 may be skipped on first reading.

1. Elliptic Operators and Regular Boundary Value Problems

1.1 Elliptic Operators

On the space $\mathbf{R}^n$, let

(1.1) $$A(D)\,u = \sum_{|p| \le l} a_p\, D^p u$$

be a linear differential operator of order l with constant coefficients; we associate to it the polynomial in $\xi = (\xi_1, \ldots, \xi_n) \in \mathbf{R}^n$ (*characteristic form of* A):

(1.2) $$A_0(\xi) = \sum_{|p| = l} a_p\, \xi^p,$$

where $\xi^p = \xi_1^{p_1}\, \xi_2^{p_2} \ldots \xi_n^{p_n}$, for $p = (p_1, \ldots, p_n)$.

Definition 1.1. *The operator* A *is said to be* ***elliptic*** *if*

(1.3) $$A_0(\xi) \neq 0 \qquad \forall \xi \in \mathbf{R}^n, \quad \xi \neq 0.$$

We have

Proposition 1.1. *For* $n > 2$, *every elliptic operator is of even order.*

Proof. Let ξ and ξ' be two linearly independent vectors in $\mathbf{R}^n$; consider the polynomial $A_0(\xi + \tau\,\xi')$ in the complex variable τ.

We have

$$A_0(\xi + \tau\,\xi') = \tau^l A_0(\xi') + \tau^{l-1} A_1(\xi, \xi') + \cdots + A_l(\xi)$$

where $A_i(\xi, \xi')$ are polynomials in ξ and ξ'.

The equation in τ

$$A_0(\xi + \tau\,\xi') = 0 \tag{1.4}$$

does not have any real roots, when for fixed ξ', ξ runs through the set J, the complement of the straight line in $\mathbf{R}^n$ joining $\{0\}$ to ξ'. Since $A_0(\xi') \neq 0$ and is independent of ξ, the roots of (1.4) depend continuously on $\xi \in J$. Furthermore since, for $n > 2$, J is connected, we see that the number of roots of (1.4) (of course, each root being counted with its multiplicity) with imaginary part > 0 (resp. < 0) is constant as ξ runs through J; we denote this number by m (resp. $l - m$).

For $\xi \in J$, we observe that $-\xi \in J$ and that

$$A_0(-\xi + \tau\,\xi') = (-1)^l A_0(\xi - \tau\,\xi').$$

Then:

(number of roots of $A_0(-\xi+\tau\xi')$ with imaginary part $<0) = l - m =$

(number of roots of $A_0(\xi-\tau\xi')$ with imaginary part $<0) =$

(number of roots of $A_0(\xi+\tau\xi')$ with imaginary part $>0) = m$,

therefore $l = 2m$. □

Proposition 1.1 no longer holds for $n = 2$: the Cauchy-Riemann operator $\dfrac{\partial}{\partial x_1} + i\dfrac{\partial}{\partial x_2}$ is elliptic in $\mathbf{R}^2$ and of order 1. However, if the coefficients of A are *real* we see that Proposition 1.1 is also true for $n = 2$.

1.2 Properly and Strongly Elliptic Operators

Although for the general properties of solutions of the equation $A\,u = f$, Definition 1.1 of ellipticity is sufficient (see for example Hörmander [6] and Section 3 of this chapter), from the point of view of the theory of boundary value problems, the property on the number of roots of (1.4) with positive imaginary part, mentioned in the proof of Proposition 1.1, is very important (see Section 8). For this reason, we shall assume, from now on, that the order of A is even ($l = 2m$) and introduce

Definition 1.2. *The operator A defined by* (1.1), *with $l = 2m$, is said to be* ***properly elliptic*** *if it is elliptic and if for every linearly independent couple of vectors ξ and ξ' belonging to $\mathbf{R}^n$, the polynomial $A_0(\xi + \tau\,\xi')$ in the complex variable τ has m roots with positive imaginary part.*

The proof of Proposition 1.1 yields

Proposition 1.2. *If $n > 2$, every elliptic operator is properly elliptic.* □

A remarkable class of properly elliptic operators is formed by the *strongly elliptic* operators:

Definition 1.3. *The operator* A *defined by* (1.1), *with* $l = 2m$, *is said to be strongly elliptic if there exists a complex number* γ *and a constant* $\alpha > 0$ *such that*

$$\mathrm{Re}\,(\gamma\, A_0(\xi)) \geqq \alpha\,|\xi|^{2m} \qquad \forall \xi \in \mathbf{R}^n. \quad \square$$

Remark 1.1. There exist properly elliptic operators which are not strongly elliptic; for example, in $\mathbf{R}^3$:

$$A = \frac{\partial^4}{\partial x_1^4} + \frac{\partial^4}{\partial x_2^4} - \frac{\partial^4}{\partial x_3^4} + \mathrm{i}\left(\frac{\partial^2}{\partial x_1^2} + \frac{\partial^2}{\partial x_2^2}\right)\frac{\partial^2}{\partial x_3^2}. \quad \square$$

Now, let Ω be an open set in $\mathbf{R}^n$ and assume that the coefficients a_p in (1.1) are functions defined on Ω (or $\bar{\Omega}$). Then the characteristic form (1.2) will also depend on x; we shall denote it by $A_0(x, \xi)$ and we shall denote the operator A by $A(x, D)$. We shall say that the operator A is elliptic, properly elliptic, strongly elliptic in Ω (or $\bar{\Omega}$) if for each $x \in \Omega$ (or $\bar{\Omega}$) the operator $A(x, D)$, considered as an operator with constant coefficients $a_p(x)$, is elliptic, properly elliptic, strongly elliptic. We shall also say that A is *uniformly strongly elliptic in* $\bar{\Omega}$ if there exists a complex number γ and $\alpha > 0$, independent of x, such that

$$\mathrm{Re}(\gamma\, A_0(x, \xi)) \geqq \alpha\,|\xi|^{2m} \qquad \forall \xi \in \mathbf{R}^n \text{ et } \forall x \in \bar{\Omega}. \quad \square$$

Remark 1.2. If the boundary Γ of Ω is connected, the ellipticity of A in $\bar{\Omega}$ and the continuity of the coefficients a_p in $\bar{\Omega}$ are sufficient to guarantee that A is properly elliptic in $\bar{\Omega}$, if the condition on the roots of $A_0(x, \xi + \tau\,\xi')$ is satisfied for only one point of Γ and for one couple of linearly independent vectors ξ and ξ'.

1.3 Regularity Hypotheses on the Open Set Ω and the Coefficients of the Operator A

Let Ω be an open set in $\mathbf{R}^n$. Since we want to study boundary value problems for the operator A in the spaces $H^s(\Omega)$, for *arbitrary* real s, we shall impose very strong regularity conditions on Ω and on the coefficients of A. We shall be able to use these conditions for all s, but we could, for each fixed s, impose more general conditions which depend on s (see Section 8).

We assume Ω to be a *bounded open set in* $\mathbf{R}^n$, *with boundary* Γ, *an* $(n-1)$*-dimensional infinitely differentiable variety,* Ω *being locally on one side of* Γ, i.e. we consider $\bar{\Omega}$ to be a compact variety with boundary of class C^∞, the boundary being Γ.

We assume the operator A *to be of order* $2m$ *and to have infinitely differentiable coefficients in* $\bar{\Omega}$; we write A in the form

$$(1.5)\qquad A\,u = A(x, D)\,u = \sum_{|p|,|q| \leqq m} (1-)^{|p|} D^p(a_{pq}(x)\, D^q u),$$

with

$$(1.6)\qquad a_{pq} \in \mathcal{D}(\bar{\Omega}).$$

Then, the characteristic form of A is

$$A_0(x, \xi) = \sum_{|p|,|q|=m} (-1)^m\, a_{pq}(x)\, \xi^{p+q}, \quad \text{where} \quad \xi^{p+q} = \xi_1^{p_1+p_1} \ldots \xi_n^{p_n+q_n}.$$

We assume A to be *properly elliptic* on $\bar{\Omega}$. Under the given hypotheses, we obtain that A is *uniformly elliptic* on $\bar{\Omega}$, i.e. there exists a constant $c > 0$, independent of x, such that

$$(1.7)\qquad c^{-1}\,|\xi|^{2m} \leqq |A_0(x, \xi)| \leqq c\,|\xi|^{2m} \quad \forall x \in \bar{\Omega} \text{ and } \forall \xi \in \mathbf{R}^n.$$

1.4 The Boundary Operators

The aim of this chapter is to study boundary value problems for the elliptic equations:

$$(1.8)\qquad \begin{cases} A\,u = f & \text{in } \Omega, \\ B_j\,u = g_j & \text{on } \Gamma, \end{cases}$$

where the B_j's are certain differential "boundary" operators in suitable finite number and f and g_j are given (we formally denote problem (1.8) by $\{A, B\}$).

However, it is well known for the simplest classical cases (for example $A = -\Delta$) that we can not arbitrarily assign the operators B_j and obtain a "well-posed" problem. Therefore, we must introduce certain *admissibility conditions with respect to the operator* A on the operators B_j. This is what we shall do in this section by introducing some definitions whose role will become clear in the sequel (see, in particular, Section 8).

Let B_j, $0 \leqq j \leqq \nu - 1$, be ν "boundary" operators defined by

$$(1.9)\qquad B_j\,\varphi = \sum_{|h| \leqq m_j} b_{jh}(x)\, D^h \varphi,$$

with

$$(1.10)\qquad b_{jh} \in \mathcal{D}(\Gamma)$$

m_j being the order of B_j. More precisely, $B_j\,\varphi$ denotes the operator

$$\varphi \to \sum_{|h| \leqq m_j} b_{jh}(x)\, \gamma_0(D^h \varphi),$$

φ being a function defined on $\bar{\Omega}$ for which $\gamma_0(D^h \varphi)$, the trace of $D^h \varphi$ on Γ, may be defined either in the classical sense or in the sense of the trace theorem of Section 9.2 of Chapter 1.

We consider a subset Γ_1 of Γ and introduce

Definition 1.4. *The system of operators* $\{B_j\}_{j=0}^{\nu-1}$ *is a normal system on* Γ_1 *if*

a) $\sum_{|h|=m_j} b_{jh}(x)\,\xi^h \neq 0 \ \forall x \in \Gamma_1$ and $\forall \xi \neq 0$ and normal to Γ at x,

b) $m_j \neq m_i$ for $j \neq i$.

Now, assume that $\nu = m$; then, we have

Definition 1.5. *The system* $\{B_j\}_{j=0}^{m-1}$ *covers the operator* A *on* Γ_1 *if for all* $x \in \Gamma_1$, *all* $\xi \in \mathbf{R}^n$, *not equal to zero and tangent to* Γ *at* x, *and all* $\xi' \in \mathbf{R}^n$, *not equal to zero and normal to* Γ *at* x, *the polynomials in the complex variable* τ: $\sum_{|h|=m_j} b_{jh}(x)\,(\xi + \tau\,\xi')^h$, $j = 0, \ldots, m-1$, *are linearly independent modulo the polynomial* $\prod_{i=1}^{m} (\tau - \tau_i^+(x, \xi, \xi'))$, *where* $\tau_i^+(x, \xi, \xi')$ *are the roots of the polynomial* $A_0(x, \xi + \tau\,\xi')$ *with positive imaginary part.*

We also impose the conditions

$$m_j \leqq 2m - 1, \quad j = 0, \ldots, m-1 \tag{1.11}$$

on the operators B_j.

We summarize the hypotheses which we shall use to obtain the final results of the theory:

1) *The operator* A *is properly elliptic in* $\bar{\Omega}$ *and has infinitely differentiable coefficients in* $\bar{\Omega}$;

2) *there are* m *operators* B_j;

3) *the coefficients of* B_j *are infinitely differentiable on* Γ;

4) *the system* $\{B_j\}_{j=0}^{m-1}$ *is normal on* Γ;

5) *the system* $\{B_j\}_{j=0}^{m-1}$ *covers the operator* A *on* Γ;

6) *the order* m_j *of* B_j *is* $\leqq 2m - 1$.

If these hypotheses are satisfied we shall sometimes refer to problem (1.8) as a *regular elliptic problem.*

We insist on the fact that, for intermediate results, only some of the hypotheses 1), ..., 6) will be used. Thus, for example, in Green's formula (see Section 2) hypothesis 5) is unnecessary; see also Section 8.4. □

Remark 1.3. Among the systems of operators $\{B_j\}$ which satisfy hypotheses 1), ..., 6) for *every properly elliptic* operator A, there is the system of Dirichlet conditions

$$B_j = \gamma_j, \quad j = 0, 1, \ldots, m-1,$$

where $\gamma_j = \dfrac{\partial^j}{\partial \nu^j}$, with ν normal to Γ, oriented towards the interior of Ω.

Problem $\{A, B\}$ is then called the *Dirichlet problem* for the operator A.

Remark 1.4. It is interesting to note that there exist strongly elliptic operators and normal systems of "boundary" operators which do not cover them (see Schechter [2], Appendix I).

Remark 1.5. The definitions given in this section are all invariant under infinitely differentiable homeomorphisms of the open set Ω.

2. Green's Formula and Adjoint Boundary Value Problems

2.1 The Adjoint of A in the Sense of Distributions or Formal Adjoint

With the operator A still given by (1.5) with (1.6), we denote by A^* the operator defined by

$$A^* u = \sum_{|p|,|q| \leq m} (-1)^{|p|} D^p(\overline{a_{qp}(x)}\, D^q u). \tag{2.1}$$

A^* is often called the formal adjoint of A: actually it is the adjoint of A in the sense of distributions on Ω, for we have

$$\int_\Omega A\, u\, \bar{v}\, dx - \int_\Omega u\, \overline{A^*\, v}\, dx = 0 \quad \forall u, v \in \mathscr{D}(\Omega). \tag{2.2}$$

It is easy to verify that A is (properly) elliptic if and only if A^* is (properly) elliptic.

2.2 The Theorem on Green's Formula

Let $\{F_i\}_{i=0}^{\nu-1}$ be a system of ν differential boundary operators defined by

$$F_i\, \varphi = \sum_{|h| \leq m_i} f_{ih}(x)\, D^h \varphi,$$

with $f_{ih}(x) \in \mathscr{D}(\Gamma)$.

Definition 2.1. *The system $\{F_i\}_{i=0}^{\nu-1}$ is a Dirichlet system of order ν on Γ_1 (subset of Γ) if it is normal on Γ_1 and if the orders m_i run through exactly the set $0, 1, \ldots, \nu - 1$, when i goes from 0 to $\nu - 1$.*

Theorem 2.1. *Let A be the operator defined by* (1.5), *with* (1.6), *and assume it to be elliptic; let $\{B_j\}_{j=0}^{m-1}$ be a normal system on Γ given by* (1.9), *with* (1.10), (1.11). *It is always possible to choose, non-uniquely, another system of boundary operators $\{S_j\}_{j=0}^{m-1}$ normal on Γ, the S_j's having infinitely differentiable coefficients on Γ and being of order $\mu_j \leqq 2m - 1$, such that the system $\{B_0, \ldots, B_{m-1}, S_0, \ldots, S_{m-1}\}$ is a*

Dirichlet system of order $2m$ *on* Γ. *Having made this choice, there exist* $2m$ *"boundary" operators* C_j, T_j, $j = 0, \ldots, m-1$, *uniquely defined, having the properties:*

a) *the coefficients of* C_j *and* T_j *are in* $\mathscr{D}(\Gamma)$;

b) *the order of* C_j *is* $2m - 1 - \mu_j$ *and the order of* T_j *is* $2m - 1 - m_j$;

c) *the system* $\{C_0, \ldots, C_{m-1}, T_0, \ldots, T_{m-1}\}$ *is a Dirichlet system of order* $2m$ *on* Γ,

such that the following Green's formula holds:

$$(2.3)\qquad \int_\Omega A u \bar{v}\, dx - \int_\Omega u \overline{A^* v}\, dx = \sum_{j=0}^{m-1} \int_\Gamma S_j u \overline{C_j v}\, d\sigma - \sum_{j=0}^{m-1} \int_\Gamma B_j u \overline{T_j v}\, d\sigma$$

for all u *and* $v \in \mathscr{D}(\Omega)$.

2.3 Proof of the Theorem

1) Via "local maps" and "partition of unity" we are brought back to the case of a half-ball. More precisely, there exists a finite covering of Γ by open sets $\mathscr{O}_i$, $i = 1, \ldots, N$, such that, for all i, there exists an infinitely differentiable diffeomorphism θ_i of $\mathscr{O}_i$ onto the unit ball in $\mathbf{R}^n$,

$$\sigma = \{x \mid |x| < 1\},$$

such that the image of $\mathscr{O}_i \cap \Omega$ under θ_i is $\sigma_+ = \{x \mid x \in \sigma, x_n > 0\}$ and the image of $\mathscr{O}_i \cap \Gamma$ is $\partial_1\sigma_+ = \{x \mid |x| < 1, x_n = 0\}$.

Via the diffeomorphism θ_i (arbitrary fixed i) the operator A is transformed into an operator $\mathscr{A}_i$ with infinitely differentiable coefficients in $\sigma_+ \cup \partial_1\sigma_+$, and we immediately see that $\mathscr{A}_i$ is still elliptic; the system $\{B_j\}_{j=0}^{m-1}$ is transformed into a system of operators $\{\mathscr{B}_{j,i}\}_{j=0}^{m-1}$ with infinitely differentiable coefficients in $\partial_1\sigma_+$ and which is normal on $\partial_1\sigma_+$.

If we recall that for $u, v \in \mathscr{D}(\bar{\Omega})$ we have formula (2.2), then we see that, by using $\{\mathscr{O}_i\}$ and an appropriate partition of unity, we may consider the problem under the following hypotheses:

in the half-ball σ_+, we define a differential operator in the form (from now on, we shall denote the point $x \in \mathbf{R}^n$ by $x = (y, t)$, with

$$y = (y_1, \ldots, y_{n-1}) \in \mathbf{R}^{n-1} \quad \text{and} \quad t \in \mathbf{R}):$$

(2.4)

$$\mathscr{A} u = \sum_{|p| \leq 2m} a_p(y, t)\, D_y^{p'} D_t^{p_n} u, \qquad p = (p', p_n), \qquad p' = (p_1, \ldots, p_{n-1}).$$

We assume that the coefficients a_p of $\mathscr{A}$ belong to $\mathscr{D}(\sigma_+ \cup \partial_1\sigma_+)$ and that $\mathscr{A}$ is *elliptic* in $\sigma_+ \cup \partial_1\sigma_+$.

We also define a system of operators $\{\mathscr{B}_j\}_{j=0}^{m-1}$ in the form

$$(2.5)\quad \mathscr{B}_j\varphi = \sum_{|h|\leqq m_j} b_{jh}(y)\, D_y^{h'} D_t^{h_n}\varphi, \qquad h = (h', h_n), \qquad h' = (h_1, \ldots, h_{n-1})$$

normal on $\partial_1\sigma_+$ and with $b_{jh} \in \mathscr{D}(\partial_1\sigma_+)$ and $m_j \leqq 2m - 1$.

From the above we immediately deduce that we can find operators $\{\mathscr{S}\}_{j=0}^{m-1}$ with infinitely differentiable coefficients in $\partial_1\sigma_+$, of order $\mu_j \leqq 2m - 1$, forming a normal system on $\partial_1\sigma_+$ such that

$$\{\mathscr{B}_0, \ldots, \mathscr{B}_{m-1}, \mathscr{S}_0, \ldots, \mathscr{S}_{m-1}\}$$

is a Dirichlet system of order $2m$ on $\partial_1\sigma_+$.

It is sufficient to take

$$\mathscr{S}_j = D_t^{\mu_j}$$

such that the numbers m_j and μ_j, $j = 0, \ldots, m - 1$, run through the set $0, 1, \ldots, 2m - 1$.

2) At this stage, we prove two lemmas before proceeding with the proof of the theorem.

Lemma 2.1. *If $\{F_j\}_{j=0}^{\nu-1}$ and $\{F'_j\}_{j=0}^{\nu-1}$ are two Dirichlet systems of order ν on $\partial_1\sigma_+$ and with coefficients belonging to $\mathscr{D}(\partial_1\sigma_+)$, then*

$$(2.6)\qquad F_j = \sum_{s=0}^{j} \Lambda'_{js} F'_s,$$

$$j = 0, \ldots, \nu - 1,$$

$$(2.7)\qquad F'_j = \sum_{s=0}^{j} \Lambda_{js} F_s,$$

where Λ_{jj} and Λ'_{jj} are non-vanishing functions on $\partial_1\sigma$, belonging to $\mathscr{D}(\partial_1\sigma_+)$, and Λ_{js} and Λ'_{js} $(s \neq j)$ are tangential differential operators (i.e. consisting of derivatives with respect to $y_1, \ldots, y_{n-1}$ only) of order $j - s$, with coefficients belonging to $\mathscr{D}(\partial_1\sigma_+)$.

Proof. It is sufficient to verify (2.7) for $F'_j = D_t^j$. Indeed, if $\{F'_j\}$ is a Dirichlet system of order ν, we may write F'_j in the form

$$(2.8)\qquad F'_j = \sum_{i=0}^{j} \Theta'_{ji} D_t^i,$$

where the Θ'_{jj}'s are non-vanishing functions, infinitely differentiable on $\partial_1\sigma_+$, and the Θ'_{ji}'s, $i \neq j$, are tangential operators of order $\leqq j - i$, with coefficients belonging to $\mathscr{D}(\partial_1\sigma_+)$. Then, if (2.7) holds for $F'_j = D_t^j$, that is if

$$(2.9)\qquad D_t^j = \sum_{s=0}^{j} \Phi_{js} F_s, \quad j = 0, \ldots, \nu - 1,$$

where the Φ_{js}'s have analogous properties to those of Λ_{js}, we have

$$F'_j = \sum_{i=0}^{j} \Theta'_{ji} D_t^i = \sum_{i=0}^{j} \Theta'_{ji} \sum_{s=0}^{j} \Phi_{is}\, F_s = \sum_{s=0}^{j} \Lambda_{js} F_s, \quad j = 0, \ldots, \nu - 1,$$

where $\Lambda_{jj} = \Theta'_{jj}\, \Phi_{jj}$ is a non-vanishing, infinitely differentiable function in $\partial_1 \sigma_+$ and $\Lambda_{js} = \sum_{i=s}^{j} \Theta'_{ji}\, \Phi_{is}$ is a tangential operator of order $\leqq j - s$, with coefficients in $\mathscr{D}(\partial_1 \sigma_+)$.

Therefore, we only need to prove formula (2.9). It holds for $j = 0$. By induction, we assume that it holds for $j < k$ and we verify that it holds for $j = k$. But $\{F_j\}_{j=0}^{\nu-1}$ being a Dirichlet system, we may write F_k in the form

$$F_k = \sum_{i=0}^{k} \Theta_{ki}\, D_t^i,$$

where Θ_{ki} has the same properties as Θ'_{ki}. Therefore

$$\begin{aligned}\Theta_{kk}\, D_t^k = F_k - \sum_{i=0}^{k-1} \Theta_{ki}\, D_t^i &= F_k - \sum_{i=0}^{k-1} \Theta_{ki} \sum_{s=0}^{i} \Phi_{is}\, F_s \\ &= F_k - \sum_{s=0}^{k-1} \left(\sum_{i=s}^{k-1} \Theta_{ki}\, \Phi_{is} \right) F_s,\end{aligned}$$

from which we immediately deduce (2.9) for $j = k$. □

Lemma 2.2. *If $\{F_j\}_{j=0}^{\nu-1}$ is a Dirichlet system of order ν and with coefficients belonging to $\mathscr{D}(\partial_1 \sigma_+)$, then, for every system $\{\varphi_j\}_{j=0}^{\nu-1}$ of functions belonging to $\mathscr{D}(\partial_1 \sigma_+)$, there exists a function $v(y, t) \in \mathscr{D}(\sigma_+ \cup \partial_1 \sigma_+)$ such that*

$$F_j\, v = \varphi_j, \quad j = 0, 1, \ldots, \nu - 1.$$

Proof. Thanks to Lemma 2.1, we have

$$(2.10) \qquad \begin{cases} F_j = \sum_{s=0}^{j} \Lambda'_{js}\, D_t^s \\ D_s^j = \sum_{s=0}^{j} \Phi_{js}\, F_s, \end{cases}$$

with the properties we have already pointed out for Λ'_{js} and Φ_{js}; we also have

$$(2.11) \qquad \sum_{s=i}^{j} \Lambda'_{js}\, \Phi_{si} = \delta_{ji}, \quad 0 \leqq i \leqq j \leqq \nu - 1.$$

Since the functions $\psi_s = \sum_{i=0}^{j} \Phi_{si}\, \varphi_i$ are infinitely differentiable on $\partial_1 \sigma_+$, there exists a function $v \in \mathscr{D}(\sigma_+ \cup \partial_1 \sigma_+)$ such that

$$D_t^s\, v = \psi_s, \quad s = 0, \ldots, \nu - 1.$$

Then, according to (2.11), v also verifies

$$F_j v = \sum_{s=0}^{j} \Lambda'_{js} \varphi_s = \sum_{s=0}^{j} \Lambda_{js} \sum_{i=0}^{s} \Phi_{si} \varphi_i = \sum_{i=0}^{j} \left(\sum_{s=i}^{j} \Lambda'_{js} \Phi_{si} \right) = \varphi_j . \quad \square$$

3) We return to the proof of the theorem under the hypotheses of 1). Let $u, v \in \mathscr{D}(\sigma_+ \cup \partial_1 \sigma_+)$ vanish in a neighborhood of

$$\partial_2 \sigma_+ = \{(y, t) \mid y_1^2 + \cdots + y_{n-1}^2 = 1, t \geqq 0\};$$

then, integrating by parts first in y and next in t, we obtain

(2.12)

$$\begin{aligned}
\int_{\sigma_+} (\mathscr{A} u) \bar{v} \, dy \, dt &= \sum_{|p| \leqq 2m} (-1)^{|p'|} \int_{\sigma_+} D_t^{p_n} u(y, t) D_y^{p'} (a_p(y, t) \overline{v(y, t)}) \, dy \\
&= \sum_{|p| \leqq 2m} (-1)^{|p|} \int_{\sigma_+} u(y, t) \overline{D^p (\overline{a_p(y, t)} v(y, t))} \, dy \, dt + \\
&\quad + \sum_{j=0}^{2m-1} \int_{\partial_1 \sigma_+} [D_t^j u(y, t) \overline{N_{2m-j} v(y,)}]_{t=0} \, dy \\
&= \int_{\sigma_+} u \overline{\mathscr{A}^* v} \, dy \, dt + \sum_{j=0}^{2m-1} \int_{\partial_1 \sigma_+} [D_t^j u(y, t) \overline{N_{2m-j} v(y, t)}]_{t=0} \, dy,
\end{aligned}$$

where

$$\mathscr{A}^* v = \sum_{|p| \leqq 2m} (-1)^{|p|} D^p (\bar{a}_p v) \quad \text{(formal adjoint of } \mathscr{A}\text{)}$$

and

$$\text{(2.13)} \quad N_{2m-j} v = \sum_{\substack{|p| \leqq 2m \\ p_n \geqq j+1}} (-1)^{|p|-j-1} D_t^{p_n - j - 1} D_y^{p'} \overline{(a_p(y, t)} v(y, t)).$$

From the ellipticity hypothesis on $\mathscr{A}$, we easily deduce that the system

$$\{N_{2m-j}\}_{j=0}^{2m-1}$$

is a Dirichlet system with infinitely differentiable coefficients on $\partial_1 \sigma_+$.

Now consider (see 1)) the system $\{\mathscr{B}_0, \ldots, \mathscr{B}_{m-1}, \mathscr{S}_0, \ldots, \mathscr{S}_{m-1}\}$, of order $2m - 1$ on $\partial_1 \sigma_+$; in order to simplify the notation, we denote it by $\{F_s\}_{s=0}^{2m-1}$. Then, according to Lemma 2.1, we have

$$\text{(2.14)} \qquad D_t^j = \sum_{s=0}^{j} \Phi_{js} F_s, \quad j = 0, \ldots, 2m-1,$$

where the Φ_{js}'s have the properties of Lemma 2.1. We denote by Φ_{js}^* the formal adjoint of Φ_{js} (adjoint in the sense of distributions on $\partial_1 \sigma_+$), i.e. the "tangential" differential operator defined by the identity

$$\text{(21.15)} \quad \int_{\partial_1 \sigma_+} \Phi_{js} \varphi \bar{\psi} \, dy = \int_{\partial_1 \sigma_+} \varphi \overline{\Phi_{js}^* \psi} \, dy, \quad \forall \varphi, \psi \in \mathscr{D}(\partial_1 \sigma_+).$$

From (2.12), (2.13), (2.14), we deduce

$$(2.16)\quad \int_{\sigma_+} (\mathscr{A}\, u)\, \bar{v}\, dy\, dt - \int_{\sigma_+} u\, \overline{\mathscr{A}^* v}\, dy\, dt$$
$$= \int_{\partial_1 \sigma_+} \sum_{s=0}^{2m-1} [F_s\, u(y, t)]_{t=0} \sum_{s=0}^{2m-1} \Phi_{js}^* [N_{2m-j}\, v(y, t)]_{t=0}\, dy .$$

It is easily verified that the operators

$$F_s' = \sum_{j=s}^{2m-1} \Phi_{js}^* N_{2m-j} = \bar{\Phi}_{ss}\, N_{2m-s} + \sum_{j=s+1}^{2m-1} \Phi_{js}^*\, N_{2m-j}$$
$$= (-1)^{|p|-s+1}\, \bar{\Phi}_{ss}\, a_{(0,0,\ldots,0,2m)}(x, 0)\, D_t^{2m-s+1} + \sum_{j=s+1}^{2m-1} \Phi_{js}^*\, N_{2m-j}$$

make up a Dirichlet system of order $2m$ with infinitely differentiable coefficients on $\partial_1 \sigma_+$.

If we denote by $\mathscr{C}_j$ the F_s''s for which $F_s = \mathscr{S}_j$ and by $\mathscr{T}_j$ the F_s''s for which $F_s = \mathscr{B}_j$ (in (2.16)), we obtain the following Green's formula

$$(2.17)\quad \int_{\sigma_+} (\mathscr{A}\, u)\, \bar{v}\, dy\, dt - \int_{\sigma_+} u\, \overline{\mathscr{A}^* v}\, dy\, dt$$
$$= \sum_{j=0}^{m-1} \int_{\partial_1 \sigma_+} \mathscr{S}_j\, u\, \overline{\mathscr{C}_j\, v}\, dy - \sum_{j=0}^{m-1} \int_{\partial_1 \sigma_+} \mathscr{B}_j\, u\, \overline{\mathscr{T}_j\, v}\, dy$$

and, according to the very construction of the F_s''s, we verify that the order of $\mathscr{C}_j$ is $2m - 1 - \mu_j$ and the order of $\mathscr{T}_j$ is $2m - 1 - m_j$. Following the remarks of 1), formula (2.17) proves the theorem. □

Remark 2.1. From (2.3), we easily deduce

Corollary 2.1. *If $u \in \mathscr{D}(\bar{\Omega})$, then $B_j\, u = 0$, $j = 0, \ldots, m - 1$, if and only if*

$$\int_{\Omega} (A\, u)\, \bar{v}\, dx = \int_{\Omega} u\, \overline{A^* v}\, dx$$

$\forall v \in \mathscr{D}(\bar{\Omega})$ verifying $C_j\, v = 0$, $j = 0, \ldots, m - 1$.

The necessary condition is obvious; conversely, if the condition holds, we have, according to (2.3):

$$\sum_{j=0}^{m-1} \int_{\Gamma} B_j\, u\, \overline{T_j\, v}\, d\sigma = 0$$

$\forall v \in \mathscr{D}(\bar{\Omega})$ verifying $C_j\, v = 0$, $j = 0, \ldots, m - 1$. Then, it is sufficient to take v such that we also have $T_j\, v = B_j\, u$, $j = 0, \ldots, m - 1$, which is possible according to Lemma 2.2. From which we obtain $B_j\, u = 0$, $j = 0, \ldots, m - 1$. □

Remark 2.2. Formula (2.3) may be extended by continuity to the case $u, v \in H^{2m}(\Omega)$. Indeed, $\mathscr{D}(\bar{\Omega})$ is dense in $H^{2m}(\Omega)$ (see Chapter 1, Theorem 8.2) and the integrals on Γ in (2.3) are well-defined thanks to the trace theorem for the elements u of $H^{2m}(\Omega)$ (see Chapter 1, Theorem 8.3).

2.4 A Variant of Green's Formula

With the elliptic operator A still defined by (1.5) with (1.6), we may associate with it the sesquilinear form given by

$$a(u, v) = \int_\Omega \sum_{|p|,|q| \leq m} a_{pq}(x) D^p u \overline{D^q v} \, dx. \tag{2.18}$$

Then, if $\{F_j\}_{j=0}^{m-1}$ is a Dirichlet system of order m, with infinitely differentiable coefficients on Γ, we may define a system $\{\Phi_j\}_{j=0}^{m-1}$ which is normal on Γ, has infinitely differentiable coefficients, with: order of F_j + order of $\Phi_j = 2m + 1$, such that

$$a(u, v) = \int_\Omega (A u) \bar{v} \, dx - \sum_{j=0}^{m-1} \int_\Gamma \Phi_j u \overline{F_j v} \, d\sigma \quad \forall u, v \in \mathscr{D}(\bar{\Omega}). \tag{2.19}$$

Indeed, as in Section 2.3, via "local maps", we may consider the case of the half-ball σ_+ and then, in the same notation and with the same reasoning as in Section 2.3, 1) and 3), we have, for $u, v \in \mathscr{D}(\sigma_+ \cup \partial_1 \sigma_+)$ and vanishing in a neighborhood of $\partial_2 \sigma_+$:

$$\int_{\sigma_+} (\mathscr{A} u) \bar{v} \, dy \, dt = \sum_{|p|,|q| \leq m} \int_{\sigma_+} a_{pq} D^p u \overline{D^q v} \, dy \, dt + \sum_{j=0}^{m-1} \int_{\partial_1 \sigma_+} [N_{m-j} u \overline{D_t^j v}]_{t=0} \, dy,$$

where the system $\{N_{m-j}\}_{j=0}^{m-1}$ is normal on $\partial_1 \sigma_+$ and has infinitely differentiable coefficients, the order of N_j being $2m - j - 1$; from which, by reasoning analogous to the case of Section 2.3, 3), we deduce (2.19). □

Remark 2.3. As in Remark 2.2, there is no difficulty in extending (2.19) to functions $u, v \in H^{2m}(\Omega)$. □

Remark 2.4. If $A = A^*$ (A is formally self-adjoint), then $a(u, v) = \overline{a(v, u)}$ and, writing (2.19) for the couples u, v and v, u, we deduce from it that

$$\int_\Omega u \overline{A v} \, dx - \int_\Omega (A u) \cdot \bar{v} \, dx = \sum_{j=0}^{m-1} \int_\Gamma F_j u \overline{\Phi_j v} \, d\sigma - \sum_{j=0}^{m-1} \int_\Gamma \Phi_j u \overline{F_j v} \, d\sigma. \tag{2.20}$$

2.5 Formal Adjoint Problems with Respect to Green's Formula

We again consider Green's formula (2.3) and introduce

Definition 2.2. *In the notation of Theorem 2.1, the system* $\{C_j\}_{j=0}^{m-1}$ *is said to be the adjoint of the system* $\{B_j\}_{j=0}^{m-1}$ *with respect to* A *and to Green's formula* (2.3).

Also, the boundary value problem $\{A^*, C\}$

$$(2.21)\qquad \begin{cases} A^* u = f & \text{in } \Omega \\ C_j u = g_j & \text{on } \Gamma, \quad j = 0, \ldots, m-1 \end{cases}$$

is called the *formal adjoint problem of* (1.8) *with respect to Green's formula* (2.3). □

Remark 2.5. Since the system $\{C_j\}_{j=0}^{m-1}$ depends on the choice of the system $\{S_j\}_{j=0}^{m-1}$, there exists an infinity of formal adjoint problems of (1.8). However, it is easy to see from Remark 2.1 that all adjoint systems of $\{B_j\}$ are *equivalent*, in the sense that if $\{C_j\}$ and $\{C'_j\}$ are two such systems and if $u \subset \mathscr{D}(\bar{\Omega})$, the conditions $C_j u = 0$, $j = 0, \ldots, m-1$, imply $C'_j u = 0$, $j = 0, \ldots, m-1$, and conversely. □

Now, assume that the operator A is *properly elliptic* and that the system $\{B_j\}_{j=0}^{m-1}$ *covers* A. Then, it is natural to ask whether the formal adjoint systems of $\{B_j\}_{j=0}^{m-1}$, in Definition 2.2, cover the formal adjoint A^* of A. The answer is yes; in fact we have

Theorem 2.2. *If* A, *as defined by* (1.5) *with* (1.6), *is properly elliptic, the system* $\{B_j\}_{j=0}^{m-1}$, *normal on* Γ, *covers* A *if and only if each normal system* $\{C_j\}_{j=0}^{m-1}$, *adjoint of the system* $\{B_j\}_{j=0}^{m-1}$ *with respect to* A *and to Green's formula* (2.3), *covers* A^*.

It is possible to give a direct, purely algebraic proof of this theorem (see Schechter [2], Appendix II); however, it will be more convenient for us to obtain this result by an indirect method in Section 4.3, below. □

Remark 2.6. It is easy to see, for example using (2.19) for A and A^*, with $F_j = \gamma_j$, that the Dirichlet problem $\{A, \gamma\}$ for A (i.e. $B_j = \gamma_j$) admits the Dirichlet problem $\{A^*, \gamma\}$ for A^* as a formal adjoint problem.

3. The Regularity of Solutions of Elliptic Equations in the Interior of Ω

3.1 Two Lemmas

The following lemmas will be used in the sequel. We recall that, if K is a compact subset of $\mathbf{R}^n$, we denote by $H^s_K(\mathbf{R}^n)$, arbitrary real s, the subspace of $H^s(\mathbf{R}^n)$ of distributions with support in K. For every $\varepsilon > 0$, the injection of $H_K^s(\mathbf{R}^n)$ into $H_K^{s-\varepsilon}(\mathbf{R}^n)$ is compact; this is an immediate consequence of Theorem 16.1 of Chapter 1.

Lemma 3.1. *For every positive real s and $\varepsilon > 0$, the norm of the injection of $H_K^s(\mathbf{R}^n)$ into $H_K^{s-\varepsilon}(\mathbf{R}^n)$ tends to zero as the diameter of K tends to zero.*

Proof. We recall that the norm under consideration is defined by

$$N(K) = \sup_{\|u\|_{H^s(\mathbf{R}^n)}=1} \|u\|_{H^{s-\varepsilon}(\mathbf{R}^n)}. \tag{3.1}$$

If the lemma was not true, there would exist a sequence of functions $u_i \in H^s(\mathbf{R}^n)$ with the properties: support of u_i contained in a ball of fixed center x_0, with radius $1/i$, $\|u_i\|_{H^s(\mathbf{R}^n)} = 1$ and $\|u_i\|_{H^{s-\varepsilon}(\mathbf{R}^n)}$ does not tend to zero. Then, according to the compactness of the injection of $H_K^s(\mathbf{R}^n)$ into $H_K^{s-\varepsilon}(\mathbf{R}^n)$, we may assume that there exists a subsequence, still denoted by u_i, such that $u_i \to u \neq 0$ in $H^{s-\varepsilon}(\mathbf{R}^n)$. Since u necessarily has support $\{x_0\}$ and $s > 0$ and since we may assume $s - \varepsilon \geqq 0$, we obtain $u = 0$, which is absurd. □

Lemma 3.2. *Let s be an arbitrary real number, l a positive integer and let a be a C^∞ function in $\mathbf{R}^n$, bounded together with all its derivatives. Let $\sigma(\varrho)$ denote the ball in $\mathbf{R}^n$ centered at the origin and of radius ϱ; we assume that $0 < \varrho \leqq \varrho_0/2$, $\varrho_0 > 0$ given. Then $\forall u \in H^s_{\sigma(\varrho)}(\mathbf{R}^n)$ and $\forall p$ such that $|p| = l$, we have*

$$\|a D^p u\|_{H^{s-l}(\mathbf{R}^n)} \leqq c\left(\max_{\sigma(2\varrho)} |a|\right) \|u\|_{H^s(\mathbf{R}^n)} + L_\varrho \|u\|_{H^{s-1}(\mathbf{R}^n)},$$

where c = constant and $L_\varrho > 0$ depends on ϱ.

Proof. Let φ_0 be a function in $\mathscr{D}(\mathbf{R}^n)$, $0 \leqq \varphi_0 \leqq 1$, $\varphi_0 = 1$ *on* $\sigma(\varrho_0)$. For every $u \in H^s_{\sigma(\varrho)}(\mathbf{R}^n)$, we have

$$a D^p u = \varphi_0 a D^p u = D^p(\varphi_0 a u) + B u,$$

where B = linear differential operator of order $< l$, with coefficients in $\mathscr{D}(\mathbf{R}^n)$ with support in $\sigma(\varrho_0)$.

Therefore, applying Lemma 7.2 of Chapter 1, we have

$$\|a D^p u\|_{H^{s-1}(\mathbf{R}^n)} \leqq c_1 \|\varphi_0 a u\|_{H^s(\mathbf{R}^n)} + c_2 \|u\|_{H^{s-1}(\mathbf{R}^n)}. \tag{3.2}$$

We now introduce the functions $\varphi_\varrho \in \mathscr{D}(\mathbf{R}^n)$ such that $0 \leqq \varphi_\varrho \leqq 1$, $\varphi_\varrho = 1$ in $\sigma(\varrho)$ and with support in $\sigma(2\varrho)$.

We apply Lemma 7.1 with $\varphi = \varphi_\varrho \varphi_0 a$; we obtain

$$\begin{aligned} \|\varphi_0 a u\|_{H^s(\mathbf{R}^n)} &= \|\varphi_\varrho \varphi_0 a u\|_{H^s(\mathbf{R}^n)} \leqq \\ &\leqq c(\max|\varphi_\varrho \varphi_0 a|) \|u\|_{H^s(\mathbf{R}^n)} + c_\varrho \|u\|_{H^{s-1}(\mathbf{R}^n)} \leqq \\ &\leqq c\left(\max_{\sigma(2\varrho)} |a|\right) \|u\|_{H^s(\mathbf{R}^n)} + c_\varrho \|u\|_{H^{s-1}(\mathbf{R}^n)}, \end{aligned} \tag{3.3}$$

with c_ϱ depending on φ_0, φ_ϱ, a and l.

The lemma follows from (3.2) and (3.3). □

We note that if $a(0) = 0$ and $s > 0$, we obtain, using Lemma 3.1, the existence of a function $\varepsilon(\varrho) > 0$ such that

$$\lim_{\varrho \to 0} \varepsilon(\varrho) = 0$$

and such that

$$\| a\, D^p\, u \|_{H^{s-l}(\mathbf{R}^n)} \leqq \varepsilon(\varrho)\, \| u \|_{H^s(\mathbf{R}^n)}, \qquad \forall u \in H^s_{\sigma(\varrho)}(\mathbf{R}^n), \qquad |p| = l. \quad \square$$

3.2 A priori Estimates in $\mathbf{R}^n$

Let

$$A = A(x, D) = \sum_{|p| \leqq l} a_p(x)\, D^p \tag{3.4}$$

be a linear differential operator of order l with *infinitely differentiable coefficients which, together with each of their derivatives, are bounded in* $\mathbf{R}^n$.

Denote by

$$A_0 = A_0(x, D) = \sum_{|p| = l} a_p(x)\, D^p$$

the homogeneous part of degree l of A. Assume that

(3.5) *the operator $A_0(0, D)$ with constant coefficients is elliptic.*

Then, we have

Theorem 3.1. *If A is defined by* (3.4) *and satisfies* (3.5), *for every integer $r \geq 0$ there exists a positive number ϱ_0 such that if $\varrho < \varrho_0$ and if $u \in L^2(\mathbf{R}^n)$ vanishes outside the ball $\sigma(\varrho)$ with center at the origin and radius ϱ, and $A\, u \in H^{-l+r}(\mathbf{R}^n)$, then $u \in H^r(\mathbf{R}^n)$ and*

$$\| u \|_{H^r(\mathbf{R}^n)} \leqq C_{r,\varrho} \{ \| A\, u \|_{H^{-l+r}(\mathbf{R}^n)} + \| u \|_{H^{r-1}(\mathbf{R}^n)} \} \tag{3.6}$$

($C_{r,\varrho}$ depending on r and ϱ, but not on u).

Proof. 1) Here, as well as in the sequel, c, C, C_r, $C_{r,\varrho}$ shall denote positive constants which may change from one inequality to another. We first prove the theorem for $r = 0$. From (3.5), we deduce that

$$1 + |\xi|^{2l} \leqq c\, |A_0(0, \xi)|^2 + 1 \qquad \forall \xi \in \mathbf{R}^n,$$

therefore also

$$1 \leqq c \frac{|A_0(0, \xi)|^2}{1 + |\xi|^{2l}} + \frac{1 + |\xi|^{2(l-1)}}{1 + |\xi|^{2l}} \qquad \forall \xi \in \mathbf{R}^n.$$

Multiplying by $|\hat{u}(\xi)|^2$ ($\hat{u}$ denoting the Fourier transform of u with respect to $x_1, \ldots, x_n$) and integrating on $\mathbf{R}^n$, we obtain

$$\| u \|_{L^2(\mathbf{R}^n)} \leqq C \{ \| A_0(0, D)\, u \|_{H^{-l}(\mathbf{R}^n)} + \| u \|_{H^{-1}(\mathbf{R}^n)} \}. \tag{3.7}$$

But, thanks to the hypotheses on the coefficients of $A(x, D)$, we may apply Lemma 3.2; therefore, there exists a $\varrho_0 > 0$ such that, for u

with support in $\sigma(\varrho)$ and $\varrho < \varrho_0$, we have

$$(3.8)\qquad \begin{aligned} &\| A_0(0, D)\, u - A(x, D)\, u \|_{H^{-l}(\mathbf{R}^n)} \leqq \\ &\leqq \| A_0(0, D)\, u - A_0(x, D)\, u \|_{H^{-l}(\mathbf{R}^n)} + \\ &+ \| A_0(x, D)\, u - A(x, D)\, u \|_{H^{-l}(\mathbf{R}^n)} \leqq \\ &\leqq \frac{1}{2c} \| u \|_{L^2(\mathbf{R}^n)} + C_\varrho \| u \|_{H^{-1}(\mathbf{R}^n)} \end{aligned}$$

and therefore from (3.7) and (3.8) we deduce

$$(3.9)\qquad \| u \|_{L^2(\mathbf{R}^n)} \leqq C \left\{ \| A(x, D)\, u \|_{H^{-l}(\mathbf{R}^n)} + \frac{1}{2c} \| u \|_{L^2(\mathbf{R}^n)} + C_\varrho \| u \|_{H^{-1}(\mathbf{R}^n)} \right\}$$

from which we obtain (3.6), with C_ϱ depending on ϱ.

2) Now, using the well-known method of "differential quotients", we shall prove the theorem for $r > 0$.

Indeed, since the theorem is valid for $r = 0$, we may proceed by induction. Thus, we assume that $u \in H^{r-1}(\mathbf{R}^n)$, $r \geqq 1$ and that (3.6) holds with $r - 1$ replacing r.

For $h \neq 0$, we set

$$(3.10)\qquad \varrho_{i,h}\, u(x) = \frac{u(x_1, \ldots, x_{i-1}, x_i + h, x_{i+1}, \ldots, x_n) - u(x_1, \ldots, x_n)}{h}, \qquad i = 1, \ldots, n.$$

If h is sufficiently small, the support of $\varrho_{i,h}\, u$ is contained in the ball $\sigma(\varrho')$ with radius

$$\varrho' = \varrho + (\varrho_0 + \varrho)/2 < \varrho_0,$$

if the support of $u(x)$ is in $\sigma(\varrho)$. Thus, we may apply (3.3) with $r - 1$ to $\varrho_{i,h}\, u$ and we obtain

$$(3.11)\qquad \begin{aligned} &\| \varrho_{i,h}\, u \|_{H^{r-1}(\mathbf{R}^n)} \leqq \\ &\leqq C_{r,\varrho'} \{ \| A(x, D)\, \varrho_{i,h}\, u \|_{H^{-l+r-1}(\mathbf{R}^n)} + \| \varrho_{i,h}\, u \|_{H^{r-2}(\mathbf{R}^n)} \}. \end{aligned}$$

We set $\tau_{i,h}\, u = u(x_1, \ldots, x_{i-1}, x_i + h, x_{i+1}, \ldots, x_n)$ and we verify that

$$A(x, D)\, \varrho_{i,h}\, u - \varrho_{i,h}\, A(x, D)\, u = \sum_{|p| \leqq l} [\varrho_{i,h}\, a_p(x)]\, D^p\, \tau_{i,h}\, u$$

and therefore

$$\| A(x, D)\, \varrho_{i,h}\, u - \varrho_{i,h}\, A(x, D)\, u \|_{H^{-l+r-1}(\mathbf{R}^n)} \leqq C_{r,\varrho} \| u \|_{H^{r-1}(\mathbf{R}^n)}.$$

Then, from (3.11) we deduce

$$(3.12)\quad \|\varrho_{i,h} u\|_{H^{r-1}(\mathbf{R}^n)} \leqq C_{r,\varrho}\{\|\varrho_{ih} A u\|_{H^{-l+r-1}(\mathbf{R}^n)} + \|\varrho_{i,h} u\|_{H^{r-2}(\mathbf{R}^n)} + \|u\|_{H^{r-1}(\mathbf{R}^n)}\} \leqq C_{r,\varrho}\{\|A u\|_{H^{-l+r}(\mathbf{R}^n)} + \|u\|_{H^{r-1}(\mathbf{R}^n)}\}$$

from which we deduce the fact that $\varrho_{i,h} u$ remains in a bounded set of $H^{r-1}(\mathbf{R}^n)$ as h varies; but then we can find a sequence $h \to 0$ such that $\varrho_{i,h} u \to \chi_i$ in $H^{r-1}(\mathbf{R}^n)$ weakly; but $\varrho_{i,h} u \to \frac{\partial u}{\partial x_i}$ in $\mathscr{D}'(\mathbf{R}^n)$, so that $\frac{\partial u}{\partial x_i} \in H^{r-1}(\mathbf{R}^n)$ for $i = 1, \ldots, n$, therefore $u \in H^r(\mathbf{R}^n)$ and we also have (3.6). □

3.2 The Regularity in the Interior of Ω and the Hypoellipticity of Elliptic Operators

From Theorem 3.1, we shall now deduce the regularity of solutions of elliptic equations in the interior of Ω.

Let Ω be an arbitrary open set in $\mathbf{R}^n$ and let

$$(3.13)\qquad A = A(x, D) = \sum_{|p| \leqq l} a_p(x) D^p$$

be a linear operator of order l *with infinitely differentiable coefficients in Ω and elliptic in Ω.*

We denote by $H^r_{\text{loc}}(\Omega)$, arbitrary interger r, the space of distributions u on Ω such that $\varphi u \in H^r(\Omega)$ for all $\varphi \in \mathscr{D}(\Omega)$. We have

Theorem 3.2. *Let A be defined by* (3.13) *and r be an arbitrary integer; if u is a distribution on Ω such that $A u \in H^{-l+r}_{\text{loc}}(\Omega)$, then $u \in H^r_{\text{loc}}(\Omega)$; in particular, if $A u$ is infinitely differentiable in Ω, then u is also infinitely differentiable in Ω.*

Proof. 1) We first show that

$$(3.14)\quad \textit{if } u \in H^{r-1}_{\text{loc}}(\Omega) \textit{ and } A u \in H^{-l+r}_{\text{loc}}(\Omega), \textit{ then } u \in H^r_{\text{loc}}(\Omega).$$

Let $\varphi \in \mathscr{D}(\Omega)$ be arbitrary and fixed; set $v = \varphi u$ and denote by $\tilde{v}$ the extension of v by zero outside of Ω.

Of course, $v \in H^{r-1}(\Omega)$ and $\tilde{v} \in H^{r-1}(\mathbf{R}^n)$.

Also, $A v \in H^{-l+r}_{\text{loc}}(\Omega)$, since

$$A v = \varphi A u + A_1 u,$$

where $A_1 u$ is an operator of order $l - 1$ and therefore $A_1 u \in H^{-l+r}_{\text{loc}}(\Omega)$ since, by hypothesis, $u \in H^{r-1}_{\text{loc}}(\Omega)$.

Now assume that $r - 1 \geqq 0$. Because of the local character of the theorem, it is sufficient to show that if φ has its support in a ball $\sigma(\varrho)$

with center at the origin and with sufficiently small radius ϱ, then $\tilde{v} \in H^r(\mathbf{R}^n)$. But then we may apply Theorem 3.1, since $r - 1 \geqq 0$ (of course, we may assume to have extended the coefficients of A to $\mathbf{R}^n$ so that Theorem 3.1 applies); which shows that $\tilde{v} \in H^r(\mathbf{R}^n)$ and therefore that (3.14) is valid.

Now, let $r - 1 < 0$. Then, we may write $\tilde{v}$ in the form

$$\tilde{v} = (I - \Delta)^k \tilde{w}, \qquad \text{with } k \text{ such that } r - 1 + 2k \geqq 1.$$

In fact, it is sufficient to set

$$\hat{\tilde{w}}(\xi) = \left(\frac{1}{1 + |\xi|^2}\right)^k \hat{\tilde{v}}(\xi) \qquad (\hat{\tilde{v}}(\xi) = \text{Fourier transform of } \tilde{v})$$

and to take $\tilde{w}$ = Fourier anti-transform of $\hat{\tilde{w}}$.

Then $\tilde{w} \in H^{r-1+2k}(\mathbf{R}^n)$ and its restriction w to Ω belongs to $H^{r-1+2k}(\Omega)$. Finally, we have

$$v = (I - \Delta)^k w.$$

But $A(I - \Delta)^k w \in H^{-l+r}_{\text{loc}}(\Omega)$ by hypothesis and the operator $A(I - \Delta)^k$ is elliptic and of order $l + 2k$.

Therefore, we may apply (3.14), for the case already demonstrated, to w; thus, $w \in H^{r+2k}_{\text{loc}}(\Omega)$ and therefore $v \in H^r_{\text{loc}}(\Omega)$ and (3.14) is proved for arbitrary r.

2) If u is an arbitrary distribution on Ω and $A u \in H^{-l+r}_{\text{loc}}(\Omega)$, we note that, because of the fact that every distribution is locally of finite order, for every ball ω such that $\bar{\omega} \subset \Omega$, there exists an integer r_0 such that $u \in H^{r_0}_{\text{loc}}(\omega)$.

So that if $r_0 \geqq r$, the theorem is obvious; if $r_0 = r - 1$, we may use (3.14). Finally, if $r_0 < r - 1$, we have

$$u \in H^{r_0}_{\text{loc}}(\omega), \qquad A u \in H^{-l+r}_{\text{loc}}(\omega) \subset H^{-l+r_0+1}_{\text{loc}}(\omega)$$

and therefore, thanks to (3.14), $u \in H^{r_0+1}_{\text{loc}}(\omega)$. Now, we may proceed in the same way a finite number of times until we have shown $u \in H^r_{\text{loc}}(\omega)$ and therefore the theorem is proved. □

Remark 3.1. Following a terminology which by now has become standard, we say that a linear differential operator A with infinitely differentiable coefficients in Ω is *hypoelliptic* if $u \in \mathscr{D}'(\Omega)$, $A u \in C^\infty(\Omega)$ implies $u \in C^\infty(\Omega)$.

Thus, we have shown that if *A is elliptic, than it is hypoelliptic.* Furthermore, we shall see in Volume 3 that if the coefficients of A and f are analytic (or belong to a Gevrey class) in Ω, then u is analytic (or belongs to a Gevrey class) in Ω.

Remark 3.2. In the preceding theory, we may also use the spaces H^r with arbitrary real r; in particular, Theorem 3.2 is valid for *all real* r (see, for example, Schwartz [7]). But, we shall only use the case integer r in the sequel.

4. A priori Estimates in the Half-Space

4.1 A new Formulation of the Covering Condition

The "a priori" estimates on the boundary shall be given first in the case of the half-space and for operators with constant coefficients.

We shall use the notation already introduced in Section 2 for the point $x \in \mathbf{R}^n$: $x = (y, t)$, with $y = (y_1, \ldots, y_{n-1}) \in \mathbf{R}^{n-1}$ and $t \in \mathbf{R}$. The "dual" variable of x will be denoted by ξ and represented by $\xi = (\eta, t')$, with $\eta = (\eta_1, \ldots, \eta_{n-1}) \in \mathbf{R}^{n-1}$ and $t' \in \mathbf{R}$.

$\mathbf{R}^n_+$ denotes the half-space of $\mathbf{R}^n$ of x's with $t > 0$.

Let

$$(4.1) \qquad \begin{cases} A(D)\, u = A(D_y, D_t) = \sum_{|p| = 2m} a_p\, D_y^{p'}\, D_t^{p_n}, \\ p = (p', p_n), \quad p' = (p_1, \ldots, p_{n-1}) \end{cases}$$

be a differential operator with constant coefficients, homogeneous of degree $2m$; let

$$(4.2) \qquad \begin{cases} B_j(D) = B_j(D_y, D_t) = \sum_{|h| = m_j} b_{jh}\, D_y^{h'}\, D_t^{h_n}, \\ h = (h', h_n), \quad h' = (h_1, \ldots, h_{n-1}), \quad j = 0, \ldots, m-1 \end{cases}$$

be m differential operators with constant coefficients, B_j being homogeneous of degree m_j, with

$$(4.3) \qquad 0 \leqq m_j \leqq 2m - 1.$$

We assume that A is properly elliptic, that is, thanks to Remark 1.2, we assume that

I) *the characteristic form* $A(\xi) = A(\eta, t')$ *is different from zero for all* $\xi \neq 0$, $\xi \in \mathbf{R}^n$; *and for all* $\eta \in \mathbf{R}^{n-1}$ *and* $\neq 0$, *the polynomial* $A(\eta, \tau)$ *in the complex variable* τ *has* m *roots with positive imaginary part.*

We denote by $\tau_i^+(\eta)$ (resp. $\tau_i^-(\eta)$) the roots of $A(\eta, \tau)$ with positive imaginary part (resp. negative imaginary part) and set

$$(4.4) \qquad M^{\pm}(\eta, \tau) = \prod_{i=1}^{m} (\tau - \tau_i^{\pm}(\eta)) = \sum_{k=0}^{m} c_k^{\pm}(\eta)\, \tau^{m-k}.$$

We have

$$M^+(\eta, \tau) = (-1)^m\, M^-(-\eta, -\tau)$$

and the coefficients $c_k^{\pm}(\eta)$ are analytic functions of $\eta \in \mathbf{R}^{n-1}$ and $\eta \neq 0$, and homogeneous of degree k.

For every $j = 0, \ldots, m-1$, we also consider the polynomial in τ

$$M_j^+(\eta, \tau) = \sum_{k=0}^{j} c_k^+(\eta)\, \tau^{j-k}.$$

Then, we have

Proposition 4.1. *For $\eta \in \mathbf{R}^{n-1}$ and $\neq 0$, we have*

(4.5)
$$\frac{1}{2\pi i}\int_\gamma \frac{M_{m-j-1}^+(\eta, \tau)}{M^+(\eta, \tau)}\,\tau^k\, d\tau = \delta_{jk}, \quad 0 \leqq j \leqq m-1, \quad 0 \leqq k \leqq m-1,$$

for every rectifiable Jordan curve γ in the complex plane, which encircles all the roots $\tau_i^+(\eta)$.

Proof. We not that, if $j \geqq k$, $M_{m-j-1}^+ \tau^k$ is a polynomial in τ of degree $m-1-j+k$ for which the term of maximum degree is given by $a_0^+(\eta)\,\tau^{m-1-j+k}$. In that case, it suffices to deform γ into a circle with center at the origin and to make the radius of the circle tend to infinity to obtain (4.5). If $j < k$, then $M_{m-j-1}\tau^k = \tau^{k-j-1} M^+ + Q$, where Q is a polynomial of degree $< k-1$ and, therefore, to obtain (4.5), it suffices to calculate $\int_\gamma \frac{Q}{M^+}\, d\tau$, and this integral vanishes because the degree of $Q \leqq m-1$ (thus, we may again deform γ into a circle of radius tending towards infinity). □

We shall also make the hypothesis that the system $\{B_j\}_{j=0}^{m-1}$ *covers* A, that is:

II) *For all $\eta \in \mathbf{R}^{n-1}$ and $\neq 0$, the polynomials $B_j(\eta, \tau)$ in τ are linearly independent modulo $M^+(\eta, \tau)$.*

Therefore, if we set

$$(4.6) \quad B_j'(\eta, \tau) = B_j(\eta, \tau), \quad \operatorname{mod} M^+(\eta, \tau), \quad (\text{i.e. } B_j = Q_j M^+ + B_j')$$

hypothesis II) is equivalent to the fact that, if $B_j'(\eta, \tau)$ is given by

$$B_j'(\eta, \tau) = \sum_{k=0}^{m-1} b_{jk}'(\eta)\, \tau^k,$$

(4.7) *the determinant of the matrix $\| b_{jk}'(\eta) \|$ is $\neq 0$ for $\eta \in \mathbf{R}^{n-1}$ and $\neq 0$.*

We shall also require another equivalent formulation of hypothesis II). For this purpose, we consider, for all $\eta \in \mathbf{R}^{n-1}$ and $\neq 0$, the ordinary differential equation

$$A\left(\eta, \frac{1}{\mathrm{i}} \frac{d}{dt}\right) \varphi(\eta, t) = 0 \tag{4.8}$$

with the conditions

$$B_j\left(\eta, \frac{1}{\mathrm{i}} \frac{d}{dt}\right) \varphi(\eta, t)\bigg|_{t=0} = c_j, \quad j = 0, \ldots, m-1, \tag{4.9}$$

where the c_j's are given arbitrary complex numbers. In agreement with the usual notation (see L. Schwartz [1]), we denote by $\mathscr{S}(\mathbf{R}_+)$ the space of functions φ, infinitely differentiable for $t \geqq 0$ and rapidly decreasing for $t \to +\infty$ (i.e. $t^k \varphi^{(j)}(t) \to 0$ as $t \to +\infty$, $\forall k$, $\forall j$).

We shall no prove

Proposition 4.2. *Assume that hypothesis* I) *holds, then hypothesis* II) *is equivalent to one of the two following conditions:*

II') *problem* (4.8) – (4.9) *admits a solution belonging to* $\mathscr{S}(\mathbf{R}_+)$, *for all* c_j, $j = 0, \ldots, m-1$;

II'') *problem* (4.8) – (4.9), *with* $c_j = 0$, *admits only the null solution in* $\mathscr{S}(\mathbf{R}_+)$.

Proof. 1) Let us first show that II') is equivalent to II''). It is a well-known fact that the solutions of (4.8) are in the form of exponential-polynomials; therefore, we deduce from I) that the space of solutions of (4.8) which belong to $\mathscr{S}(\mathbf{R}_+)$ has dimension m. Thus, we may construct a mapping of this space into $\mathbf{C}^m$ by making correspond to each solution φ of (4.8) which belongs to $\mathscr{S}(\mathbf{R}_+)$, the vector in $\mathbf{C}^m$ having components $B_j\left(\eta, \frac{1}{\mathrm{i}} \frac{d}{dt}\right) \varphi\Big|_{t=0}$, $j = 0, \ldots, m-1$. This mapping is a linear mapping of an m-dimensional vector space into $\mathbf{C}^m$; therefore, it is surjective (i.e. II') is satisfied) if and only if it is injective (i.e. if II'') is satisfied).

2) Now, we show that II) is equivalent to II').

Assume that II) is satisfied and let c_j, $j = 0, 1, \ldots, m-1$, be m given complex numbers and $\eta \in \mathbf{R}^{n-1}$, $\neq 0$; then, because of (4.7), the system

$$\sum_{s=0}^{m-1} b'_{js}(\eta)\, q_s(\eta) = c_j, \quad j = 0, \ldots, m-1 \tag{4.10}$$

admits a unique solution $\{q_s(\eta)\}_{s=0}^{m-1}$ which depends on η. We set

$$u(\eta, t) = \frac{1}{2\pi \mathrm{i}} \int_\gamma \sum_{s=0}^{m-1} q_s(\eta) \frac{M^+_{m-s-1}(\eta, \tau)}{M^+(\eta, \tau)} \mathrm{e}^{\mathrm{i}t\tau}\, d\tau, \tag{4.11}$$

where γ is a rectifiable Jordan curve which encircles the roots $\tau^+_i(\eta)$ of $M^+(\eta, \tau)$. We may assume that γ is in the complex half-plane corresponding to positive imaginary parts, since the $\tau_i^+(\eta)$'s are in this half-plane. Then, the function $t \to u(\eta, t) \in \mathscr{S}(\mathbf{R}_+)$ and we easily verify that it satisfies equation (4.8); furthermore, for $t = 0$, we have

$$B_j\left(\eta, \frac{1}{\mathrm{i}}\frac{d}{dt}\right)u(\eta, t)|_{t=0} = \frac{1}{2\pi\mathrm{i}}\int_\gamma \sum_{s=0}^{m-1} q_s(\eta)\, B_j(\eta, \tau)\, \frac{M^+_{m-s-1}(\eta, \tau)}{M^+(\eta, \tau)}\, d\tau$$

$$= \frac{1}{2\pi\mathrm{i}}\int_\gamma \sum_{s=0}^{m-1}\sum_{k=0}^{m-1} b'_{jk}(\eta)\, \tau^k\, q_s(\eta)\, \frac{M^+_{m-s-1}(\eta, \tau)}{M^+(\eta, \tau)}\, d\tau$$

$$= (\text{because of } (4.5)) = \sum_{s=0}^{m-1} b'_{js}(\eta)\, q_s(\eta) = c_j,$$

$$j = 0, \ldots, m-1$$

and therefore u satisfies II').

Conversely, we assume that II) does not hold; this is equivalent to the fact that the determinant of $\| b'_{jk}(\eta) \|$ vanishes and therefore there exist non-null solutions $\{q_s(\eta)\}_{s=0}^{m-1}$ of system (4.10) with

$$c_j = 0, \quad j = 0, \ldots, m-1.$$

We show that u, as given by (4.11), is not identical to zero. Indeed, there exists s_0, $0 \leqq s_0 \leqq m-1$, such that $q_{s_0}(\eta) \neq 0$ and therefore

$$\left.\frac{d^{s_0} u(\eta, t)}{dt^{s_0}}\right|_{t=0} = \frac{1}{2\pi\mathrm{i}}\int_\gamma \sum_{s=0}^{m-1} q_s(\eta)\, (\mathrm{i}\,\tau)^{s_0}\, \frac{M^+_{m-s-1}(\eta, \tau)}{M^+(\eta, \tau)}\, d\tau$$

$$= (\text{because of } (4.5)) = c^{s_0}\, q_{s_0}(\eta) \neq 0;$$

so that u is not identically zero. But, this contradicts II'') and therefore II'). □

4.2 A Lemma on Ordinary Differential Equations

The following lemma is important for the sequel.

Lemma 4.1. *Under the hypotheses* I) *and* II) *of Section* 4.1, *for all* $\eta \in \mathbf{R}^{n-1}$ *and* $\neq 0$, *the operator*

$$\mathscr{P}_\eta : \varphi \to \mathscr{P}_\eta \varphi$$

$$= \left\{A\left(\eta, \frac{1}{\mathrm{i}}\frac{d}{dt}\right)\varphi,\ B_0\left(\eta, \frac{1}{\mathrm{i}}\frac{d}{dt}\right)\varphi\bigg|_{t=0}, \ldots, B_{m-1}\left(\eta, \frac{1}{\mathrm{i}}\frac{d}{dt}\right)\varphi\bigg|_{t=0}\right\}$$

is an (algebraic and topological) isomorphism of $H^{2m}(\mathbf{R}_+)$ *onto* $L^2(\mathbf{R}_+) \times \mathbf{C}^m$.

Proof. Using the Banach theorem, it is sufficient to show that $\mathscr{P}_\eta$ is an algebraic isomorphism; thanks to Proposition 4.2 and the fact that the solutions of (4.8) in $H^{2m}(\mathbf{R}_+)$ (and even in $\mathscr{S}'(\mathbf{R}_+)$) are in

$\mathscr{S}(\mathbf{R}_+)$, it suffices to show that $\mathscr{P}_\eta$ is surjective, i.e. that there exists $\varphi \in H^{2m}(\mathbf{R}_+)$, solution of

$$(4.12)\qquad \begin{cases} A\left(\eta, \dfrac{1}{\mathrm{i}}\dfrac{d}{dt}\right)\varphi(t) = f(t) \quad t > 0, \\ B_j\left(\eta, \dfrac{1}{\mathrm{i}}\dfrac{d}{dt}\right)\varphi(t)\Big|_{t=0} = c_j, \quad j = 0, \ldots, m-1, \end{cases}$$

where arbitrary $f \in L^2(\mathbf{R}_+)$ is given and the c_j's are arbitrary complex numbers.

Let $\tilde{f}$ be the extension of f by zero for all $t < 0$ (i.e. $\tilde{f}(t) = f(t)$ if $t \geqq 0$ and $\tilde{f}(t) = 0$ if $t < 0$). There exists $\psi(t) \in H^{2m}(\mathbf{R})$, solution of

$$(4.13)\qquad A\left(\eta, \frac{1}{\mathrm{i}}\frac{d}{dt}\right)\psi(t) = \tilde{f}(t).$$

By Fourier transform in t, it suffices to take

$$\hat{\psi}(t') = \frac{\hat{\tilde{f}}(t')}{A(\eta, t')} \qquad (\hat{v} = \text{Fourier transform of } v)$$

and to note that $A(\eta, t') \neq 0$ for real t'; then the Fourier anti-transform ψ of $\hat{\psi}$ belongs to $H^{2m}(\mathbf{R})$, for $\dfrac{(t')^j}{A(\eta, t')}$ is bounded if $t' \in \mathbf{R}$ for $j \leqq 2m$. Furthermore, ψ satisfies (4.13).

Consider the restriction ψ_1 of ψ to $\mathbf{R}_+$; $\psi_1 \in H^{2m}(\mathbf{R}_+)$ and therefore we may calculate

$$B_j\left(\eta, \frac{1}{\mathrm{i}}\frac{d}{dt}\right)\psi_1(t)\Big|_{t=0} = \gamma_j, \quad j = 0, \ldots, m-1.$$

Now we apply Proposition 4.2 to solve the problem

$$A\left(\eta, \frac{1}{\mathrm{i}}\frac{d}{dt}\right)\psi_2(t) = 0, \quad t > 0$$

$$B_j\left(\eta, \frac{1}{\mathrm{i}}\frac{d}{dt}\right)\psi_2(t)\Big|_{t=0} = c_j - \gamma_j, \quad j = 0, \ldots, m-1$$

in $\mathscr{S}(\mathbf{R}_+)$.

Then $\psi_2 \in H^{2m}(\mathbf{R}_+)$ and $\varphi(t) = \psi_1(t) + \psi_2(t)$ solves problem (4.12). □

Remark 4.1. Setting $C_\eta = \| \mathscr{P}_\eta^{-1} \|_{\mathscr{L}(L^2(\mathbf{R}_+)\times \mathbf{C}^m; H^{2m}(\mathbf{R}_+))}$, we deduce from Lemma 4.1 that

$$(4.14)\qquad \|\varphi\|_{H^{2m}(\mathbf{R}_+)} \leqq C_\eta \|\mathscr{P}_\eta \varphi\|_{L^2(\mathbf{R}_+)\times \mathbf{C}^m} \qquad \forall \varphi \in H^{2m}(\mathbf{R}_+).$$

We shall see that the function $\eta \to C_\eta$ is continuous in $\mathbf{R}^{n-1} - \{0\}$.

Let

$$G = \mathscr{L}(L^2(\mathbf{R}_+) \times \mathbf{C}^m; H^{2m}(\mathbf{R}_+));$$

the function $\eta \to \| \mathscr{P}_\eta^{-1} \|_G$ is bounded on every compact set of $\mathbf{R}^{n-1} - \{0\}$.

Then, for $\eta, \eta_0 \in \mathbf{R}^{n-1} - \{0\}$:

$$\mathscr{P}_\eta^{-1} - \mathscr{P}_{\eta_0}^{-1} = -\mathscr{P}_\eta^{-1} (\mathscr{P}_\eta - \mathscr{P}_{\eta_0}) \mathscr{P}_{\eta_0}^{-1}$$

yields, as $\eta \to \eta_0$:

$$\| \mathscr{P}_\eta^{-1} - \mathscr{P}_{\eta_0}^{-1} \|_G \leqq c \| \mathscr{P}_\eta - \mathscr{P}_{\eta_0} \|_F, \quad F = \mathscr{L}(H^{2m}(\mathbf{R}_+); L^2(\mathbf{R}_+) \times \mathbf{C}^m),$$

and $\| \mathscr{P}_\eta - \mathscr{P}_{\eta_0} \|_F \to 0$ as $\eta \to \eta_0$ (for A and the B_j's are polynomials in η). □

Remark 4.2. In order to point out the role played in the theory of this chapter by the hypothesis that A be *properly* elliptic and by the hypothesis that $\{B_j\}_{j=0}^{m-1}$ *covers* A (see also Section 8.4, further on), we note that the hypothesis on the roots of $A(\eta, \tau)$ in I) and hypothesis II) are also *necessary* for the validity of Lemma 4.1.

Indeed, assume first that I) is satisfied, but that II) is not satisfied for some $\eta \neq 0$; then, as we have seen in the proof of Proposition 4.2, there exists a $u \in \mathscr{S}(\mathbf{R}_+)$, which satisfies (4.8) and (4.10) with $c_j = 0$, $j = 0, \ldots, m-1$ and which is not identically zero; which contradicts the lemma.

Next, if we assume A to be elliptic, but not properly elliptic, then we may assume that, for some $\eta \neq 0$, there are $m' > m$ roots $\tau_i^+(\eta)$ of $A(\eta, \tau)$. We set $M^+(\eta, \tau) = \prod^{m'} (\tau - \tau_i^+(\eta))$ and, because $m < m'$, there exists a function $u(t) \in \mathscr{S}(\mathbf{R}_+)$ and $\neq 0$ which satisfies the equation $M^+\left(\eta, \dfrac{1}{i}\dfrac{d}{dt}\right) u(t) = 0$, $t > 0$, and consequently also (4.8) and the boundary conditions (4.10) with $c_j = 0$, q.e.d. □

Further on, we shall make use of inequality (4.14) to obtain a priori estimates in the spaces $H^s(\mathbf{R}^n_+)$ with $s \geqq 2m$. But, we shall also need certain "dual" estimates; for this reason we shall now study the adjoint mapping $\mathscr{P}_\eta^*$ of $\mathscr{P}_\eta$.

$\mathscr{P}_\eta^*$ is a continuous linear mapping of $L^2(\mathbf{R}_+) \times \mathbf{C}^m$ into $(H^{2m}(\mathbf{R}_+))'$, defined by the formula

(4.15)

$$\langle \varphi, \overline{\mathscr{P}_\eta^* \Psi} \rangle = \langle \mathscr{P}_\eta \varphi, \overline{\Psi} \rangle, \quad \forall \varphi \in H^{2m}(\mathbf{R}_+) \text{ and } \Psi \in L^2(\mathbf{R}_+) \times \mathbf{C}^m,$$

where Ψ denotes the element $\{\psi; c_0, \ldots, c_{m-1}\}$ of $L^2(\mathbf{R}_+) \times \mathbf{C}^m$ and the brackets denote the appropriate dualities.

But $(H^{2m}(\mathbf{R}_+))'$ may be identified (see Chapter 1, Section 12.6) with $H^{-2m}_{\overline{\mathbf{R}}_+}(\mathbf{R})$. Then, we denote by φ_0 a fixed extension of the functions φ

of $H^{2m}(\mathbf{R}_+)$ to $H^{2m}(\mathbf{R})$ (such an extension exists and depends linearly and continuously on φ; see Chapter 1, Section 2.2), $\tilde{\psi}$ the extension of $\psi \in L^2(\mathbf{R}_+)$ by zero for $t < 0$, δ the Dirac measure at the origin; we may write (4.15) in the form

$$(4.16) \quad \langle \varphi, \overline{\mathscr{P}_\eta^* \Psi} \rangle$$

$$= \left\langle A\left(\eta, \frac{1}{i}\frac{d}{dt}\right) \varphi_0, \bar{\tilde{\psi}} \right\rangle + \sum_{j=0}^{m-1} \left\langle B_j\left(\eta, \frac{1}{i}\frac{d}{dt}\right) \varphi_0(t)\Big|_{t=0}, \bar{c}_j \right\rangle^{((1))}$$

$$= \left\langle \varphi_0, \overline{A^*\left(\eta, \frac{1}{i}\frac{d}{dt}\right) \tilde{\psi}} \right\rangle + \sum_{j=0}^{m-1} \left\langle B_j\left(\eta, \frac{1}{i}\frac{d}{dt}\right) \varphi_0, \overline{c_j \delta} \right\rangle$$

$$= \left\langle \varphi_0, \overline{A^*\left(\eta, \frac{1}{i}\frac{d}{dt}\right) \tilde{\psi}} \right\rangle + \sum_{j=0}^{m-1} \left\langle \varphi_0, \overline{B_j^*\left(\eta, \frac{1}{i}\frac{d}{dt}\right) c_j \delta} \right\rangle,$$

where A^* and B_j^* are the formal adjoints of A and B_j (as differential operators in t) and the brackets denote the appropriate dualities. Thus, $\mathscr{P}_\eta^*$ may be defined by

$$(4.17) \quad \mathscr{P}_\eta^* \Psi = A^*\left(\eta, \frac{1}{i}\frac{d}{dt}\right) \tilde{\psi}(t) + \sum_{j=0}^{m-1} B_j^*\left(\eta, \frac{1}{i}\frac{d}{dt}\right) c_j \delta.$$

From Lemma 4.1, we deduce the estimate

$$(4.18) \quad \|\Psi\|_{L^2(\mathbf{R}_+)\times \mathbf{C}^m} \leqq C_\eta \|\mathscr{P}_\eta^* \Psi\|_{H^{-2m}(\mathbf{R})} \qquad \forall \Psi \in L^2(\mathbf{R}_+) \times \mathbf{C}^m.$$

4.3 First Application: Proof of Theorem 2.2

Before going on with the study of a priori estimates in the half-space, we shall give an application of Lemma 4.1 by proving Theorem 2.2.

Let A and $\{B_j\}_{j=0}^{m-1}$ satisfy the hypotheses of Theorem 2.2. To prove the theorem, *it is obviously sufficient to prove that the system* $\{C_j\}_{j=0}^{m-1}$ *covers* A^*, where $\{C_j\}_{j=0}^{m-1}$ is given by Green's formula (2.3).

Via "local maps" and "partition of unity", as in Section 2.3, 1) (the notion of *covering* having a *local* character), we are brought back to the case of the half-ball σ_+ and to Green's formula (2.17), using the notation of Section 2.3.

Therefore, for every u and v belonging to $\mathscr{D}(\overline{\mathbf{R}_+^n})$ and having support in $\sigma_+ \cup \partial_1 \sigma_+$, we have

$$(4.19) \quad \int_{\mathbf{R}_+^n} (\mathscr{A} u)\, \bar{v}\, dy\, dt - \int_{\mathbf{R}_+^n} u\, \overline{\mathscr{A}^* v}\, dy\, dt$$

$$= \sum_{j=0}^{m-1} \int_{\mathbf{R}^{n-1}} \mathscr{S}_j u\, \overline{\mathscr{C}_j v}\, dy - \sum_{j=0}^{m-1} \int_{\mathbf{R}^{n-1}} \mathscr{B}_j u\, \overline{\mathscr{T}_j v}\, dy$$

((1)) We use the fact that $c\,\delta \in H^{-2m+m_j}(\mathbf{R})$ and that if $v \in H^{2m-m_j}(\mathbf{R})$, then $\langle v, c\,\delta\rangle = \langle v(0), c\rangle$, the first bracket denoting the duality between $H^{2m-m_j}(\mathbf{R})$ and $H^{-2m+m_j}(\mathbf{R})$ and the second bracket denoting the product $v(0)\,c$.

and the problem is to show that $\{\mathscr{C}_j\}_{j=0}^{m-1}$ covers $\mathscr{A}^*$, under the hypothesis that $\{\mathscr{B}_j\}_{j=0}^{m-1}$ covers $\mathscr{A}$.

Let λ be a real number, $\lambda > 1$: then the functions $u(\lambda y, \lambda t)$ and $v(\lambda y, \lambda t)$ are still in $\mathscr{D}(\overline{\mathbf{R}^n_+})$ and have support in $\sigma_+ \cup \partial_1 \sigma_+$; therefore, we may apply (4.19) to $u(\lambda y, \lambda t)$ and $v(\lambda y, \lambda t)$ and, after some easy calculations, we find, in the obvious notation,

$$\sum_{|p|=2m} \int_{\mathbf{R}^n_+} a_p\left(\frac{y}{\lambda}, \frac{t}{\lambda}\right)(D^p u(y,t))\,\overline{v(y,t)}\,dy\,dt -$$

$$-\sum_{j=0}^{m-1} \int_{\mathbf{R}^{n-1}} \left\{ \sum_{|h|=\mu_j} s_{jh}\left(\frac{y}{\lambda}\right) D^h u(y,t)\Big|_{t=0} \times \right.$$

$$\left. \times \overline{\sum_{|h|=2m-1-\mu_j} c_{jh}\left(\frac{y}{\lambda}\right) D^h v(y,t)}\Big|_{t=0} \right\} dy +$$

$$+\sum_{j=0}^{m-1} \int_{\mathbf{R}^{n-1}} \left\{ \sum_{|h|=m_j} b_{jh}\left(\frac{y}{\lambda}\right) D^h u(y,t)\Big|_{t=0} \times \right.$$

$$\left. \times \overline{\sum_{|h|=2m-1-m_j} t_{jh}\left(\frac{y}{\lambda}\right) D^h v(y,t)}\Big|_{t=0} \right\} dy = 0\left(\frac{1}{\lambda}\right).$$

Then, letting $\lambda \to +\infty$, we obtain

$$\text{(4.20)} \quad \int_{\mathbf{R}^n_+} (\mathscr{A}_0(0,D)\,u(y,t))\,\overline{v(y,t)}\,dy\,dt - \int_{\mathbf{R}^n_+} u(y,t)\,\overline{\mathscr{A}_0^*\,v(y,t)}\,dy\,dt =$$

$$= \sum_{j=0}^{m-1} \int_{\mathbf{R}^{n-1}} \mathscr{S}_{j,0}(0,D)\,u(y,t)\Big|_{t=0} \overline{\mathscr{C}_{j,0}(0,D)\,v(y,t)}\Big|_{t=0} dy -$$

$$- \sum_{j=0}^{m-1} \int_{\mathbf{R}^{n-1}} \mathscr{B}_{j,0}(0,D)\,u(y,t)\Big|_{t=0} \overline{\mathscr{T}_{j,0}(0,D)\,v(y,t)}\Big|_{t=0} dy,$$

where $\mathscr{A}_0$ and $\mathscr{A}_0^*$, $\mathscr{B}_{j,0}$, $\mathscr{C}_{j,0}$, $\mathscr{S}_{j,0}$, $\mathscr{T}_{j,0}$ have constant coefficients and are homogeneous of suitable degree: "by homogeneity", we obtain without difficulty that (4.20) is also valid for arbitrary u and v in $\mathscr{D}(\overline{\mathbf{R}^n_+})$.

Now, let $\varphi(t)$ and $\psi(t)$ be given functions in $\mathscr{D}(\overline{\mathbf{R}}_+)$, $\chi(y)$ be given in $\mathscr{D}(\mathbf{R}^{n-1})$ and real $\lambda > 0$; we apply (4.20) to the functions

$$u(y,t) = e^{i\langle \eta, y\rangle}\,\chi(\lambda y)\,\varphi(t)$$

and

$$v(y,t) = e^{i\langle \eta, y\rangle}\,\chi(\lambda y)\,\psi(y)$$

for all $\eta \in \mathbf{R}^{n-1}$ and $\neq 0$. It follows (write

$$\mathscr{A}_0(0, D) = \mathscr{A}_0(0, D_y, D_t), \ldots)$$

that

$$\Bigg\{ \int_0^{+\infty} \left(\mathscr{A}_0\left(0, \mathrm{i}\,\eta, \frac{d}{dt}\right) \varphi(t) \right) \overline{\psi(t)}\, dt - $$

$$ - \overline{\int_0^{+\infty} \varphi(t)\, \mathscr{A}_0^*\left(0, \mathrm{i}\,\eta, \frac{d}{dt}\right) \psi(t)\, dt} \Bigg\} \int_{\mathbf{R}^{n-1}} |\chi(y)|^2\, dy - $$

$$ - \Bigg\{ \sum_{j=0}^{m-1} \mathscr{S}_{j,0}\left(0, \mathrm{i}\,\eta, \frac{d}{dt}\right) \varphi(t)\Big|_{t=0} \overline{\mathscr{C}_{j,0}\left(0, \mathrm{i}\,\eta, \frac{d}{dt}\right) \psi(t)\Big|_{t=0}} - $$

$$ - \sum_{j=0}^{m-1} \mathscr{B}_{j,0}\left(0, \mathrm{i}\,\eta, \frac{d}{dt}\right) \varphi(t)\Big|_{t=0} \overline{\mathscr{T}_{j,0}\left(0, \mathrm{i}\,\eta, \frac{d}{dt}\right) \psi(t)\Big|_{t=0}} \Bigg\} \times $$

$$ \times \int_{\mathbf{R}^{n-1}} |\chi(y)|^2\, dy = 0(\lambda). $$

Letting $\lambda \to 0$, we obtain

$$\int_0^{+\infty} \left(\mathscr{A}_0\left(0, \mathrm{i}\,\eta, \frac{d}{dt}\right) \varphi(t) \right) \overline{\psi(t)}\, dt - \overline{\int_0^{+\infty} \varphi(t)\, \mathscr{A}_0^*\left(0, \mathrm{i}\,\eta, \frac{d}{dt}\right) \psi(t)\, dt}$$

$$(4.21) \qquad = \sum_{j=0}^{m-1} \mathscr{S}_{j,0}\left(0, \mathrm{i}\,\eta, \frac{d}{dt}\right) \varphi(t)\Big|_{t=0} \overline{\mathscr{C}_{j,0}\left(0, \mathrm{i}\,\eta, \frac{d}{dt}\right) \psi(t)\Big|_{t=0}} - $$

$$ - \sum_{j=0}^{m-1} \mathscr{B}_{j,0}\left(0, \mathrm{i}\,\eta, \frac{d}{dt}\right) \varphi(t)\Big|_{t=0} \overline{\mathscr{T}_{j,0}\left(0, \mathrm{i}\,\eta, \frac{d}{dt}\right) \psi(t)\Big|_{t=0}} $$

for all $\varphi, \psi \in \mathscr{D}(\overline{\mathbf{R}_+})$ and therefore also for all $\varphi, \psi \in H^{2m}(\mathbf{R}_+)$, since $\mathscr{D}(\overline{\mathbf{R}^+})$ is dense in $H^{2m}(\mathbf{R}_+)$.

At this point, to show that $\{\mathscr{C}_j\}_{j=0}^{m-1}$ covers $\mathscr{A}^*$, it is sufficient (for conditions II″) and II) of Section 4.1 are equivalent (Proposition 4.2)) to show that the differential problem

$$(4.22) \qquad \begin{cases} \mathscr{A}^*\left(0, \mathrm{i}\,\eta, \dfrac{d}{dt}\right) \psi(t) = 0, & t > 0, \\ \mathscr{C}_{j,0}\left(0, \mathrm{i}\,\eta, \dfrac{d}{dt}\right) \psi(t)\Big|_{t=0} = 0, & j = 0, \ldots, m-1, \end{cases}$$

admits only the solution $\psi(t) = 0$ in $H^{2m}(\mathbf{R}_+)$ (and therefore also in $\mathscr{S}(\mathbf{R}_+)$). Therefore, let $\psi \in H^{2m}(\mathbf{R}_+)$ be a solution of (4.22). We may apply (4.21) with arbitrary $\varphi \in H^{2m}(\mathbf{R}_+)$ and obtain

$$\text{(4.23)} \quad \int_0^{+\infty} \left(\mathscr{A}_0\left(0, \mathrm{i}\eta, \frac{d}{dt}\right)\varphi(t)\right)\overline{\psi(t)}\,dt$$
$$= -\sum_{j=0}^{m-1} \mathscr{B}_{j,0}\left(0, \mathrm{i}, \eta, \frac{d}{dt}\right)\varphi(t)\Big|_{t=0} \overline{\mathscr{T}_{j,0}\left(0, \mathrm{i}\eta, \frac{d}{dt}\right)\psi(t)\Big|_{t=0}}.$$

But the system $\{B_j\}_{j=0}^{m-1}$ covers $\mathscr{A}$ and therefore $\mathscr{A}_0$ and $\{\mathscr{B}_{j,0}\}$ satisfy hypotheses I) and II) of Section 4.1; therefore, we may apply Lemma 4.1, which shows that the problem

$$\mathscr{A}_0\left(0, \mathrm{i}\eta, \frac{d}{dt}\right)\varphi(t) = f(t), \quad t > 0,$$

$$\mathscr{B}_{j,0}\left(0, \mathrm{i}\eta, \frac{d}{dt}\right)\varphi(t)\Big|_{t=0} = 0, \quad j = 0, \ldots, m-1,$$

admits, for every $f \in L^2(\mathbf{R}_+)$, a unique solution in $H^{2m}(\mathbf{R}_+)$. Therefore, from (4.23), we deduce that

$$\int_0^{+\infty} \overline{f\psi}\,dt = 0 \qquad \forall f \in L^2(\mathbf{R}_+)$$

and therefore $\psi = 0$. □

4.4 A priori Estimates in the Half-Space for the Case of Constant Coefficients

Now, let us again consider the operators A and $\{B_j\}_{j=0}^{m-1}$ given by (4.1) and (4.2) under the hypotheses of Section 4.1 [i.e. (4.3), I) and II)]; we consider the operator

$$\text{(4.24)} \quad \mathscr{P}\colon u \to \mathscr{P}u = \{A(D)u, B_0(D)u|_{t=0}, \ldots, B_{m-1}(D)u|_{t=0}\},$$

which according to the trace theorem of Chapter 1, Section 8 is a continuous linear mapping of $H^{2m}(\mathbf{R}^n_+)$ into

$$L^2(\mathbf{R}^n_+) \times \prod_{j=0}^{m-1} H^{2m-m_j-1/2}(\mathbf{R}^{n-1}).$$

We denote by $\mathscr{P}^*$ the adjoint of $\mathscr{P}$ which maps

$$L^2(\mathbf{R}^n_+) \times \prod_{j=0}^{m-1} H^{-2m+m_j+1/2}(\mathbf{R}^{n-1}) \quad \text{into} \quad (H^{2m}(\mathbf{R}^n_+))'.$$

Then, we proceed in the same way as we did to obtain (4.17). If

$$F = \{f; g_0, \ldots, g_{m-1}\} \in L^2(\mathbf{R}^n_+) \times \prod_{j=0}^{m-1} H^{-2m+m_j+1/2}(\mathbf{R}^{n-1})$$

and

$$\varphi \in H^{2m}(\mathbf{R}^n_+),$$

we have

$$\langle \mathscr{P}^* F, \bar{\varphi}\rangle = \langle F, \overline{\mathscr{P}\varphi}\rangle = \langle f, \overline{A\varphi}\rangle + \sum_{j=0}^{m-1} \langle g_j, \overline{B_j \varphi|_{t=0}}\rangle$$

$$= \langle \tilde{f}, \overline{A\varphi_0}\rangle + \sum_{j=0}^{m-1} \langle g_j \otimes \delta(t), \overline{B_j \varphi_0}\rangle$$

$$= \langle A^* \tilde{f}, \overline{\varphi_0}\rangle + \sum_{j=0}^{m-1} \langle B_j^*(g_j \otimes \delta(t)), \overline{\varphi_0}\rangle$$

where $\tilde{f}$ is the extension of f by zero for $t < 0$, $\delta(t)$ is the Dirac measure at zero, φ_0 is an extension of φ, $\varphi \to \varphi_0$ being a continuous linear mapping of $H^{2m}(\mathbf{R}^n_+)$ into $H^{2m}(\mathbf{R}^n)$.

Then

$$\mathscr{P}^* F = A^*(D)\tilde{f} + \sum_{j=0}^{m-1} B_j^*(D)(g_j(y) \otimes \delta(t)), \tag{4.25}$$

where A^* and B_j^* are the formal adjoints of A and B_j.

Then, we have

Theorem 4.1. *Under hypotheses* I) *and* II), *the following inequalities hold:*

$$\|u\|_{H^{2m}(\mathbf{R}^n_+)} \leqq C\{\|\mathscr{P}u\|_{L^2(\mathbf{R}^n_+)\times\prod_{j=0}^{m-1} H^{2m-m_j-1/2}(\mathbf{R}^{n-1})} + \|u\|_{H^{2m-1}(\mathbf{R}^n_+)}\} \quad \forall u \in H^{2m}(\mathbf{R}^n_+) \tag{4.26}$$

and, with $\mathscr{P}^*$ *given by* (4.25),

$$\|F\|_{L^2(\mathbf{R}^n_+)\times\prod_{j=0}^{m-1} H^{-2m+m_j+1/2}(\mathbf{R}^{n-1})} \leqq C\left\{\|\mathscr{P}^* F\|_{H^{-2m}(\mathbf{R}^n)} + \|\tilde{f}\|_{H^{-1}(\mathbf{R}^n)} + \sum_{j=0}^{m-1} \|g_j\|_{H^{-2m+m_j-1/2}(\mathbf{R}^{n-1})}\right\} \tag{4.27}$$

$$\forall F = \{f; g, \ldots, g_{m-1}\} \in L^2(\mathbf{R}^n_+) \times \prod_{j=0}^{m-1} H^{-2m+m_j+1/2}(\mathbf{R}^{n-1}),$$

C denoting a constant (independent of u and F).

Proof. 1) C shall denote various constants.

Again, we use the notation of Lemma 4.1 and the inequality (4.14) for $|\eta| = 1$; C_η being a continuous function of η, we obtain, writing out the norms explicitly:

$$\sum_{l=0}^{2m} \int_0^{+\infty} |\varphi^{(l)}(t)|^2\, dt \leqq C\left\{ \int_0^{+\infty} \left| A\left(\eta, \frac{1}{\mathrm{i}}\frac{d}{dt}\right)\varphi(t)\right|^2 dt + \right.$$

$$\left. + \sum_{j=0}^{m-1} \left| B_j\left(\eta, \frac{1}{\mathrm{i}}\frac{d}{dt}\right)\varphi(t)\Big|_{t=0}\right|^2 \right\} \qquad \forall \varphi \in H^{2m}(\mathbf{R}_+), \quad |\eta| = 1,$$

C being independent (of φ and) of η.

Since the operators A and B_j are homogeneous of degree $2m$ and m_j respectively, we obtain, *for arbitrary* $\eta \neq 0$,

$$\sum_{l=0}^{2m} |\eta|^{2(2m-l)} \int_0^{+\infty} |\varphi^{(l)}(t)|^2\, dt \leqq C\left\{ \int_0^{+\infty} \left| A\left(\eta, \frac{1}{\mathrm{i}}\frac{d}{dt}\right)\varphi(t)\right|^2 dt + \right.$$

$$\left. + \sum_{j=0}^{m-1} |\eta|^{2(2m-m_j-1/2)} \left| B_j\left(\eta, \frac{1}{\mathrm{i}}\frac{d}{dt}\right)\varphi(t)\Big|_{t=0}\right|^2 \right\},$$

C being independent of η.

We apply this inequality with $\varphi(t)$ replaced by $\hat{u}(\eta, t)$ (Fourier transform of $u(y, t)$ with respect to y), with $u \in H^{2m}(\mathbf{R}^n_+)$; then, setting $A(D)\, u(y, t) = f(y, t)$ and $B_j(D)\, u(y, t)\,|_{t=0} = g_j(y)$ for the sake of simplicity, we obtain

$$\sum_{l=0}^{m} |\eta|^{2(2m-l)} \int_0^{+\infty} \left| \frac{\partial^l \hat{u}(\eta, t)}{\partial t^l}\right|^2 dt$$

$$\leqq C\left\{ \int_0^{+\infty} |\hat{f}(\eta, t)|^2\, dt + \sum_{j=0}^{m-1} |\eta|^{2(2m-m_j-1/2)} |\hat{g}_j(\eta)|^2 \right\}$$

from which, integrating over $|\eta| \geqq 1$:

$$\sum_{l=0}^{2m} \int_{|\eta|\geqq 1} \int_0^{+\infty} (1 + |\eta|^2)^{2m-l} \left| \frac{\partial^l u(\eta, t)}{\partial t^l}\right|^2 dt\, d\eta \leqq \tag{4.28}$$

$$\leqq C\left\{ \int_{|\eta|\geqq 1} \int_0^{+\infty} |\hat{f}(\eta, t)|^2\, d\eta\, dt + \right.$$

$$\left. + \sum_{j=0}^{m-1} \int_{|\eta|\geqq 1} (1 + |\eta|^2)^{2m-m_j-1/2} |\hat{g}_j(\eta)|^2 \right\}.$$

Now, consider the integrals over $|\eta| \leqq 1$. We obviously have

$$(4.29) \quad \sum_{l=0}^{2m-1} \int_{|\eta|\leqq 1} \int_0^{+\infty} (1 + |\eta|^2)^{2m-l} \left| \frac{\partial^l \hat{u}(\eta, t)}{\partial t^l} \right|^2 dt\, d\eta$$

$$\leqq C \sum_{l=0}^{2m-1} \int_{|\eta|\leqq 1} \int_0^{+\infty} \left| \frac{\partial^l \hat{u}(\eta, t)}{\partial t^l} \right|^2 dt\, d\eta \leqq C \, \|u\|^2_{H^{2m-1}(\mathbf{R}^n_+)} .$$

Furthermore, since A is elliptic, we may write it in the form

$$A\left(\eta, \frac{\partial}{\partial t}\right) = \alpha_0 \frac{\partial^{2m}}{\partial t^{2m}} + \sum_{l=0}^{2m-1} \alpha_{2m-l}(\eta) \frac{\partial^l}{\partial t^l},$$

with $\alpha_0 \neq 0$ and the $\alpha_{2m-l}(\eta)$'s being continuous functions of η; therefore, we have

$$\alpha_0 \frac{\partial^{2m} \hat{u}(\eta, t)}{\partial t^{2m}} = \hat{f}(\eta, t) - \sum_{l=0}^{2m-1} \alpha_{2m-l}(\eta) \frac{\partial^l \hat{u}(\eta, t)}{\partial t^l},$$

whence

$$(4.30) \quad \int_{|\eta|\leqq 1} \int_0^{+\infty} \left| \frac{\partial^{2m} \hat{u}(\eta, t)}{\partial t^{2m}} \right|^2 dt\, d\eta \leqq C \left\{ \int_{|\eta|\leqq 1} \int_0^{+\infty} |\hat{f}(\eta, t)|^2 \, dt\, d\eta + \right.$$

$$\left. + \sum_{l=0}^{2m-1} \int_{|\eta|\leqq 1} \int_0^{+\infty} \left| \frac{\partial^l \hat{u}(\eta, t)}{\partial t^l} \right|^2 dt\, d\eta \right\}$$

$$\leqq C \{ \|f\|_{L^2(\mathbf{R}^n_+)} + \|u\|_{H^{2m-1}(\mathbf{R}^n_+)} \} .$$

Then (4.26) follows from (4.28) and (4.30).

2) Now, we prove (4.27). In the notation of Lemma 4.1, we deduce from (4.17) and (4.18) that

$$\int_0^{+\infty} |\psi(t)|^2 \, dt + \sum_{j=0}^{m-1} |c_j|^2 \leqq C \int_{-\infty}^{+\infty} (1 + |t'|^2)^{-2m} \Big| \mathrm{i}^{2m} A^*(\eta, t') \times$$

$$\times \int_0^{+\infty} \mathrm{e}^{-\mathrm{i}t't} \psi(t) \, dt + \sum_{j=0}^{m-1} \mathrm{i}^{m_j} B_j^*(\eta, t') \, c_j \Big|^2 dt'$$

$$\forall \Psi = \{\psi; c_0, \ldots, c_{m-1}\} \in L^2(\mathbf{R}_+) \times \mathbf{C}^m \quad \text{and} \quad \text{for } |\eta| = 1,$$

where C is independent of η.

The operators A^* and B_j^* also being homogeneous, we obtain, *for arbitrary* $\eta \neq 0$,

$$\int_0^{+\infty} |\psi(t)|^2 \, dt + \sum_{j=0}^{m-1} |\eta|^{2(m_j - 2m + 1/2)} |c_j|^2 \leqq C \left\{ \int_{-\infty}^{+\infty} (|\eta| + |t'|^2)^{-2m} \times \right.$$

$$\left. \times \left| i^{2m} A^*(\eta, t') \int_0^{+\infty} e^{-it't} \psi(t) \, dt + \sum_{j=0}^{m-1} i^{m_j} B_j^*(\eta, t') c_j \right|^2 dt' \right\}.$$

We apply this inequality with $\psi(t)$ replaced by $\hat{f}(\eta, t)$ and c_j by $\hat{g}_j(\eta)$, with $f \in L^2(\mathbf{R}_+^n)$ and $g_j \in H^{m_j - 2m + 1/2}(\mathbf{R}^{n-1})$; setting

$$v(\eta, t') = \frac{1}{\sqrt{2\pi}} \int_0^{+\infty} e^{-it't} \hat{f}(\eta, t) \, dt = \frac{1}{\sqrt{2\pi}} \int_{-\infty}^{+\infty} e^{it't} \hat{\tilde{f}}(\eta, t) \, dt,$$

we have

$$\int_0^{+\infty} |\hat{f}(\eta, t)|^2 \, dt + \sum_{j=0}^{m-1} |\eta|^{2(m_j - 2m + 1/2)} |\hat{g}_j(\eta)|^2$$

$$\leqq C \left\{ \int_{-\infty}^{+\infty} (|\eta|^2 + |t'|^2)^{-2m} \cdot \left| i^{2m} A^*(\eta, t') v(\eta, t') + \right. \right.$$

$$\left. \left. + \sum_{j=0}^{m-1} i^{m_j} B_j^*(\eta, t') \hat{g}_j(\eta) \right|^2 dt' \right\}.$$

But the expression

$$w(\eta, t') = i^{2m} A^*(\eta, t') v(\eta, t') + \sum_{j=0}^{m-1} i^{m_j} B_j^*(\eta, t') \hat{g}_j(\eta)$$

is the Fourier transform in the variables (y, t) of $\mathscr{P}^* F$ for

$$F = \{f; g_0, \ldots, g_{m-1}\};$$

thus integrating over $|\eta| \geqq 1$, we obtain

$$(4.31) \quad \int_{|\eta| \geqq 1} \int_0^{+\infty} |\hat{f}(\eta, t)^2| \, dt \, d\eta + \sum_{j=0}^{m-1} \int_{|\eta| \geqq 1} (1 + |\eta|^2)^{m_j - 2m + 1/2} |g_j(\eta)|^2 \, d\eta$$

$$\leqq C \int_{|\eta| \geqq 1} \int_0^{+\infty} (1 + |\eta|^2 + |t'|^2)^{-2m} |w(\eta, t')|^2 \, dt' \, d\eta$$

$$\leqq C \, \| \mathscr{P}^* F \|^2_{H^{-2m}(\mathbf{R}^n)} .$$

Now, we consider the integrals over $|\eta| \leqq 1$.

First of all, since $|\eta| \leqq 1$,

(4.32)
$$\sum_{j=0}^{m-1} \int_{|\eta|\leqq 1} (1 + |\eta|^2)^{m_j - 2m + 1/2} |\hat{g}_j(\eta)|^2 \, d\eta$$
$$\leqq C \sum_{j=0}^{m-1} \int_{|\eta|\leqq 1} (1 + |\eta|^2)^{m_j - 2m - 1/2} |\hat{g}_j(\eta)|^2 \, d\eta$$
$$\leqq C \sum_{j=0}^{m-1} \| g_j \|^2_{H^{m_j - 2m - 1/2}(\mathbf{R}^n)}.$$

By Fourier transform in t, we immediately see that $\left(\dfrac{d}{dt} + I\right)^{2m}$ is an isomorphism of $L^2(\mathbf{R})$ onto $H^{-2m}(\mathbf{R})$; therefore, we have

$$\int_0^{+\infty} |\hat{f}(\eta, t)|^2 \, dt \leq C \left\| \left(\frac{\partial}{\partial t} + I\right)^{2m} \tilde{\hat{f}}(\eta, t) \right\|^2_{H^{-2m}(\mathbf{R})}$$
$$\leqq C \left\{ \left\| \frac{\partial^{2m} \tilde{\hat{f}}(\eta, t)}{\partial t^{2m}} \right\|^2_{H^{-2m}(\mathbf{R})} + \| \tilde{\hat{f}}(\eta, t) \|^2_{H^{-1}(\mathbf{R})} \right\},$$

where, as always, $\tilde{\hat{f}}(\eta, t)$ denotes the extension of $\hat{f}(\eta, t)$ by zero for $t < 0$. Hence

(4.33)
$$\int_{|\eta|\leqq 1} \int_0^{+\infty} |\hat{f}(\eta, t)|^2 \, dt \, d\eta$$
$$\leqq C \left\{ \int_{|\eta|\leqq 1} \int_{-\infty}^{+\infty} (1 + |t'|^2)^{-1} |v(\eta, t')|^2 \, dt' \, d\eta + \right.$$
$$\left. + \int_{|\eta|\leqq 1} \left\| \frac{\partial^{2m} \tilde{\hat{f}}(\eta, t)}{\partial t^{2m}} \right\|^2_{H^{-2m}(\mathbf{R})} d\eta \right\}$$
$$\leqq C \left\{ \int_{|\eta|\leqq 1} \int_{-\infty}^{+\infty} (1 + |\eta|^2 + |t'|^2)^{-1} |v(\eta, t')|^2 \, dt' \, d\eta + \right.$$
$$\left. + \int_{|\eta|\leqq 1} \left\| \frac{\partial^{2m} \tilde{\hat{f}}(\eta, t)}{\partial t^{2m}} \right\|^2_{H^{-2m}(\mathbf{R})} d\eta \right\}$$
$$\leqq C \left\{ \| \hat{f} \|^2_{H^{-1}(\mathbf{R}^n)} + \int_{|\eta|\leqq 1} \left\| \frac{\partial^{2m} \tilde{\hat{f}}(\eta, t)}{\partial t^{2m}} \right\|^2_{H^{-2m}(\mathbf{R})} d\eta \right\}.$$

In order to estimate the last term in (4.33), we express $\frac{\partial^{2m}\tilde{f}}{\partial t^{2m}}$ with the aid of $\mathscr{P}^* F$; indeed, we may write $A^*(\eta, t')$ in the form

$$A^*(\eta, t') = \alpha_0^* t'^{2m} + \sum_{l=0}^{2m-1} \alpha_{2m-l}^*(\eta)\, t'^l,$$

with $\alpha_0^* \neq 0$ and $\alpha_{2m-l}^*(\eta)$ continuous functions of η. Therefore

$$\begin{aligned}\alpha_0^* (\mathrm{i}\, t')^{2m} v(\eta, t') &= \mathrm{i}^{2m} A^*(\eta, t')\, v(\eta, t') - \sum_{l=0}^{2m-1} \mathrm{i}^{2m} \alpha_{2m-l}^*(\eta)\, t'^l\, v(\eta, t') \\ &= w(\eta, t') - \sum_{j=0}^{m-1} \mathrm{i}^{m_j} B_j^*(\eta, t')\, \hat{g}_j(\eta) - \\ &\qquad - \sum_{l=0}^{2m-1} \mathrm{i}^{2m} \alpha_{2m-l}^*(\eta)\, t'^l\, v(\eta, t'),\end{aligned}$$

from which we obtain

$$\begin{aligned}(4.34)\quad &\int_{|\eta| \leqq 1} \left\| \frac{\partial^{2m}\tilde{f}(\eta, t)}{\partial t^{2m}} \right\|^2_{H^{-2m}(\mathbf{R})} d\eta \\ &= \int_{|\eta| \leqq 1} \int_{-\infty}^{+\infty} (1 + |t'|^2)^{-2m} |(\mathrm{i}\, t')^{2m} v(\eta, t')|^2\, dt'\, d\eta \\ &\leqq C \left\{ \int_{|\eta| \leqq 1} \int_{-\infty}^{+\infty} (1 + |\eta|^2 + |t'|^2)^{-2m} |w(\eta, t')|^2\, dt'\, d\eta + \right. \\ &\quad + \sum_{l=0}^{2m-1} \int_{|\eta| \leqq 1} \int_{-\infty}^{+\infty} (1 + |\eta|^2 + |t'|^2)^{-2m} |t'^l\, v(\eta, t')|^2\, dt'\, d\eta + \\ &\quad \left. + \sum_{j=0}^{m-1} \int_{|\eta| \leqq 1} \int_{-\infty}^{+\infty} (1 + |t'|^2)^{m_j - 2m} |\hat{g}_j(\eta)|^2\, d\eta\, dt' \right\} \\ &\leqq C \left\{ \| \mathscr{P}^* F \|^2_{H^{-2m}(\mathbf{R}^n)} + \| \tilde{f} \|^2_{H^{-1}(\mathbf{R}^n)} + \sum_{j=0}^{m-1} \| g_j \|^2_{H^{-2m+m_j-1/2}(\mathbf{R}^{n-1})} \right\}.\end{aligned}$$

So that (4.27) follows from (4.31), (4.32), (4.33) and (4.34). □

4.5 A priori Estimates in the Half-Space for the Case of Variable Coefficients

Let us now consider the case of operators with variable coefficients, still in the half-space $\mathbf{R}^n_+$.

Therefore, still with the notation $x = (y, t)$ for the elements of $\mathbf{R}^n$, let

$$A(x, D) = \sum_{|p| \leq 2m} a_p(x) D^p \tag{4.35}$$

be a linear differential operator of order $2m$ with *infinitely differentiable coefficients which, together with all their derivatives, are bounded in* $\mathbf{R}^n$. We consider $A(x, D)$ as an operator in $\mathbf{R}^n_+$, but it is convenient to extend the coefficients of $A(x, D)$ to all of $\mathbf{R}^n$.

We denote by

$$A_0(x, D) = \sum_{|p| = 2m} a_p(x) D^p u$$

the homogeneous part of degree $2m$ of $A(x, D)$.

Let

$$B_j(y, D) = \sum_{|h| \leq m_j} b_{j,h}(y) D^h, \quad j = 0, \ldots, m-1, \tag{4.36}$$

be differential operators of order m_j, with $0 \leq m_j < 2m$ and with *infinitely differentiable coefficients which, together with all their derivatives, are bounded in* $\mathbf{R}^{n-1}$.

We denote by

$$B_{j,0}(y, D) = \sum_{|h| = m_j} b_{j,h}(y) D^h, \quad j = 0, \ldots, m-1,$$

the homogeneous part of degree m_j of B_j.

We shall assume that the *differential operators* $A_0(0, D)$ *and* $B_{j,0}(0, D)$ *with constant coefficients satisfy hypotheses* I) *and* II) *of Section* 4.1.

We denote by $\mathscr{P}$ the operator

$$\mathscr{P}: u \to \mathscr{P}u = \{A(x, D)u; B_0(y, D)u|_{t=0}, \ldots, B_{m-1}(y, D)u|_{t=0}\} \tag{4.37}$$

which, thanks to the hypotheses on the coefficients of A and B_j, is a continuous linear mapping of

$$H^{2m}(\mathbf{R}^n_+) \quad \text{into} \quad L^2(\mathbf{R}^n_+) \times \prod_{j=0}^{m-1} H^{2m-m_j-1/2}(\mathbf{R}^{n-1}).$$

We denote by $\mathscr{P}^*$ the adjoint of $\mathscr{P}$, which maps

$$L^2(\mathbf{R}^n_+) \times \prod_{j=0}^{m-1} H^{-2m+m_j+1/2}(\mathbf{R}^{n-1}) \quad \text{into} \quad (H^{2m}(\mathbf{R}^n_+))' = H^{-2m}_{\overline{\mathbf{R}}^n_+}(\mathbf{R}^n)$$

and which may be defined, as for the case of constant coefficients, by

$$\mathscr{P}^* F = A^*(x, D)\tilde{f}(x) + \sum_{j=0}^{m-1} B_j^*(y, D)(g_j(y) \otimes \delta(t)) \tag{4.38}$$

$$\forall F = \{f; g_0, \ldots, g_{m-1}\} \in L^2(\mathbf{R}^n_+) \times \prod_{j=0}^{m-1} H^{2m-m_j-1/2}(\mathbf{R}^{n-1}).$$

A^* and B_j^* being the formal adjoints of A and B_j, and $\tilde{f}$ still having the same meaning.

Theorem 4.3. *Let A and $\{B_j\}_{j=0}^{m-1}$ be defined by* (4.35) *and* (4.36). *Assume that the operators $A_0(0, D)$ and $\{B_j(0, D)\}_{j=0}^{m-1}$ with constant coefficients satisfy hypotheses* I) *and* II) *of Section* 4.1; *then, for every fixed integer $r \geqq 0$, there exists a positive number ϱ_0 such that if $\varrho < \varrho_0$ we have*

i) *if $u \in H^{2m}(\mathbf{R}^n_+)$ and vanishes outside the ball $\sigma(\varrho)$ with center at the origin and radius ϱ and*

$$\mathscr{P} u \in H^r(\mathbf{R}_n^+) \times \prod_{j=0}^{m-1} H^{2m+r-m_j-1/2}(\mathbf{R}^{n-1})$$

then

$$u \in H^{2m+r}(\mathbf{R}^n_+)$$

and

$$(4.39)\quad \|u\|_{H^{2m+r}(\mathbf{R}^n_+)} \leqq C_{r,\varrho} \{\|\mathscr{P} u\|_{H^r(\mathbf{R}^n_+)\times \prod_{j=0}^{m-1} H^{2m+r-m_j-1/2}(\mathbf{R}^{n-1})} + \|u\|_{H^{2m+r-1}(\mathbf{R}^n_+)}\}$$

($C_{r,\varrho}$ *depending on r and ϱ*);

ii) *if $F = \{f; g_0, \ldots, g_{m-1}\} \in L^2(\mathbf{R}^n_+) \times \prod_{j=0}^{m-1} H^{-2m+m_j+1/2}(\mathbf{R}^{n-1})$ and vanishes outside $\sigma(\varrho)$* (i.e. *f vanishes outside $\sigma(\varrho)$ and $g_j(y)$ vanishes for $|y| > \varrho$*), *then*

$$(4.40)\quad \|F\|_{L^2(\mathbf{R}^n_+)\times \prod_{j=0}^{m-1} H^{-2m+m_j+1/2}(\mathbf{R}^{n-1})}$$
$$\leqq C_{0,\varrho} \left\{\|\mathscr{P}^* F\|_{H^{-2m}(\mathbf{R}^n)} + \|\tilde{f}\|_{H^{-1}(\mathbf{R}^n)} + \sum_{j=0}^{m-1} \|g_j\|_{H^{-2m+m_j-1/2}(\mathbf{R}^{n-1})}\right\},$$

where $C_{0,\varrho}$ is independent of F.

Proof. 1) We prove i). First we verify (4.39) for $r = 0$.

If $u \in H^{2m}(\mathbf{R}^n_+)$, we may apply Theorem 4.1 to the operators $A_0(0, D)$ and $B_{j,0}(0, D)$ and therefore we have

$$(4.41)\quad \|u\|_{H^{2m}(\mathbf{R}^n_+)} \leqq C \left\{\|A_0(0, D) u\|_{L^2(\mathbf{R}^n_+)} + \right.$$
$$\left. + \sum_{j=0}^{m-1} \|B_{j,0}(0, D) u\|_{H^{2m-m_j-1/2}(\mathbf{R}^{n-1})} + \|u\|_{H^{2m-1}(\mathbf{R}^n_+)}\right\}.$$

But, thanks to the hypotheses on the coefficients of $A(x, D)$ and $B_j(y, D)$, there exists $\varrho_0 > 0$ such that, for u with support in $\sigma(\varrho)$, $\varrho < \varrho_0$, we have

$$(4.42)\quad \|A_0(0, D) u - A(x, D) u\|_{L^2(\mathbf{R}^n_+)}$$
$$\leqq \|A_0(0, D) u - A_0(x, D) u\|_{L^2(\mathbf{R}^n_+)} +$$
$$+ C'_\varrho \|u\|_{H^{2m-1}(\mathbf{R}^n_+)} \leqq \sum_{|p|=2m} \|[a_p(0) - a_p(x)] D^p u\|_{L^2(\mathbf{R}^n_+)} +$$
$$+ C'_\varrho \|u\|_{H^{2m-1}(\mathbf{R}^n_+)} \leqq \frac{1}{4C} \|u\|_{H^{2m}(\mathbf{R}^n_+)} + C'_\varrho \|u\|_{H^{2m-1}(\mathbf{R}^n_+)}$$

and, using the trace theorem of Chapter 1, Section 8, also:

(4.43) $$\sum_{j=0}^{m-1} \| B_{j,0}(0, D)\, u - B_{j,0}(y, D)\, u \|_{H^{2m-m_j-1/2}(\mathbf{R}^{n-1})}$$
$$\leqq \frac{1}{4C} \| u \|_{H^{2m}(\mathbf{R}^n_+)} + C'_\varrho \| u \|_{H^{2m-1}(\mathbf{R}^n_+)}.$$

Therefore, from (4.41) we deduce

$$\| u \|_{H^{2m}(\mathbf{R}^n_+)} \leqq C'_\varrho \{ \| \mathscr{P} u \|_{L^2(\mathbf{R}^n_+) \times \prod_{j=0}^{m-1} H^{2m-m_j-1/2}(\mathbf{R}^{n-1})} +$$
$$+ \| u \|_{H^{2m-1}(\mathbf{R}^n_+)} \} + \tfrac{1}{2} \| u \|_{H^{2m}(\mathbf{R}^n_+)},$$

whence (4.39) with $r = 0$.

Now, we show (4.39) for $r > 0$ by induction on r. We know by hypothesis that $u \in H^{2m}(\mathbf{R}^n_+)$ and, as we have just seen, we also know that (4.39) holds for $r = 0$.

Therefore, we assume that $u \in H^{2m+r-1}(\mathbf{R}^n_+)$ and that (4.39) holds with r replaced by $r - 1$, and use the method of "differential quotients".

For $h \neq 0$, we set

$$\varrho_{i,h}\, u(x) = \frac{u(x_1, \ldots, x_{i-1}, x_i + h, x_{i+1}, \ldots, x_n) - u(x_1, \ldots, x_n)}{h}.$$

For sufficiently small h, the support of $\varrho_{i,h}\, u$, $i = 1, \ldots, n-1$, is in the ball $\sigma(\varrho')$ with radius $\varrho' = \varrho + (\varrho_0 - \varrho)/2 < \varrho_0$, if the support of u is in $\sigma(\varrho)$. Therefore, applying (4.39) with $r - 1$, we have

(4.44) $$\| \varrho_{i,h}\, u \|_{H^{2m+r-1}(\mathbf{R}^n_+)} \leqq C_{r,\varrho'} \{ \| A(x, D)\, \varrho_{i,h}\, u \|_{H^{r-1}(\mathbf{R}^n_+)} +$$
$$+ \sum_{j=0}^{m-1} \| B_j(x, D)\, \varrho_{i,h}\, u \|_{H^{2m+r-m_j-3/2}(\mathbf{R}^{n-1})} + \| \varrho_{i,h}\, u \|_{H^{2m+r-2}(\mathbf{R}^n_+)} \}.$$

Set $\tau_{i,h}\, u = u(x_1, \ldots, x_{i-1}, x_i + h, x_{i+1}, \ldots, x_n)$; we have

$$A(x, D)\, \varrho_{i,h}\, u - \varrho_{i,h}\, A(x, D)\, u = \sum_{|p| \leqq 2m} [\varrho_{i,h}\, a_p(x)]\, D^p\, \tau_{i,h}\, u$$

and therefore

$$\| A(x, D)\, \varrho_{i,h}\, u - \varrho_{i,h}\, A(x, D)\, u \|_{H^{r-1}(\mathbf{R}^n_+)} \leqq C_{r,\varrho} \| u \|_{H^{2m+r-1}(\mathbf{R}^n_+)}.$$

Also

$$\| B_j(y, D)\, \varrho_{i,h} u - \varrho_{i,h} B_j(y, D)\, u \|_{H^{2m+r-m_j-1/2}(\mathbf{R}^{n-1})} \leqq C_{r,\varrho} \| u \|_{H^{2m+r-1}(\mathbf{R}^n_+)}.$$

Finally, from (4.44), we deduce

(4.45) $$\| \varrho_{i,h}\, u \|_{H^{2m+r-1}(\mathbf{R}^n_+)}$$
$$\leqq C_{r,\varrho'} \{ \| \mathscr{P} u \|_{H^r(\mathbf{R}^n_+) \times \prod_{j=0}^{m-1} H^{2m+r-m_j-1/2}(\mathbf{R}^{n-1})} + \| u \|_{H^{2m+r-1}(\mathbf{R}^n_+)} \}.$$

This inequality proves that $\varrho_{i,h} u$ remains in a bounded set of $H^{2m+r-1}(\mathbf{R}^n_+)$ as h varies, with $0 < |h| < (\varrho_0 - \varrho)/2$; but then we can find a sequence $h \to 0$ such that $\varrho_{i,h} u \to \chi$ in $H^{2m+r-1}(\mathbf{R}^n_+)$ weakly; but $\varrho_{i,h} u \to \dfrac{\partial u}{\partial x_i}$ in $\mathscr{D}'(\mathbf{R}^n_+)$, so that

$$\frac{\partial u}{\partial x_i} \in H^{2m+-1}(\mathbf{R}^n_+) \quad \text{for} \quad i = 1, \ldots, n-1$$

and furthermore its norm satisfies (4.45).

To show that $\dfrac{\partial u}{\partial x_n}$ is also in $H^{2m+r-1}(\mathbf{R}^n_+)$, we use the fact that $A\, u \in H^r(\mathbf{R}^n_+)$ and the ellipticity of A; indeed, we have $a_{(0,\ldots,0,2m)}(0) \neq 0$ and therefore also $|a_{(0,\ldots,0,2m)}(x)| > \varepsilon > 0$ for $|x| < \varrho_0$ if ϱ_0 is small enough. Consequently,

$$\frac{\partial^{2m} u}{\partial x_n^{2m}} = \frac{1}{a_{(0,\ldots,0,2m)}(x)} \left\{ A(x, D)\, u - \sum_{\substack{|p| \leqq 2m \\ p_n < 2m}} a_p(x)\, D^p u \right\} \in H^r(\mathbf{R}^n_+),$$

from which we deduce that $\dfrac{\partial u}{\partial x_n} \in H^{2m+r-1}(\mathbf{R}^n_+)$ and furthermore that its norm in this space is less than or equal to

$$C \left\{ \| A\, u \|_{H^r(\mathbf{R}^n_+)} + \sum_{i=1}^{n-1} \left\| \frac{\partial u}{\partial x_i} \right\|_{H^{2m+r-1}(\mathbf{R}^n_+)} \right\}$$

and therefore also less than or equal to the second member of (4.45). Therefore i) is proved. □

2) To show ii), let us assume that

$$F = \{f; g_0, \ldots, g_{m-1}\} \in L^2(\mathbf{R}^n_+) + \prod_{j=0}^{m-1} H^{-2m+m_j+1/2}(\mathbf{R}^{n-1}).$$

We consider the operator

$$\mathscr{P}_{(0)} = \{A_0(0, D), B_{0,0}(0, D)\,|_{t=0}, \ldots, B_{m-1,0}(0, D)\,|_{t=0}\}$$

as a continuous linear mapping of

$$H^{2m}(\mathbf{R}^n_+) \quad \text{into} \quad L^2(\mathbf{R}^n_+) \times \prod_{j=0}^{m-1} H^{2m-m_j-1/2}(\mathbf{R}^{n-1}),$$

and its adjoint $\mathscr{P}^*_{(0)}$ defined by

$$\mathscr{P}^*_0 F = A^*_0(0, D)\, \tilde{f} + \sum_{j=0}^{m-1} B^*_{j,0}(0, D)\, (g_j(y) \otimes \delta(t)),$$

where $A^*_0(x, D)$ and $B^*_{j,0}(y, D)$ denote the formal adjoints of $A_0(x, D)$ and $B_{j,0}(y, D)$, $j = 0, \ldots, m-1$.

We apply Theorem 4.1 and obtain

$$(4.46) \quad \|F\|_{L^2(\mathbf{R}^n_+)\times\prod_{j=0}^{m-1} H^{-2m+m_j+1/2}(\mathbf{R}^{n-1})} \leqq \Big\{C\,\|A_0^*(0,D)\tilde{f} + \sum_{j=0}^{m-1} B_{j,0}^*(0,D)\,g_j(y)\otimes\delta(t)\|_{H^{-2m}(\mathbf{R}^n)} + \|\tilde{f}\|_{H^{-1}(\mathbf{R}^n)} + \sum_{j=0}^{m-1}\|g_j\|_{H^{-2m+m_j-1/2}(\mathbf{R}^{n-1})}\Big\}.$$

But we see that the homogeneous part of degree m of $A^*(x,D)$ is $A_0^*(x,D)$ and the homogeneous part of degree m_j of $B_j^*(y,D)$ is $B_{j,0}^*(y,D)$. Thus we obtain that

$$(4.47) \quad \Big\|A_0^*(0,D)\tilde{f} - A^*(x,D)\tilde{f} + \sum_{j=0}^{m-1} B_{j,0}^*(0,D)(g_j(y)\otimes\delta(t)) - \sum_{j=0}^{m-1} B_j^*(y,D)(g_j(y)\otimes\delta(t))\Big\|_{H^{-2m}(\mathbf{R}^n)} \leqq$$

$$\leqq \Big\|[A_0^*(0,D) - A_0^*(x,D)]\tilde{f} + \sum_{j=0}^{m-1}[B_{j,0}^*(0,D) - B_{j,0}(y,D)](g_j(y)\otimes\delta(t))\Big\|_{H^{-2m}(\mathbf{R}^n)} +$$

$$+ C_\varrho\Big\{\|\tilde{f}\|_{H^{-1}(\mathbf{R}^n)} + \sum_{j=0}^{m-1}\|g_j(y)\otimes\delta(t)\|_{H^{-2m+m_j-1}(\mathbf{R}^n)}\Big\} \leqq$$

(if ϱ_0 is sufficiently small, we may apply Lemma 2.3, thanks to the hypotheses on the coefficients of A and B_j)

$$\leqq \varepsilon\Big\{\|\tilde{f}\|_{L^2(\mathbf{R}^n)} + \sum_{j=0}^{m-1}\|g_j(y)\otimes\delta(t)\|_{H^{-2m+m_j}(\mathbf{R}^n)}\Big\} +$$

$$+ C'_\varrho\Big\{\|\tilde{f}\|_{H^{-1}(\mathbf{R}^n)} + \sum_{j=0}^{m-1}\|g_j(y)\otimes\delta(t)\|_{H^{-2m+m_j-1}(\mathbf{R}^n)}\Big\} \leqq$$

(here we use the fact that, for $s > 0$,

$$\|g_j(y)\otimes\delta(t)\|_{H^{-s}(\mathbf{R}^n)} \leqq C\,\|g_j\|_{H^{-s+1/2}(\mathbf{R}^{n-1})})$$

$$\leqq \varepsilon\Big\{\|f\|_{L^2(\mathbf{R}_+{}^n)} + \sum_{j=0}^{m-1}\|g_j\|_{H^{-2m+m_j+1/2}(\mathbf{R}^{n-1})}\Big\} +$$

$$+ C''_\varrho\Big\{\|\tilde{f}\|_{H^{-1}(\mathbf{R}^n)} + \sum_{j=0}^{m-1}\|g_j\|_{H^{-2m+m_j-1/2}(\mathbf{R}^{n-1})}\Big\}.$$

Then (4.40′) follows from (4.46) and (4.47) (choosing $\varepsilon = \frac{1}{2}$). □

5. A priori Estimates in the Open Set Ω and the Existence of Solutions in $H^s(\Omega)$-Spaces, with Real $s \geqq 2m$

5.1 A priori Estimates in the Open Set Ω

We return now to a boundary value problem $\{A, B\}$:

$$(5.1) \quad \begin{cases} A\,u = f & \text{in } \Omega \\ B_j\,u = g_j & \text{on } \Gamma, \quad j = 0, 1, \ldots, m-1, \end{cases}$$

under the hypotheses of Section 1 on the open set Ω and on the operators A and B_j, that is:

(i) *Ω is a bounded open set in $\mathbf{R}^n$, with boundary Γ, an $n-1$ dimensional infinitely differentiable variety, Ω being locally on one side of Γ;*

(ii) *the operator A is defined by*

$$A\,u = \sum_{|p|,\,|q| \leqq m} (-1)^{|p|}\, D^p(a_{pq}(x)\, D^q u),$$

with $a_{pq} \in \mathscr{D}(\bar{\Omega})$ and is properly elliptic in $\bar{\Omega}$;

(iii) *the operators B_j are defined by*

$$B_j\,u = \sum_{|h| \leqq m_j} b_{jh}(x)\, D^h u$$

with $b_{jh} \in \mathscr{D}(\Gamma)$, $0 \leqq m_j \leqq 2m - 1$, the system $\{B_j\}_{j=0}^{m-1}$ being normal on Γ and covering A on Γ.

We consider the operator $\mathscr{P}$, defined by

$$\mathscr{P}\colon\ u \to \mathscr{P}u = \{A\,u;\, B_0\,u, \ldots, B_{m-1}\,u\},$$

a continuous linear mapping of $H^{2m}(\Omega)$ into

$$L^2(\Omega) \times \prod_{j=0}^{m-1} H^{2m-m_j-1/2}(\Gamma).$$

We denote by $\mathscr{P}^*$ the adjoint of $\mathscr{P}$, which maps

$$L^2(\Omega) \times \prod_{j=0}^{m-1} H^{-2m+m_j+1/2}(\Gamma)$$

into $(H^{2m}(\Omega))' \equiv H_{\bar{\Omega}}^{-2m}(\mathbf{R}^n)$ and is defined by

$$(5.2) \quad \langle u, \overline{\mathscr{P}^* F}\rangle = \langle \mathscr{P}\,u, \bar{F}\rangle \left(= \langle A\,u, \bar{f}\rangle + \sum_{j=0}^{m-1} \langle B_j\,u, \bar{g}_j\rangle \right)$$

$\forall u \in H^{2m}(\Omega)$ and $\forall F = \{f; g_0, \ldots, g_{m-1}\} \in L^2(\Omega) \times \prod_{j=0}^{m-1} H^{-2m+m_j+1/2}(\Gamma)$, where the brackets in (5.2) denote the appropriate dualities.

Theorem 5.1. *Under hypotheses* (i), (ii), (iii), *for every integer* $r \geqq 0$, *we have:*

1°) *if*

$$u \in H^{2m}(\Omega) \quad \textit{and} \quad \mathscr{P} u \in H^r(\Omega) \times \prod_{j=0}^{m-1} H^{2m+r-m_j-1/2}(\Gamma),$$

then $u \in H^{2m+r}(\Omega)$ *and*

$$(5.3) \quad \| u \|_{H^{2m+r}(\Omega)} \leqq C_r \{ \| \mathscr{P} u \|_{H^r(\Omega) \times \prod_{j=0}^{m-1} H^{2m+r-m_j-1/2}(\Gamma)} + \| u \|_{H^{2m+r-1}(\Omega)} \}$$

(C_r *depending on* r);

2°) *if*

$$F = \{f; g_0, \ldots, g_{m-1}\} \in L^2(\Omega) \times \prod_{j=0}^{m-1} H^{-2m+m_j+1/2}(\Gamma),$$

then there exists a C_0, *independent of* F, *such that*

$$(5.4) \quad \| F \|_{L^2(\Omega) \times \prod_{j=0}^{m-1} H^{-2m+m_j+1/2}(\Gamma)} \leqq$$

$$\leqq C_0 \left\{ \| \mathscr{P}^* F \|_{H_{\bar{\Omega}}^{-2m}(\mathbf{R}^n)} + \| \tilde{f} \|_{H^{-1}(\mathbf{R}^n)} + \sum_{j=0}^{m-1} \| g_j \|_{H^{-2m+m_j-1/2}(\Gamma)} \right\},$$

where $\tilde{f}$ *denotes the extension of* f *by zero outside of* Ω.

Proof. 1) Via "local maps" and "partition of unity", the usual arguments bring us back to Theorems 3.1 and 4.3. Indeed, we may take a finite covering of $\bar{\Omega}$ by open sets $\{\mathcal{O}_i\}_{i=1}^N$, whose diameter will be suitably fixed, and an associated partition of unity $\{\theta_i\}_{i=1}^N$, $\theta_i \in \mathscr{D}(\mathbf{R}^n)$, the support of θ_i being in $\mathcal{O}_i$ and $\sum_{i=1}^N \theta_i = 1$ in $\bar{\Omega}$.

Now, we show 1°). We have $u = \sum_{i=1}^N \theta_i u$; therefore it suffices to consider $\theta_i u$ for each i. Two possibilities exist: $\mathcal{O}_i \cap \Omega = \mathcal{O}_i$ or $\mathcal{O}_i \cap \cap \Omega \subset \mathcal{O}_i$, strictly.

In the first case, we apply Theorem 3.1; indeed if the diameter of $\mathcal{O}_i$ is sufficiently small, there exists an infinitely differentiable diffeomorphism ψ_i of $\mathcal{O}_i$ into the ball $\sigma(\varrho)$ of $\mathbf{R}^n$; this diffeomorphism transforms the spaces $H^s(\mathcal{O}_i)$, $s \geqq 0$, into the spaces $H^s(\sigma(\varrho))$ (see Chapter 1, Section 12.9), that is every element $v \in H^s(\mathcal{O}_i)$ is transformed into an element $v' = v \circ \psi_i \in H^s(\sigma(\varrho))$.

In the same way, A is transformed into an elliptic operator $\mathscr{A}$ of order $2m$ in $\sigma(\varrho)$.

Then, if ϱ is sufficiently small, we may apply Theorem 3.1; it follows that $\theta_i' u'$ (transform of $\theta_i u$) belongs to $H^{2m+r}(\sigma(\varrho))$ and

$$(5.5) \quad \| \theta_i' u' \|_{H^{2m+r}(\sigma(\varrho))} \leqq C_r \{ \| \mathscr{A}(\theta_i' u') \|_{H^r(\sigma(\varrho))} + \| \theta_i' u' \|_{H^{2m+r-1}(\sigma(\varrho))} \}.$$

Coming back to the open set $\mathcal{O}_i$, we see that $\theta_i u \in H^{2m+r}(\mathcal{O}_i)$ and therefore, thanks to the property of the support of θ_i, $\theta_i u \in H^{2m+r}(\Omega)$ and

$$\|\theta_i u\|_{H^{2m+r}(\Omega)} \leqq C_r \{\|A(\theta_i u)\|_{H^r(\Omega)} + \|\theta_i u\|_{H^{2m+r-1}(\Omega)}\}.$$

But

$$A(\theta_i u) = \theta_i A u + \sum_{\substack{|p|<2m \\ |q|\leqq 2m}} \gamma_{pq} D^p u D^q \theta_i \quad \text{with} \quad \gamma_{pq} \in \mathscr{D}(\bar{\Omega}),$$

and therefore

$$(5.6) \qquad \|\theta_i u\|_{H^{2m+r}(\Omega)} \leqq C_r \{\|\theta_i A u\|_{H^r(\Omega)} + \|\theta_i u\|_{H^{2m+r-1}(\Omega)}\}.$$

In the second case ($\mathcal{O}_i \cap \Omega \subset \mathcal{O}_i$, strictly), the argument is analogous; we only need to note that we may define the diffeomorphism ψ_i so that $\mathcal{O}_i \cap \Omega$ is transformed into the half-ball $\sigma_+(\varrho) = \{x \mid x \in \sigma(\varrho), x_n > 0\}$ and $\mathcal{O}_i \cap \Gamma$ into $\partial_1 \sigma_+(\varrho) = \{x \mid x \in \sigma(\varrho), x_n = 0\}$, and that we must also consider the operators $\mathscr{B}_j$, the transforms of B_j, $j = 0, \ldots, m-1$. We may also assume that the coefficients of $\mathscr{A}$ and $\mathscr{B}_j$ are extended to $\mathbf{R}^n$ and $\mathbf{R}^{n-1}$ respectively, into infinitely differentiable functions which, together with all their derivatives, are bounded. Then (if the diameter of $\mathcal{O}_i$ is sufficiently small), we may apply Theorem 4.3. It follows that $\theta_i u \in H^{2m+r}(\Omega)$ and furthermore that

$$(5.7) \quad \|\theta_i u\|_{H^{2m+r}(\Omega)} \leqq$$
$$\leqq C_r \left\{\|\theta_i A u\|_{H^r(\Omega)} + \sum_{j=0}^{m-1} \|\theta_i B_j u\|_{H^{2m+r-m_j-1/2}(\Gamma)} + \|\theta_i u\|_{H^{2m+r-1}(\Omega)}\right\} \leqq$$
$$\leqq C_r \left\{\|A u\|_{H^r(\Omega)} + \sum_{j=0}^{m-1} \|B_j u\|_{H^{2m+r-m_j-1/2}(\Gamma)} + \|u\|_{H^{2m+r-1}(\Omega)}\right\}.$$

Summing over i, (5.6) and (5.7) yield

$$\|u\|_{H^{2m+r}(\Omega)} \leqq \sum_{i=1}^{N} \|\theta_i u\|_{H^{2m+r}(\Omega)} \leqq$$
$$\leqq C_r \{\|\mathscr{P} u\|_{H^r(\Omega) \times \prod_{j=0}^{m-1} H^{2m+r-m_j-1/2}(\Gamma)} + \|u\|_{H^{2m+r-1}(\Omega)}\}. \quad \square$$

2) Now we prove 2°) in analogous fashion. We use the same notation as in 1) and the results of Chapter 1, Section 12.9 on the spaces H^s, with $s < 0$.

If $F = \{f; g_0, \ldots, g_{m-1}\}$, then

$$F = \sum_{i=1}^{N} \theta_i F, \quad \text{i.e.} \quad f = \sum_{i=1}^{N} \theta_i f \quad \text{and} \quad g_j = \sum_{i=1}^{N} \theta_i g_j, \quad j = 0, \ldots, m-1.$$

Assume that $\mathcal{O}_i \cap \Omega = \mathcal{O}_i$; then ψ_i transforms $\theta_i F = \{\theta_i f; 0, \ldots, 0\}$ into $\theta_i' F' = \{\theta_i' f'; 0, \ldots, 0\}$. If $\mathscr{A}^*$ denotes the formal adjoint of $\mathscr{A}$, then $\mathscr{A}^*$ is the transform of A and we may apply Theorem 3.1; we obtain

$$\| \theta_i' f' \|_{L^2(\sigma(\varrho))} \leqq C \{ \| \mathscr{A}^* (\theta_i' f') \|_{H^{-2m}(\sigma(\varrho))} + \| \theta_i' f \|_{H^{-1}(\sigma(\varrho))} \},$$

from which we deduce that

$$\| \theta_i f \|_{L^2(\Omega)} \leqq C \{ \| A^* (\theta_i f) \|_{H^{-2m}(\Omega)} + \| \theta_i f \|_{H^{-1}(\Omega)} \}. \tag{5.8}$$

Now, assume that $\mathcal{O}_i \cap \Omega \subset \mathcal{O}_i$, strictly. Then ψ_i transforms $\theta_i F$ into $\theta_i' F' = \{\theta_i' f'; \theta_i' g_0', \ldots, \theta_i' g_{m-1}'\}$. We denote by $\mathscr{P}^{*\prime}(\theta_i' F')$ the transform of $\mathscr{P}^*(\theta_i F)$ and seek an explicit representation of $\mathscr{P}^{*\prime}(\theta_i' F')$; we recall that $\mathscr{P}^*(\theta_i F)$ is defined (see (5.2)) by

$$\langle u, \overline{\mathscr{P}^*(\theta_i F)} \rangle = \langle A u, \overline{\theta_i f} \rangle + \sum_{j=0}^{m-1} \langle B_j u, \overline{\theta_i g_i} \rangle \tag{5.9}$$

for all $u \in H^{2m}(\Omega)$. Taking into account the property of the support of θ_i and therefore also of θ_i', we see that $\mathscr{P}^{*\prime}(\theta_i' F')$ is defined by

$$\langle u', \overline{\mathscr{P}^{*\prime}(\theta_i' F')} \rangle = \langle \mathscr{A} u', \overline{\theta_i' f'} \rangle + \sum_{j=0}^{m-1} \langle \mathscr{B}_j u' |_{t=0}, \overline{\theta_i' g_j} \rangle \quad \forall u' \in H^{2m}(\mathbf{R}^n_+),$$

the brackets of the first member denoting the duality between $H^{2m}(\mathbf{R}^n_+)$ and $(H^{2m}(\mathbf{R}^n_+))'$, and the brackets of the second member denoting the duality between $L^2(\mathbf{R}^n_+)$ and itself and between $H^{2m-m_j-1/2}(\mathbf{R}^{n-1})$ and $H^{-2m+m_j+1/2}(\mathbf{R}^{n-1})$, respectively. But, following the method of Section 4, we may identify $(H^{2m}(\mathbf{R}^n_+))'$ with $H^{-2m}_{\overline{\mathbf{R}}^n_+}(\mathbf{R}^n)$ and, denoting by u_0' a fixed extension of $u' \in H^{2m}(\mathbf{R}^n_+)$ to $H^{2m}(\mathbf{R}^n)$ and by $\bar{v}$ the extension by zero for $t < 0$ of $v \in L^2(\mathbf{R}^n_+)$ to $L^2(\mathbf{R}^n)$, we have

$$\begin{aligned} \langle u', \overline{\mathscr{P}^{*\prime}(\theta_i' F')} \rangle &= \langle \mathscr{A} u_0', \overline{\overline{\theta_i' f'}} \rangle + \sum_{j=0}^{m-1} \langle \mathscr{B}_j u_0', \overline{\theta_i' g_j' \otimes \delta(t)} \rangle \\ &= \langle u_0', \overline{\mathscr{A}^* (\overline{\theta_i' f'})} \rangle + \sum_{j=0}^{m-1} \langle u_0', \overline{\mathscr{B}_j^* (\theta_i' g_j' \otimes \delta(t))} \rangle, \end{aligned}$$

the brackets, as usually, denoting the appropriate dualities. Therefore $\mathscr{P}^{*\prime}(\theta_i' F')$ is defined by

$$\mathscr{A}^* (\overline{\theta_i' f'}) + \sum_{j=0}^{m-1} \mathscr{B}_j^* (\theta_i' g_j' \otimes \delta(t))$$

and therefore coincides with the adjoint $\mathscr{P}'^*$ of the operator

$$\mathscr{P}' u = \{\mathscr{A} u; \mathscr{B}_0 u |_{t=0}, \ldots, \mathscr{B}_{m-1} u |_{t=0}\},$$

calculated on $\theta_i' F'$.

Then we may apply Theorem 4.3, which shows that

$$\|\theta_i' F'\|_{L^2(\mathbf{R}^n_+)\times\prod_{j=0}^{m-1} H^{-2m+m_j+1/2}(\mathbf{R}^{n-1})} \leqq$$

$$\leqq C\left\{\|\mathscr{P}^{*\prime}(\theta_i' F')\|_{H^{-2m}_{\overline{\mathbf{R}}^n_+}(\mathbf{R}^n)} + \|\theta_i' f'\|_{H^{-1}(\mathbf{R}^n_+)} + \right.$$

$$\left. + \sum_{j=0}^{m-1} \|\theta_i' g_i'\|_{H^{-2m+m_j-1/2}(\mathbf{R}^{n-1})}\right\}.$$

But then, coming back to the open set $\mathcal{O}_i \cap \Omega$, we see that

$$\|\theta_i F\|_{L^2(\Omega)\times\prod_{j=0}^{m-1} H^{-2m+m_j+1/2}(\Gamma)} \leqq$$

$$\leqq C\left\{\|\mathscr{P}^*(\theta_i F)\|_{H^{-2m}_{\bar\Omega}(\mathbf{R}^n)} + \|\widetilde{\theta_i f}\|_{H^{-1}(\mathbf{R}^n)} + \right.$$

$$\left. + \sum_{j=0}^{m-1} \|\theta_i g_j\|_{H^{-2m+m_j-1/2}(\Gamma)}\right\},$$

and noting that

$$\|\mathscr{P}^*(\theta_i F)\|_{H^{-2m}_{\bar\Omega}(\mathbf{R}^n)} \leqq C\left\{\|\mathscr{P}^* F\|_{H^{-2m}_{\bar\Omega}(\mathbf{R}^n)} + \|\tilde f\|_{H^{-1}(\mathbf{R}^n)} + \right.$$

$$\left. + \sum_{j=0}^{m-1} \|g_j\|_{H^{-2m+m_j-1/2}(\Gamma)}\right\},$$

it also follows that (5.4) holds. □

Remark 5.1. The estimates (5.4) are sometimes called the "dual estimates" of the estimates (5.3) (see also the Comments at the end of this chapter). □

5.2 Existence of Solutions in $H^s(\Omega)$-Spaces, with Integer $s \geqq 2m$

We can now study the boundary value problem (5.1) from the point of view of the existence of solutions in the spaces $H^{2m+r}(\Omega)$, with integer $r \geqq 0$. The essential aim is to show that the operator

$$\mathscr{P}: u \to \mathscr{P} u = \{A u; B_0 u, \ldots, B_{m-1} u\} \tag{5.10}$$

is an indexed operator, mapping $H^{2m+r}(\Omega)$ into

$$H^r(\Omega) \times \prod_{j=0}^{m-1} H^{2m+r-m_j-1/2}(\Gamma)$$

and then to express the *compatibility conditions of the problem* using an *adjoint boundary value problem.*

We recall that $\mathscr{P}$ is an *indexed operator* if the dimension of the kernel $\ker(\mathscr{P})$ of $\mathscr{P}$ is finite and if $\operatorname{Im}(\mathscr{P})$ is closed and its codimension is also finite; then the index $\chi(\mathscr{P})$ of $\mathscr{P}$ is given by

$$\chi(\mathscr{P}) = \dim\ker(\mathscr{P}) - \operatorname{codim}\operatorname{Im}(\mathscr{P}).$$

To show that $\mathscr{P}$, as defined by (5.10), is an indexed operator, we shall use Theorem 5.1; from the "direct estimates" (5.3), we deduce that $\ker(\mathscr{P})$ is finite-dimensional and that $\operatorname{Im}(\mathscr{P})$ is closed; from the "dual estimates" (5.4) we shall deduce that the codimension of $\operatorname{Im}(\mathscr{P})$ is finite.

We shall make use of the following lemma, due to Peetre [2]:

Lemma 5.1. *Let E, F, G be three reflexive Banach spaces such that $E \subset F$ with compact injection and let $\mathscr{C}$ be a continuous linear operator of E into G. Then the following conditions are equivalent:*

I) *the image under $\mathscr{C}$ in G is closed and the kernel of $\mathscr{C}$ is finite-dimensional;*

II) *there exists a constant C such that*

$$(5.11) \qquad \|u\|_E \leqq C\{\|\mathscr{C}u\|_G + \|u\|_F\} \qquad \forall u \in E^{((1))}.$$

We are now in a position to prove

Theorem 5.2. *Under the hypotheses* (i), (ii), (iii), *the operator $\mathscr{P}$, defined by* (5.10), *mapping $H^{2m+r}(\Omega)$ into*

$$H^r(\Omega) \times \prod_{j=0}^{m-1} H^{2m+r-m_j-1/2}(\Gamma), \qquad r = 0, 1, 2, \ldots,$$

admits an index $\chi(\mathscr{P})$ which is independent of r; the kernel of $\mathscr{P}$ is equal to the finite-dimensional space

$$(5.12) \qquad N = \{u \mid u \in \mathscr{D}(\bar{\Omega}),\ \mathscr{P}u = 0\}$$

((1)) Proof of the lemma: 1) Condition II) implies I). First of all, (5.11) implies that the unit ball in the kernel $E_0 = \ker(\mathscr{C})$ of $\mathscr{C}$ is compact; therefore E_0 is finite dimensional. We decompose E into a direct sum $E = E_0 \oplus E_1$.

The restriction of $\mathscr{C}$ to E_1 is injective and we can show that

$$(0) \qquad \|u\|_E \leqq C\|\mathscr{C}u\|_G \qquad \forall u \in E_1$$

(by contradiction, using (5.11) and the hypothesis that the injection of E into F is compact). It follows that $\operatorname{Im}(\mathscr{C})$ is closed in G, for, if $v_j \in \operatorname{Im}(\mathscr{C})$ and tends to v in G, then there exists $u_j \in E_1$ such that $\mathscr{C}u_j = v_j$ and we deduce from (0) that u_j tends to $u \in E_1$ and obviously $\mathscr{C}u = v$, therefore $v \in \operatorname{Im}(\mathscr{C})$.

2) Condition I) implies II). Again, decompose E into a direct sum $E = E_0 \oplus E_1$, where

$$E_0 = \ker(\mathscr{C}).$$

The restriction of $\mathscr{C}$ to E_1 is injective and surjective; therefore, thanks to the closed graph theorem, we have (0). We can also show that

$$(00) \qquad \|u\|_E \leqq C\|u\|_F \qquad \forall u \in E_0$$

(still by contradiction, using the hypothesis that the injection of E into F is compact). Then, if $u \in E$ and $u = u_0 + u_1$ with $u_i \in E_i$, (5.11) follows from (0) and (00).

and the image under $\mathscr{P}$ *is given by the elements* $\{f; g_0, \ldots, g_{m-1}\}$ *of*

$$H^r(\Omega) \times \prod_{j=0}^{m-1} H^{2m+r-m_j-1/2}(\Gamma)$$

such that

$$(5.13)\quad \begin{cases} \displaystyle\int_\Omega f\bar{v}\,dx + \sum_{j=0}^{m-1} \langle g_j, \bar{\varphi}_j\rangle = 0 \\ \text{for every element } \Phi = \{v; \varphi_0, \ldots, \varphi_{m-1}\} \text{ of the space} \\ \displaystyle\mathscr{N} = \left\{\Phi \in L^2(\Omega) \times \prod_{j=0}^{m-1} H^{-2m+m_j+1/2}(\Gamma), \quad \mathscr{P}^*\Phi = 0\right\}, \end{cases}$$

where $\mathscr{P}^*$ *is the adjoint of* $\mathscr{P}$ *considered for* $r = 0$ *(therefore* $\mathscr{P}^*$ *is given by* (5.2)).

Proof. For the time being, we denote by $\mathscr{P}_{(r)}$ the operator $\mathscr{P}$ considered as a mapping of

$$H^{2m+r}(\Omega) \quad \text{into} \quad H^r(\Omega) \times \prod_{j=0}^{m-1} H^{2m+r-m_j-1/2}(\Gamma).$$

It follows from 1°) of Theorem 5.1 that the kernel of $\mathscr{P}_{(r)}$ is N, as given in (5.12), for $\bigcap_{r\geq 0} H^{2m+r}(\Omega) = \mathscr{D}(\bar{\Omega})$ (see Chapter 1, Section 9.4), and that

$$(5.14)\quad \operatorname{Im}(\mathscr{P}_{(r)}) = \operatorname{Im}(\mathscr{P}_{(0)}) \cap \left\{H^r(\Omega) \times \prod_{j=0}^{m-1} H^{2m+r-m_j-1/2}(\Gamma)\right\}.$$

Applying Lemma 5.1 to $\mathscr{P}_{(r)}$ (the injection of $H^{2m+r}(\Omega)$ into $H^{2m-r-1}(\Omega)$ being compact, see Chapter 1, Section 16), we deduce from (5.3) that N is finite-dimensional and that $\operatorname{Im}(\mathscr{P}_{(r)})$ is closed for all r.

We first consider $\operatorname{Im}(\mathscr{P}_{(0)})$; since $\operatorname{Im}(\mathscr{P}_{(0)})$ is closed, the equation

$$\mathscr{P}_{(0)}u = F, \quad \text{with } F = \{f; g_0, \ldots, g_{m-1}\} \in L^2(\Omega) \times \prod_{j=0}^{m-1} H^{2m-m_j-1/2}(\Gamma)$$

admits a solution if and only if

$$\langle F, \Phi\rangle = 0,$$

($\langle\ ,\ \rangle$ denoting the duality between

$$L^2(\Omega) \times \prod_{j=0}^{m-1} H^{2m-m_j-1/2}(\Gamma)$$

and its dual) for every solution of the equation $\mathscr{P}^*_{(0)}\Phi = 0$, where $\mathscr{P}^*_{(0)}$ is the adjoint of $\mathscr{P}_{(0)}$ and therefore coincides with the operator $\mathscr{P}^*$ defined by (5.2).

Therefore, we have shown that $\operatorname{Im}\mathscr{P}_{(0)}$ is the polar of the kernel of $\mathscr{P}^*$.

We note that this follows only from 1°) of Theorem 5.1.

Now, we show that the kernel of $\mathscr{P}^*$ is finite-dimensional and coincides with the space $\mathscr{N}$. For this purpose, we use 2°) of Theorem 5.1. In fact, thanks to (5.4), we can apply Lemma 5.1 to the operator $\mathscr{P}^*$, for the injection of

$$L^2(\Omega) \times \prod_{j=0}^{m-1} H^{-2m+m_j+1/2}(\Gamma) \quad \text{into} \quad H^{-1}(\Omega) \times \prod_{j=0}^{m-1} H^{-2m+m_j-1/2}(\Gamma)$$

is compact (see Chapter 1, Section 16).

Therefore, the kernel of $\mathscr{P}^*$ $(= \mathscr{P}^*_{(0)})$ is finite-dimensional.

Finally, we have to show that

$$\operatorname{codim} \operatorname{Im}(\mathscr{P}_{(r)}) = \operatorname{codim} \operatorname{Im}(\mathscr{P}_{(0)}).$$

But

$$\operatorname{codim} \operatorname{Im}(\mathscr{P}_{(r)}) = \dim \frac{H^r(\Omega) \times \prod_{j=0}^{m-1} H^{2m+r-m_j-1/2}(\Gamma)}{\operatorname{Im}(\mathscr{P}_{(r)})}, \qquad r = 0, 1, \ldots,$$

and we can easily see that the injection of

$$H^r(\Omega) \times \prod_{j=0}^{m-1} H^{2m+r-m_j-1/2}(\Gamma) \quad \text{into} \quad L^2(\Omega) \times \prod_{j=0}^{m-1} H^{2m-m_j-1/2}(\Gamma)$$

induces an isomorphism between the quotient spaces (with respect to $\operatorname{Im}(\mathscr{P}_{(r)})$ and $\operatorname{Im}(\mathscr{P}_{(0)})$ respectively), for we already know that $\operatorname{Im}(\mathscr{P}_{(r)})$ is closed, that (5.14) is valid and that $\operatorname{codim} \operatorname{Im}(\mathscr{P}_{(0)})$ is finite. Therefore $\operatorname{codim} \operatorname{Im}(\mathscr{P}_{(r)}) = \operatorname{codim} \operatorname{Im}(\mathscr{P}_{(0)}) < +\infty$ and the theorem is proved. □

5.3 Precise Statement of the Compatibility Conditions for Existence

We shall now state the compatibility conditions of the problem more precisely by using *Green's formula* and the *formal adjoint problems* introduced in Section 2.

Given A and the B_j's, $j = 0, \ldots, m-1$, under the hypotheses (i), (ii) and (iii), we may apply Theorem 2.1; therefore let us choose the system of "boundary" operators $\{S_j\}_{j=0}^{m-1}$ in such a way that Green's formula (2.3) holds, that is (see also Remark 2.2):

$$(5.15) \qquad \int_\Omega (A\,u)\,\bar{v}\,dx - \int_\Omega u\,\overline{A^* v}\,dx = \sum_{j=0}^{m-1} \int_\Gamma S_j u\,\overline{C_j v}\,d\sigma - \sum_{j=0}^{m-1} \int_\Gamma B_j u\,\overline{T_j v}\,d\sigma, \qquad \forall u, v \in H^{2m}(\Omega),$$

where the operators C_j and T_j, $j = 0, \ldots, m-1$, depend on A, $\{B_j\}_{j=0}^{m-1}$ and $\{S_j\}_{j=0}^{m-1}$, according to Theorem 2.1.

We first prove a "regularity" result which is analogous to Theorem 5.1 i). We denote the scalar product on the space $H^{2m-m_j-1/2}(\Gamma)$ by $(\, , \,)_j$ and introduce the form

$$[u, v] = \int_\Omega A\,u\,\overline{A\,v}\,dx + \sum_{j=0}^{m-1} (B_j\,u, B_j\,v)_j \tag{5.16}$$

which is continuous on $H^{2m}(\Omega) \times H^{2m}(\Omega)$.

Applying (5.3) for $r = 0$ and Theorem 16.3, Chapter 1, we easily obtain the existence of a constant c such that

$$c^{-1}\,\|u\|^2_{H^{2m}(\Omega)} \leqq [u, u] + \|u\|^2_{L^2(\Omega)} \leqq c\,\|u\|^2_{H^{2m}(\Omega)}, \qquad \forall v \in H^{2m}(\Omega). \tag{5.17}$$

Let us then prove

Proposition 5.1. *Let* $u \in H^{2m}(\Omega)$ *and* $f \in H^r(\Omega)$ *be given* (r being a fixed positive integer) *and satisfy*

$$[u, v] = \int_\Omega f\,\bar{v}\,dx \qquad \forall v \in H^{2m}(\Omega); \tag{5.18}$$

then $u \in H^{4m+r}(\Omega)$.

Proof. Choosing $v \in \mathscr{D}(\Omega)$ in (5.18), we obtain $A^*\,A\,u = f$; since $A^*\,A$ is elliptic and of order $4m$, we obtain from Theorem 3.1 that $u \in H^{2m+r}(\omega)$ for any open set ω such that $\bar{\omega} \subset \Omega$.

We now use (as in the proof of Theorem 5.1 and in the same notation) local maps and a partition of unity to reduce (5.18) to

$$\{u', \theta'\,v'\} = \int_{\sigma_+(\varrho)} f'\,\overline{\theta'\,v'}\,dx, \qquad \forall v' \in H^{2m}(\sigma_+(\varrho)), \tag{5.19}$$

where θ' is a C^∞-function in $\sigma_+(\varrho)$ which vanishes in a neighborhood of $\partial_2\sigma_+(\varrho) = \{x \mid |x| = \varrho, x_n \geqq 0\}$, and where $\{u', v'\}$ is defined by

$$\{u', v'\} = \int_{\sigma_+(\varrho)} \mathscr{A}\,u'\,\overline{\mathscr{A}\,v'}\,dx + \sum_{j=0}^{m-1} \{\mathscr{B}_j\,u', \mathscr{B}_j\,v'\}_j, \tag{5.20}$$

$\{ \, , \, \}_j$ denoting the scalar product in

$$H^{2m-m_j-1/2}(\partial_1\sigma_+(\varrho)).$$

Thanks to (5.17), $\mathscr{A}$ and $\mathscr{B}_j$ satisfy the inequalities

$$\begin{cases} (c')^{-1}\,\|v'\|^2_{H^{2m}(\sigma_+(\varrho))} \leqq \{v', v'\} + \|v'\|^2_{L^2(\sigma_+(\varrho))} \leqq \\ \qquad \leqq c'\,\|v'\|^2_{H^{2m}(\sigma_+(\varrho))}, \qquad \forall v' \in H^{2m}(\sigma_+(\varrho)). \end{cases} \tag{5.21}$$

Using these inequalities, we can prove that $\theta' u'$ belongs to $H^{4m+r}(\sigma_+(\varrho))$ by the method of difference quotients. To simplify the notation we shall write $u, v, \theta, \ldots$, and set

$$(u, v) = \int_{\sigma_+(\varrho)} u \bar{v} \, dx$$

and furthermore use the notations $\varrho_{i,h} u$ and $\tau_{i,h} u$ introduced in Section 4.5.

For the time being, we admit

Lemma 5.2. *Let s be a positive integer. Then, for all q such that* $|q| = s$, *we have*

(5.22) $$|\{\varrho_{i,h}(D_y^q(\theta u)), v\} - (-1)^{s+1}\{u, D_y^q(\theta \varrho_{i,-h} v)\}| \leqq$$
$$\leqq K_s \|v\|_{H^{2m}(\sigma_+(\varrho))} \left(\sum_{|q| \leqq s} \|D_y^q u\|_{H^{2m}(\sigma_+(\varrho))} \right),$$
$$\forall u, v \in H^{2m}(\sigma_+(\varrho)), \quad D_y^p u \in H^{2m}(\sigma_+(\varrho)), \quad |p| \leqq s,$$

where $|q| = s$ *and* K_s *is independent of* u, v *and* h.

From (5.19), for $1 \leqq i \leqq n - 1$ and h sufficiently small, we obtain

(5.23) $$\{u, \theta \varrho_{i,-h} v\} = (f, \theta \varrho_{i,-h} v) \quad \forall v \in H^{2m}(\sigma_+(\varrho)),$$

hence

(5.24) $$|\{u, \theta \varrho_{i,-h} v\}| \leqq K \|v\|_{H^1(\sigma_+(\varrho))},$$

with K independent of v and h.

Using (5.22), with $s = 0$, we get

(5.25) $$|\{\varrho_{i,h}(\theta u), v\}| \leqq K_0^* \|v\|_{H^{2m}(\sigma_+(\varrho))} \quad \forall v \in H^{2m}(\sigma_+(\varrho)),$$

where K_0^* is independent of h and v. Hence, thanks to (5.21) and taking $v = \varrho_{i,h}(\theta u)$, we obtain

(5.26) $$\frac{\partial(\theta u)}{\partial x_1} \in H^{2m}(\sigma_+(\varrho)), \quad i = 1, \ldots, n - 1.$$

By iteration, we can show in an analogous manner that every tangential derivative $D_y^q(\theta u)$, $|q| \leqq r + 2m$, belongs to $H^{2m}(\sigma_+(\varrho))$. Indeed, assume that this is true for $1 \leqq |q| < r + 2m$. Then, using (5.19) for $i = 1, \ldots, n - 1$ and $h \neq 0$, h sufficiently small, and $|p| \leqq r$, we get

$$\{u, D_y^q(\theta \varrho_{i,-h} v)\} = (f, D_y^q(\theta \varrho_{i,-h} v)) = (-1)^{|p|} (D_y^p f, D_y^{q-p}(\theta \varrho_{i,-h} v)),$$
$$\forall v \in H^{2m}(\sigma_+(\varrho));$$

since $f \in H^r(\sigma_+(\varrho))$ and $|q| < 2m + r$, we have

$$|\{u, D_y^q(\theta\, \varrho_{i,-h}\, v)\}| \leqq K\, \|v\|_{H^{2m}(\sigma_+(\varrho))};$$

consequently, using (5.22) for $s = |q|$, we obtain

$$|\{\varrho_{i,h}(D_y^q(\theta\, u)), v| \leqq K_s^*\, \|v\|_{H^{2m}(\sigma_+(\varrho))},$$

where K_s^* is independent of h and v. Therefore, using (5.21) again, we have

$$\frac{\partial D_y^q(\theta\, u)}{\partial x_i} \in H^{2m}(\sigma_+(\varrho)), \quad i = 1, \ldots, n-1, \tag{5.27}$$

hence

$$D_y^q(\theta\, u) \in H^{2m}(\sigma_+(\varrho)) \quad \text{for} \quad |q| \leqq r + 2m. \tag{5.28}$$

Let us now use the fact that $\mathscr{A}^* \mathscr{A}(\theta\, u) = f$ in $\sigma_+(\varrho)$; writing $\mathscr{A}\, w$ in the form

$$\mathscr{A}\, w = \sum_{|p| \leqq 2m} \alpha_p(x)\, D^p u, \tag{5.29}$$

we can always assume, thanks to the ellipticity of $\mathscr{A}$, that the coefficient of $\dfrac{\partial^{2m}}{\partial x_n^{2m}}$ in $\mathscr{A}$ equals 1. Therefore

$$\mathscr{A}^* \mathscr{A}(\theta\, u) = \frac{\partial^{4m}(\theta\, u)}{\partial x_n^{4m}} + \sum_{|p| < 2m} D^p \left(\sum_{i=1}^{n-1} \frac{\partial}{\partial x_i} \left(\sum_{|q| \leqq 2m} D^q(\beta_{p,i,q}(\theta\, u)) \right) \right), \tag{5.30}$$

where $\beta_{p,i,q}$ is of class C^∞ in $\sigma_+(\varrho)$. Then we have

$$\frac{\partial^{4m}(\theta\, u)}{\partial x_n^{4m}} = f - \sum_{|p| < 2m} D^p g_p, \tag{5.31}$$

with

$$\begin{cases} f \in H^r(\sigma_+(\varrho)) \text{ and, thanks to (5.28),} \\ D_y^q g_p \in L^2(\sigma_+(\varrho)) \quad \text{for} \quad |q| \leqq 2m + r - 1. \end{cases} \tag{5.32}$$

It follows that

$$\frac{\partial^{4m}}{\partial x_n^{4m}}(D_y^q(\theta\, u)) = D_y^q f - \sum_{|p| < 2m} D^p D_y^q g_p \in H^{-2m+1}(\sigma_+(\varrho)), \tag{5.33}$$

$$\text{for} \quad |q| \leqq 2m + r - 1.$$

Therefore (5.18), (5.33) and Lemma 12.3, Chapter 1, applied to $D_y^q(\theta\, u)$, $|q| \leqq 2m + r - 1$, imply that

$$D_y^q(\theta\, u) \in H^{2m+1}(\sigma_+(\varrho)), \quad \text{for} \quad |q| \leqq 2m + r - 1. \tag{5.34}$$

But then (5.30) yields

$$D_y^q g_p \in H^1(\sigma_+(\varrho)), \quad |q| \leqq m + r - 2.$$

Then using (5.33) again, we obtain

$$\frac{\partial^{4m}}{\partial x_n^{4m}} (D_y^q(\theta\, u)) \in H^{-2m+2}(\sigma_+(\varrho)), \quad \text{for} \quad |q| \leqq 2m + r - 2$$

and therefore, thanks to Lemma 12.3, Chapter 1, we have

$$D_y^q(\theta\, u) \in H^{2m+2}(\sigma_+(\varrho)), \quad |q| \leqq 2m + r - 2;$$

by iteration, we conclude that

$$\theta\, u \in H^{4m+r}(\sigma_+(\varrho)), \quad \text{which proves the theorem.} \quad \square$$

We still have to show Lemma 5.2.
We first show that if $|q| = s$, then

(5.35) $$|(\mathscr{A}(\varrho_{i,h}(D_y^q(\theta\, u))), \mathscr{A}\, v) - (-1)^{s+1}(\mathscr{A}\, u, \mathscr{A}(D_y^q(\theta\, \varrho_{i,-h}\, v)))| \leqq$$
$$\leqq K_s' \, \| v \|_{H^{2m}(\sigma_+(\varrho))} \left(\sum_{|p| \leqq s} \| D_y^p u \|_{H^{2m}(\sigma_+(\varrho))} \right),$$

where K_s' is independent of u, v and h.
We recall that for every φ and ψ in $L^2(\sigma_+(\varrho))$, we have

(5.36) $$(\varrho_{i,h}(\theta\, \varphi), \psi) = -(\theta\, \varphi, \varrho_{i,-h}\, \psi).$$

Using (5.29), we obtain

(5.37) $$\mathscr{A}(\varrho_{i,h}(D_y^q(\theta\, u))) = \varrho_{i,h}\, \mathscr{A}(D_y^q(\theta\, u)) -$$
$$- \sum_{|p| \leqq 2m} (\varrho_{i,h}\, \alpha_p)\, D^p\, \tau_{i,h}(D_y^q(\theta\, u))$$
$$= \varrho_{i,h}(\theta\, D_y^p\, \mathscr{A}\, u) + \varrho_{i,h}(\mathscr{A}(D_y^q(\theta\, u)) - \theta\, D_y^q\, \mathscr{A}\, u) -$$
$$- \sum_{|p| \leqq 2m} (\varrho_{i,h}\, \alpha_p)\, D^p \tau_{i,h}(D_y^q(\theta\, u))$$
$$= \varrho_{i,h}(\theta\, D_y^q\, \mathscr{A}\, u) + \cdots$$

and therefore

(5.38) $$(\mathscr{A}(\varrho_{i,h}(D_y^q(\theta\, u))), \mathscr{A}\, v) = (\varrho_{i,h}(\theta\, D_y^q\, \mathscr{A}\, u), \mathscr{A}\, v) + (\ldots, \mathscr{A}\, v),$$

and it is easy to see that we have

(5.39) $$|(\ldots, \mathscr{A}\, v)| \leqq K_s'' \, \| v \|_{H^{2m}(\sigma_+(\varrho))} \left(\sum_{|p| \leqq s} \| D_y^p u \|_{H^{2m}(\sigma_+(\varrho))} \right),$$

where K_s'' is independent of u, v and h.

Moreover, we have

$$(5.40)\quad (\mathscr{A}\, u, \mathscr{A}(D_y^q(\theta\, \varrho_{i,-h}\, v))) = (\mathscr{A}\, u, D_y^q(\theta\, \varrho_{i,-h}\, \mathscr{A}\, v)) + $$
$$+ (\mathscr{A}\, u, \mathscr{A}(D_y^q(\theta\, \varrho_{i,-h}\, v)) - D_y^q(\theta\, \varrho_{i,-h}\, \mathscr{A}\, v))$$
$$= (-1)^{|q|}(\theta\, D_y^q\, \mathscr{A}\, u, \varrho_{i,-h}\, \mathscr{A}\, v) + (\mathscr{A}\, u, \ldots)$$
$$= (-1)^{|q|+1}(\varrho_{i,h}(\theta\, D_y^q\, \mathscr{A}\, u), \mathscr{A}\, v) + (\mathscr{A}\, u, \ldots);$$

by integration by parts in the term $(\mathscr{A}\, u, \ldots)$ we get

$$(5.41)\quad |(\mathscr{A}\, u, \ldots)| \leqq K_s'''\, \|v\|_{H^{2m}(\sigma_+(\varrho))} \left(\sum_{|p| \leqq s} \|D u\|_{H^{2m}(\sigma_+(\varrho))} \right).$$

Then (5.36) follows from (5.38), ..., (5.41).

We use similar arguments for the terms

$$(5.42)\quad \{\mathscr{B}_j(\varrho_{i,h}(D_y^q u)), \mathscr{B}_j\, v\}_j - (-1)^{|q|+1}\{\mathscr{B}_j\, u, \mathscr{B}_j(D_y^q(\theta\, \varrho_{i,-h}\, v))\}_j,$$

taking into account the formulas

$$\{\varphi, D_y^q(\theta\, \psi)\}_j = (-1)^{|q|}\{\theta\, D_y^q \varphi, \psi\}_j,$$
$$\{\varrho_{i,h}(\theta\, \varphi), \psi\}_j = -\{\theta\, \varphi, \varrho_{i,-h}\, \psi\}_j$$

and also using the trace theorem of Section 8, Chapter 1.

This completes the proof of the Lemma. □

From Proposition 5.1, we obtain

Corollary 5.1. *If in Proposition* 5.1, *we assume that* $f \in \mathscr{D}(\bar{\Omega})$, *then* $u \in \mathscr{D}(\bar{\Omega})$.

Let us now recall that (see Theorem 5.2) the kernel of $\mathscr{P}$ is finite-dimensional and is given by

$$N = \{u \mid u \in \mathscr{D}(\bar{\Omega}), A\, u = 0, B_j\, u = 0, j = 0, \ldots, m-1\}.$$

We introduce the vector spaces

$$M = \left\{v \mid v \in H^{2m}(\Omega), \int_\Omega v\, \bar{u}\, dx = 0, \forall u \in N\right\},$$

$$H_B^{2m}(\Omega) = \{u \mid u \in H^{2m}(\Omega), B_j\, u = 0, j = 0, \ldots, m-1\},$$
$$H_C^{2m}(\Omega) = \{u \mid u \in H^{2m}(\Omega), C_j\, u = 0, j = 0, \ldots, m-1\}.$$

We have

Proposition 5.2. *If* $f \in \mathscr{D}(\bar{\Omega})$, *then there exists at least one solution* $u \in \mathscr{D}(\bar{\Omega})$ *of the problem*

$$(5.43)\quad \begin{cases} A^*\, u = f & \text{in } \Omega \\ C_j\, u = 0 & \text{on } \Gamma, \quad j = 0, \ldots, m-1, \end{cases}$$

if and only if $f \in M$.

Proof. Green's formula (5.15) implies that the condition $f \in M$ is necessary. Let us show that it is also sufficient.

First, we note that N is a closed subspace of $H^{2m}(\Omega)$ and of $L^2(\Omega)$; M is also closed a subspace of $H^{2m}(\Omega)$ and moreover

(5.44) $\quad \| u \|^2_{H^{2m}(\Omega)} \leqq C[u, u], \quad \forall u \in M, \quad C$ independent of u.

We can indeed show (5.44) by contradiction. If we assume that (5.44) is false, then there exist a constant C_1 and a sequence $u_n \in M$ such that

$$\| u_n \|_{H^{2m}(\Omega)} \to +\infty \quad \text{and} \quad [u_n, u_n] \leqq C_1 \qquad \forall n.$$

Set

$$w_n = \frac{u_n}{\| u_n \|_{H^{2m}(\Omega)}}.$$

Then we have

(5.45) $$\| w_n \|_{H^{2m}(\Omega)} = 1, \quad \forall n,$$

(5.46) $$[w_n, w_n] \to 0 \text{ if } n \to +\infty.$$

Thanks to Theorem 16.1 of Chapter 1 and to (5.15), there exists a subsequence of w_n, still denoted by w_n, such that

$$\| w_m - w_n \|_{L^2(\Omega)} \to 0 \text{ if } m, n \to +\infty;$$

but using (5.17) and (5.46) it follows that

$$\| w_m - w_n \|_{H^{2m}(\Omega)} \to 0 \text{ if } m, n \to +\infty$$

and therefore there exists a $w \in M$ such that

(5.47) $$\| w_n - w \|_{H^{2m}(\Omega)} \to 0 \text{ if } n \to +\infty.$$

But (5.46) implies $[w, w] = 0$, hence $w \in \ker \mathscr{P} = N$; but we have also shown that $w \in M$, hence $w = 0$, which is absurd since (5.45), (5.47) imply that $\| w \|_{H^{2m}(\Omega)} = 1$.

From (5.44) and (5.17) it follows that in M, $[u, u]^{1/2}$ is equivalent to $\| u \|_{H^{2m}(\Omega)}$. But since $v \to \int_\Omega f \bar{v} \, dx$ is continuous on M, there exists a $g \in M$ such that

$$[g, v'] = \int_\Omega f \bar{v}' \, dx, \qquad \forall v' \in M.$$

We note that every $v \in H^{2m}(\Omega)$ may be written

$$v = v' + v'', \quad v' \in M, \quad v'' \in N$$

(choose v'' = projection of v on N in $L^2(\Omega)$).

But $[v'', v''] = 0$ and $\int_\Omega f\,\bar{v}''\,dx = 0$. Then

$$[g, v] = [g, v'] = \int_\Omega f\,\bar{v}'\,dx = \int_\Omega f\,\bar{v}\,dx, \qquad \forall v \in H^{2m}(\Omega) \tag{5.48}$$

and Corollary 5.1, shows that $g \in \mathscr{D}(\bar{\Omega})$.

Set $u = A\,g$ and integrate by parts in (5.48) with $v \in \mathscr{D}(\Omega)$; we obtain $A^*\,u = f$ and

$$\int_\Omega u\,\overline{A\,v}\,dx = \int_\Omega A^*\,u\,\bar{v}\,dx, \qquad \forall v \in H_B^{2m}(\Omega). \tag{5.49}$$

But $\{B_0\,u, \ldots, B_{m-1}\,u, S_0\,u, \ldots, S_{m-1}\,u\}$ is a Dirichlet system of order $2m$ on Γ. We can therefore apply Lemma 2.2 extended to Ω and show that as v describes $H_B^{2m}(\Omega) \cap \mathscr{D}(\bar{\Omega})$, $S_j\,v$ describes $\mathscr{D}(\Gamma)$; therefore (5.49) and (5.15) imply $C_j\,u = 0$, $0 \leqq j \leqq m-1$. □

Of course Propositions 5.1, 5.2 and Theorems 5.1, 5.2 remain valid if, instead of problem $\{A^*, B\}$, we consider the adjoint problem $\{A^*, C\}$. In particular, let us state the analogue to Proposition 5.2. We introduce

$$N^* = \{v \mid v \in \mathscr{D}(\bar{\Omega}),\ A^*\,v = 0,\ C_j\,v = 0,\ j = 0, \ldots, m-1\}$$

and

$$M^* = \left\{w \mid w \in H^{2m}(\Omega),\ \int_\Omega w\,\bar{v}\,dx = 0,\ \forall v \in N^*\right\}.$$

We have

Proposition 5.3. *If $f \in \mathscr{D}(\bar{\Omega})$, then there exists at least one solution $u \in \mathscr{D}(\bar{\Omega})$ of*

$$\begin{cases} A\,u = f & \text{in } \Omega \\ B_j\,u = 0 & \text{on } \Gamma, \quad j = 0, \ldots, m-1, \end{cases} \tag{5.50}$$

if and only if $f \in M^*$.

We can now show

Proposition 5.4. *The vector space $\mathscr{N}$ defined in (5.13) coincides with the set described by $\{v, T_0\,v, \ldots, T_{m-1}\,v\}$ as v describes N^*.*

Proof. Let us first show that if $\Phi = \{v, \varphi_0, \ldots, \varphi_{m-1}\} \in \mathscr{N}$, then $v \in \mathscr{D}(\bar{\Omega})$. Indeed, if $\Phi \in \mathscr{N}$, then by definition we have

$$\int_\Omega A\,u\,\bar{v}\,dx + \sum_{j=0}^{m-1} \langle B_j\,u, \bar{\varphi}_j\rangle = 0, \qquad \forall u \in H^{2m}(\Omega), \tag{5.51}$$

hence in particular

$$\int_\Omega A\,u\,\bar{v}\,dx = 0, \qquad \forall u \in H_B^{2m}(\Omega). \tag{5.52}$$

But (5.51), (5.52) (where $u \in \mathscr{D}(\Omega)$) imply

$$A^* v = 0. \tag{5.53}$$

We decompose $v = v_1 + v_2$, $v_1 \in N^* =$ closed subspace of $L^2(\Omega)$; we have

$$\int_\Omega v_2 \bar{w} \, dx = 0, \quad \forall w \in N^*. \tag{5.54}$$

Applying (5.15) to $u \in H_B^{2m}(\Omega)$ and to v_1 we obtain

$$\int_\Omega A u \bar{v}_1 \, dx = 0, \quad \forall u \in H_B^{2m}(\Omega).$$

Then (5.52) implies

$$\int_\Omega A u \bar{v}_2 \, dx = 0, \quad \forall u \in H_B^{2m}(\Omega). \tag{5.55}$$

We shall now show that $v_2 = 0$. Indeed, let us take $h \in \mathscr{D}(\bar{\Omega})$; we may write

$$h = A u_1 + h_1 \quad \text{with} \quad u_1 \in H_B^{2m}(\Omega) \quad \text{and} \quad h_1 \in N^*,$$

since, if $h_1 =$ projection in $L^2(\Omega)$ of h on N^*, we have

$$\int_\Omega (h - h_1) \bar{w} \, dx = 0, \quad \forall w \in N^*.$$

Then Proposition 5.3 implies the existence of $u_1 \in H_B^{2m}(\Omega)$ such that $A u_1 = h - h_1$.

Furthermore, using (5.54), (5.55), we have

$$\int_\Omega h \bar{v}_2 \, dx = \int_\Omega A u_1 \bar{v}_2 \, dx + \int_\Omega h_1 \bar{v}_2 \, dx = 0.$$

Since h is arbitrary in $\mathscr{D}(\bar{\Omega})$, it follows that $v_2 = 0$.

Consequently $v = v_1 \in \mathscr{D}(\bar{\Omega})$.

We can now apply Green's formula (5.15) to $\int_\Omega (A u) \bar{v} \, dx$ in (5.51); using (5.53), we obtain

$$\sum_{j=0}^{m-1} \langle B_j u, \bar{\varphi}_j - \overline{T_j v} \rangle + \sum_{j=0}^{m-1} \int_\Gamma S_j u \overline{C_j v} \, d\sigma = 0, \quad \forall u \in H^{2m}(\Omega). \tag{5.56}$$

Since the system $\{B_0 u, \ldots, B_{m-1} u, S_0 u, \ldots, S_{m-1} u\}$ is a Dirichlet system of order $2m$ on Γ, we may use Lemma 2.2 extended to the open set Ω and therefore, if u describes $\mathscr{D}(\bar{\Omega})$, $\{B_j u, S_j u\}_{j=0}^{m-1}$ describes $(\mathscr{D}(\Gamma))^{2m}$; then from (5.56) we deduce

$$T_j v = \varphi_j, \quad C_j v = 0, \quad j = 0, 1, \ldots, m-1,$$

which proves the proposition. □

From Proposition 5.4 and Theorem 5.2, we finally obtain

Theorem 5.3. *Under the hypotheses* (i), (ii), (iii) *and having chosen a formal adjoint problem* $\{A^*, C\}$ *of* $\{A, B\}$ *with respect to Green's formula* (5.15), *the operator* $\mathscr{P} = \{A; B_0, \ldots, B_{m-1}\}$, *considered as an operator of* $H^{2m+r}(\Omega)$ *into*

$$H^r(\Omega) \times \prod_{j=0}^{m-1} H^{2m+r-m_j-1/2}(\Gamma), \qquad r = 0, 1, 2, \ldots,$$

has a finite-dimensional kernel given by the space

$$N = \{u \mid u \in \mathscr{D}(\overline{\Omega}), A\, u = 0, B_0\, u = 0, \ldots, B_{m-1}\, u = 0\}$$

and the image under $\mathscr{P}$ *is the subspace of elements* $\{f; g_0, \ldots, g_{m-1}\}$ *of*

$$H^r(\Omega) \times \prod_{j=0}^{m-1} H^{2m+r-m_j-1/2}(\Gamma)$$

which satisfy

$$\int_\Omega f\,\bar{v}\, dx + \sum_{j=0}^{m-1} \int_\Gamma g_j\, \overline{T_j\, v}\, d\sigma = 0 \tag{5.57}$$

for every v *belonging to the finite-dimensional space*

$$N^* = \{v \mid v \in \mathscr{D}(\overline{\Omega}), A^*\, v = 0, C_0\, v = 0, \ldots, C_{m-1}\, v = 0\}.$$

Therefore, the operator $\mathscr{P}$ *is an indexed operator and its index* $\chi(\mathscr{P})$ *is given by*

$$\chi(\mathscr{P}) = \dim N - \dim N^*.$$

Remark 5.2. The functions v and the operators T_j depend on the *choice* of Green's formula, in such a way that (5.57) does not mean that the image-space (which is obviously independent of Green's formula!) depends on the choice of Green's formula. □

Remark 5.3. Denoting by $H^{2m+r}(\Omega)/N$ the quotient space of $H^{2m+r}(\Omega)$ by N and by

$$\left\{H^r(\Omega) \times \prod_{j=0}^{m-1} H^{2m+r-m_j-1/2}(\Gamma); N^*, \mathscr{T}\right\}$$

the subspace of

$$H^r(\Omega) \times \prod_{j=0}^{m-1} H^{2m+r-m_j-1/2}(\Gamma)$$

defined by (5.57), and still denoting by $\mathscr{P}$ the operator after passage to the quotient by N, Theorem 5.3 may be stated in the following form:

(5.58) *the operator* $\mathscr{P}$ *defines an (algebraic and topological) isomorphism of* $H^{2m+r}(\Omega)/N$ *onto*

$$\left\{H^r(\Omega) \times \prod_{j=0}^{m-1} H^{2m+r-m_j-1/2}(\Gamma); N^*, \mathscr{T}\right\}.$$

Remark 5.4. For elliptic boundary value problems, Theorem 5.3 yields the theorem of the alternative of Riesz-Fredholm in the usual form for elliptic equations of the second order (see Miranda [1]), for the functions v of N^* are the solutions in $\mathscr{D}(\bar{\Omega})$ of the homogeneous formal adjoint problem $\{A^*, C\}$ with respect to Green's formula (5.15)

$$A^* v = 0 \quad \text{in} \quad \Omega, \qquad C_j v = 0, \qquad j = 0, \ldots, m-1.$$

5.4 Existence of Solutions in $H^s(\Omega)$-Spaces, with Real $s \geqq 2m$

We shall give a first application of interpolation between Hilbert spaces (Chapter 1) in order to study problem (5.1) in the spaces $H^s(\Omega)$ with *real* $s \geqq 2m$.

In fact, it is sufficient to interpolate between r and $r - 1$ in (5.58) for fixed $r \geqq 0$; also applying the results of Chapter 1, Sections 7 and 9, we immediately obtain

Theorem 5.4. *Under the hypotheses* (i), (ii), (iii), *the operator* $\mathscr{P}$ *defines an (algebraic and topological) isomorphism of*

$$H^s(\Omega)/N \quad \textit{onto} \quad \left\{H^{s-2m}(\Omega) \times \prod_{j=0}^{m-1} H^{s-m_j-1/2}(\Gamma);\, N^*, \mathscr{T}\right\}$$

for every real $s \geqq 2m$.

Remark 5.5. The use of interpolation in Theorem 5.4 is not indispensable; in fact, the same result can be obtained by applying the method used in Sections 4 and 5 for the case integer s. A more interesting application of interpolation theory will be given in Section 7. □

Remark 5.6. Even in the case $s \geqq 2m$, we may still say that, in $H^s(\Omega)$, $\mathscr{P}$ is an indexed operator and its index is still given by

$$\chi(\mathscr{P}) = \dim N - \dim N^*$$

(invariance of the index by interpolation: Proposition 5.1 of Geymonat [2]). □

Remark 5.7. From the preceding theorems we deduce the fact that if $f \in \mathscr{D}(\bar{\Omega})$ and $g_j \in \mathscr{D}(\Gamma)$, then the solutions u of problem (5.1) belong to $\mathscr{D}(\bar{\Omega})$. In Volume 3, we shall see that there is an even stronger regularity: if Γ is an analytic (or Gevrey) variety, if the coefficients of A and of the B_j's are analytic (or of Gevrey class), f is analytic (or of Gevrey class) in $\bar{\Omega}$, g_j is analytic (or of Gevery class) on Γ, then u is analytic (or of Gevrey class) in $\bar{\Omega}$. □

6. Application of Transposition: Existence of Solutions in $H^s(\Omega)$-Spaces, with Real $s \leqq 0$

6.1 The Transposition Method; Generalities

In Section 5, we studied the boundary value problem (5.1) in "regular" function spaces: the spaces $H^s(\Omega)$, with $s \geqq 2m$. Now we want to investigate what can be obtained from these results by applying "transposition". We have already made use of the idea of transposition in the search for "dual estimates" (see Section 4).

In this section we shall systematically develop this idea in order to study problem (5.1) in "non-regular" spaces of solutions.

Many different exploitations of this idea of transposition are possible. In the continuation of this chapter, we shall give a "natural" meaning to problem (5.1) for u belonging to $H^s(\Omega)$-spaces, with arbitrary real s, with the g_j's belonging to $H^{s-m_j-1/2}(\Gamma)$ and *f to a suitable, sufficiently general space of distributions on Ω.* □

We still assume that the hypotheses (i), (ii), (iii) *of Section* 5 *are satisfied.* According to Theorem 2.2, the system $\{C_j\}_{j=0}^{m-1}$ covers A^* and therefore we may apply Theorem 5.3, taking A^* instead of A and $\{C_j\}_{j=0}^{m-1}$ instead of $\{B_j\}_{j=0}^{m-1}$. In particular, denoting by $H_C^{2m+r}(\Omega)$ the space

$$H_C^{2m+r}(\Omega) = \{u \mid u \in H^{2m+r}(\Omega),\, C_j v = 0, j = 0, \ldots, m-1\}, \quad r \geqq 0$$

and by $\{H^r(\Omega); N\}$ the space

$$\{H^r(\Omega); N\} = \left\{f \mid f \in H^r(\Omega),\, \int_\Omega f \bar{u}\, dx = 0 \quad \forall u \in N\right\}, \quad r \geqq 0$$

and still denoting by A^* the operator after passage to the quotient by N^*, we have

(6.1) *the operator A^* defines an (algebraic and topological) isomorphism of $H_C^{2m+r}(\Omega)/N^*$ onto $\{H^r(\Omega); N\}$ for all real $s \geqq 0$.*

Therefore, we may consider transposing the isomorphism (6.1); but for $r > \frac{1}{2}$, the dual of $H^r(\Omega)$ is not $H^{-r}(\Omega)$ (it is not even a space of distributions on Ω). Thus, in order to obtain the desired result, we must restrict the operator A^* to a subspace of $H_C^{2m+r}(\Omega)$; we choose the space

$$(6.2) \quad X^r(\Omega) = \{v \mid v \in H^{2m+r}(\Omega),\, C_j v = 0, j = 0, \ldots, m-1;\; A^* v \in H_0^r(\Omega)\}, \quad \text{real } r \geqq 0.$$

With the norm of the graph

$$\| v \|^2_{X^r(\Omega)} = \| v \|^2_{H^{2m+r}(\Omega)} + \| A\, v \|^2_{H^r(\Omega)},$$

it is easy to see that *$X^r(\Omega)$ is a Hilbert space.*

From definition (6.2) and (6.1), with A^* still denoting the operator A^* after passage to the quotient by N^* and $\{H_0^r(\Omega); N\}$ denoting the space $\{H^r(\Omega); N\} \cap H_0^r(\Omega)$, we deduce:

$$(6.3)\quad \begin{cases} \textit{the operator } A^* \textit{ defines an (algebraic and topological) isomorphism} \\ \textit{of } X^r(\Omega)/N^* \textit{ onto } \{H_0^r(\Omega); N\},\ r \geqq 0. \end{cases}$$

This result is our starting point for the application of the method of transposition. Indeed, by transposition, we deduce from (6.3):

Proposition 6.1. *Assume that* (i), (ii), (iii) *of Section 5 are satisfied and that r is real and $\geqq 0$. Then, for every continuous antilinear form $v^\bullet \to L(v^\bullet)$ on $X^r(\Omega)/N^*$ there exists one and only one element $u^\bullet$ in the space $\{H_0^r(\Omega); N\}'$ such that*

$$(6.4)\qquad \langle u^\bullet, \overline{A^* v^\bullet}\rangle = L(v^\bullet) \qquad \forall v^\bullet \subset X^r(\Omega)/N^*$$

and $u^\bullet$ depends continuously on L (for the strong dual topologies), the brackets in (6.4) *denoting the duality between $\{H_0^r(\Omega); N\}'$ and $\{H_0^r(\Omega); N\}$.* □

It is immediate to interprete the space $\{H_0^r(\Omega); N\}'$; indeed, we have

$$(6.5)\qquad \{H_0^r(\Omega); N\}' \cong H^{-r}(\Omega)/N.$$

Therefore, also noting that (using (5.15))

$$\int_\Omega w\, \overline{A^* v}\, dx = 0 \qquad \forall w \in N \quad \text{and} \quad v \in X^r(\Omega),$$

we may say: *given a continuous antilinear form L on $X^r(\Omega)/N^*$, there exists $u \in H^{-r}(\Omega)$, determined up to addition of a function of N, such that*

$$(6.6)\qquad \langle u, \overline{A^* v}\rangle = L(v^\bullet) \qquad \forall v^\bullet \in X^r(\Omega)/N^*$$

and

$$\forall v \in X^r(\Omega) \ \textit{belonging to } v^\bullet,$$

the brackets denoting the duality between $H^{-r}(\Omega)$ and $H_0^r(\Omega)$. □

Remark 6.1. Proposition 6.1, specified by (6.6), in a certain sense, gives a solution of problem (5.1) in $H^{-r}(\Omega)$; but in (6.6) we have "mixed" the equation $A\,u = f$ and the boundary conditions $B_j\, u = g_j$, $j = 0, \ldots, m-1$. *Therefore, we now have to "separate" the two items by choosing the form L in a suitable way and giving an interpretation of the solved problem, once L has been fixed.*

6.2 Choice of the Form L

Concerning the choice of L, it is natural to decompose it into two forms, $L = L_1 + L_2$, in such a way that L_1 gives rise to equation $A\,u = f$ in the sense of distributions on Ω and L_2 gives rise to the boundary conditions $B_j u = g_j, \ldots$ in a sense to be specified in the *most natural possible way*.

There is an "optimal" choice for L_2, but a different situation is encountered for the choice of L_1.

For the choice of L_2, we *must* (recall Green's formula (5.14) and (6.6)!) first consider the mapping

$$v \to \mathcal{T} v = \{T_0 v, \ldots, T_{m-1} v\} \tag{6.7}$$

as v describes $X^r(\Omega)$. We can characterize the image $\mathcal{T}(X^r(\Omega))$ of $X^r(\Omega)$ under $\mathcal{T}$: for, we have

Theorem 6.1. *Under hypotheses* (i), (ii) *of Section 5 and if the system* $\{B_j\}_{j=0}^{m-1}$ *is normal on* Γ, $v \to \mathcal{T} v$ *is a continuous linear mapping of*

$$X^r(\Omega) \quad \textit{onto} \quad \prod_{j=0}^{m-1} H^{r+m_j+1/2}(\Gamma) \quad \textit{for all real } r \geqq 0$$

and there exists a continuous "right-inverse" of $\mathcal{T}$.

Proof. 1) Applying the trace theorem for the spaces $H^{2m+r}(\Omega)$ (Chapter 1, Theorem 8.3), we immediately see, T_j being of order $2m - m_j - 1$, that $v \to \mathcal{T} v$ is a continuous linear mapping of

$$X^r(\Omega) \quad \text{into} \quad \prod_{j=0}^{m-1} H^{r+m_j+1/2}(\Gamma).$$

2) To show that $\mathcal{T}$ is surjective and the existence of a continuous right-inverse of $\mathcal{T}$, we take an arbitrary $\varphi_j \in H^{r+m_j+1/2}(\Gamma)$, $j = 0, \ldots, m-1$. We have to solve the problem: construct $v \in H^{2m+r}(\Omega)$ such that

$$C_j v = 0, \quad T_j v = \varphi_j, \quad j = 0, \ldots, m-1, \tag{6.8}$$

$$A^* v \in H_0^r(\Omega), \tag{6.9}$$

and such that $\{\varphi_0, \ldots, \varphi_{m-1}\} \to v$ is a continuous linear mapping of

$$\prod_{j=0}^{m-1} H^{r+m_j+1/2}(\Gamma) \quad \text{into} \quad H^{2m+r}(\Omega).$$

If $r \leqq \frac{1}{2}$, then condition (6.9) is a consequence of the fact that $v \in H^{2m+r}(\Omega)$ and that $H_0^r(\Omega) = H^r(\Omega)$ (Chapter 1, Theorem 11.1). If $r > \frac{1}{2}$, then (6.9) is equivalent (Chapter 1, Theorem 11.5) to

$$\gamma_i(A^* v) = 0, \quad i = 0, 1, \ldots, [r - \tfrac{1}{2}]^- \text{ ((1))}. \tag{6.10}$$

Since A^* is elliptic, the system

$$\{C_j, T_j, \gamma_i(A^*)\}, \quad j = 0, \ldots, m-1, \quad i = 0, \ldots, [r - \tfrac{1}{2}]^-,$$

((1)) $[\alpha]^{-1}$ denotes the greatest integer less than α.

is a Dirichlet system of order $2m + [r - \frac{1}{2}]^{-1} + 1$, with infinitely differentiable coefficients on Γ; thus, using Lemma 2.1 and the arguments of Lemma 2.2 (in an obvious formulation, with Ω instead of σ_+) it all comes down to constructing $w \in H^{2m+r}(\Omega)$ with

$$\gamma_j w = \psi_j, \quad j = 0, 1, \ldots, 2m + [r - \tfrac{1}{2}]^-,$$

with ψ_j given in $H^{2m+r-j-1/2}(\Gamma)$ and w depending continuously on the ψ_j's. This is possible thanks to Theorem 9.4 of Chapter 1. □

Choice of L_2. Therefore, according to Theorem 6.1, we have an "optimal" choice for L_2; more precisely

$$(6.11) \qquad L_2(v) = \sum_{j=0}^{m-1} \langle g_j, \overline{T_j v} \rangle, \quad \text{with } g_j \in H^{-r-m_j-1/2}(\Gamma),$$

where the brackets denote the duality between $H^{-r-m_j-1/2}(\Gamma)$ and $H^{r+m_j+1/2}(\Gamma)$; L_2, as defined by (6.11), is, for *all* $r \geqq 0$, a continuous antilinear form on $X^r(\Omega)$, the g_j's being arbitrarily chosen in $H^{-r-m_j-1/2}(\Gamma)$. □

Choice of L_1. For L_1, we may in general proceed as follows: we choose a Hilbert space (or, more generally, a topological vector space) $K^r(\Omega)$ of distributions on Ω such that

$$(6.12) \qquad \begin{cases} X^r(\Omega) \subset K^r(\Omega) \subset L^2(\Omega) \\ \textit{with continuous injection}; \end{cases}$$

(6.13) $\mathscr{D}(\Omega)$ *is dense in* $K^r(\Omega)$ (and therefore $K^r(\Omega)$ is a normal space of distributions on Ω). □

Remark 6.2. Such spaces $K^r(\Omega)$ *always* exist. For example, we may take $K^r(\Omega) = L^2(\Omega)$ or $K^r(\Omega) = H^s(\Omega)$, $0 \leqq s \leqq \frac{1}{2}$ (for $\mathscr{D}(\Omega)$ is dense in $H^s(\Omega)$ if $0 \leqq s \leqq \frac{1}{2}$; see Chapter 1, Theorem 11.1). For further comments on the choice of $K^r(\Omega)$, we refer the reader to Section 6.3. See also Problem 13.10 in Chapter 3. □

Since $\mathscr{D}(\Omega) \subset X^r(\Omega)$, $X^r(\Omega)$ is also dense in $K^r(\Omega)$. The dual $K^{-r}(\Omega)$ of $K^r(\Omega)$ therefore may be identified to a subspace of $\mathscr{D}'(\Omega)$ and, if $f \in K^{-r}(\Omega)$, the form

$$(6.14) \qquad L_2(v) = \langle f, \bar{v} \rangle$$

(where the brackets denote the duality between $K^{-r}(\Omega)$ and $K^r(\Omega)$) also defines a continuous antilinear form on $X^r(\Omega)$.

Thus, consider the form

$$(6.15) \qquad L(v) = L_1(v) + L_2(v) = \langle f, \bar{v} \rangle + \sum_{j=0}^{m-1} \langle g_j, \overline{T_j v} \rangle.$$

If we make the convention that $L(v^\bullet) = L(v)$ for every $v \in X^r(\Omega)$, element of the class $v^\bullet \in X^r(\Omega)/N^*$, we see that (6.15) defines *a con-*

tinuous antilinear form $v^\bullet \to L(v^\bullet)$ *on* $X^r(\Omega)/N^*$ *if and only if*

(6.16) $$\langle f, \bar{v}\rangle + \sum_{j=0}^{m-1} \langle g_j, \overline{T_j v}\rangle = 0 \qquad \forall v \in N^*.$$

Finally, we have obtained

Theorem 6.2. *Under hypothesis* (i), (ii), (iii) *of Section 5, for all fixed real* $r \geqq 0$, *if* $f \in K^{-r}(\Omega)$ *and* $g_j \in H^{-r-m_j-1/2}(\Gamma)$, $j = 0, \ldots, m-1$, *with* (6.12), (6.13) *and* (6.16), *there exists* $u \in H^{-r}(\Omega)$, *determined up to addition of a function of* N, *such that*

(6.17) $$\langle u, \overline{A^* v}\rangle = \langle f, \bar{v}\rangle + \sum_{j=0}^{m-1} \langle g_j, \overline{T_j v}\rangle \qquad \forall v \in X^r(\Omega).$$

Furthermore $\{f; g_0, \ldots, g_{m-1}\} \to u^\bullet = u + N$ *is a continuous linear mapping of the subspace of*

$$K^{-r}(\Omega) \times \prod_{j=0}^{m-1} H^{-r-m_j-1/2}(\Gamma)$$

made up of elements satisfying (6.16), *into* $H^{-r}(\Omega)/N$. □

Writing (6.17) for all $v \in \mathscr{D}(\Omega)$, we obtain

$$\langle u, \overline{A^* v}\rangle = \langle f, \bar{v}\rangle \qquad \forall v \in \mathscr{D}(\Omega)$$

and therefore u *satisfies the equation*

(6.18) $$A u = f$$

in the sense of distributions on Ω.

We still have to specify the choice of f and to interpret the boundary conditions "contained" in (6.17).

6.3 The Spaces $\Xi^s(\Omega)$ and $D^s_A(\Omega)$

Problem statement

The problem solved by Theorem 6.2 is the more general as f is taken in a "larger" space, therefore as $K^r(\Omega)$ is "smaller".

We have seen (Remark 6.2) that spaces $K^r(\Omega)$ exist, but we ignore if there exists an optimum space $K^r(\Omega)$ (i.e. the smallest possible). [The difficulty is that if $\mathscr{D}(\Omega)$ is dense in $F_1(\Omega)$ and $F_2(\Omega)$, it is not necessarily dense in $F_1(\Omega) \cap F_2(\Omega)$ provided with the sup-norm topology.]

Also note that the space $K^r(\Omega)$ *may depend on the boundary conditions*; thus, with the Dirichlet boundary conditions we may take $K^r(\Omega) = H^m_0(\Omega)$ (or, more generally, $H^{m+1/2}_0(\Omega)$) which is no longer possible for other boundary value problems.

In what is to follow, we shall give a "universal" (i.e. independent of the boundary conditions) choice for $K^r(\Omega)$, while limiting ourselves (in order to simplify the presentation) to the case where $K^r(\Omega)$ is a

Hilbert space (which is not mandatory!), which seems to be very "close" to the "minimal" space valid for all boundary conditions (if such a space exists!).

For this purpose we need the spaces $\Xi^s(\Omega)$.

$\Xi^s(\Omega)$-Spaces, arbitrary real s

We first introduce the definition for $s = 0, 1, 2, \ldots$

Let $\varrho(x) \in \mathscr{D}(\bar{\Omega})$ and be positive in Ω and vanishing on Γ of the same order as the distance $d(x, \Gamma)$ of x to Γ

$$\left(\text{i.e.} \quad \lim_{x \to x_0 \in \Gamma} \frac{\varrho(x)}{d(x, \Gamma)} = d \neq 0\right);$$

there exist such functions, for Γ is an infinitely differentiable variety (we have already introduced a function of this type in Chapter 1, Section 11.2).

Definition 6.1. *If $s = 0, 1, \ldots$, we set*

$$\Xi^s(\Omega) = \{u \mid \varrho^{|\alpha|} D^\alpha u \in L^2(\Omega), |\alpha| \leqq s\},$$

provided with the norm

$$\| u \|_{\Xi^s(\Omega)} = \left(\sum_{|\alpha| \leqq s} \| \varrho^{|\alpha|} D^\alpha u \|_{L^2(\Omega)} \right)^{1/2}.$$

$\Xi^s(\Omega)$ is a Hilbert space and we have

(6.19) $$\Xi^0(\Omega) = L^2(\Omega), \quad H^s(\Omega) \subset \Xi^s(\Omega) \subset L^2(\Omega)$$

with continuous injection.

Proposition 6.2. *$\mathscr{D}(\Omega)$ is dense in $\Xi^s(\Omega)$.*

Proof. Let $\delta_\nu(x)$ be a sequence of functions of $\mathscr{D}(\Omega)$ such that $\delta_\nu(x) = 1$ if $d(x, \Gamma) \geqq 2/\nu$ and $\delta_\nu(x) = 0$ if $d(x, \Gamma) \leqq 1/\nu$ and

$$|D^\alpha \delta_\nu(x)| \leqq \frac{C_\alpha}{|d(x, \Gamma)|^{|\alpha|}}$$

(C_α depending on α, but not on ν); such a sequence exists. Now, let $u \in \Xi^s(\Omega)$; then $\delta_\nu u \in H^s(\Omega)$ and has compact support in Ω. We verify that

$$\delta_\nu u \to u \quad \text{in} \quad \Xi^s(\Omega) \quad \text{as} \quad \nu \to +\infty.$$

For this purpose we have to show that $\varrho^{|\alpha|} D^\alpha(\delta_\nu u) \to \varrho^{|\alpha|} D^\alpha u$ in $L^2(\Omega)$; but

$$\delta_\nu \varrho^{|\alpha|} D^\alpha u \to \varrho^{|\alpha|} D^\alpha u$$

in $L^2(\Omega)$; therefore, it is sufficient to show that

$$\varrho^{|\alpha|} (D^\beta \delta_\nu) (D^\gamma u) \to 0$$

in $L^2(\Omega)$, $|\beta| \geqq 1$, $|\gamma| + |\beta| = |\alpha|$, $\nu \to +\infty$. But $\varrho^{|\alpha|}(D^\beta \delta_\nu)(D^\gamma u)$ vanishes for $d(x, \Gamma) \geqq 2/\nu$ (for $|\beta| \geqq 1$) so that, according to Lebesgue's theorem, we have the desired result if we note that

$$|\varrho^{|\alpha|}(D^\beta \delta_\nu)(D^\gamma u)| = |\varrho^{|\beta|}(D^\beta \delta_\nu)|\,|\varrho^{|\gamma|} D^\gamma u| \leqq C_\beta |\varrho^{|\gamma|} D^\gamma u| \in L^2(\Omega).$$

Therefore now it suffices to approach $\delta_\nu u$ (fixed ν) in $H^s(\Omega)$ with functions of $\mathscr{D}(\Omega)$; this is possible via the usual regularization procedure, for the support of $\delta_\nu u$ (ν being fixed) belongs to a compact set in Ω. □

Definition 6.2. *Let real $s > 0$ not be an integer, $s = k + \theta$, with integer $k \geqq 0$ and $0 < \theta < 1$; we set*

$$\Xi^s(\Omega) = [\Xi^{k+1}(\Omega), \Xi^k(\Omega)]_{1-\theta}.$$

From this definition, (6.19) and the inclusion properties of $H^s(\Omega)$, there results that

(6.20) $H^s(\Omega) \subset \Xi^s(\Omega) \subset \Xi^{s'}(\Omega) \subset L^2(\Omega)$, $\quad s, s'$ real and >0, $\quad s' < s$.

From Proposition 6.2 and the density theorem (Chapter 1, Section 2), we also deduce that

(6.21) $\quad \mathscr{D}(\Omega)$ *is dense in* $\Xi^s(\Omega)$ *for all real* $s \geqq 0$.

Therefore, we see that $\Xi^s(\Omega)$ is a *normal space of distributions on* Ω; its dual may be identified to a space of distributions on Ω. We denote this dual space by $\Xi^{-s}(\Omega)$, i.e. we set

(6.22) $$\Xi^{-s}(\Omega) = (\Xi^s(\Omega))', \quad s > 0.$$

Proposition 6.4. *For integer $s > 0$, every $f \in \Xi^{-s}(\Omega)$ may be represented (non-uniquely) in the form*

(6.23) $$f = \sum_{|\alpha| \leqq s} D^\alpha(\varrho^{|\alpha|} f_\alpha), \quad \text{with } f_\alpha \in L^2(\Omega).$$

Proof. According to the Hahn-Banach theorem, every continuous linear form on $\Xi^s(\Omega)$ may be written

$$M(\varphi) = \sum_{|\alpha| \leqq s} \int_\Omega g_\alpha \varrho^{|\alpha|} D^\alpha \varphi \, dx \qquad g_\alpha \in L^2(\Omega).$$

According to Proposition 6.2, M is defined by its values for each $\varphi \in \mathscr{D}(\Omega)$, from which (6.23) follows if we set $f_\alpha = (-1)^{|\alpha|} g_\alpha$. □

Choice of the space $K^r(\Omega)$

For all real $r \geqq 0$, we take

$$K^r(\Omega) = \Xi^{2m+r}(\Omega).$$

This choice is permissible, for (6.12) is verified according to (6.20) and (6.13) according to Proposition 6.2. □

Then Theorem 6.2 may be stated as follows: denote by $D_A^{-r}(\Omega)$ the space

(6.24) $D_A^{-r}(\Omega) = \{u \mid u \in H^{-r}(\Omega), A\,u \in \Xi^{-2m-r}(\Omega)\}, \quad r \geqq 0,$

with the norm of the graph

$$\|u\|^2_{D_A^{-r}(\Omega)} = \|u\|^2_{H^{-r}(\Omega)} + \|A\,u\|^2_{\Xi^{-2m-r}(\Omega)}.$$

Then, we have

Theorem 6.3. *Under hypotheses* (i), (ii), (iii) *of Section 5, for every fixed real* $r \geqq 0$, *if* $f \in \Xi^{-2m-r}(\Omega)$ *and* $g_j \in H^{-r-m_j-1/2}(\Gamma)$, $j = 0, \ldots, m-1$, *with* (6.16), *there exists* $u \in D_A^{-r}(\Omega)$, *determined up to the addition of a function of* N, *such that*

(6.25) $$\langle u, \overline{A^* v}\rangle = \langle f, \bar{v}\rangle + \sum_{j=0}^{m-1} \langle g_j, \overline{T_j v}\rangle \qquad \forall v \in X^r(\Omega).$$

Furthermore, $\{f; g_0, \ldots, g_{m-1}\} \to u^{\bullet} = u + N$ *is a continuous linear mapping of the subspace of*

(6.26) $$\Xi^{-2m-r}(\Omega) \times \prod_{j=0}^{m-1} H^{-r-m_j-1/2}(\Gamma)$$

made of the elements satisfying (6.16), *into* $H^{-r}(\Omega)/N$.

To go further with the interpretation of the result we must now give *trace theorems* for the elements of $D_A^{-r}(\Omega)$; we shall do this in the following sections.

6.4 Density Theorem

Theorem 6.4. *Under hypotheses* (i), (ii) *of Section 5, the space* $\mathscr{D}(\bar{\Omega})$ *is dense in* $D_A^{-r}(\Omega)$ *for all real* $r \geqq 0$ *with* $r - \frac{1}{2}$ *not an integer.*

Proof. Let $u \to M(u)$ be a continuous linear form on $D_A^{-r}(\Omega)$: it may be written

(6.27) $M(u) = \langle f, u\rangle + \langle g, A\,u\rangle$, with $f \in H_0^r(\Omega)$ and $g \in \Xi^{2m+r}(\Omega)$,

since the intervening spaces are Hilbert spaces and therefore reflexive.

Suppose that we have

(6.28) $$M(\varphi) = 0 \quad \text{for all } \varphi \in \mathscr{D}(\bar{\Omega}).$$

We have to show that under these conditions

(6.29) $$M(u) = 0 \quad \text{for all } u \in D_A^{-r}(\Omega).$$

But every $\varphi \in \mathscr{D}(\bar{\Omega})$ is a restriction to $\bar{\Omega}$ of a function $\Phi \in \mathscr{D}(\mathbf{R}^n)$. We denote by $\tilde{f}$ and $\tilde{g}$ the extension to $\mathbf{R}^n$ of f and g by zero outside Ω and by $\mathscr{A}$ an operator extending A to $\mathbf{R}^n$ in the following sense: $\mathscr{A}$ is a linear operator of order $2m$ with infinitely differentiable coefficients in $\mathbf{R}^n$, which coincides with the operator A in $\bar{\Omega}$ and which is *properly elliptic* in $\bar{\mathcal{O}}$, where $\mathcal{O}$ is a bounded open set with infinitely differentiable boundary $\partial\mathcal{O}$, with $\bar{\Omega} \subset \mathcal{O}$. Of course, there exists such an extension of A. Also note that $\tilde{f}$ and $\tilde{g}$ belong to $L^2(\mathbf{R}^n)$.

Therefore (6.28) may be written

$$M(\varphi) = \langle \tilde{f}, \Phi \rangle + \langle \tilde{g}, \mathscr{A}\,\Phi \rangle = 0 \tag{6.30}$$

where the brackets are taken in the sense of distributions on $\mathbf{R}^n$; so if $\mathscr{A}^*$ denotes the formal adjoint of $\mathscr{A}$, we have

$$\mathscr{A}^*\,\tilde{g} = -\tilde{f} \tag{6.31}$$

in the sense of distributions on $\mathbf{R}^n$.

But, since $r - \frac{1}{2}$ is not an integer and $f \in H_0^r(\Omega)$, according to Theorem 11.3 of Chapter 1, we have $\tilde{f} \in H^r(\mathbf{R}^n)$ and therefore its restriction to $\mathcal{O}$ belongs to $H^r(\mathcal{O})$. We would like to show that $\tilde{g}\,|_{\mathcal{O}} \in H^{2m+r}(\mathcal{O})$ follows.

Let us show that there exists $w \in H^{2m+r}(\mathcal{O}) \cap H_0^m(\mathcal{O})$ such that

$$\mathscr{A}^*\,w = -\tilde{f}. \tag{6.32}$$

Using Theorem 5.4, applied to the open set $\mathcal{O}$ and to the Dirichlet problem $\mathscr{A}^*\,w = -\tilde{f}$, $\gamma_j\,w = 0$, $j = 0, \ldots, m-1$, such a w exists if and only if

$$\int_{\mathcal{O}} \tilde{f}\,\bar{z}\,dx = 0 \tag{6.33}$$

for all $z \in \mathscr{D}(\bar{\mathcal{O}})$ such that $\mathscr{A}\,z = 0$ in $\mathcal{O}$ and $\gamma_j\,z = 0$, $j = 0, \ldots, m-1$ on $\partial\mathcal{O}$.

We shall prove that this condition is satisfied: let θ be a function of $\mathscr{D}(\mathcal{O})$ equal to 1 on $\bar{\Omega}$; then

$$\int_{\mathcal{O}} \tilde{f}\,\bar{z}\,dx = \int_{\mathcal{O}} \tilde{f}\,\overline{\theta\,z}\,dx = -\int_{\mathcal{O}} (\mathscr{A}^*\,\tilde{g})\,\overline{\theta\,z}\,dx = -\int_{\mathcal{O}} \tilde{g}\,\overline{\mathscr{A}(\theta\,z)}\,dx = 0,$$

for $\tilde{g} \equiv 0$ outside Ω and $\theta = 1$ on $\bar{\Omega}$.

Therefore w exists and

$$w - \tilde{g} \in L^2(\mathcal{O}) \quad \text{with} \quad \mathscr{A}^*(w - \tilde{g}) = 0 \quad \text{in} \quad \mathcal{O}.$$

According to the hypoellipticity of $\mathscr{A}^*$ (Theorem 3.2), it follows that $w - \tilde{g}$ is infinitely differentiable in $\mathcal{O}$ and therefore, since $w \in H^{2m+r}(\mathcal{O})$ and $\tilde{g}$ vanishes outside Ω, we see that $\tilde{g} \in H^{2m+r}(\mathcal{O})$, whence (since g vanishes outside Ω) $\gamma_j\,g = 0$, $j = 0, \ldots, [2m + r - \frac{1}{2}]^-$, on Γ and therefore (Chapter 1, Section 11) $g \in H_0^{2m+r}(\Omega)$.

Now, we compute $\langle g, A\,u \rangle$ with $u \in D_A^{-r}(\Omega)$; since $\mathscr{D}(\Omega)$ is dense in $H_0^{2m+r}(\Omega)$, there exists a sequence $\psi_k \in \mathscr{D}(\Omega)$ with $\psi_k \to g$ in $H_0^{2m+r}(\Omega)$;

so that we have

$$\langle g, A u\rangle = \lim_{k\to\infty} \langle g_k, A u\rangle = \lim_{k\to\infty} \langle A^* g_k, u\rangle$$
$$= \langle A^* g, u\rangle,$$

for $u \in H^{-r}(\Omega)$ and $A^* g_k \to A g$ in $H_0^r(\Omega)$. Therefore

$M(u) = \langle A^* g + f, u\rangle = 0$, for $A^* g + f = 0$ according to (6.31). □

Remark 6.3. In the proof of Theorem 6.4, an exception appears for the parameter r: $r - \frac{1}{2}$ = integer; but we shall see further on (Remark 6.5) that the theorem holds also when $r - \frac{1}{2}$ is an integer. □

6.5 Trace Theorem and Green's Formula for the Space $D_A^s(\Omega)$, $s \leq 0$

Theorem 6.5. *Under hypotheses* (i), (ii) *of Section 5 and if the system* $\{B_j\}_{j=0}^{m-1}$ *is normal on* Γ, *for all real* $r \geqq 0$, *with* $r - \frac{1}{2}$ *not an integer, the mapping* $u \to B u = \{B_0 u, \ldots, B_{m-1} u\}$ *of* $\mathscr{D}(\bar{\Omega})$ *into* $\{\mathscr{D}(\Gamma)\}^m$ *extends by continuity to a continuous linear mapping, still denoted* $u \to B u$, *of*

$$D_A^{-r}(\Omega) \quad \text{into} \quad \prod_{j=0}^{m-1} H^{-r-m_j-1/2}(\Gamma).$$

Furthermore, for $u \in D_A^{-r}(\Omega)$ *and* $v \in X^r(\bar{\Omega})$, *we have "Green's formula":*

$$\text{(6.34)} \qquad \langle A u, \bar{v}\rangle - \langle u, \overline{A^* v}\rangle = -\sum_{j=0}^{m-1} \langle B_j u, \overline{T_j v}\rangle,$$

where the first brackets denote the duality between $\Xi^{-2m-r}(\Omega)$ *and* $\Xi^{2m+r}(\Omega)$, *the second between* $H^{-r}(\Omega)$ *and* $H_0^r(\Omega)$ *and* $\langle B_j u, \overline{T_j v}\rangle$ *the duality between*

$$H^{-r-m_j-1/2}(\Gamma) \quad \text{and} \quad H^{r+m_j+1/2}(\Gamma).$$

Proof. Let u be given in $D_A^{-r}(\Omega)$ and let φ_j, $j = 0, \ldots, m-1$, be given in $H^{r+m_j+1/2}(\Gamma)$. Consider the right-inverse of $\mathscr{T}$ introduced in Theorem 6.1 and apply it to $\{\varphi_0, \ldots, \varphi_{m-1}\}$; we obtain a function $v_\varphi \in H^{2m+r}(\Omega)$ such that

$$C_j v_\varphi = 0, \quad T_j v_\varphi = \varphi_j, \quad j = 0, \ldots, m-1 \quad \text{and} \quad A^* v_\varphi \in H_0^r(\Omega),$$

v_φ depending continuously on the φ_j's. Then we may consider the expression

$$\text{(6.35)} \qquad Z(v_\varphi) = \langle u, \overline{A^* v_\varphi}\rangle - \langle A u, \bar{v}_\varphi\rangle,$$

where the first brackets denote the duality between $H^{-r}(\Omega)$ and $H_0^r(\Omega)$ and the second between $\Xi^{-2m-r}(\Omega)$ and $\Xi^{2m+r}(\Omega)$ (note that $v_\varphi \in X^r(\Omega) \subset \subset \Xi^{2m+r}(\Omega)$).

$Z(v_\varphi)$ is independent of the "right-inverse" used; indeed, if v_1 and v_2 are two functions such as v_φ, then $\chi = v_1 - v_2$ satisfies the conditions

$$\begin{aligned} \gamma_j \chi = 0, \quad & j = 0, \ldots, 2m-1, & \text{if } r \leqq \tfrac{1}{2}, \\ & j = 0, \ldots, 2m + [r - \tfrac{1}{2}]^- & \text{if } r > \tfrac{1}{2} \end{aligned}$$

(we use the same considerations as in the proof of Theorem 6.1, 2)). Therefore (Chapter 1, Theorem 11.5)

$$\chi \in H_0^{2m+r}(\Omega), \quad \text{so that} \quad \langle u, \overline{A^* \chi} \rangle = \langle A\, u, \bar{\chi} \rangle$$

and therefore $Z(v_1) = Z(v_2)$.

Therefore (6.35) depends only on φ and we may write $Z(\varphi)$ instead of $Z(v_\varphi)$. Using (6.35) for $Z(\varphi)$ we see that $\varphi \to Z(\varphi)$ is a continuous antilinear form on

$$\prod_{j=0}^{m-1} H^{r+m_j+1/2}(\Gamma);$$

therefore

$$Z(\varphi) = \sum_{j=0}^{m-1} \langle \tau_j\, u, \bar{\varphi}_j \rangle, \tag{6.36}$$

where $\tau_j\, u \in H^{-r-m_j-1/2}(\Gamma)$.

Still using (6.35), we easily see that the mappings $u \to \tau_j\, u$ are continuous linear mappings of $D_A^{-r}(\Omega)$ into $H^{-r-m_j-1/2}(\Gamma)$.

Now, we shall verify that

$$\tau_j\, u = B_j\, u \quad \text{for} \quad u \in \mathscr{D}(\bar{\Omega}). \tag{6.37}$$

If we take $\varphi_j \in \mathscr{D}(\Gamma)$, $j = 0, \ldots, m-1$, we may construct the right-inverse v in $\mathscr{D}(\bar{\Omega})$ (see the proof of Theorem 6.1 and Lemma 2.2); then, according to Green's formula (5.15):

$$Z(\varphi) = \sum_{j=0}^{m-1} \int_\Gamma B_j\, u\, \overline{T_j\, v}\, d\sigma = \sum_{j=0}^{m-1} \int_\Gamma (B_j\, u)\, \bar{\varphi}_j\, d\sigma \qquad \forall \varphi_j \in \mathscr{D}(\Gamma),$$

whence (6.37), comparing with (6.36).

Also note that the mapping τ_j is uniquely determined by the operator B_j, for, thanks to Theorem 6.4, $\mathscr{D}(\bar{\Omega})$ is dense in $D_A^{-r}(\Omega)$.

Finally, Green's formula (6.34) results from the preceding considerations: for if $v \in X^r(\Omega)$, then $\varphi_j = T_j\, v \in H^{r+m_j+1/2}(\Gamma)$ and we may take $v_\varphi = v$ in (6.35) and therefore (6.34) follows from (6.35) and (6.36). □

Remark 6.4. If $r - \frac{1}{2}$ is an integer, the proof may still be carried out: we obtain the existence of a continuous linear mapping $u \to \tau\, u$

$= (\tau_0 u, \ldots, \tau_{m-1} u)$ of $D_A^{-r}(\Omega)$ into

$$\prod_{j=0}^{m-1} H^{-r-m_j-1/2}(\Gamma),$$

which coincides with $B u = \{B_0 u, \ldots, B_{m-1} u\}$ on $\mathscr{D}(\bar{\Omega})$. But we can not say that τu is the extension by continuity of $B u$ and therefore that it is uniquely determined by $B u$, for we do not know whether $\mathscr{D}(\bar{\Omega})$ is dense or not in $D_A^{-r}(\Omega)$. Nevertheless, we shall see further on (Remark 6.5) that Theorem 6.5 also holds for $r - \frac{1}{2} =$ integer.

In this case we may still write Green's formula (6.34), but with the operator τ_j instead of B_j.

6.6 Existence of Solutions in $D_A^s(\Omega)$-Spaces, with Real $s \leqq 0$

We are now in a position to interpret Theorem 6.3 in a more precise way.

Indeed, we may write either formula (6.25) or (6.34) for the solution u obtained in Theorem 6.3. But we already know (see (6.18)) that $A u = f$; therefore, it follows that

$$\sum_{j=0}^{m-1} \langle B_j u - g_j, T_j v\rangle = 0 \qquad \forall v \in X^r(\Omega)$$

and, thanks to Theorem 6.1:

$$\sum_{j=0}^{m-1} \langle B_j u - g_j, \varphi_j\rangle = 0 \qquad \forall \varphi_j \in H^{r+m_j+1/2}(\Gamma)$$

and therefore $B_j u = g_j$, $j = 0, \ldots, m-1$.

Therefore, we have shown

Theorem 6.6. *Under hypotheses* (i), (ii), (iii) *of Section 5, for all real* $s \leqq 0$, *with* $s - \frac{1}{2}$ *not an integer, the operator*

$$\mathscr{P}: u \to \mathscr{P} u = \{A u, B_0 u, \ldots, B_{m-1} u\},$$

defines an (algebraic and topological) isomorphism of $D_A^s(\Omega)/N$ *onto the space*

$$\left\{\Xi^{s-2m}(\Omega) \times \prod_{j=0}^{m-1} H^{s-m_j-1/2}(\Gamma); N^*, \mathscr{T}\right\}$$

of elements $\{f; g_0, \ldots, g_{m-1}\}$ *belonging to*

$$\Xi^{s-2m}(\Omega) \times \prod_{j=0}^{m-1} H^{s-m_j-1/2}(\Gamma)$$

and satisfying (6.16).

Remark 6.5. Let us look at the case $s - \frac{1}{2} = k$, integer $k \leqq -1$. Using Remark 6.4 and the same arguments as for Theorem 6.6, we obtain:

$$(6.38)\quad \left\{\begin{array}{l} \textit{under the hypotheses (i), (ii), (iii) of Section 5, the operator} \\ \qquad u \to \mathscr{P}_\tau u = \{A u; \tau_0 u, \ldots, \tau_{m-1} u\} \\ \textit{defines an (algebraic and topological) isomorphism of } D_A^{k+1/2}(\Omega)/N \\ \textit{onto the space} \\ \qquad \left\{\Xi^{k+1/2-2m}(\Omega) \times \prod_{j=0}^{m-1} H^{k-m_j}(\Gamma); N^*, \mathscr{T}\right\} \\ \textit{of elements } \{f; g_0, \ldots, g_{m-1}\} \textit{ belonging to} \\ \qquad \Xi^{k+1/2-2m}(\Omega) \times \prod_{j=0}^{m-1} H^{k-m_j}(\Gamma) \\ \textit{and satisfying } (6.16). \end{array}\right.$$

On the other hand, using Theorem 6.6, we obtain that $u \to \mathscr{P} u$ is also an isomorphism of

$$(6.39)\quad D_A^k(\Omega)/N \quad \text{onto} \quad \left\{\Xi^{k-2m}(\Omega) \times \prod_{j=0}^{m-1} H^{k-m_j-1/2}(\Gamma); N^*, \mathscr{T}\right\}$$

and of

$$(6.40)\quad D_A^{k+1}(\Omega)/N \quad \text{onto} \quad \left\{\Xi^{k+1-2m}(\Omega) \times \prod_{j=0}^{m-1} H^{k+1-m_j-1/2}(\Gamma); N^*, \mathscr{T}\right\}.$$

By interpolation, there results that $\mathscr{P}$ is an isomorphism of

$$(6.41)\quad [D_A^k(\Omega)/N, D_A^{k+1}(\Omega)/N]_{1/2} \quad \text{onto}$$
$$\left[\left\{\Xi^{k-2m}(\Omega) \times \prod_{j=0}^{m-1} H^{k-m_j-1/2}(\Gamma); N^*, \mathscr{T}\right\}, \left\{\Xi^{k+1-2m}(\Omega) \times \prod_{j=0}^{m-1} H^{k+1-m_j-1/2}(\Gamma); N^*, \mathscr{T}\right\}\right]_{1/2}.$$

It is easy to verify, by one of the (equivalent) definitions of the spaces $[X, Y]_\theta$ given in Chapter 1, that

$$(6.42)\quad [X_1 \times X_2, Y_1 \times Y_2]_\theta = [X_1, X_2]_\theta \times [Y_1, Y_2]_\theta, \quad 0 < \theta < 1,$$

if $\{X_i, Y_i\}$ is a couple of Hilbert spaces having analogous properties to the couple $\{X, Y\}$. Of course, this formula is valid for a product of any finite number of spaces. Therefore, using the results of Chapter 1, Section 13.4, the definition of $\Xi^s(\Omega)$-spaces, the duality theorem (Chapter 1, Section 6.2) and the interpolation between the spaces $H^s(\Gamma)$, we deduce from (6.41) that $\mathscr{P}$ is an isomorphism of

$$([D_A^k(\Omega), D_A^{k+1}(\Omega)]_{1/2})/N \text{ onto } \left\{\Xi^{k+1/2-2m}(\Omega) \times \prod_{j=0}^{m-1} H^{k-m_j}(\Gamma); N^*, \mathscr{T}\right\}.$$

Therefore, we have the following situation:

$\mathscr{P}$ and $\mathscr{P}_\tau$ are two isomorphisms of

$$([D_A^k(\Omega), D_A^{k+1}(\Omega)]_{1/2})/N \quad \text{and} \quad D_A^{k+1/2}(\Omega)/N$$

respectively, onto the same space $\left\{\Xi^{k+1/2-2m}(\Omega) \times \prod_{j=0}^{m-1} H^{k-m_j}(\Gamma); N^*, \mathscr{T}\right\}$.

But it can be shown, directly and easily, that

$$[D_A^k(\Omega), D_A^{k+1}(\Omega)]_{1/2} \subset D_A^{k+1/2}(\Omega). \tag{6.43}$$

Furthermore, $\mathscr{D}(\bar{\Omega})$ is dense in $[D_A^k(\Omega), D_A^{k+1}(\Omega)]_{1/2}$, thanks to the density Theorem 6.4 and to Chapter 1, Section 2.1, and on $\mathscr{D}(\bar{\Omega})$

$$\mathscr{P} u = \mathscr{P}_\tau u \qquad \text{(see Remark 6.4).}$$

Therefore

$$[D_A^k(\Omega), D_A^{k+1}(\Omega)]_{1/2} = D_A^{k+1/2}(\Omega) \tag{6.44}$$

and $\mathscr{P} u = \mathscr{P}_\tau u$ on this space.

Therefore, even if $s - \frac{1}{2}$ is an integer, the density, trace and existence theorems (Theorems 6.4, 6.5 and 6.6) are valid.

Finally, we may state

Theorem 6.7. *Under hypotheses* (i), (ii), (iii) *of Section 5, for all real* $s \leqq 0$, *the operator*

$$\mathscr{P}: u \to \mathscr{P} u = \{A u, B_0 u, \ldots, B_{m-1} u\}$$

defines an (*algebraic and topological*) *isomorphism of* $D_A^s(\Omega)/N$ *onto the space* $\left\{\Xi^{s-2m}(\Omega) \times \prod_{j=0}^{m-1} H^{s-m_j-1/2}(\Gamma); N^*, \mathscr{T}\right\}$ *of elements* $\{f; g_0, \ldots, g_{m-1}\}$ *belonging to* $\Xi^{s-2m} \times \prod_{j=0}^{m-1} H^{s-m_j-1/2}(\Gamma)$ *and satisfying* (6.16).

This is the extension of Theorem 5.4 to $H^s(\Omega)$-spaces, with $s \leqq 0$.

We may again say that, in $D_A^s(\Omega)$, $\mathscr{P}$ is an indexed operator and that its index is given by the formula $\chi(\mathscr{P}) = \dim N - \dim N^*$.

Note that we have solved problem (5.1), that is

$$A u = f \quad \text{in } \Omega, \tag{6.45}$$

$$B_j u = g_j \quad \text{on } \Gamma, \quad j = 0, \ldots, m-1, \tag{6.46}$$

the equation (6.45) being satisfied in the *sense of distributions* on Ω and the boundary conditions (6.46) in the *sense of the trace Theorem* 6.5; therefore, we have given a "natural" meaning to problem (5.1) in the spaces $H^s(\Omega)$ with real $s \leqq 0$ (see also Section 8.1, below, for the interpretation of (6.46)). □

Remark 6.6. Theorem 6.7 solves the boundary value problem (6.45), (6.46) with *arbitrary distributions* g_j on Γ (and f, for example, belonging to a space $\Xi^s(\Omega)$, arbitrary and fixed).

Indeed, Γ being compact, we have

$$\bigcup_{s \geq 0} H^{-s}(\Gamma) = \mathcal{D}'(\Gamma).$$

Therefore, if we want to solve problem $A u = f$, $B_j u = g_j$, $j = 0, \ldots, m-1$, with f given in a space $\Xi^s(\Omega)$, arbitrary and fixed, and g_j given in $\mathcal{D}'(\Gamma)$, there exists $\mu > 0$, sufficiently large, such that $f \in \Xi^{-\mu-2m}(\Omega)$ and $g_j \in H^{-\mu-m_j-1/2}(\Gamma)$ and therefore we may apply Theorem 6.6 and find $u \in H^{-\mu}(\Omega)$.

Note that $\bigcup_{s>0} H^{-\mu}(\Omega) \subset \mathcal{D}'(\Omega)$, strictly. But we shall see in Volume 3 how this may be further extended.

7. Application of Interpolation: Existence of Solutions in $H^s(\Omega)$-Spaces, with Real s, $0 < s < 2m$

7.1 New Properties of $\Xi^s(\Omega)$-Spaces

We shall first introduce some new properties of the spaces $\Xi^s(\Omega)$ of Section 6.2.

Proposition 7.1. *For integer* $s > 0$, $\Xi^s(\Omega)$ *coincides with the space of* u*'s belonging to* $\mathcal{D}'(\Omega)$ *such that*

$$\varrho^s u \in H_0^s(\Omega). \tag{7.1}$$

Proof. 1) Let $u \in \Xi^s(\Omega)$; then we see that $\varrho^s u \in H^s(\Omega)$. Indeed, using the formula of Leibniz, we have

$$D^\alpha(\varrho^s u) = \sum_{\substack{\gamma \\ \{0 \leq \gamma_i \leq \alpha_i\}}} L_\gamma(x)\, \varrho^{s-|\gamma|} D^{\alpha-\gamma} u \tag{7.2}$$

where $L_\gamma(x)$ is a continuous function on $\bar{\Omega}$; therefore for $|\alpha| \leq s$, we have

$$D^\alpha(\varrho^s u) \in L^2(\Omega), \quad \text{i.e.} \quad \varrho^s u \in H^s(\Omega).$$

Furthermore, $\varrho^s u$ may be approximated with functions of $\mathcal{D}(\Omega)$ in $H^s(\Omega)$; indeed, there exists (Proposition 6.1) a sequence $\varphi_\nu \in \mathcal{D}(\Omega)$ such that $\varphi_\nu \to u$ in $\Xi^s(\Omega)$, i.e.

$$\varrho^{|\alpha|} D^\alpha \varphi_\nu \to \varrho^{|\alpha|} D^\alpha u$$

in $L^2(\Omega)$, for $|\alpha| \leq s$; again using (7.2), it follows that

$$\varrho^s \varphi_\nu \to \varrho^s u$$

in $H^s(\Omega)$; but $\varrho^s \varphi_\nu \in \mathcal{D}(\Omega)$, therefore $\varrho^s u \in H_0^s(\Omega)$.

2) Conversely, let $u \in \mathscr{D}'(\Omega)$ with (7.1), we show that

$$\varrho^{|\alpha|} D^\alpha u \in L^2(\Omega) \qquad \forall \alpha \quad \text{with} \quad |\alpha| \leqq s. \tag{7.3}$$

Via "local maps" and "partition of unity" we are led to the following situation: we have a function $v(y, t)$, which is locally in $L^2(\mathbf{R}^n_+)$ and has compact support in $\overline{\mathbf{R}}^n_+$, such that, for fixed r, with $r = 0, 1, \ldots, s$, and fixed D^γ_y, with $|\gamma| \leqq r$, the function $w(y, t) = t^s D^\gamma_y v(y, t)$ belongs to $L^2(\mathbf{R}^n_+)$ as well as its derivatives $D^j_t w$ for $j = 1, \ldots, r - |\gamma|$ and furthermore $D^j_t w(y, t)|_{t=0} = 0$, $j = 0, \ldots, r - |\gamma| - 1$. Then, we may apply Lemma 10.1 of Chapter 1 to obtain

$$\begin{aligned}
\int_0^{+\infty} t^{-2(r-|\gamma|)} \| w(y, t) \|^2_{L^2(\mathbf{R}^{n-1})} \, dt &\leqq \\
\leqq C \int_0^{+\infty} t^{-2(r-|\gamma|)} \| t \, D_t w(y, t) \|^2_{L^2(\mathbf{R}^{n-1})} \, dt & \\
= C \int_0^{+\infty} t^{-2(r-|\gamma|-1)} \| D_t w(y, t) \|^2_{L^2(\mathbf{R}^{n-1})} \, dt &\leqq \\
\leqq C^2 \int_0^{+\infty} t^{-2(r-|\gamma|-1)} \| t \, D_t w(y, t) \|^2_{L^2(\mathbf{R}^{n-1})} \, dt &\leqq \\
\leqq \cdots \leqq C^{s-1} \int_0^{+\infty} t^{-2} \| D_t^{r-|\gamma|-1} w(y, t) \|^2_{L^2(\mathbf{R}^{n-1})} \, dt &\leqq \\
\leqq C^s \int_0^{+\infty} t^{-2} \| t \, D_t^{r-|\gamma|} w(y, t) \|^2_{L^2(\mathbf{R}^{n-1})} \, dt & \\
= \| D^{r-|\gamma|} w \|^2_{L^2(\mathbf{R}^n_+)} < +\infty . &
\end{aligned} \tag{7.4}$$

Therefore

$$t^{|\gamma|-r+j} D^j_t (t^s D^\gamma_y v) \in L^2(\mathbf{R}^n_+) \tag{7.5}$$

for $r = 0, 1, \ldots, s$, $|\gamma| \leqq r$, $j = 0, 1, \ldots, r - |\gamma|$, from which we obtain

$$t^{|\gamma|+j} D^j_t D^\gamma_y v \in L^2(\mathbf{R}^n_+) \quad \text{for} \quad j + |\gamma| \leqq s,$$

which proves (7.3). □

We are now in a position to prove

Theorem 7.1. *If integer $s > 0$, we have*

$$[\Xi^s(\Omega), L^2(\Omega)]_\theta = \{u \mid u \in \mathscr{D}'(\Omega), \varrho^{s(1-\theta)} u \in [H^s_0(\Omega), L^2(\Omega)]_\theta\}. \tag{7.6}$$

Proof. — 1) Let $u \in [\Xi^s(\Omega), L^2(\Omega)]_\theta$; then according to the holomorphic interpolation (Chapter 1, Section 14), there exists f such that

$u = f(\theta)$, where $\zeta \to f(\zeta)$ is a continuous function taking its values in $L^2(\Omega)$ in the strip

$$\mathscr{B} = \{\zeta \mid \zeta = \xi + i\eta, 0 \leqq \xi \leqq 1\},$$

of polynomial growth in η (Chapter 1, Corollary 14.1), and holomorphic taking its values in $L^2(\Omega)$ in the interior of $\mathscr{B}$, $\eta \to f(i\eta)$ being continuous, of polynomial growth in η, taking its values in $\Xi^s(\Omega)$, $\eta \to \to f(\xi + i\eta)$ being continuous, of polynomial growth in η, taking its values in $L^2(\Omega)$.

Consider the function

(7.7) $$\zeta \to g(\zeta) = \varrho^{s(1-\zeta)} f(\zeta), \quad \zeta \in \mathscr{B}$$

and let us verify that

(7.8) $\begin{cases} \zeta \to g(\zeta) \text{ is a continuous function, of polynomial growth, of } \mathscr{B} \\ \text{into } L^2(\Omega) \text{ and analytic in the interior of } \mathscr{B}. \end{cases}$

(7.9) $\begin{cases} \eta \to g(i\eta) \text{ is a continuous function, of polynomial growth, of } \mathbf{R} \\ \text{into } H_0^s(\Omega). \end{cases}$

First of all g is continuous, for

$$\| g(\zeta) \|_{L^2(\Omega)} \leqq C \, \| f(\zeta) \|_{L^2(\Omega)}.$$

It is also analytic in the interior of $\mathscr{B}$; in fact, it suffices to apply a theorem of Grothendieck on holomorphic functions taking their values in a topological vector space (see Grothendieck [1]) and to verify that

$$\zeta \to \langle g(\zeta), \varphi \rangle = \int_\Omega g(\zeta) \varphi \, dx$$

for all $\varphi \in \mathscr{D}(\Omega)$, is a scalar holomorphic function in the interior of $\mathscr{B}$. But

$$\langle g(\zeta), \varphi \rangle = \langle f(\zeta), \varrho^{s(1-\zeta)} \varphi \rangle \quad \text{and} \quad \zeta \to f(\zeta) \quad \text{and} \quad \zeta \to \varrho^{s(1-\zeta)} \varphi$$

are holomorphic with values in $L^2(\Omega)$; therefore, $\zeta \to \langle g(\zeta), \varphi \rangle$ is holomorphic and (7.8) is proved.

To verify (7.9), using Theorem 11.8 of Chapter 1, it suffices to verify first that

(7.10) $$\varrho^{-s+|\alpha|} D^\alpha g(i\eta) \in L^2(\Omega) \qquad \forall \eta \in \mathbf{R}, \quad |\alpha| \leqq s.$$

But, by the formula of Leibniz, we have

$$\varrho^{-s+|\alpha|} D^\alpha g(i\eta) = \varrho^{-s+|\alpha|} \sum_{\substack{\beta \\ \{0 \leqq \beta_i \leqq \alpha_i\}}} L_\beta(x, y) \, \varrho^{s(1-i\eta)-|\beta|} D^{\alpha-\beta} f(i\eta)$$

where the $L_\beta(x, \eta)$ are polynomials in η, with coefficients which are continuous functions of x in $\bar{\Omega}$. But $f(i\eta) \in \Xi^s(\Omega)$ and therefore

$\varrho^{|\alpha|-|\beta|} D^{\alpha-\beta} f(\mathrm{i}\,\eta) \in L^2(\Omega)$ and therefore (7.10) is proved. Furthermore, we obtain

$$\| \varrho^{-s+|\alpha|} D^\alpha g(\mathrm{i}\,\eta) \|_{L^2(\Omega)} \leqq P(\eta) \, \| f(\mathrm{i}\,\eta) \|_{\Xi^s(\Omega)}$$

with $P(\eta)$ a polynomial in η of degree $|\alpha|$. Therefore (7.9) is also proved. It follows that

$$g(\theta) \in [H_0^s(\Omega), L^2(\Omega)]_\theta,$$

that is

$$\varrho^{s(1-\theta)} f(\theta) = \varrho^{s(1-\theta)} u \in [H_0^s(\Omega), L^2(\Omega)]_\theta.$$

2) Conversely, let $u \in \mathscr{D}'(\Omega)$ with $\varrho^{s(1-\theta)} u \in [H_0^s(\Omega), L^2(\Omega)]_\theta$; then, still using the method of holomorphic interpolation, we see that $\varrho^{s(1-\theta)} u = \psi(\theta)$, where $\zeta \to \psi(\zeta)$ is a function taking its values in $L^2(\Omega)$, continuous in $\mathscr{B}$, of polynomial growth in η and holomorphic in the interior of $\mathscr{B}$, $\eta \to \psi(\mathrm{i}\,\eta)$ being continuous and of polynomial growth in η, taking its values in $H_0^s(\Omega)$.

Consider the function

$$\zeta \to \chi(\zeta) = \varrho^{s(\zeta-1)} \psi(\zeta), \quad \zeta \in \mathscr{B} \tag{7.11}$$

and verify that

(7.12) $\zeta \to \chi(\zeta)$ is a continuous function of $\mathscr{B}$ into $L^2(\Omega)$ of polynomial growth in η;

(7.13) $\zeta \to \chi(\zeta)$ is holomorphic in the interior of $\mathscr{B}$ (with values in $L^2(\Omega)$);

(7.14) $\eta \to \chi(\mathrm{i}\,\eta)$ is a continuous function of $\mathbf{R}$ into $\Xi^s(\Omega)$ and of polynomial growth in η.

We verify that (7.12) holds. We note that not only $\psi(\theta)$ belongs to $[H_0^s(\Omega), L^2(\Omega)]_\theta$, but also

$$\psi(\theta + \mathrm{i}\,\eta_0) \in [H_0^s(\Omega), L^2(\Omega)]_\theta, \quad \forall \eta_0.$$

Indeed (and this is a general property on $[X, Y]_\theta$-spaces) it suffices to not that the function $\tilde{\psi}(\zeta) = \psi(\zeta + \mathrm{i}\,\eta_0)$ satisfies exactly the same conditions as $\psi(\zeta)$ (i.e. is a continuous function of $\mathscr{B}$ into $L^2(\Omega)$, of polynomial growth in η, holomorphic in the interior of $\mathscr{B}$, $\eta \to \tilde{\psi}(\mathrm{i}\,\eta)$ being a continuous function of $\mathbf{R}$ into $H_0^s(\Omega)$ of polynomial growth in η); therefore $\tilde{\psi}(\theta) = \psi(\theta + \mathrm{i}\,\eta_0) \in [H_0^s(\Omega), L^2(\Omega)]_\theta$. Using Remark 11.9 of Chapter 1, for $\psi(\zeta)$ with $\zeta = \theta + \mathrm{i}\,\eta$, we obtain

$$\varrho^{-s(1-\mathrm{Re}\,\zeta)} \psi(\zeta) \in L^2(\Omega) \qquad \forall \zeta \in \mathscr{B}$$

and therefore that $\chi(\zeta) \in L^2(\Omega)$ and

$$\| \chi(\zeta) \|_{L^2(\Omega)} = \| \varrho^{s(\mathrm{Re}\,\zeta - 1)} \psi(\zeta) \|_{L^2(\Omega)} < +\infty, \quad \forall \zeta \in \mathscr{B},$$

whence (7.12), thanks to the properties of ψ.

To verify (7.13) we use the theorem of Grothendieck [1] again; the scalar function $\zeta \to \langle \chi(\zeta), \varphi \rangle$, $\forall \varphi \in \mathscr{D}(\Omega)$, is holomorphic in the interior of $\mathscr{B}$, for $\langle \chi(\zeta), \varphi \rangle = \langle \psi(\zeta), \varrho^{s(\zeta-1)} \varphi \rangle$ and $\zeta \to \psi(\zeta)$ and $\zeta \to \to \varrho^{s(\zeta-1)} \varphi$ are holomorphic and take their values in $L^2(\Omega)$. Finally, we verify (7.14); for $|\alpha| \leqq s$, we have

$$\begin{aligned}\varrho^{|\alpha|} D^\alpha \chi(\mathrm{i}\,\eta) &= \varrho^{|\alpha|} D^\alpha(\varrho^{s(\mathrm{i}\eta-1)} \psi(\mathrm{i}\,\eta)) \\ &= \varrho^{|\alpha|} \sum_{\substack{\beta \\ 0 \leqq \beta_i \leqq \alpha_i}} L'_\beta(x, \eta)\, \varrho^{\mathrm{i}\eta s - s - |\beta|} D^{\alpha-\beta} \psi(\mathrm{i}\,\eta),\end{aligned}$$

where the $L'_\beta(x, \eta)$'s are polynomials with respect to η with coefficients which are continuous functions of χ in $\bar{\Omega}$. But $\psi(\mathrm{i}\,\eta) \in H_0^s(\Omega)$ and therefore (Theorem 11.8, Chapter 1) $\varrho^{-s+|\alpha|-|\beta|} D^{\alpha-\beta} \psi(\mathrm{i}\,\eta) \in L^2(\Omega)$, therefore

$$\varrho^{|\alpha|} D^\alpha \chi(\mathrm{i}\,\eta) \in L^2(\Omega), \quad \text{i.e.} \quad \chi(\mathrm{i}\,\eta) \in \Xi^s(\Omega).$$

Furthermore

$$\| \varrho^{|\alpha|} D^\alpha \chi(\mathrm{i}\,\eta) \|_{L^2(\Omega)} \leqq P(\eta)\, \| \psi(\mathrm{i}\,\eta) \|_{H_0{}^s(\Omega)} \quad \text{for} \quad |\alpha| \leqq s,$$

where $P(\eta)$ is a polynomial in η and therefore (7.14) is verified.

From (7.12), (7.13) and (7.14) we deduce that

$$\chi(\theta) = \varrho^{s(\theta-1)} \psi(\theta) = \varrho^{s(\theta-1)} \varrho^{s(1-\theta)} u = u \in [\Xi^s(\Omega), L^2(\Omega)]_\theta$$

and therefore the theorem is proved. □

From Theorem 7.1 with the results of Chapter 1, Section 11.5, we deduce

Corollary 7.1. *If integer* $s > 0$, *we have*

$$[\Xi^s(\Omega), L^2(\Omega)]_\theta = \{u \mid u \in \mathscr{D}'(\Omega), \varrho^{s(1-\theta)} u \in H_0^{s(1-\theta)}(\Omega)\}$$

for $0 < \theta < 1$, *with* $s(1-\theta) + \frac{1}{2} \neq$ *integer.*

Using Proposition 7.1, we obtain

Corollary 7.2. *If integer* $s > 0$ *and* $s(1-\theta)$ *is an integer, we have*

$$[\Xi^s(\Omega), L^2(\Omega)]_\theta = \Xi^{s(1-\theta)}(\Omega).$$

Using the reiteration theorem (Chapter 1, Section 6.1) we also obtain

Corollary 7.3. *If integer* $s > 0$, *we have*

$$[\Xi^s(\Omega), L^2(\Omega)]_\theta = \Xi^{s(1-\theta)}(\Omega) \qquad \forall \theta, \quad 0 < \theta < 1.$$

Indeed, let $k < s(1-\theta) < k+1$, integer k, then by Definition 6.2, we have

$$\Xi^{s(1-\theta)}(\Omega) = [\Xi^{k+1}(\Omega), \Xi^k(\Omega)]_{\theta'}, \qquad \text{with} \quad \theta' = 1 - [s(1-\theta) - k].$$

But from Corollary 7.2 and the reiteration theorem (Chapter 1, Section 6.1) we deduce

$$[\Xi^{k+1}(\Omega), \Xi^k(\Omega)]_{\theta'} = [[\Xi^s(\Omega), L^2(\Omega)]_{\theta_0}, [\Xi^s(\Omega), L^2(\Omega)]_{\theta_1}]_{\theta'}$$
$$= [\Xi^s(\Omega), L^2(\Omega)]_{\theta^*},$$

with

$$\theta_0 = \frac{s-k-1}{s}, \quad \theta_1 = \frac{s-k}{s}, \quad \theta^* = (1-\theta')\,\theta_0 + \theta'\,\theta_1$$

and therefore

$$\theta^* = [s(1-\theta) - k]\frac{s-k-1}{s} + \{1 - [s(1-\theta)-k]\}\frac{s-k}{s} = \theta. \quad \square$$

Finally, from Corollaries 7.1 and 7.3, we deduce

Corollary 7.4. For real $s > 0$, $s - \frac{1}{2} \neq$ integer, we have

$$\Xi^s(\Omega) = \{u \mid u \in \mathscr{D}'(\Omega),\ \varrho^s u \in H_0^s(\Omega)\}.$$

7.2 Use of Interpolation; First Results

We have already applied the theory of interpolation in Section 5.4 to study the problem (5.1) in the spaces $H^s(\Omega)$ with real $s > 2m$ and in Remark 6.5. We shall use it again now, in an essential way, for the case real s, $0 < s < 2m$, which is still missing. We assume that hypotheses (i), (ii), (iii) of Section 5.1 are satisfied.

From Theorems 5.4 ($s = 2m$) and 6.6 ($s = 0$), we deduce that the operator $\mathscr{P} = \{A; B_0, \ldots, B_{m-1}\}$ is an (algebraic and topological) isomorphism of

$$(7.15)\quad H^{2m}(\Omega)/N \quad \text{onto} \quad \left\{L^2(\Omega) \times \prod_{j=0}^{m-1} H^{2m-m_j-1/2}(\Gamma); N^*, \mathscr{T}\right\}$$

and of

$$(7.16)\quad D_A^0(\Omega)/N \quad \text{onto} \quad \left\{\Xi^{-2m}(\Omega) \times \prod_{j=0}^{m-1} H^{-m_j-1/2}(\Gamma); N^*, \mathscr{T}\right\}.$$

By interpolation, it follows that

(7.17) $\quad \mathscr{P}$ is an isomorphism of $[H^{2m}(\Omega)/N, D_A^0(\Omega)/N]_\theta$

onto

$$\Bigg[\left\{L^2(\Omega) \times \prod_{j=0}^{m-1} H^{2m-m_j-1/2}(\Gamma); N^*, \mathscr{T}\right\},$$
$$\left\{\Xi^{-2m}(\Omega) \times \sum_{j=0}^{m-1} H^{-m_j-1/2}(\Gamma); N^*, \mathscr{T}\right\}\Bigg]_\theta \qquad \forall\theta, \quad 0 < \theta < 1.$$

Now, we have to interpret the interpolation spaces we have obtained. $\square$

First, thanks to Chapter 1, Section 13.4 and to (6.42), we have

$$(7.18)\quad \begin{cases} \Big[\Big\{L^2(\Omega) \times \prod_{j=0}^{m-1} H^{2m-m_j-1/2}(\Gamma); N^*, \mathscr{T}\Big\}, \\ \qquad \Big\{\Xi^{-2m}(\Omega) \times \prod_{j=0}^{m-1} H^{-m_j-1/2}(\Gamma); N^*, \mathscr{T}\Big\}\Big]_\theta \\ = \Big\{[L^2(\Omega), \Xi^{-2m}(\Omega)]_\theta \times \prod_{j=0}^{m-1} H^{2m(1-\theta)-m_j-1/2}(\Gamma); N^*, \mathscr{T}\Big\}. \end{cases}$$

Furthermore, thanks to the duality theorem (Chapter 1, Section 6.2) and to Corollary 7.3, we have

$$(7.19)\quad [L^2(\Omega), \Xi^{-2m}(\Omega)]_\theta = ([\Xi^{2m}(\Omega), L^2(\Omega)]_{1-\theta})' = (\Xi^{2m\theta}(\Omega))' = \Xi^{-2m\theta}(\Omega) \quad \forall\theta,\ 0 < \theta < 1.$$

Also, thanks to Chapter 1, Section 13.4:

$$(7.20)\quad [H^{2m}(\Omega)/N, D_A^0(\Omega)/N]_\theta = ([H^{2m}(\Omega), D_A^0(\Omega)]_\theta)/N, \quad 0 < \theta < 1.$$

Therefore, in short

(7.21) *$\mathscr{P}$ is an isomorphism of* $([H^{2m}(\Omega), D_A^0(\Omega)]_\theta)/N$ *onto*

$$\Big\{\Xi^{-2m\theta}(\Omega) \times \prod_{j=0}^{m-1} H^{2m(1-\theta)-m_j-1/2}(\Gamma); N^*, \mathscr{T}\Big\}, \quad 0 < \theta < 1. \quad \square$$

The space $D_A^s(\Omega)$, $0 < s < 2m$

There remains to interpret the space $[H^{2m}(\Omega), D_A^0(\Omega)]_\theta$. For this purpose, we introduce the space

$$D_A^s(\Omega) = \{u \mid u \in H^s(\Omega), A\,u \in \Xi^{s-2m}(\Omega)\}, \quad 0 < s < 2m,$$

provided with the norm of the graph

$$\|u\|_{D_A^s(\Omega)} = (\|u\|^2_{H^s(\Omega)} + \|A\,u\|^2_{\Xi^{s-2m}(\Omega)})^{1/2},$$

which makes it a Hilbert space.

Theorem 7.2. *Under hypotheses* (i) *and* (ii) *of Section 5, we have*

$$(7.22)\quad [H^{2m}(\Omega), D_A^0(\Omega)]_\theta = D_A^{2m(1-\theta)}(\Omega) \qquad \forall\theta,\ 0 < \theta < 1.$$

Proof. — 1) First, we note that A^* is properly elliptic in $\bar{\Omega}$; therefore we may apply Remark 1.3 according to which the Dirichlet problem for A^* satisfies conditions (i), (ii), (iii) of Section 5. Therefore, we may apply the a priori estimate (5.3) with $r = 0$ and obtain the fact that for all functions $u \in H^{2m}(\Omega) \cap H_0^m(\Omega)$ and therefore a fortiori for all $u \in H_0^{2m}(\Omega)$, we have

$$\|u\|^2_{H^{2m}(\Omega)} \leqq C\{\|A^*u\|^2_{L^2(\Omega)} + \|u\|^2_{H^{2m-1}(\Omega)}\}.$$

Applying Theorem 16.3 of Chapter 1 we obtain that for all $u \in H_0^{2m}(\Omega)$ we have

$$\| u \|^2_{H^{2m}(\Omega)} \leqq C' \{ \| A^* u \|^2_{L^2(\Omega)} + \| u \|^2_{L^2(\Omega)} \}.$$

The sesquilinear form

$$\pi(u, v) = \int_\Omega A^* u \overline{A^* v} \, dx + \int_\Omega u \bar{v} \, dx$$

defines on $H_0^{2m}(\Omega)$ a scalar product which is *equivalent* to the "natural" scalar product. Let $f \in H^{-2m}(\Omega)$; $v \to \langle f, \bar{v} \rangle$ is a continuous antilinear form on $H_0^{2m}(\Omega)$, therefore it may be expressed by a scalar product (in the scalar product $\pi(u, v)$):

$$\text{(7.23)} \qquad \begin{cases} \langle f, \bar{v} \rangle = \pi(G f, v), & G f \in H_0^{2m}(\Omega) \\ G \in \mathscr{L}(H^{-2m}(\Omega); H_0^{2m}(\Omega)). \end{cases}$$

Setting $G f = u$, (7.23) is equivalent to

$$(A A^* + I) u = f, \qquad u \in H_0^{2m}(\Omega).$$

2) We shall now use Theorem 14.3 of Chapter 1. Using 1), we see that this theorem may be applied if we set

$$\begin{array}{lll} X = H^{2m}(\Omega), & Y = L^2(\Omega) & (\Phi = L^2(\Omega)) \\ \mathscr{X} = L^2(\Omega), & \mathscr{Y} = \Xi^{-2m}(\Omega) & (\Psi = H^{-2m}(\Omega)) \\ \tilde{\mathscr{X}} = L^2(\Omega), & \tilde{\mathscr{Y}} = H^{-2m}(\Omega), & \\ \partial = A, & \mathscr{G} = A^*(A A^* + I)^{-1} & \end{array}$$

(we note that thanks to 1) $(A A^* + I)^{-1}$, inverse of $A A^* + I$, exists and operates from $H^{-2m}(\Omega)$ onto $H_0^{2m}(\Omega)$),

$$r = -(A A^* + I)^{-1}.$$

The application of Theorem 14.3 of Chapter 1, under the preceding conditions, yields Theorem 7.2. □

Remark 7.1. From Theorem 7.2, we deduce that $\mathscr{D}(\bar{\Omega})$ is dense in $D^{2m(1-\theta)}(\Omega)$, for it is dense in $H^{2m}(\Omega)$ (Chapter 1, Section 8.1), and in $D_A^0(\Omega)$ (Theorem 6.4). □

7.3 The Final Results

Applying (7.22), we can now obtain the trace theorem for the space $D_A^s(\Omega)$, $0 < s < 2m$, and the final result on the boundary value problems.

Theorem 7.3. *Under hypothesis* (i), (ii) *of Section* 5 *and if the system* $\{B_j\}_{j=0}^{m-1}$ *is normal on* Γ, *for all real* s *such that* $0 < s < 2m$, *the mapping*

$u \to B u = \{B_0 u, \ldots, B_{m-1} u\}$ of $\mathscr{D}(\bar{\Omega})$ into $\{\mathscr{D}(\Gamma)\}^m$ extends by continuity to a continuous linear mapping, still denoted by $u \to B u$, of $D^s_A(\Omega)$ into

$$\prod_{j=0}^{m-1} H^{s-m_j-1/2}(\Gamma).$$

Proof. $u \to B u$ is a continuous linear mapping of $H^{2m}(\Omega)$ into $\prod_{j=0}^{m-1} H^{2m-m_j-1/2}(\Gamma)$ (Chapter 1, Section 8.2) and of $D^0_A(\Omega)$ into $\prod_{j=0}^{m-1} H^{-m_j-1/2}(\Gamma)$ (Theorem 6.5); therefore, by interpolation, using Theorem 7.2 and the properties of the spaces $H^s(\Gamma)$ (Chapter 1, Section 7), it is a continuous linear mapping of $D^s_A(\Omega)$ into $\prod_{j=0}^{m-1} H^{s-m_j-1/2}(\Gamma)$ for real s, $0 < s < 2m$. Furthermore, this mapping is obtained by extension by continuity of $\mathscr{D}(\bar{\Omega})$ into $D^s_A(\Omega)$, for $\mathscr{D}(\bar{\Omega})$ is dense in $D^s_A(\Omega)$ (see Remark 7.1). □

Finally, from (7.21) and Theorem 7.2, we deduce

Theorem 7.4. *Under hypotheses* (i), (ii), (iii) *of Section 5, the operator $\mathscr{P} = \{A; B_0, \ldots, B_{m-1}\}$ is an algebraic and topological isomorphism of*

$$D^s_A(\Omega)/N \quad \textit{onto} \quad \left\{\Xi^{s-2m}(\Omega) \times \prod_{j=0}^{m-1} H^{s-m_j-1/2}(\Gamma); N^*, \mathscr{T}\right\}$$

for all real s such that $0 < s < 2m$. □

Again we note that, in $D^s_A(\Omega)$, $\mathscr{P}$ is an operator with index

$$\chi(\mathscr{P}) = \dim N - \dim N^*.$$

Remark 7.2. Thus having concluded the study of boundary value problem (5.1) in the spaces $H^s(\Omega)$ with arbitrary real s, we believe that it might be useful to the reader if we summarize the various results in one statement.

Under hypotheses (i), (ii) *and* (iii) *of Section 5.1 (which, as we recall, define the "regular elliptic problems"), we consider the boundary value problem*

$$(7.24) \qquad A u = f \quad \textit{in } \Omega$$

$$(7.25) \qquad B_j u = g_j \quad \textit{on } \Gamma, \quad j = 0, 1, \ldots, m-1.$$

Let s be an arbitrary real number; let

$$f \in H^{s-2m}(\Omega) \ \textit{if} \ s \geqq 2m, \quad \textit{or} \quad f \in \Xi^{s-2m}(\Omega) \ \textit{if} \ s < 2m,$$

$$g_j \in H^{s-m_j-1/2}(\Gamma), \quad j = 0, 1, \ldots, m-1.$$

Then, these exists $u \in H^s(\Omega)$, *"solution" of* (7.24), (7.25), *if and only if*

$$\langle f, \bar{v}\rangle + \sum_{j=0}^{m-1} \langle g_j, \overline{T_j v}\rangle = 0, \quad \forall v \in N^* \tag{7.26}$$

where

$$N^* = \{v \mid v \in \mathscr{D}(\bar{\Omega}),\ C_j v = 0, j = 0, \ldots, m-1,\ A^* v = 0\}.$$

Such a "solution" u is determined up to addition of a function of N (*i.e. if* $w \in N$, $u + w$ *is again a "solution"*), *where*

$$N = \{w \mid w \in \mathscr{D}(\bar{\Omega}), B_j w = 0, j = 0, \ldots, m-1, A w = 0\}.$$

The term "solution" of (7.24), (7.25) *is to be understood in the following sense:*

a) (7.24) *is verified in the sense of distributions on* Ω;

b) (7.25) *is verified in the sense of the various trace theorems* (for $s \geqq 2m$, Chapter 1, Section 9.2, for $0 < s < 2m$, Chapter 2, Section 7, for $s < 0$, Chapter 2, Section 6), *which all depend on an extension by continuity (and therefore "natural")*[1] *of the classical trace operator on* Γ *for sufficiently regular functions in* Ω. *Finally, the solutions depend continuously on the data in the sense of the following estimates:*

$$\inf_{z \in N} \|u + z\|_{H^s(\Omega)} \leqq C \left\{\|A u\|_{H^{s-2m}(\Omega)} + \sum_{j=0}^{m-1} \|B_j u\|_{H^{s-m_j-1/2}(\Gamma)}\right\} \tag{7.27}$$

if $s \geqq 2m$;

$$\inf_{z \in N} \|u + z\|_{H^s(\Omega)} \leqq C \left\{\|A u\|_{\Xi^{s-2m}(\Omega)} + \sum_{j=0}^{m-1} \|B_j u\|_{H^{s-m_j-1/2}(\Gamma)}\right\} \tag{7.28}$$

if $s < 2m$.

Remark 7.3. As we have already pointed out, the space $\Xi^{s-2m}(\Omega)$ intervening in Sections 6 and 7 (when $s < 2m$) is *not optimal*. We shall give an example for which the space $\Xi^{s-2m}(\Omega)$ may be improved upon (and for which we can obtain the optimal space).

We go back to Section 6.1 and, *in order to simplify, we assume that* $N = N^* = \{0\}$; then, with the definitions introduced in Section 6.1, we have

(7.29) A (resp. A^*) *is an isomorphism of* $H^{2m}_B(\Omega)$ (resp. $H^{2m}_C(\Omega)$) *onto* $H^0(\Omega)$.

By transposition (this is exactly what we have done in Section 6.1, see Proposition 6.1 for $r = 0$):

(7.30) $(A^*)^*$ *is an isomorphism of* $H^0(\Omega)$ *onto* $(H^{2m}_C(\Omega))'$;

[1] See also Section 8.1 below.

Thanks to the formula (2.3) we immediately verify that

$$(A^*)^* \varphi = A \varphi, \quad \forall \varphi \in H_B^{2m}(\Omega).$$

Therefore $(A^*)^*$ is an *extension* of A and we set

$$(A^*)^* = A.$$

Then, interpolating between (7.29) and (7.30), we have:

$$(7.31) \quad \begin{cases} A \text{ is an isomorphism of } [H_B^{2m}(\Omega), H^0(\Omega)]_\theta \text{ onto} \\ \qquad [H^0(\Omega)\,(H_C^{2m}(\Omega))']_\theta, \quad 0 < \theta < 1. \end{cases}$$

The second space coincides (duality theorem, Chapter 1, Section 6.2) with $([H_C^{2m}(\Omega), H^0(\Omega)]_{1-\theta})'$. But, according to Theorem 14.4 of Chapter 4 (the proof of which does not use this remark!), we have

$$(7.32) \qquad [H_C^{2m}(\Omega), H^0(\Omega)]_{1-\theta} = H^{2m\theta}(\Omega)$$

if

$$2m - \mu_j - 1 \ (= \text{order of } C_j) > 2m\theta - \tfrac{1}{2}.$$

On the other hand, according to P. Grisvard [8] (see also Remark 14.5, Chapter 4, Volume 2), we have, if

$$m_j < 2m(1-\theta) - \tfrac{1}{2}, \quad j = 0, \ldots, m-1,$$

$$[H_B^{2m}(\Omega), H^0(\Omega)]_\theta = H_B^{2m(1-\theta)}(\Omega)$$
$$= \{v \mid v \in H^{2m(1-\theta)}(\Omega), B_j v = 0 \text{ on } \Gamma, 0 \leq j \leq m-1\}.$$

Consequently:

(7.33) *A is an isomorphism of $H_B^{2m(1-\theta)}(\Omega)$ onto $(H^{2m\theta}(\Omega))'$ if $2m\theta < \frac{1}{2}$.*

But if $2m\theta < \frac{1}{2}$, we have (Chapter 1, Section 11):

$$(H^{2m\theta}(\Omega))' = H^{-2m\theta}(\Omega).$$

Therefore:

Theorem 7.5. *Assume that hypotheses* (i), (ii), (iii) *are satisfied and that* (7.29) *holds. Let* $0 < s < \frac{1}{2}$. *Then A is an isomorphism of $H_B^{2m-s}(\Omega)$ onto $H^{-s}(\Omega)$.*

We combine this result with Remark 7.2 to obtain: in (7.24), (7.25), assume that

$$f \in H^{-s}(\Omega), \quad g_j \in H^{2m-s-m_j-1/2}(\Gamma), \quad 0 < s < \tfrac{1}{2}.$$

Then the solution $u \in H^{2m-s}(\Omega)$ (compare also with Section 8.3). □

Remark 7.4. The preceding remark also applies to the evolution equations treated in Chapters 4 and 5, Volume 2; see, for example, Chapter 4, Remark 15.1. □

8. Complements and Generalizations

In this section, we provide some complements to the preceding theory.

8.1 Continuity of Traces on Surfaces Neighbouring Γ

We first reconsider the trace theorems.

Let $\{\Gamma_\varrho\}$, $0 \leqq \varrho \leqq \varrho_0 < 1$, be a family of surfaces which are *parallel* to Γ and *tend towards* Γ when $\varrho \to 0$.

More precisely, we assume, as hypothesis (i) of Section 5 allows us to do, that:

(8.1)

there exists a finite family of open sets $O_1, \ldots, O_N$ in $\mathbf{R}^n$ such that

$$\Gamma_\varrho \quad \text{and} \quad \Gamma \subset \bigcup_{i=1}^{N} O_i$$

and that, for each i, there exists an infinitely differentiable homeomorphism θ_i, with non-null jacobian, of O_i onto the cylinder

$$Q = \left\{(y, t), \sum_{i-1}^{m-1} y_i^2 < 1, -1 < t < 1\right\}$$

satisfying

(a) θ_i maps $O_i \cap \Omega$ onto $Q_+ = \{(y, t) \in Q, t > 0\}$,

(b) θ_i maps $O_i \cap \Gamma$ onto $Q_0 = \{(y, t) \in Q, t = 0\}$

(c) θ_i maps $O_i \cap \Gamma_\varrho$ onto $Q_\varrho = \{(y, t) \in Q, t = \varrho\}$

(d) if $O_i \cap O_j \neq \emptyset$, then θ_i and θ_j satisfy the usual compatibility conditions on $O_i \cap O_j$ (i.e. there exists an infinitely differentiable homeomorphism J_{ij}, with positive jacobian, of $\theta_i(O_i \cap O_j)$ onto $\theta_j(O_i \cap O_j)$ such that

$$\theta_j(x) = J_{ij}(\theta_i(x)) \; (\forall x \in O_i \cap O_j)).$$

With the help of $\{\theta_i\}$, we can define a homeomorphism

$$x \to \psi(x, \varrho) \quad \text{of} \quad \Gamma_\varrho \text{ onto } \Gamma, \tag{8.2}$$

which is infinitely differentiable both ways and where ψ and ψ^{-1} and all their derivatives are bounded by constants independent of ϱ (but dependent on the order of the derivative). Finally, we let Ω_ϱ be the open set with boundary Γ_ϱ, contained in Ω.

We assume, as we are allowed to do, that the coefficients of the operators $\{B_j\}$, $\{C_j\}$, $\{S_j\}$, $\{T_j\}$ are defined and infinitely differentiable not only on Γ but in $\bar{\Omega} - \Omega_{\varrho_0}$, so that the systems $\{B_j\}$, $\{C_j\}$, $\{S_j\}$, $\{T_j\}$ are normal on Γ_ϱ for all ϱ with $0 < \varrho \leqq \varrho_0$.

We need to consider three cases:

1) $s \geqq 2m$.

We consider the space $H^s(\Omega)$; if $u \in H^s(\Omega)$, its restriction to Ω_ϱ (still denoted by u) belongs to $H^s(\Omega_\varrho)$ and therefore we may define $B_j u|_{\Gamma_\varrho}$ on Γ_ϱ, $j = 0, \ldots, m-1$ (Chapter 1, Section 9). With the help of (8.2), we may, by transfer of structure, define the *image* of $B_j u|_{\Gamma_\varrho}$ on Γ_ϱ; we set:

(8.3) $$B_j^{(\varrho)} u = \text{image under (8.2) of } B_j u|_{\Gamma_\varrho}.$$

We have $B_j^{(\varrho)} u \in H^{s-m_j-1/2}(\Gamma)$ and, using the definition and the properties of $H^s(\Omega)$-spaces (Chapter 1, Sections 7–9), it is easy to see that

(8.4) $$B_j^{(\varrho)} u \to B_j u \ \textit{in} \ H^{s-m_j-1/2}(\Gamma) \ \textit{as} \ \varrho \to 0, \quad j = 0, \ldots, m-1. \quad \square$$

2) $s \leqq 0$.

Note that in this case ($s \leqq 0$), if $u \in D_A^s(\Omega)$, its restriction to Ω_ϱ does not in general belong to $D_A^s(\Omega_\varrho)$ (defined in analogous manner as $D_A^s(\Omega)$), for the mapping "restriction" to Ω_ϱ does not map $\Xi^{s-2m}(\Omega)$ into $\Xi^{s-2m}(\Omega_\varrho)$. Therefore $B_j u|_{\Gamma_\varrho}$ can not be defined for arbitrary $u \in D_A^s(\Omega)$. This leads to constrain the class of u's for which we shall prove (8.4). For example, we can introduce the spaces

$$\tilde{D}_A^s(\Omega) = \{u \mid u \in H^s(\Omega), A u \in L^2(\Omega)\}$$

and

$$\tilde{D}_A^s(\Omega_\varrho) = \{u \mid u \in H^s(\Omega_\varrho), A u \in L^2(\Omega_\varrho)\},$$

provided with the norm of the graph and note that the restriction to Ω_ϱ of $u \in \tilde{D}_A^s(\Omega)$ belongs to $\tilde{D}_A^s(\Omega_\varrho)$. By the same arguments as in Sections 6.4, 6.5, 6.6, we verify that $\mathscr{D}(\bar{\Omega})$ (resp. $\mathscr{D}(\bar{\Omega}_\varrho)$) is dense in $\tilde{D}_A^s(\Omega)$ (resp. $\tilde{D}_A^s(\Omega_\varrho)$) and that, for $u \in \tilde{D}_A^s(\Omega)$ (resp. $\tilde{D}_A^s(\Omega_\varrho)$), we have a trace theorem analogous to Theorem 6.5. Therefore, we may define $B_j u|_\Gamma$ on Γ and $B_j u|_{\Gamma_\varrho}$ on Γ_ϱ for $u \in \tilde{D}_A^s(\Omega)$. Using (8.2), we then define $B_j^{(\varrho)} u$ by (8.3).

We show that we still have (8.4).

Clearly (8.4) holds if $u \in \mathscr{D}(\bar{\Omega})$. Therefore, it is sufficient to show that

(8.5) $$\begin{cases} B_j^{(\varrho)} \ \textit{remains in a bounded set of} \ \mathscr{L}(\tilde{D}_A^s(\Omega), H^{s-m_j-1/2}(\Gamma)), \\ j = 0, \ldots, m-1, \ \textit{as} \ \varrho \to 0. \end{cases}$$

Therefore, we have to show that for fixed $\varphi = (\varphi_0, \ldots, \varphi_{m-1})$ in

$$\prod_{j=0}^{m-1} H^{-s+m_j+1/2}(\Gamma),$$

we have

$$\left| \sum_{j=0}^{m-1} \langle B_j^{(\varrho)} u, \bar{\varphi}_i \rangle \right| \leqq \text{constant, independent of } \varrho .$$

After a transfer of structure by (8.2), this amounts to showing that

$$\left| \sum_{j=0}^{m-1} \langle B_j u |_{\Gamma_\varrho}, \bar{\varphi}_{j,\varrho} \rangle \right| \leqq \text{constant}$$

when $\varphi_\varrho = (\varphi_{0,\varrho}, \ldots, \varphi_{m-1,\varrho})$ belongs to a bounded set of

$$\prod_{j=0}^{m-1} H^{-s+m_j+1/2}(\Gamma_\varrho) .$$

But according to the definition of $B_j u|_{\Gamma_\varrho}$, we have

$$\sum_{i=0}^{m-1} \langle B_j u |_{\Gamma_\varrho}, \bar{\varphi}_{j,\varrho} \rangle = \langle u, \overline{A^* v_{\varphi_\varrho}} \rangle - \langle A u, \bar{v}_{\varphi_\varrho} \rangle ,$$

where $\langle u, A^* v_{\varphi_\varrho} \rangle$ denotes the duality between $H^s(\Omega_\varrho)$ and $H_0^{-s}(\Omega_\varrho)$, $\langle A u, \bar{v}_{\varphi_\varrho} \rangle$ the scalar product in $L^2(\Omega_\varrho)$ and v_{φ_ϱ} is the right-inverse of φ_ϱ, analogous to the one in Theorem 6.1 but passing from Γ_ϱ to Ω_ϱ (therefore

$$v_{\varphi_\varrho} \in X^{-s}(\Omega_\varrho) = \{v \mid v \in H^{2m-s}(\Omega_\varrho),\ C_j v |_{\Gamma_\varrho} = 0, j = 0, \ldots, m-1, \\ A^* v \in \Pi_0^{-s}(\Omega_\varrho)\}).$$

Then it suffices to verify that we may choose v_{φ_ϱ} so that it remains in a ball of $X^{-s}(\Omega_\varrho)$ whose radius is independent of ϱ, as $\varrho \to 0$.

But this follows from the proof of Theorem 6.1, since all the "maps" are realized by functions which together with their derivatives are bounded in ϱ, thanks to (8.2). □

3) $0 < s < 2m$.

We consider $\tilde{D}_A^s(\Omega)$, again defined as for the case $s \leqq 0$. With the same notation as before, we show (8.4) for $0 < s < 2m$, by interpolation between the cases $s = 2m$ and $s = 0$ and using Theorem 7.2 of this chapter and Theorem 5.2 of Chapter 1.

Therefore, we have obtained

Theorem 8.1. *Under the hypotheses of the various trace theorems* (Chapter 1, Theorem 9.4 if $s \geqq m$, Chapter 2, Theorem 7.3 if $0 < s < 2m$, Chapter 2, Theorem 6.5 and Remark 6.5 if $s \leqq 0$), *if* $\{\Gamma_\varrho\}$ *is a family of parallel surfaces to* Γ *which tends to* Γ *when* $\varrho \to 0$ (in the sense of (8.1)), *then* $B_j^{(\varrho)} u \to B_j u$ *in* $H^{s-m_j-1/2}(\Gamma)$ *as* $\varrho \to 0$, $j = 0, \ldots, m-1$, *for* $u \in H^s(\Omega)$ *if* $s \geqq 2m$ *and for* $u \in \tilde{D}_A^s(\Omega)$ *if* $s < 2m$. □

8.2 A Generalization; Application to Dirichlet's Problem

The starting point of Section 4 to obtain the *direct* a priori estimates (see (4.39)) was the study of the operator

$$\mathscr{P}_\eta = \left\{A\left(\eta, \frac{1}{i}\frac{d}{dt}\right); B_0\left(\eta, \frac{1}{i}\frac{d}{dt}\right)\bigg|_{t=0}, \ldots, B_{m-1}\left(\eta, \frac{1}{i}\frac{d}{dt}\right)\bigg|_{t=0}\right\}.$$

We have considered $\mathscr{P}_\eta$ (see (4.2)) as an operator of $H^{2m}(\mathbf{R}_+)$ into

$$L^2(\mathbf{R}_+) \times \mathbf{C}^m.$$

But suppose that $\max_j m_j < 2m - 1$ and set $l = 2m - 1 - \max_j m_j$. Then we can see that the theory of Section 4 for the direct estimates may be developed again if we consider $\mathscr{P}_\eta$ as an operator of $H^{2m-l}(\mathbf{R}_+)$ into $H^{-l}(\mathbf{R}_+) \times \mathbf{C}^m$. We only need to use an extension $f \to f^*$ of $H^{-l}(\mathbf{R}_+)$ into $H^{-l}(\mathbf{R})$ instead of $f \to \tilde{f}$ of $L^2(\mathbf{R}_+)$ into $L^2(\mathbf{R})$ (extension which exists, see Chapter 1, Section 2) in the proof of Lemma 4.1. We may then carry on as in Sections 4 and 5 and obtain the following generalization of formula (5.3) of Theorem 5.1:

Theorem 8.2. *Under hypotheses* (i), (ii) *and* (iii) *of Section 5 and if*

$$l = 2m - 1 - \max_j m_j > 0,$$

then, for $r = -l, -l+1, \ldots, 0, 1, \ldots,$ *we have: if* $u \in H^{2m-l}(\Omega)$ *and*

$$\mathscr{P} u = \{A u; B_0 u, \ldots, B_{m-1} u\} \in H^r(\Omega) \times \prod_{j=0}^{m-1} H^{2m+r-m_j-1/2}(\Gamma),$$

then $u \in H^{2m+r}(\Omega)$ *and*

$$\text{(8.6)} \quad \|u\|_{H^{2m+r}(\Omega)} \leqq c_r \{\|\mathscr{P} u\|_{H^r(\Omega) \times \prod_{j=0}^{m-1} H^{2m+r-m_j-1/2}(\Gamma)} + \|u\|_{H^{2m+r-1}(\Omega)}\}. \ \square$$

We point out the consequences of (8.6). Using the same arguments as in the proof of Theorem 5.2, we see again that $\mathscr{P}_{(r)}$ (notation of Section 5 extended to the case integer $r \geqq -l$) has kernel N, that $\operatorname{Im}(\mathscr{P}_{(r)})$ is closed and that

$$\operatorname{Im}(\mathscr{P}_{(r)}) = \operatorname{Im}(\mathscr{P}_{(-l)}) \cap H^r(\Omega) \times \prod_{j=0}^{m-1} H^{2m+r-m_j-1/2}(\Gamma),$$

$$r = -l, \ldots, -1, 0, \ldots.$$

It can also be shown that

$$\text{(8.7)} \quad \operatorname{codim} \operatorname{Im}(\mathscr{P}_{(r)}) = \operatorname{codim} \operatorname{Im}(\mathscr{P}_{(0)}), \quad r = -l, \ldots, -1,$$

by still using the fact that $\operatorname{codim} \operatorname{Im}(\mathscr{P}_{(0)})$ is finite and by dual arguments to those used for Theorem 5.2 to prove (8.7) for $r > 0$. Therefore $\mathscr{P}_{(r)}$ is an indexed operator and its index χ is independent of r; and *therefore Theorem* 5.2 *is still valid for* $r = -l, -l+1, \ldots, -1$. $\square$

It follows that the theory developed in Sections 5.3 and 5.4 is also valid under these new conditions: in particular, we have Theorem 5.3 for $r = -l, \ldots, -1$ and Theorem 5.4 for $s \geqq 2m - l$, with $s - \frac{1}{2} \neq$ $\neq$ integer if $s < 2m$ (if $s < 2m$, the exception occurs because of Section 12.8 of Chapter 1). □

Starting from this point, it is possible to develop the theory of transposition and of interpolation as in Sections 6 and 7. We shall not insist on this point. □

Applied to the Dirichlet problem (for which $l = m$), the preceding remarks yield

Theorem 8.3. *Under hypotheses* (i) *and* (ii) *of Section 5, the operator*

$$\{A; \gamma_0, \ldots, \gamma_{m-1}\}$$

defines an isomorphism of $H^s(\Omega)/N$ *onto*

$$\left\{H^{s-2m}(\Omega) \times \prod_{j=0}^{m-1} H^{s-j-1/2}(\Gamma); N^*, \mathscr{T}\right\}$$

for all real $s \geqq m$ *and different from* $m + \frac{1}{2}, m + \frac{3}{2}, \ldots, 2m - \frac{1}{2}$. □

Remark 8.1. For the generalizations obtained in this section we have used the *direct* estimates for $r \geqq -l$. □

Remark 8.2. We could make a more precise examination of the regularity hypotheses on the coefficients of the operators A and B_j and on the open set Ω. We have assumed that the coefficients and the boundary Γ are infinitely differentiable so as to be able to develop the theory for all $H^s(\Omega)$ with arbitrary real s. But the hypotheses could be specified for each s separately (see, for example, Geymonat [2]).

But, from this point of view, the use of interpolation and transposition is not very *economical*, since, for example in interpolation, to obtain a result in $H^s(\Omega)$ we must start from the corresponding result in $H^{[s]+1}$ ($[s]$ = greatest integer less than or equal to s). □

8.3 Remarks on the Hypotheses on A and B_j

Let us now take a closer look at the different hypotheses made on the operators A and B_j.

The four essential steps in Sections 4 and 5 were the following:

I. proof of the direct estimates (5.3);
II. proof of the dual estimates (5.4);
III. proof of the existence of the index of $\mathscr{P}$ (Theorem 5.2);
IV. proof of the theorem on the alternative (Theorem 5.3).

A first remark is evident: the hypothesis on the normality of the system $_{j=0}^{m-1}\{B_j\}$ was not used in Section 4 and we see immediately that in Section 5 it was used only starting from 5.3, when it was required to introduce *Green's formula* (5.24).

Therefore, in order to obtain I, II, III, we only used the hypotheses that A be *properly elliptic*, that the B_j's *cover* A and that $m_j \leqq 2m - 1$.

But it is possible to go further: the method of Peetre [2], which we have followed, may be used in a more general setting by eliminating the hypothesis that $m_j \leqq 2m - 1$ and obtaining I, II, III starting from the spaces $H^s(\Omega)$ with real $s > \max_j m_j + \frac{1}{2}$ (still more generally, with real $s > \max \mu_j + \frac{1}{2}$, where μ_j is the *normal* order of B_j) and $s \neq$ integer $+ \frac{1}{2}$ if $s < 2m$. A partial example was given in Section 8.2; for the general case, we refer the reader to the original work of Peetre [2]. Thus, we see that the hypothesis on the normality of $\{B_j\}$ and the hypothesis on the order of B_j are essentially connected to the use of *Green's formula*; therefore, they play an essential role in IV and in the use of interpolation and transposition in Sections 6 and 7. But for I, II, III, only the proper ellipticity of A and the condition that $\{B_j\}$ covers A are required.

We note that *these last hypotheses are necessary*: more precisely, it can be shown that the *proper ellipticity* of A and the condition that $\{B_j\}$ *covers* A are necessary conditions to obtain I (by reduction to the case $\mathbf{R}_+$ and use of Remark 4.2, see Agmon-Douglis-Nirenberg [1]). □

8.4 The Realization of A in $L^2(\Omega)$

In Section 6, we have already considered the differential operators A and A^* as (bounded) operators in certain Hilbert spaces. It is also important to consider them as unbounded operators in $L^2(\Omega)$.

We shall denote by A_2 the unbounded operator in $L^2(\Omega)$ defined by

$$D(A_2) = H_B^{2m}(\Omega) = \{u \mid u \in H^{2m}(\Omega), B_j u = 0, j = 0, \ldots, m-1\},$$

$$A_2 u = A(x, D) u \quad \text{for} \quad u \in D(A_2).$$

Similarly, we denote by A_2^* the operator defined by

$$D(A_2^*) = H_C^{2m}(\Omega) = \{u \mid u \in H^{2m}(\Omega), C_j u = 0, j = 0, \ldots, m-1\},$$

$$A_2^* u = A^*(x, D) u \quad \text{for} \quad u \in D(A_2^*).$$

A_2 (A_2^*) is called the *realization of* A (A^*) in $L^2(\Omega)$ under the boundary conditions $B_j u = 0$ $(C_j u = 0)$, $j = 0, \ldots, m-1$. In the sequel, we shall sometimes denote the operator $A_2(A_2^*)$ by $A(A^*)$ if there is no danger of confusion.

Theorem 8.4. *Under hypotheses* (i), (ii), (iii) *of Section 5,* A_2^* *is the adjoint of the operator* A_2, *in the sense of unbounded operators in* $L^2(\Omega)$, *i.e.* $A_2^* = (A_2)^*$.

Proof. We apply Theorem 5.3 to problems $\{A, B\}$ and $\{A^*, C\}$; there results that A_2 and A_2^* are closed operators in $L^2(\Omega)$, with closed image,

$$\operatorname{Ker} A_2 = N, \quad \operatorname{Ker} A_2^* = N^*, \quad \operatorname{Im}(A_2) = \{L^2(\Omega); N^*\}$$

(orthogonal subspace of N^* in $L^2(\Omega)$) and $\operatorname{Im}(A_2^*) = \{L^2(\Omega); N\}$.

It is also obvious that $D(A_2^*)$ and $D(A_2)$ are dense in $L^2(\Omega)$.

But the adjoint $(A_2)^*$ of A_2 in $L^2(\Omega)$ is also closed, has domain $D((A_2)^*)$ dense in $L^2(\Omega)$ and closed image; it is defined by

$$(A_2 u, v) = (u, (A_2)^* v) \quad \forall u \in D(A_2), \quad v \in D((A_2)^*).$$

Therefore, we have

$$\operatorname{Ker}(A_2)^* = \text{orthogonal subspace of } \operatorname{Im}(A_2) \text{ in } L^2(\Omega)$$

and therefore

$$\operatorname{Ker}(A_2)^* = N^* = \operatorname{Ker} A_2^*. \tag{8.8}$$

We also have $\operatorname{Im}((A_2)^*) =$ orthogonal subspace of $\operatorname{Ker} A_2 = \{L^2(\Omega); N\}$ in $L^2(\Omega)$; therefore

$$\operatorname{Im}((A_2)^*) = \operatorname{Im}(A_2^*). \tag{8.9}$$

Applying Green's formula (5.14) we see that $A_2^* \subset (A_2)^*$. Therefore, it is sufficient to verify that $D((A_2)^*) \subset D(A_2^*)$. Let $u \in D((A_2)^*)$: thus $f = (A_2)^* u \in \operatorname{Im}((A_2)^*) = \operatorname{Im}(A_2^*)$; consequently, there exists $v \in D(A_2^*)$ such that $A_2^* v = f$ and therefore $v \in D(A_2^*) \subset D((A_2)^*)$ and $(A_2)^* v = f$. It follows that $u - v \in \operatorname{Ker}(A_2)^* = \operatorname{Ker} A_2^* \subset D(A_2^*)$; from which we deduce that $u = (u - v) + v \in D(A_2^*)$. □

Of course, we may also consider A as an unbounded operator in $H^r(\Omega)$ with real $r \geqq 0$; more precisely, we consider in $H^r(\Omega)$ the operator (*realization* of A in $H^r(\Omega)$) $A_{2,r}$ defined by

$$D(A_{2,r}) = H_B^{2m+r}(\Omega) = \{u \mid u \in H^{2m+r}(\Omega), B_j u = 0, j = 0, \ldots, m-1\}$$

$$A_{2,r} u = A(x, D) u \quad \text{for} \quad u \in D(A_{2,r}).$$

Similarly, we define $A_{2,r}^*$.

Thanks to Theorems 5.3 and 5.4, it is easy to see that $A_{2,r}$ is the restriction of A_2 to $H^r(\Omega)$ and that $\operatorname{Ker} A_{2,r} = \operatorname{Ker} A_2 = N$ and

$$\operatorname{Im}(A_{2,r}) = \{H^r(\Omega); N^*\}. \quad \square$$

Remark 8.3. From what we have seen for A_2 there also results that A_2 is an indexed operator in $L^2(\Omega)$ and that

$$\chi(A_2) = \chi(\mathscr{P}) = \dim N - \dim N^*, \tag{8.10}$$

where $\mathscr{P}$ is the operator $\mathscr{P} = \{A; B_0, \ldots, B_{m-1}\}$ defined in Section 5. □

8.5 Some Remarks on the Index of $\mathscr{P}$

We shall now make some remarks concerning the index of the operator $\mathscr{P} = \{A; B_0, \ldots, B_{m-1}\}$ (or of the operator A_2, thanks to (8.10)). We have seen in Sections 5, 6, 7 that $\mathscr{P}$, as an operator in $H^s(\Omega)$, if $s \geqq 2m$, or in $D^s_A(\Omega)$, if $s < 2m$, always admits an index $\chi(\mathscr{P})$ which is independent of s and which is given by

$$\chi(\mathscr{P}) = \dim N - \dim N^*. \quad \square \tag{8.11}$$

Remark 8.4. It is known (see, for example, Kato [5]) that the index of an operator between Hilbert spaces (for example!) does not change if we add a compact operator to the operator. Therefore, according to the compactness theorem of Section 16 of Chapter 1, if we consider the operator

$$\mathscr{Q} = \{A + Q; B_0 + F_0, \ldots, B_{m-1} + F_{m-1}\}, \quad \text{with } Q \text{ (resp. } F_j)$$

a linear differential operator of order $< 2m$ (resp. $< m_j$) with coefficients in $\mathscr{D}(\bar{\Omega})$ (resp. $\mathscr{D}(\Gamma)$), we have

$$\chi(\mathscr{Q}) = \chi(\mathscr{P});$$

therefore the index of $\mathscr{P}$ *does not change if we add operators of smaller order to* A *and* B_j. $\square$

Remark 8.5. It is also interesting to know when

$$\chi(\mathscr{P}) = 0, \quad \text{i.e.} \quad \dim N = \dim N^* \tag{8.12}$$

(then $\mathscr{P}$ is sometimes called a Fredholm operator).

A sufficient condition is

$$H^{2m}_B(\Omega) = \overline{H^{2m}_C(\Omega)} \quad (\text{i.e. } D(A_2) = \overline{D(A_2^*)}) \tag{8.13}$$

(here $\bar{E}$ denotes the set of conjugate complex functions of a function space E).

Indeed, in this case we can consider the differential operators $\bar{A}^*$ and $\bar{C}_j$ ($j = 0, \ldots, m-1$) as being deduced from A^* and C_j by replacing the coefficients with their complex conjugates. Let $\bar{A}_2^*$ be the realization of $\bar{A}^*$ in $L^2(\Omega)$ under the boundary conditions $\bar{C}_j u = 0$, $j = 0, \ldots, m-1$; we see that $D(\bar{A}_2^*) = D(A_2)$. But $A - \bar{A}^*$ is an operator of order $\leqq 2m - 1$, therefore $A_2 - \bar{A}_2^*$ is a compact operator of $D(A_2)$ into $L^2(\Omega)$; it follows that $\chi(A_2) = \chi(\bar{A}_2^*)$. But $\chi(\bar{A}_2^*) = \chi(A_2^*)$ and, thanks to Theorem 8.4,

$$\chi(A_2^*) = -\chi(A_2).$$

Therefore

$$\chi(A_2) = -\chi(A_2) = \chi(\mathscr{P}) = 0.$$

In particular, condition (8.13) is satisfied for the Dirichlet problem, since then

$$H_B^{2m}(\Omega) = H_C^{2m}(\Omega) = H^{2m}(\Omega) \cap H_0^m(\Omega).$$

Therefore

$$\text{(8.14)} \quad \begin{cases} \textit{for every properly elliptic operator } A\textit{, the index of the Dirichlet} \\ \textit{problem is zero.} \quad \Box \end{cases}$$

We also note that, because of Remark 8.3, to have $\chi(\mathscr{P}) = 0$, it is sufficient to find $Q, F_0, \ldots, F_{m-1}$ under the conditions of Remark 8.3, such that $\chi(\mathscr{Q}) = 0$. In particular it suffices to find $\lambda \in \mathbf{C}$ such that $\chi(A_2 + \lambda I) = 0$, where I is the identity in $L^2(\Omega)$. $\Box$

8.6 Uniqueness and Surjectivity Theorems

Two other important questions are:

1) when is $\dim N = 0$?

In this case, there is a *unique* solution to boundary value problem $\{A, B\}$;

2) when is $\dim N^* = 0$?

In this case, there exists a solution of problem $\{A, B\}$ for all given f and g_j (*surjectivity* of $\mathscr{P}$).

The two questions are evidently of the same nature: surjectivity for $\{A, B\}$ is uniqueness for the adjoint problem $\{A^*, C\}$. $\Box$

In most applications these questions are asked in a little more general way, that is: to give sufficient conditions for uniqueness (or surjectivity) for problem $\{A + \lambda I, B\}$ for at least *one* complex number λ.

Of course, we shall have uniqueness for $\{A + \lambda I, B\}$ if we can show an inequality of the type

$$\text{(8.15)} \qquad \|u\|_{H^{2m}(\Omega)} \leqq c\,\|A u + \lambda u\|_{L^2(\Omega)}, \qquad \forall u \in H_B^{2m}(\Omega).$$

In Chapter 4, Section 4, we shall prove a sufficient condition for (8.15) to be satisfied for all λ with $\operatorname{Re}\lambda > \xi_0$ (ξ_0 = suitable real number); it is the following: in $\Omega \times \mathbf{R}_y$, consider the operator

$$\Lambda_\theta = A(x, D_x) + e^{i\theta}(-1)^m D_y^{2m} \qquad \forall\theta \in \left[-\frac{\pi}{2}, \frac{\pi}{2}\right],$$

with the boundary conditions

$$B_j(x, D_x) \text{ given on } \Gamma \times \mathbf{R}_y, \qquad j = 0, \ldots, m-1;$$

and *assume that for all* $\theta \in \left[-\frac{\pi}{2}, \frac{\pi}{2}\right]$, Λ_θ *is properly elliptic in* $\bar{\Omega} \times \mathbf{R}_y$ *and that the system* $\{B_j\}_{j=0}^{m-1}$ *covers* Λ_θ *on* $\Gamma \times \mathbf{R}_y$.

Under this condition, we have uniqueness for $\{A + \lambda I, B\}$ for all λ with $\operatorname{Re}\lambda > \xi_0$. For example, the condition is satisfied for the Dirichlet problem, if A is *strongly elliptic* in $\bar{\Omega}$.

Other sufficient conditions are provided by *variational theory* (coerciveness conditions, see Section 9).

9. Variational Theory of Boundary Value Problems

In this section we shall give the main outline of the "variational" theory of boundary value problems and compare the results with those obtained in the preceding sections.

9.1 Variational Problems

The linear elliptic variational problems correspond to the minimization of positive definite quadratic forms with homogeneous part of degree 2 on a Hilbert space – which we shall call V. The most classical example is the Dirichlet problem for the Laplacian: on the space $V = H_0^1(\Omega)$, we consider the quadratic form

$$Q(v) = a(v, v) - 2\int_\Omega f\bar{v}\,dx,$$

where

$$a(u, v) = \sum_{i=1}^{n} \int_\Omega \frac{\partial u}{\partial x_i}\,\overline{\frac{\partial v}{\partial x_i}}\,dx$$

and where f is given in $H^{-1}(\Omega)$ $\left(\int_\Omega f\bar{v}\,dx = \langle f, \bar{v}\rangle\right)$; then *the inf. of* $Q(v)$, *as* v *describes* V, *is reached by the unique element* u *of* V *satisfying*

(9.1) $$a(u, v) = \langle f, \bar{v}\rangle \quad \forall v \in V$$

or by the unique element of V satisfying

$$-\Delta u = f \quad \text{in } \Omega,$$

which may also be written:

(9.2) $$\begin{cases} u \in H^1(\Omega), \\ -\Delta u = f \quad \text{in } \Omega \\ u = 0 \quad \text{on } \Gamma. \end{cases}$$

Formulation (9.1) leads to the "abstract" problem:

let $u, v \to a(u, v)$ be a continuous sesquilinear form on $V \times V$ and let $f \in V'$ = antidual of $V((f, v)$ denoting the value of f at $v)$; *we seek* $u \in V$, *solution of*

(9.3) $$a(u, v) = (f, v) \quad \forall v \in V. \quad \square$$

Remark 9.1. In (9.3), we do *not necessarily assume* that $a(u, v) = \overline{a(u, v)}$; problem (9.3) therefore no longer corresponds to a problem of minimizing a quadratic form. Nevertheless, because of its close relation to the problems of the calculus of variations, problem (9.3) is called a "*variational problem*". □

Remark 9.2. $a(u, v)$ may be represented in the form

$$a(u, v) = (A\,u, v), \quad A \in \mathscr{L}(V; V') \tag{9.4}$$

and, *for the moment*, A is *arbitrary* in $\mathscr{L}(V; V')$. Then (9.3) is equivalent to the most general linear equation

$$A\,u = f \tag{9.5}$$

for $A \in \mathscr{L}(V; V')$. But we shall give convenient *sufficient* conditions for (9.5) to admit a unique solution. □

If $a(u, v) = \overline{a(u, v)}$, then we associate the form

$$Q(v) = a(v, v) - 2(f, v)$$

to (9.3) and we are led to assume $a(v, v)$ to be positive definite on V. In the non-symmetric case, we consider the *real part* of $a(u, v)$ and we are led to the *following definition*:

Definition 9.1. *The sesquilinear form* $a(u, v)$ *is said to be* V*-elliptic if*

$$\operatorname{Re} a(v, v) \geqq \alpha\,\|v\|^2 \qquad \forall v \in V, \quad \alpha < 0, \tag{9.6}$$

$$\|v\| = \text{norm of } v \text{ in } V.$$

Then, we have

Theorem 9.1. *Under hypothesis* (9.6), *problem* (9.3) *admits a unique solution, the mapping* $f \to u$ *being continuous from* $V' \to V$ (or: A is an isomorphism of V onto V').

Proof. — 1) Denote by $\|\ \|_*$ the norm (dual of $\|\ \|$) in V'. We have

$$\alpha\,\|v\|^2 \leqq \operatorname{Re} a(v, v) = \operatorname{Re}(A\,v, v) \leq \|A\,v\|_*\,\|v\|,$$

therefore

$$\|A\,v\|_* \geqq \alpha\,\|v\| \qquad \forall v \in V. \tag{9.7}$$

2) We introduce the adjoint form $a^*(u, v)$ by

$$a^*(u, v) = \overline{a(u, v)} \qquad \forall u, v \in V. \tag{9.8}$$

$a^*(u, v)$ may be represented by

$$a^*(u, v) = (A^*\,u, v), \quad A^* \in \mathscr{L}(V\,; V') \tag{9.9}$$

and we immediately verify that A^* *is the adjoint of* A.

Since $\operatorname{Re} a^*(v, v) = \operatorname{Re} a(v, v)$ we have, in the same fashion as for (9.7):

$$\|A^* v\|_* \geqq \alpha \|v\| \qquad \forall v \in V. \tag{9.10}$$

3) The theorem follows from (9.7) and (9.10) (the operator A is one-to-one and the image under it is dense and closed in V'). □

Remark 9.3. The preceding result is valid, with the same proof, under the hypothesis

$$|a(v, v)| \geqq \alpha \|v\|^2, \quad \alpha > 0, \quad \forall v \in V. \quad \square \tag{9.11}$$

Remark 9.4. For many applications – and in particular, as we shall see in the following chapters, for all the theory of *evolution equations* – we must consider, not the problem (we use the notation of Sections 1 to 8):

$$A u = f, \quad B_j u = 0, \quad 0 \leqq j \leqq m - 1,$$

but the problem

$$A u + \lambda u = f, \quad B_j u = 0, \quad 0 \leqq j \leqq m - 1,$$

where λ is a complex parameter. This leads to a variational formulation in which two Hilbert spaces intervene, instead of only the space V (since V' is intrinsically connected with V): we consider two *Hilbert spaces V and H, with*

(9.12) $V \subset H$, *V dense in H, the injection of V into H is continuous.*

We identify H to its antidual, and if V' is the antidual of V, we may identify H to a subspace of V', since V is dense in H, therefore

$$V \subset H \subset V'. \tag{9.13}$$

We denote by (,) the scalar product in H and by | | the norm in H; if $f \in V'$ and $v \in V$, their scalar product is also denoted by (f, v), which is permissible thanks to the identifications (9.13).

Then, instead of problem (9.3), we consider the problem: *find $u \in V$, solution of*

$$a(u, v) + \lambda(u, v) = (f, v) \qquad \forall v \in V, \tag{9.14}$$

where λ is given in $\mathbf{C}$.

Definition 9.1 is now generalized to:

Definition 9.2. *The sesquilinear form $a(u, v)$ is said to be V-coercive (relative to H) if there exist $\lambda_0 \in \mathbf{R}$ and $\alpha > 0$ such that*

$$\operatorname{Re} a(v, v) + \lambda_0 |v|^2 \geqq \alpha \|v\|^2 \qquad \forall v \in V. \tag{9.15}$$

Of course, Theorem 9.1 yields:

Corollary 9.1. *If the form* $a(u, v)$ *is V-coercive* (in the sense of (9.15)), *then problem* (9.14) *admits a unique solution for all* $\lambda \in \mathbf{C}$ *satisfying*

(9.16) $$\operatorname{Re}\lambda \geqq \lambda_0 . \quad \square$$

9.2 The Problem

We take up problem (5.1) again, in a *formal* way (i.e. without worrying about the data and solution spaces).

If φ is a "function" such that

$$B_j \varphi = g_j, \quad 0 \leqq j \leqq m-1,$$

then

$$u - \varphi = w$$

satisfies

$$\begin{cases} A w = f - A \varphi \\ B_j w = 0, \quad 0 \leqq j \leqq m-1. \end{cases}$$

Therefore, at least in a formal way, we are led back (changing the notation) to the problem

(9.17) $$\begin{cases} A u = f \\ B_j u = 0, \quad 0 \leqq j \leqq m-1, \end{cases}$$

to which we also associate (Remark 9.4) the problems

(9.18) $$\begin{cases} A u + \lambda u = f \\ B_j u = 0, \quad 0 \leqq j \leqq m-1. \end{cases}$$

The question is: *which* "regular elliptic" *problems* (in the sense of Sections 1 to 8) *of type* (9.17) (resp. (9.18)) *belong to the class of V-elliptic* (resp. *V-coercive*) *variational problems?*

We start with a counter-example which is due to Seeley [3].

9.3 A Counter-Example

In $\mathbf{R}^2$, consider (we use polar coordinates (r, θ)):

$$\Omega = \{(r, \theta) \mid \pi < r < 2\pi\}.$$

Take the elliptic operator

(9.19) $$A = -\left(e^{i\theta} \frac{\partial}{\partial \theta}\right)^2 - e^{2i\theta}\left(1 + \frac{\partial^2}{\partial r^2}\right)$$

and associate problem (9.18) to it, with the Dirichlet condition:

(9.19′) $$B_0 = \text{trace on } \Gamma \ (= \text{boundary of } \Omega).$$

For the choice (9.19), (9.19)', *problem* (9.18) *is never V-coercive* (and no matter what the possible choice of V is). Indeed, we shall verify that, *no matter what* λ *is*, problem

$$(9.20)\qquad (A+\lambda)\,u = 0, \quad u\,|_{\Gamma} = 0$$

admits non-null solutions (which, according to Corollary 9.1, would be impossible if the problem was V-coercive). Indeed, if $\mu \in \mathbf{C}$ satisfies $\mu^2 = \lambda$, the functions

$$u = \sin r \cos(\mu\, e^{-i\theta}) \quad \text{and} \quad u = \sin r \sin(\mu\, e^{-i\theta}) \quad (\lambda \neq 0)$$

and

$$u = \sin r \quad \text{and} \quad u = e^{-i\theta} \sin r \quad (\lambda = 0)$$

are solutions of (9.20). □

9.4 Variational Formulation and Green's Formula

Let A be defined by (1.5) and $a(u, v)$ by (2.18). We consider problem (9.17) and *apply Green's formula* (2.19).

Among the B_j's, choose those which are of order $m_j < m$; we may always assume, with an eventual permutation of indices, that they are: $B_0, \ldots, B_{p-1}$, $0 \leqq p \leqq m-1$ ($p = 0$ means that all B_j's are of order $\geqq m$). There certainly exist "boundary" operators $B'_p, \ldots, B'_{m-1}$, with infinitely differentiable coefficients on Γ, such that the system $\{B_0, \ldots, B_{p-1}, B'_p, \ldots, B'_{m-1}\}$ is a Dirichlet system of order m on Γ.

We can apply *Green's formula* (2.19), taking $\{B_0, \ldots, B_{p-1}, B'_p, \ldots, B'_{m-1}\}$ for the system $\{F_j\}$. *Formally*, we can say that, if v satisfies

$$(9.21)\qquad B_0 v = 0, \ldots, B_{p-1} v = 0,$$

then for all u, we have

$$(9.22)\qquad a(u, v) = \int_{\Omega} (A\,u)\,\bar{v}\,dx - \sum_{j=p}^{m-1} \int_{\Gamma} \Phi_j\, u\, \overline{B'_j v}\, d\sigma.$$

It follows that, if u satisfies

$$(9.23)\qquad a(u, v) = \int_{\Omega} f\,\bar{v}\,dx, \quad \forall v \text{ satisfying (9.21)},$$

then

$$(9.24)\qquad A\,u = f \quad \text{in } \Omega$$

and

$$(9.25)\qquad \sum_{j=p}^{m-1} \int_{\Gamma} \Phi_i\, u\, \overline{B'_j v}\, d\sigma = 0, \quad \forall v \text{ satisfying (9.21)}.$$

Since v is arbitrary and since $\{B_0, \ldots, B_{p-1}, B'_p, \ldots, B'_{m-1}\}$ is a Dirichlet system of order m on Γ, it follows that (see Lemma 2.2):

$$\Phi_j u = 0, \quad j = p, \ldots, m-1. \tag{9.26}$$

Conversely, if u satisfies (9.24) and (9.26), then it follows from (9.22) that u must also satisfy (9.23).

Thus, we see how we can *formally* put problem (9.1) into a *variational formulation* (we do *not yet* worry about the coerciveness): *assume that we can choose* $B'_p, \ldots, B'_{m-1}$ *so that*

$$\Phi_j = B_j, \quad j = p, \ldots, m-1. \tag{9.27}$$

Then (9.17) *is equivalent to* (9.3), *taking for V the space of v's in $H^m(\Omega)$ which satisfy* (9.21). □

Remark 9.5. In the variational formulation we must therefore divide the boundary conditions into two groups:

$$B_j u = 0, \quad j = 0, \ldots, p-1, \quad \text{with} \quad m_j < m,$$

which are sometimes called *stable conditions,* and

$$B_j u = 0, \quad j = p, \ldots, m-1, \quad \text{with} \quad m_j \geqq m,$$

which are called *natural* or *transversality conditions.* □

Remark 9.6. Therefore, we can reduce problem (9.17) to the variational formulation if we can choose $B'_p, \ldots, B'_{m-1}$ so that (9.27) is valid. But this *imposes restrictions.* Indeed, in Green's formula (2.19), A being elliptic, we have: order of $\Phi_j = 2m - 1 -$ order of F_j. Therefore, if $\mu_p, \ldots, \mu_{m-1}$ are numbers between 0 and $m-1$ so that $m_1, \ldots, m_{p-1}$, $\mu_p, \ldots, \mu_{m-1}$ yield all the numbers $0, 1, \ldots, m-1$ (in arbitrary order), then μ_j is the order of B'_j and therefore the order of B_j, with $j = p, \ldots, m-1$, is given by $2m - 1 - \mu_j$, that is

$$m_j = 2m - 1 - \mu_j, \quad j = p, \ldots, m-1. \tag{9.28}$$

Therefore, *we have found a restriction on the order of the natural conditions* (in the sense of Remark 9.5). This condition is not necessarily satisfied for a *regular elliptic problem* (i.e. satisfying conditions (i), (ii), (iii) of Section 5). □

Here is an example. Let

$$A\,u = \Delta^2 u + u = \sum_{i,j=1}^{n} \frac{\partial^2}{\partial x_i^2}\left(\frac{\partial^2 u}{\partial x_j^2}\right) + u \tag{9.29}$$

with the boundary conditions given by the *boundary* operators

$$B_0\,u = u, \quad B_1\,u = \frac{\partial \Delta u}{\partial \nu} \quad (\nu = \text{normal to } \Gamma, \text{ directed towards the interior}). \tag{9.30}$$

Then, we have $m = 2$, $p = 1$, $m_0 = 0$, $\mu_1 = 1$, $m_1 = 3$ and therefore (9.28) is not satisfied for $j = 1$.

Nevertheless, problem (9.29)–(9.30) satisfies conditions (i), (ii), (iii) of Section 5. Indeed, we can reduce it to the case of the half-space $\Omega = \mathbf{R}^n_+$. In the notation of Section 4.1, the polynomial $A_0(\eta, \tau)$ becomes

$$A_0(\eta, \tau) = (\eta^2 + \tau^2)^2;$$

therefore $\tau^+ = \mathrm{i}\,\eta$, double root, $M^+(\eta, \tau) = (\tau - \mathrm{i}\,\eta)^2$. The polynomials corresponding to B_0 and B_1 are

$$B_0(\eta, \tau) = 1, \quad B_1(\eta, \tau) = \tau(\eta^2 + \tau^2).$$

Condition I) of Section 4 is satisfied; to verify II) it suffices to verify that if $a + b\,\tau(\eta^2 + \tau^2)$ is divisible by $(\tau - \mathrm{i}\,\eta)^2$, then $a = b = 0$, which is immediate.

Finally, B_0 and B_1 form a normal system on Γ. Therefore *problem* (9.29)–(9.30) *fits the theory of Sections* 1–7, but not the *variational formulation.* □

Remark 9.7. In fact, all the considerations of Section 9.4 depend on a *choice*; starting from $A\,u = f$, one multiplies by $\bar{v}$ and integrates on Ω, then one applies integration by parts formulas; in other words one takes the scalar product *in* $L^2(\Omega)$. But, more generally, we can "replace" the equation $A\,u = f$ by the equation

$$\int_\Omega A\,u\,\overline{\Lambda\,v}\,dx = \int_\Omega f\,\overline{\Lambda\,v}\,dx, \tag{9.31}$$

where *the operator* Λ *is at our disposal* and is *to be chosen in a suitable way.*

This will eventually give us a space V *depending on the choice of* Λ.

So that what we have in fact shown in Remark 9.6 is that, *for the choice* Λ = *identity, there exist regular elliptic problems which do not fit the variational setting*. But we shall see that, for example for (9.29), (9.30), we can choose Λ in such a way that the problem fits the variational, V-elliptic setting. □

State of the problem

We started by examining when a regular elliptic problem can be put into variational form.

Now, *conversely*, we shall see which *boundary value problems* correspond to variational problems (Section 9.5). Then, we shall briefly study (Section 9.6) the *coerciveness* of variational problems.

9.5 "Concrete" Variational Problems

Let Ω be an arbitrary bounded open set in $\mathbf{R}^n$ and

$$(9.32)\qquad \begin{cases} A\,u = \sum\limits_{|p|,|q|\leq m} (-1)^{|p|} D^p(a_{pq}(x)\, D^q u), \\ a_{pq} \in L^\infty(\Omega). \end{cases}$$

In the notation of Section 9.1, we choose V and H in the following manner:

$H = L^2(\Omega)$,

V = closed subspace of $H^m(\Omega)$ such that $H_0^m(\Omega) \subset V$, with continuous injection.

We also consider a Hilbert space K (in order to stay in the hilbertian setting, but we could also consider a non-hilbertisable topological vector space) which is a *normal space of distributions on* Ω and such that $V \subset K$, with continuous injection (for example $K = L^2(\Omega)$). Let K' be the dual of K; note that K' is a space of distributions on Ω.

We consider the continuous sesquilinear form on V:

$$(9.33)\qquad a(u, v) = \sum_{|p|,|q|\leq m} \int_\Omega a_{pq}\, D^q u\, \overline{D^p v}\, dx.$$

The variational problem is the following:

$$(9.34)\qquad \begin{cases} \textit{with given } f \in K', \textit{ we seek } u \in V \textit{ such that} \\ \qquad a(u, v) = \langle f, \bar{v}\rangle \qquad \forall v \in V, \end{cases}$$

where the brackets denote the duality between K and K'.

If the form $a(u, v)$ is V-elliptic, then we may apply Theorem 9.1, since $\langle f, \bar{v}\rangle$ is a continuous antilinear form on V, according to the hypotheses on K. □

Let us give an interpretation of the problem.

First, since $\mathscr{D}(\Omega) \subset V$, we deduce from (9.34) that

$$(9.35)\qquad A\,u = f,$$

in the sense of distributions on Ω.

As far as the "boundary conditions" for u are concerned, they are *partly* contained in the condition $u \in V$ (*stable conditions*) and *partly* in equation (9.34) (*natural conditions*).

If V is determined by differential conditions such as (9.21), we may also interpret the natural conditions by using Green's formula (2.19) (see Section 9.4); but this is purely *formal* if we do not have regularity conditions on the data and on the solution u which enable us to justify the use of formula (2.19). Therefore, in general, we may say that the *natural conditions* are taken in the sense of equation (9.34).

Usual examples:

1) $V = H_0^m(\Omega)$. Then we have Dirichlet's problem; all the boundary conditions are *stable*: we have $\gamma_j u = 0$, $j = 0, \ldots, m-1$ (and the traces $\gamma_j u$ on Γ have meaning if Ω is sufficiently regular).

In this case, we may take $K = H_0^m(\Omega)$ and therefore $K' = H^{-m}(\Omega)$. □

2) $V = H^m(\Omega)$. There are no stable boundary conditions. Formally, using (2.19), we obtain the *natural* conditions

$$\Phi_j u = 0, \quad j = 0, \ldots, m-1.$$

This problem is called the *Neumann problem* (with respect to Green's formula (2.19)).

Note that we may have *different Neumann problems* relative to a given operator A. Indeed, if we decompose the operator A in different ways with respect to the *elementary* operators D^p, we may associate different forms $a(u, v)$ and therefore different Green's formulas (2.19) to it; for example, we may decompose the Laplace operator Δ in the usual way (take $n = 2$ for the sake of simplicity):

$$\Delta u = \frac{\partial}{\partial x_1}\left(\frac{\partial u}{\partial x_1}\right) + \frac{\partial}{\partial x_2}\left(\frac{\partial u}{\partial x_2}\right)$$

or

$$\Delta u = \frac{\partial}{\partial x_1}\left(\frac{\partial u}{\partial x_1}\right) + \frac{\partial}{\partial x_2}\left(\frac{\partial u}{\partial x_2}\right) + \frac{\partial}{\partial x_1}\left(c\frac{\partial u}{\partial x_2}\right) - \frac{\partial}{\partial x_2}\left(c\frac{\partial u}{\partial x_1}\right) + \\ + \frac{\partial c}{\partial x_2}\frac{\partial u}{\partial x_1} - \frac{\partial c}{\partial x_1}\frac{\partial u}{\partial x_2},$$

where c is a real function belonging to $C^1(\bar{\Omega})$. Then we have the two Green's formulas

$$\int_\Omega \left(\frac{\partial u}{\partial x_1}\frac{\overline{\partial v}}{\partial x_1} + \frac{\partial u}{\partial x_2}\frac{\overline{\partial v}}{\partial x_2}\right) dx = -\int_\Omega \Delta u\, \bar{v}\, dx - \int_\Gamma \frac{\partial u}{\partial \nu}\bar{v}\, d\sigma,$$

$$\int_\Omega \left(\frac{\partial u}{\partial x_1}\frac{\overline{\partial v}}{\partial x_1} + \frac{\partial u}{\partial x_2}\frac{\overline{\partial v}}{\partial x_2} + c\frac{\partial u}{\partial x_1}\frac{\overline{\partial v}}{\partial x_2} - c\frac{\partial u}{\partial x_2}\frac{\overline{\partial v}}{\partial x_1} + \frac{\partial c}{\partial x_2}\frac{\partial u}{\partial x_1}\bar{v} - \\ - \frac{\partial c}{\partial x_1}\frac{\partial u}{\partial x_2}\bar{v}\right) dx = -\int_\Omega \Delta u\, \bar{v}\, dx - \int_\Gamma \left(\frac{\partial u}{\partial \nu} + c\frac{\partial u}{\partial \sigma}\right)\bar{v}\, d\sigma,$$

where $\dfrac{\partial}{\partial \nu}$ and $\dfrac{\partial}{\partial \sigma}$ denote the derivative along the interior normal to Γ and along the tangent to Γ, respectively. Corresponding to these two decompositions we have the two Neumann problems:

$$-\Delta u = f \quad \text{in } \Omega, \qquad \frac{\partial u}{\partial \nu} = 0 \quad \text{on } \Gamma,$$

$$-\Delta u = f \quad \text{in } \Omega, \qquad \frac{\partial u}{\partial \nu} + c\frac{\partial u}{\partial \sigma} = 0 \quad \text{on } \Gamma.$$

Usually, the second problem is called the "regular oblique derivative" problem for Δ. □

3) Let $\Gamma_0, \ldots, \Gamma_{m-1}$ be m subsets of Γ (open sets on Γ) and set (still with respect to Green's formula (2.19) and assuming Ω to be sufficiently regular)

$$V = \{v \mid v \in H^m(\Omega), F_j v = 0 \text{ on } \Gamma_j, j = 0, \ldots, m-1\}.$$

Then, formally, we solve the *mixed problem*

$$A u = f,$$

$$F_j u = 0 \quad \text{on} \quad \Gamma_j, \qquad \Phi_j u = 0 \quad \text{on} \quad \Gamma - \Gamma_j, \qquad j = 0, \ldots, m-1.$$

This problem *does not belong* to the class of *regular boundary value problems* in the sense of Section 1. □

4) If $\Omega = \Omega_1 \cup \Omega_2 \cup \Sigma$ is divided into two parts, Ω_1 and Ω_2, by a surface Σ and if we assume the coefficients of A to be regular in $\bar{\Omega}_1$ and $\bar{\Omega}_2$ *separately*, we have a *transmission problem*: u satisfies the equation $A u = f$ in Ω_1 and Ω_2 separately, and *junction* conditions (or *transmission* conditions) on Σ. □

5) We also note that V may be determined by *integral* conditions (instead of pointwise); then we have new types of "boundary value problems" (see also the *Appendix* to Volume 2). □

9.6 Coercive Forms and Problems

Now, the fundamental problem is to give algebraic necessary and sufficient conditions for the problem to be coercive. We shall limit ourselves to the main results.

1) For the classical case of second order elliptic equations with real coefficients, it is easy to see that the ellipticity of A is sufficient for the coerciveness of $a(u, v)$ on $H^1(\Omega)$ and therefore also on each subspace V; therefore, in particular, the problems of Dirichlet, of Neumann, of mixed type and of transmission for second order equations can be solved.

2) For equations of higher order than the second, the first basic result is due to Gårding [1] and Vishik [1] for the Dirichlet problem:

Theorem 9.2. *Let Ω be a bounded open set in $\mathbf{R}^n$ and A be given by* (9.23); *Assume that $a_{pq} \in C^0(\bar{\Omega})$ if $|p| = |q| = m$ and that A is uniformly strongly elliptic in $\bar{\Omega}$. Then $a(u, v)$ is coercive on $H_0^m(\Omega)$ (i.e. the Dirichlet problem is coercive).*

For the proof, which by now is classical, we refer the reader to Yosida [2], for example. □

3) The coerciveness of the form $a(u, v)$, given by (9.33), on spaces V which are determined by a number $p \leq m$ of differential conditions on Γ of order $< m$:

$$V = \{v \mid v \in H^m(\Omega), B_j v = 0, j = 0, \ldots, m - 1\}$$

has been studied by many authors, following Aronszajn [1]. One may use methods analogous to those we have described in Sections 4 and 5 for the direct a priori estimates.

The most general theorem (at least from the point of view of the operators A and B_j), which is due to Agmon [2], is the following:

Theorem 9.3. *Assume that*

a) *Ω is a bounded open set in $\mathbf{R}^n$, of class C^m, $A(x, D)$ is given by* (9.32) *with $a_{pq} \in C^0(\bar{\Omega})$ if $|p| = |q| = m$;*

b) *B_j, $j = 0, \ldots, p - 1$, are p differential boundary operators*

$$B_j(x, D) = \sum_{|h| \leq m_j} b_{jh}(x) D^h,$$

with $0 \leq m_j \leq m - 1$, $b_{jh} \in C^{m-m_j}(\Gamma)$;

c) $\operatorname{Re} \sum_{|p|=|q|=m} a_{pq}(x) \xi^{p+q} > 0 \quad \forall x \in \bar{\Omega}, \quad \forall \xi \in \mathbf{R}^n - \{0\}$

(therefore A is strongly elliptic in $\bar{\Omega}$);

d) *for all $x \in \Gamma$, all $\xi \in \mathbf{R}^n - \{0\}$ and tangent to Γ at x, all $\xi' \in \mathbf{R}^n - \{0\}$ and normal to Γ at x:*

$$\operatorname{Re} \int_0^{+\infty} \sum_{|p|=|q|=m} a_{pq}(x) \left(\xi + \xi' \frac{1}{i} \frac{d}{dt}\right)^p v(t) \overline{\left(\xi + \xi' \frac{1}{i} \frac{d}{dt}\right)^q v(t)}\, dt > 0$$

for every function $v(t) \in \mathscr{S}(\mathbf{R}_t)$ and $\not\equiv 0$, satisfying the ordinary differential equation

$$\tilde{A}_0\left(x, \xi + \xi' \frac{1}{i} \frac{d}{dt}\right) v(t) = 0$$

and the conditions

$$B_{j,0}\left(x, \xi + \xi' \frac{1}{i} \frac{d}{dt}\right) v(t)\Big|_{t=0} = 0, \quad j = 0, \ldots, p - 1,$$

where $\tilde{A}_0$, $B_{j,0}$ are the characteristic forms of $\tilde{A} = \frac{1}{2}(A + A^*)$ and B_j (see Section 4.5).

Then the form $a(u, v)$ given by (9.33) is coercive on the space V:

$$V = \{u \mid u \in H^m(\Omega), B_j u = 0, j = 0, \ldots, p-1\}. \quad \square$$

The hypotheses on the open set Ω can be considerably weakened in case the form $a(u, v)$ is *formally positive* (i.e. of the type

$$a(u, v) = \int_\Omega \sum_{k=1}^{n} A_k(x, D)\, u\, \overline{A_k(x, D)\, v}\, dx$$

with A_k a linear differential operator of order $\leqq m$) and $V = H^m(\Omega)$ (see K. T. Smith [2]). $\square$

Remark 9.8. Here is an example of a *non-coercive* variational problem. Let $A u$ be given by

(9.36) $$\Delta^2 u = \sum_{i,j=1}^{N} \frac{\partial^2}{\partial x_i^2}\left(\frac{\partial^2 u}{\partial x_j^2}\right).$$

The form $a(u, v)$ is

$$a(u, v) = \sum_{i,j=1}^{n} \int_\Omega \frac{\partial^2 u}{\partial x_i^2}\, \overline{\frac{\partial^2 v}{\partial x_j^2}}\, dx$$

and problem (9.34), choosing $V = H^2(\Omega)$, is formally equivalent to the boundary value problem

$$\Delta^2 u = f \text{ in } \Omega, \quad B_0 u = 0, \quad B_1 u = 0 \text{ on } \Gamma,$$

with

(9.37) $$B_0 u = \Delta u, \quad B_1 u = \frac{\partial \Delta u}{\partial \nu}.$$

The form $a(u, v)$ is not coercive on $H^2(\Omega)$; it is sufficient to note that $a(u, v)$ vanishes if $u = v =$ an arbitrary harmonic function in Ω. $\square$

Remark 9.9. We can add a suitable "tangential" form to the form $a(u, v)$, for example of the type

$$\sum_{j=0}^{m-1} \langle \mathscr{B}_j u, \overline{F_j v}\rangle,$$

where $\mathscr{B}_j \in \mathscr{L}(H^m(\Omega); H^{-m+m_j+1/2}(\Gamma))$ and F_j is the operator of formula (2.19), m_j being the order of F_j, and impose the coerciveness condition on the new form

$$a(u, v) + \sum_{j=0}^{m-1} \langle \mathscr{B}_j u, \overline{F_j v}\rangle.$$

In this way, we obtain *regular oblique derivative* problems.

Similarly, we can add a coercive form on Γ, *of higher order than* a, to $a(u, v)$; thus, we obtain variational boundary value problems for which the boundary conditions contain derivatives *of higher order than* A. $\square$

9.7 Regularity of Solutions

Another very important question concerning the preceding theory is the regularity of the solution. The solution of problem (9.34) belongs to V and therefore to $H^m(\Omega)$; we would like to know the regularity properties of this solution as a function of the regularity properties of the data of the problem (open set Ω, space V, coefficients of A, f); this is a problem of *a priori estimates*.

If Ω and the coefficients of A are sufficiently regular and if problem (9.34) *is equivalent to a regular elliptic problem* (in the sense of Sections 1 and 8), we realize that the regularity of the solution can be obtained by the same methods as were used in Sections 4 and 5 and that the results are of the same type; in particular, we have: if $f \in H^r(\Omega)$, $r \geqq 0$, then $u \in H^{2m+r}(\Omega)$.

But (9.34) also contains non-regular elliptic problems, as we have already seen; and here many different situations exist.

(i) There may be a *local regularity* of the same type as for regular elliptic problems: such is the case, for example, when the boundary conditions determined by (9.34) are equivalent to regular elliptic conditions, in the neighbourhood of a point of Γ.

Example 3 of Section 9.5 (*mixed problems*) admits a local regularity in the neighbourhood of interior points of Γ_j (with respect to Γ). But we can not obtain a global regularity, of the same type, in Ω; that is, if $f \in H^r(\Omega)$, we do not in general have $u \in H^{2m+r}(\Omega)$ (see Magenes-Stampacchia [1] and the bibliography of this work.)

(ii) There are also coercive variational problems which are "irregular" everywhere in $\bar{\Omega}$. For example, we have seen (Theorem 9.2) that the Dirichlet problem may be solved in $H^m(\Omega)$ under very weak hypotheses on Ω and on the coefficients of A; under these hypotheses, the solution u, even if f is very regular, may not belong to $H^{2m}(\omega)$, for *any* set ω contained in $\bar{\Omega}$. See the Comments to this chapter for references. □

9.8 Generalizations (I)

Until now, we have considered the differential operator A as a linear combination of the *elementary operators* D^p, derivatives of order p. It seemed natural to take u in Sobolev spaces $H^s(\Omega)$. But, often the operator A is decomposed into elementary operators *different from the derivatives* D^p and then u must be taken in spaces which no longer are Sobolev spaces. For example, we may write the operator $\Delta^2 u$ as

$$\sum_{i,j=1}^{n} \frac{\partial^2}{\partial x_i \partial x_j}\left(\eta \frac{\partial^2 u}{\partial x_i \partial x_j}\right) + \sum_{i,j=1}^{N} \frac{\partial^2}{\partial x_i^2}\left(\mu \frac{\partial^2 u}{\partial x_j^2}\right)$$

$$\eta, \mu \in \mathbf{R}, \quad \text{with} \quad \eta + \mu = 1,$$

or as

$$\Delta(\Delta u).$$

In the first case, we consider the second derivatives as elementary operators and it is natural to have $u \in H^2(\Omega)$; in the second case, the elementary operator is the Laplacian and it is natural to consider the space of u's $\in L^2(\Omega)$ such that $\Delta u \in L^2(\Omega)$, which is *different* from $H^2(\Omega)$.

Problem

$$(9.38) \quad \Delta^2 u + \lambda u = f \ \text{ in } \Omega, \quad \Delta u = 0, \quad \frac{\partial \Delta u}{\partial \nu} = 0 \ \text{ on } \Gamma,$$

as we have seen (Remark 9.8), is not coercive on $H^2(\Omega)$. But obviously, if we take $H = L^2(\Omega)$ and $V = \{v \mid v \in L^2(\Omega), \Delta v \in L^2(\Omega)\}$, provided with the norm

$$(\|u\|^2_{L^2(\Omega)} + \|\Delta u\|^2_{L^2(\Omega)})^{1/2},$$

then the form

$$a(u, v) = \int_\Omega \Delta u \overline{\Delta v}\, dx + \lambda \int_\Omega u \bar{v}\, dx$$

is V-elliptic for all λ with $\operatorname{Re} \lambda > 0$ and therefore Theorem 9.1 applies. Therefore for all $f \in L^2(\Omega)$ there exists a unique $u \in V$ such that

$$a(u, v) = \int_\Omega f \bar{v}\, dx \qquad \forall v \in V.$$

It follows that $\Delta^2 u + \lambda u = f$ in the sense of distributions on Ω, the boundary conditions being taken in the *formal* sense (but note that $w = \Delta u$ is such that $w \in L^2(\Omega)$, $\Delta w \in L^2(\Omega)$ and therefore if Ω is sufficiently regular we may apply the trace Theorem 6.5 and obtain

$$\gamma_0 w \in H^{-1/2}(\Gamma), \quad \gamma_1 w \in H^{-3/2}(\Gamma)). \quad \square$$

Remark 9.10. The variational theory may be applied to non-elliptic equations (or not even hypoelliptic!); for example, we may study the problem

$$\frac{\partial^4 u}{\partial x^2\, \partial y^2} + \lambda u = f \ \text{ in a square } \Omega \text{ of } \mathbf{R}^2, \quad u = 0 \ \text{ on } \Gamma,$$

in which the operator A is hyperbolic, by taking $H = L^2(\Omega)$ and V = closure of $\mathscr{D}(\Omega)$ in the space

$$\mathscr{H} = \left\{u \mid u \in L^2(\Omega), \frac{\partial^2 u}{\partial x\, \partial y} \in L^2(\Omega)\right\}.$$

Another interesting application can be made to a remarkable class of hypoelliptic operators, which are obtained by using the spaces

$H^{\alpha,\beta}(\Omega)$ and their closed subspaces, where $\Omega = \Omega_x \times \Omega_y$ (Ω_x: open set of $\mathbf{R}_x^n$ and Ω_y: open set of $\mathbf{R}_y^m$; therefore Ω is an open set of $\mathbf{R}^{m+n}$) and

$$H^{\alpha,\beta}(\Omega) = \{u \mid u \in L^2(\Omega),\, D_x^p u \in L^2(\Omega),\, |p| \leqq \alpha,\, D_y^q u \in L^2(\Omega),\, |q| \leqq \beta\}$$

and as elementary operators the derivatives of the type D_x^p and D_y^q, with $|p| \leqq \alpha$, $|q| \leqq \beta$. For example, if, in $\mathbf{R}^2$, we take

$$\Omega_x = \{x \mid a < x < b\}, \qquad \Omega_y = \{y \mid c < y < d\},$$

$$\Omega = \Omega_x \times \Omega_y \qquad \alpha = 2, \qquad \beta = 1,$$

V = closure of $\mathscr{D}(\Omega)$ in $H^{2,1}(\Omega)$, $H = L^2(\Omega)$,

$$A\,u = \frac{\partial^4 u}{\partial x^4} - \frac{\partial^2 u}{\partial y^2}$$

$$\left(\text{and therefore } a(u, v) = \int_\Omega \left(\frac{\partial^2 u}{\partial x^2}\overline{\frac{\partial^2 v}{\partial x^2}} + \frac{\partial u}{\partial y}\overline{\frac{\partial v}{\partial y}}\right) dx\,dy\right),$$

we can solve the problem (of Dirichlet):

$$A\,u = f \quad \text{in } \Omega$$

$$u = \frac{\partial u}{\partial x} = 0 \quad \text{on the sides } x = a \text{ and } x = b \text{ of } \Omega,$$

$$u = 0 \quad \text{on the sides} \quad y = c \quad \text{and} \quad y = d \quad \text{of } \Omega. \quad \square$$

9.9 Generalizations (II)

We come back to problem (9.29) – (9.30) and Remark 9.7.
Let $H = H_0^1(\Omega)$, with the norm

$$\left(\sum_{i=1}^n \left\|\frac{\partial v}{\partial x_i}\right\|_{L^2(\Omega)}^2\right)^{1/2}$$

and

$$V = \left\{v \mid v \in H_0^1(\Omega), \frac{\partial \Delta v}{\partial x_i} \in L^2(\Omega),\, i = 1, \ldots, n\right\},$$

with the norm

$$\left(\sum_{i=1}^n \left(\left\|\frac{\partial v}{\partial x_i}\right\|_{L^2(\Omega)}^2 + \left\|\frac{\partial \Delta v}{\partial x_i}\right\|_{L^2(\Omega)}^2\right)\right)^{1/2}.$$

We define

$$a(u, v) = \sum_{i=1}^n \int_\Omega \frac{\partial \Delta u}{\partial x_i}\overline{\frac{\partial \Delta v}{\partial x_i}}\,dx + \sum_{i=1}^n \int_\Omega \frac{\partial u}{\partial x_i}\overline{\frac{\partial v}{\partial x_i}}\,dx.$$

Then $a(u, v)$ is V-elliptic and Theorem 9.1 applies; therefore, for all $f \in H_0^1(\Omega)$ there exists $u \in V$ such that

$$(9.39) \qquad \sum_{i=1}^{n} \int_{\Omega} \frac{\partial \Delta u}{\partial x_i} \overline{\frac{\partial \Delta v}{\partial x_i}} \, dx + \sum_{i=1}^{n} \int_{\Omega} \frac{\partial u}{\partial x_i} \overline{\frac{\partial v}{\partial x_i}} \, dx = \sum_{i=1}^{n} \int_{\Omega} \frac{\partial f}{\partial x_i} \overline{\frac{\partial v}{\partial x_i}} \, dx \qquad \forall v \in V.$$

We recall that Δ is an isomorphism of $H_0^1(\Omega)$ onto $H^{-1}(\Omega)$ and that for φ and $\psi \in H_0^1(\Omega)$ we have (Green's) formula

$$\sum_{i=1}^{n} \int_{\Omega} \frac{\partial \varphi}{\partial x_i} \overline{\frac{\partial \psi}{\partial x_i}} \, dx = -\langle \varphi, \overline{\Delta \psi} \rangle,$$

the brackets denoting the duality between $H^{-1}(\Omega)$ and $H_0^1(\Omega)$. Then (9.39) becomes

$$(9.40) \qquad \sum_{i=1}^{n} \int_{\Omega} \frac{\partial \Delta u}{\partial x_i} \overline{\frac{\partial \Delta v}{\partial x_i}} \, dx = \langle u, \overline{\Delta v} \rangle - \langle f, \overline{\Delta v} \rangle = \int_{\Omega} u \overline{\Delta v} \, dx - \int_{\Omega} f \overline{\Delta v} \, dx \qquad \forall v \in V.$$

But as v describes V, Δv describes a space W which, in particular, contains $H^1(\Omega)$; therefore, it follows that $w = \Delta u$ is a solution of

$$\sum_{i=1}^{n} \int_{\Omega} \frac{\partial w}{\partial x_i} \overline{\frac{\partial z}{\partial x_i}} \, dx = \int_{\Omega} (u - f) \bar{z} \, dx \qquad \forall z \in H^1(\Omega),$$

that is of the (variational) Neumann problem

$$\begin{cases} -\Delta w = u - f & \text{in } \Omega, \\ \dfrac{\partial w}{\partial \nu} = 0 & \text{on } \Gamma. \end{cases}$$

But then for u, which belongs to V, we have

$$(9.41) \qquad \Delta^2 u + u = f \quad \text{in } \Omega,$$

$$(9.42) \qquad \frac{\partial \Delta u}{\partial \nu} = 0 \quad \text{on } \Gamma, \qquad u = 0 \quad \text{on } \Gamma.$$

Note that (9.41) is taken in the sense of distributions on Ω, and the boundary conditions (9.42) are taken in the variational sense, if Ω is not regular, and in the concrete sense of trace theorems, if Ω is regular (see Theorem 7.3). *Therefore, we have solved problem* (9.29) – (9.30) *which did not fit the variational formulation of Section 9.5, with Λ = identity* (see Remark 9.7). Here, we have taken $\Lambda = -\Delta$. □

Remark 9.11. If Ω is sufficiently regular, $v \in V$ implies that $\Delta v \in \in L^2(\Omega)$ and we may take $f \in L^2(\Omega)$ in (9.41). □

10. Comments

We shall give few bibliographical references for equations of the second order and refer the reader to the books of Bitzadze [4], Courant-Hilbert [1] and C. Miranda [1].

For equations of higher order than the second, boundary value problems have only been studied recently (except for some particular cases such as, for example, the iterated Laplace operator or equations in two variables, for which the reader can consult E. E. Levi [1]), and it is mainly following the works of Garding [1] and Vishik [1] that the variational theory has been developed: various accounts of this theory can be found in Agmon [7], Berezanski [4], Browder [3, 9], Kato [5], Lions [2], Magenes-Stampacchia [1], Necas [2], Vishik-Ladyzenskaia [1]. We also note the work of Hestenes [1]. Here, we have only sketched the outline of the theory.

Theorem 9.1 has been introduced and used by many authors: Lax-Milgram [1], Vishik [1], . . .; for various extensions to "convex sets", see Stampacchia [1] and then Lions-Stampacchia [1].

For the coerciveness problem pointed out in Section 9.6, we add the works of Necas [2, 3], Schechter [1], Smith [1] to the references to Agmon, Aronszajn and Smith made in the text.

The question of the regularity at the boundary of solutions of "regular" variational problems (see Section 9.7) was resolved by Nirenberg [2] (see also Guseva [1]). Also note the "compensation method" of Aronszajn-Smith, an account of which is given in Lions [4], for example. It concerns problems for which there is a certain regularity of the data (coefficients of A, boundary conditions, open set Ω).

If these conditions of regularity are not satisfied, the question of the regularity of the solutions is much more difficult; for example, if the coefficients of A belong to $L^\infty(\Omega)$ the case of *second order* equations is resolved in a satisfactory manner, thanks to the results of De Giorgi [1] and Nash [1] and many other authors: H. O. Cordes, D. Gilbarg, O. Ladyzenskaia, W. Littman, C. Miranda, C. B. Morrey, J. Moser, J. Servin, G. Stampacchia, N. N. Uraltzeva, H. F. Weinberger . . .; we refer the reader to Ladyzenskaia-Uraltzeva [1], Miranda [1], Morrey [2], Stampacchia [2] and to the bibliographies of these works; for "mixed" and "transmission" problems, see also further on. For the case of higher order equations we note the recent results of De Giorgi [2], Morrey [5]. When the coefficients of A belong to $L^\infty(\Omega)$, the solutions can not be generalized to spaces of functions with locally integrable first derivatives without losing the fundamental properties; see Serrin [1].

Still in the setting of variational theory, we note that the idea described in Section 9.8 was introduced in a general abstract manner in Lions [2] and developed for many concrete problems by various authors. For example, it is a well-known situation for the equations of elasticity. It is also a natural situation for boundary value problems for the hypoelliptic operators associated with the spaces $H^{\alpha,\beta}(\Omega)$ and pointed out in Section 9.8; see the works of H. Marcinkowska, V. P. Milkhailov, P. P. Mosolov, S. N. Nikolski, M. Pagni, B. Pini cited in Pagni [1], and also Pagni [2], Pini [12], Ramazanov [2].

For the problems of Section 9.8, see also Magenes-Stampacchia [1], Pulvirenti [1, 2].

We call attention to Remark 9.7 and Section 9.9, which, we believe, may be developed more systematically to obtain more precise information on boundary value problems which may fit a variational theory. The idea of replacing "$A\,u = f$" by (9.31), with Λ to be *chosen*, has been used in many situations. For *hyperbolic* evolution equations this idea plays an absolutely fundamental role; see the method called "method a-b-c" of Friedrichs [3] and the obtainment of the energy inequalities in Leray [1]. For elliptic equations, the notion of conditional ellipticity (inspired by Leray [1]), which rests on a neighboring idea, was introduced in Lions [8]; see also an analogous idea for k-positive definite operators: Martyniuk [1], Petryshin [1]. Along closely related lines, see Chapter 4 of Berezanski [4] and the works [1, 2] of this author. See also Dezin [1].

The example given in Section 9.9 reconsiders, in a clearer form, one of the examples given in Lions [3].

We also note the following: if A (resp. A^*) is an isomorphism of $D(A)$ (resp. $D(A^*)$) onto H, then A is (by transposition) an isomorphism of H onto $D(A^*)'$ and therefore (by interpolation) of $[D(A), H]_{1/2}$ onto $[D(A^*), H]'_{1/2}$. In the case where $[D(A), H]_{1/2} = [D(A^*), H]_{1/2}$ (for the study of this question see Kato [4, 5], Lions [17], Shimakura [1]), setting $V = [D(A), H]_{1/2}$, we see that A is an isomorphism of V onto V' and therefore A is associated to the sesquilinear form

$$a(u, v) = (A\,u, v) = \text{scalar product of } A\,u \text{ and } v \text{ in the anti-duality between } V' \text{ and } V.$$

We also point out the use made by Necas [2, 4] of the equality of Rellich for elliptic boundary value problems.

For relations between variational theory and non-variational theory, see also Agmon [4], Shimakura [2].

We come now to the *non-variational theory* developed in Sections 1 – 7 of this chapter for regular elliptic problems.

The conditions of *proper ellipticity* of A and of *system* $\{B_j\}$ *covering* A (Definitions 1.2 and 1.5) were introduced by Lopatinski [1] and Shapiro [1] and by Agmon-Douglis-Nirenberg [1]; the terminology "proper ellipticity" and "covers", which we have used, is that of Schechter [2].

Definition 1.4 of *normal system* and Theorem 2.1 on Green's formula are due to Aronszajn-Milgram [1]; see also Schechter [2] for the proof as we have given it. The proof of Theorem 2.2, given in Section 4.3, is due to Agmon.

For the existence of local solutions, the regularity at the interior and the hypoellipticity of the elliptic operators (Section 3), starting from the lemma of Cacciopoli [1] – Weyl [1], see John [1], Friedrichs [1], Petrowski [1], Schwartz [7] (whose idea of the proof we have followed) etc.

The characterization of hypoelliptic operators with constant coefficients was given by Hörmander [1]; for sufficient conditions for the hypoellipticity of operators with variable coefficients see Friberg [1], Hörmander [2, 6], Malgrange [1], Mizohata [2], Trèves [3]; a very general sufficient condition for second order operators was obtained by Hörmander [11].

The *direct* a priori estimates (of the type of formula (4.39)) were obtained by numerous authors: Agmon-Douglis-Nirenberg [1], Browder [1, 2], Hörmander [3], Koselev [1], Peetre [1], Schechter [1], Slobodetski [2, 3]; the use of the Fourier transform to obtain them, as we did in Section 4, is classical by now.

To prove the existence of the index of the operator $\mathscr{P}$ in Sections 4 and 5, we have used the method of Peetre [2] which depends on the obtainment of the "dual" estimates (formulas (4.40) and (5.4)) (see also the account of Grisvard [3]). For Section 5.3, see Schechter [11]. We have pointed out the advantages of this method in Section 8.4. But here, we have to recall the other methods which may be used to obtain the final results of Section 5, by also generalizing them to Sobolev spaces of L^p type.

The "direct" a priori estimates may be obtained in L^p by the method of Poisson kernels which uses the theory of singular integrals of Calderon-Zygmund [1] (see Agmon-Douglis-Nirenberg [1], Agmon [3], Browder [1, 2], Nirenberg [3]) and also by a method of Peetre [13] and Arkeryd [1]. We also call attention, for equations of variational type, to the estimates in the spaces $\mathscr{L}^{2,\lambda}(\Omega)$ of Morrey type (see Stampacchia [3] and the bibliography of this work, for theses spaces) given by Campanato [3, 4, 6] and from which the estimates in the spaces $L^p(\Omega)$ can be deduced (see Campanato-Stampacchia [1], Giusti [1]), as well as certain results for problems with discontinuous coefficients (see Campanato [6], Kadlec-Necas [1]), which connects with the work of Morrey [1, 4] and Cordes [1].

The existence of the index of $\mathscr{P}$ may also be shown by the construction of a *parametrix*: see, for example, Chapter X of Hörmander [6] and Agranovich-Vishik [1]. For the case of problems with *normal* boundary conditions, the existence of the index of $\mathscr{P}$ may be proved by using the formal adjoint problem $\{A^*, C\}$ directly and certain results of variational theory (see Berezanski [4], Schechter [2, 3]).

In the case of the plane, we need to recall the methods which, taking their inspiration from the classical theory for second order equations, have studied the boundary value problems by using a *global* representation of the solutions by "double or single layer potentials" and the theory of singular integral equations in one variable (see Agmon [1], Fichera [2], Muskhelisvili [1], Vekua [1], Volpet [1], etc.).

Finally, the theory of pseudo-differential operators (which generalize both the differential operators and the singular integral operators), developed and applied by Agranovich [1], Agranovich-Dynin [1], Boutet de Monvel [1, 2], Calderon [4, 5], Calderon-Zygmund [2], Dynin [1, 2], Grusin-Vainberg [1], Hörmander [7, 8, 9], Kohn-Nirenberg [1, 2], Mihlin [1], Shamir [4, 5], Seeley [1, 4, 5], Unterberger-Bokobza [1], Vishik [5,) Vishik-Eskin [1–4], etc., also contains the "regular" elliptic boundary value problems; then, the results of Section 5 become particular cases of the results on pseudo-differential operators. Even the boundary value problems for elliptic equations which are not "regular" enter this theory; for example the general problem of the "oblique derivative" and the "mixed" problem (see also below). But we have not touched upon the questions pertaining to pseudo-differential operators. This would have implied some very considerable supplementary developments; and a great number of questions still seem to be open in this direction (see the problems for this chapter).

Sections 6 and 7 of this chapter systematically develops and completes (we have eliminated the exceptional values of the parameter s and introduced the spaces $\Xi^s(\Omega)$) the work of Lions-Magenes [1].

The method of transposition for boundary value problems was introduced by Sobolev-Vishik [1] and Fichera [1] and also used in Lions [30], Magenes-Stampacchia [1], Berezanski [5]; it was systematically applied to elliptic boundary value problems, furthermore using interpolation and Sobolev spaces of L^p type $(1 < p < \infty)$, by Lions-Magenes [1], then by Baiocchi [1], Barkoski-Roitberg [1], Berenzanksi [3, 4], Berenzanski-Krein-Roitberg [1], Berenzanski-Roitberg [1], Geymonat [1], Roitberg [1, 2], Schechter [4, 8, 9].

In order to separate the equation $A u = f$ from the boundary conditions $B_j u = g_j$ in the functional equation (6.6), we have been led (see Lions-Magenes [1]) to the density and trace theorems (Theorems 6.4 and 6.5).

This enables us to give an interpretation "of the classical type" for the resolved problems.

In Roitberg [3], the author obtains isomorphism results in spaces obtained by completion of $\mathscr{D}(\bar{\Omega})$ for suitable norms. In order to "separate" the various data (in Ω and on Σ), he uses the trace theorems (Lions-Magenes [1] and Section 6 of this chapter), which enables him to obtain our results again (up to technical variants) and other results of a more abstract nature.

In Section 7, the use of Theorem 14.3 of Chapter 1, due to Baiocchi [5], enabled us to considerably simplify the account we had given in previous works. Let us also mention that we have arrived at Proposition 7.1 and Theorem 7.1 after discussions with G. Geymonat. For other results of the type of Remark 7.3, see Lions-Magenes [3].

The property of continuity of traces on surfaces neighboring Γ (Section 8.1) draws upon the formulation of boundary value problems for elliptic equations of the second order given by Cimmino [1, 2], and for polyharmonic equations by Sobolev [1] (see also Magenes [1], Pini [4]).

The realization of A in $L^2(\Omega)$ (Section 8.4) is due to Browder [1, 5]; our proof follows Grisvard [3]. Concerning the vanishing of the index (Fredholm operators, see result (8.14)) see Agmon-Douglis-Nirenberg [1], Agranovich [1], Browder [6], Geymonat-Grisvard [1], Kaniel-Schechter [1], etc.

For uniqueness theorems different from those of Section 8.3, see Vishik [1] and Pini [3] (case of sufficiently "small" domains) and Agmon-Douglis-Nirenberg [1] (case of "weakly positive semi-definite" operators).

The technique of trace theorems given in this chapter is equally useful for *unilateral* problems of which we present a simple example: in an open set Ω with regular boundary Γ, there exists one and only one function $u \in H^1(\Omega)$ such that

$$-\Delta u + u = f, \quad f \in L^2(\Omega)$$

$$u \geqq 0 \quad \text{on } \Gamma \quad (\text{in the sense: } \gamma_0 u \geqq 0 \text{ in } H^{1/2}(\Gamma))$$

$$\frac{\partial u}{\partial \nu} \leqq 0 \quad \text{on } \Gamma \quad (\text{in the sense: } \gamma_1 u \leqq 0 \text{ in } H^{-1/2}(\Gamma))$$

$$u \frac{\partial u}{\partial \nu} = 0 \quad \text{on } \Gamma \quad (\text{the multiplication } u, v \to u \cdot v \text{ is}$$

is a continuous mapping of $H^{1/2}(\Gamma) \times H^{-1/2}(\Gamma) \to \mathscr{D}'(\Gamma)$, for example, so that $u\dfrac{\partial u}{\partial \nu}$ *has meaning*$\Big)$; for these problems, see Lions-Stampacchia [1] and the bibliography of this work.

Furthermore, the techniques of this chapter enable us to prove *uniqueness theorems* of the following type: let $\mathbf{R}^n_+$ be the half-space $\{x_n > 0\}$ and u an element of $H^{-k}(\mathbf{R}^n_+)$, k an arbitrary positive real number, solution of

$$-\varDelta u + u = 0$$

which satisfies

$$\frac{1}{x_n}(u(x', x_n) - u(x', 0)) \to 0 \quad \text{in} \quad \mathscr{D}'(\mathbf{R}^{n-1}) \quad \text{as} \quad x_n \to 0.$$

Then $u = 0$ $\Big($indeed, it follows that $\dfrac{\partial u}{\partial x_n} = 0$ in $H^{-k-3/2}(\mathbf{R}^{n-1})$, whence the result$\Big)$. This type of result extends to elliptic operators with variable coefficients of arbitrary order and in the spaces L^p by using Lions-Magenes [1] (V). For results of this type via different methods, see Butzer [1].

Finally, we point out some other interesting questions pertaining to elliptic boundary value problems which we have not studied in this book:

1) The "topological" questions pertaining to the theory of boundary value problems (study of invariants by "homotopy", computation of the index, relations with certain classical problems of algebraic topology) which have been posed and developed in the last few years and for which we refer the reader to Agranovich [1], Atiyah [1], Atiyah-Bott [1], Atiyah-Singer [1], Calderon [8], Gelfand [1, 2], Gohberg-Krein [1], Palais [1], Seeley [1, 4], Volpert [2].

2) The "mixed" problems (pointed out in Section 9.5, example 3, in the setting of variational theory): see Peetre [3], Schechter [5], Shamir [1, 3–7], Vishik [5], Vishik-Eskin [1–3], etc.

3) The "transmission" problems (pointed out in Section 9.5, example 4, in the setting of variational theory): see Campanato [1], Lions [31], Roitberg-Sheftel [2], Schechter [6], Sheftel [1–3], Stampacchia [4], Troisi [1–3], etc.

4) The general *oblique derivate* problem in any dimension: see Bitzadze [2, 3], Egorov-Kondrat'ev [1, 2], Hörmander [8], Vishik-Eskin [5], etc.

5) The boundary value problems in domains with "angular points"; see Hanna-Smith [1], Kondratiev [2], Necas [2], Volkov [1] and also 10) and 11) below.

6) The axiomatic theory of the potential and the theory of capacity according to Brelot, Choquet, Deny; see the seminar notes of Brelot-Choquet-Deny [1] and the "connection" made between this theory and the boundary value problems for elliptic equations of the second order with measurable coefficients (R. M. Hervé [1], Littman-Stampacchia-Weinberger [1], Stampacchia [2]).

7) Generalizations of the theory of Sections 3–8 to $L^p(\Omega)$ type spaces, $p \neq 2$ (Sobolev spaces $W^{s,p}(\Omega)$, Besov spaces $B^{s,p}(\Omega)$, Lebesgue spaces $H^{s,p}(\Omega)$; we follow the notations of Magenes [3], for example); see Lions-Magenes [1] and the accounts of Magenes [2, 3] and Geymonat-Grisvard [1].

8) Spectral theory (study of the spectrum of problems, eigenfunction expansions, asymptotic distribution of eigenvalues, ...): we refer the reader to the books of Agmon [7] and Berezanski [4] and to the works of Agmon [6, 8], Agmon-Kannai [1], Aronszajn [5], Browder [5, 8], Fichera [4], Gårding [2], Geymonat-Grisvard [3], Mizohata-Arima [1], Peetre [11], Plejel [1], Seeley [6], Weinstein [1], etc.

9) Fundamental solutions, Green's functions; see Berezanski [4], John [1], Hörmander [6], Miranda [1], Courrége [1], Trèves [5] ...

In this regard, we call attention to the study of Green's function of fractional powers of elliptic operators: see Kotake-Narasimhan [1], Sobolevski [2]. Moreover, this question is tied to interpolation spaces (see Kato [3–5], Lions [17], Fujiwara [1]).

10) "Variational" equations of the second order

$$\left(\text{i.e. of the type } \sum_{|p|,|q| \leq 1} D^p(a_{pq} D^q u)\right)$$

with discontinuous coefficients a_{pq}: problem of the regularity of the solution (which we have already mentioned in these "Comments" and in Section 9.7), maximum principle, Harnack's inequality, "regular" and "non-regular" points of the boundary, "removable" singularities of the solution, "isolated" singularities of the solution ... (see the authors and the bibliography already cited in connection with regularity problems).

11) "Non-variational" equations of the second order

$$\left(\text{i.e. of the type } \sum_{|p| \leq 2} a_p D^p\right)$$

with discontinuous coefficients a_p, for which analogous problems to those of 10) exist: see Alexandrov [1, 2], Bers-Nirenberg [1], Miranda [3], Nirenberg [1], Pucci [1], ..., [4], Finn and Serrin [1], Gilbarg and Serrin [1], Talenti [1–3], Cordes [1], Stampacchia [5] ...

12) Boundary value problems in weighted Sobolev spaces and for equations which "degenerate" on the boundary: see Baouendi [1],

Geymonat-Grisvard [2], Kudryavcev [1], Morel [1], Murthy-Stampacchia [1], Necas [2], Oleinik [5], Vishik [3], Vishik-Eskin [2], and the results of Baouendi-Goulaouic [2], elucidating, as a consequence, the topological structure of $\mathscr{D}(\bar{\Omega})$.

13) "Non-local" boundary value problems; see Bade-Freeman [1], Beals [1], Browder [7], Fishel [1], Freeman [2], Grubb [1], Peetre [1], Schechter [10], Vishik [3]; see also the Appendix to Volume 2 of this book and the works on the infinitesimal generators of Markov semigroups (see Dynkin [1] and the bibliography of this work). Also, the integro-differential operators of the second order, which come up in "non-local" problems, appear as infinitesimal generators of Feller semigroups on a variety with boundary; see Bony-Courregè-Priouret [1] and the bibliography to this note.

14) Boundary value problems in unbounded open sets: see Baiocchi [1], Barros-Neto [1], Freeman [1], Kudrjavcev [2], Lax [2], Lions-Magenes [1] (I), Miranda [4], Peetre [1, 2], etc.: in this regard, particularly Beppo-Levi type spaces (see Deny-Lions [1]) and the completion of $\mathscr{D}(\Omega)$ in these spaces (see Hörmander-Lions [1]) are used.

15) Singular perturbations: see Friedrichs [2], Huet [1] ... [5], Oleinik [6], Peetre [4], Vishik-Ljusternik [1], etc.

16) Study of boundary value problems in the spaces $C^{k,\alpha}(\bar{\Omega})$ of "holderian" functions (Schauder type estimates, ...); see Agmon [9], Agmon-Douglis-Nirenberg [1], Miranda [2], etc.

17) Uniqueness and unique extension problems: see Aronszajn [6], Cordes [2], Heinz [1], Landis [1], Müller [1], Pederson [1]; for equations of general type, see Calderon [7], Hörmander [6].

18) *Generalizations of the preceding questions to systems of elliptic operators* (containing, in particular, the classical systems of elasticity and of Stokes).

We recall the main definitions pertaining to elliptic systems.

Consider the *matrix operator* given by

$$A = A(x; D) = \| l_{ij}(x; D) \|, \quad i, j = 1, \ldots, m,$$

where the l_{ij}'s are linear differential operators with coefficients defined in $\bar{\Omega}$.

A is said to be *elliptic* in $\bar{\Omega}$, according to Douglis-Nirenberg [1], if there exist integers s_i, t_i, $i = 1, \ldots, m$, such that the order of l_{ij} is $s_i + t_j$ (where we take $l_{ij} \equiv 0$, if $s_i + t_j < 0$) and if, denoting by $l^0_{ij}(x, D)$ the principal (or characteristic) part of l_{ij} and by $A^0(x, D)$ the matrix $\| l^0_{ij}(x, D) \|$, we have, for all $x \in \bar{\Omega}$ and all $\xi \in \mathbf{R}^n$ and $\neq 0$,

$$L^0(x, \xi) = \det \| A^0(x, \xi) \| \neq 0.$$

The system of operators A is said to be *properly elliptic* if, in addition,

$$\sum_{i=1}^{m} (s_i + t_i) = 2r,$$

integer $r > 0$, and if, for all $x \in \bar{\Omega}$ and every couple of linearly independent vectors ξ, ξ' in $\mathbf{R}^n$, the polynomial $L^0(x, \xi + \tau\xi')$ in τ has r roots with positive imaginary parts: $\tau_1^+(x, \xi, \xi'), \ldots, \tau_r^+(x, \xi, \xi')$.

If $n \geqq 3$, every elliptic system is properly elliptic. We could also consider more particular classes of elliptic systems (Petrowski systems, strongly elliptic systems ..., see, for example, Volevich [2]). We just recall the definition of *strongly elliptic* systems: A is said to be *strongly elliptic* in $\bar{\Omega}$ if $t_i = s_i > 0$ and if for every $x \in \bar{\Omega}$ and for every complex vector $\lambda = (\lambda_1, \ldots, \lambda_m)$ and every $\xi \in \mathbf{R}^n$, with $\lambda \neq 0$ and $\xi \neq 0$, we have

$$\operatorname{Re} \sum_{i,j=1}^{m} (-1)^{s_i} l_{ij}^0(x, \xi)\, \lambda_i \bar{\lambda}_j \geqq k \sum_{i=1}^{m} |\xi|^{2s_i} |\lambda_i|^2,$$

with k a positive constant.

We also recall the definition of *systems of boundary operators covering A* (generalization of Definition 1.5 of this chapter). Given a properly elliptic matrix operator A, we denote by $\mathscr{A}(x, \xi)$ the adjoint of the matrix $A^0(x, \xi)$ (i.e. such that $A^0(x, \xi)\, \mathscr{A}(x, \xi) = L^0(x, \xi)\, I$, I = identity matrix). We consider a matrix of linear differential operators with coefficients defined in Γ:

$$B = B(x, D) = \| B_{qj}(x, D) \|, \quad q = 1, \ldots, r, \quad j = 1, \ldots, m.$$

We say that B *covers* A if there exist integers σ_q, $q = 1, \ldots, r$, such that B_{qj} is of order $\sigma_q + t_j$ (if $\sigma_q + t_j < 0$, we assume that $B_{qj} \equiv 0$) and if, denoting by $B_0^{qj}(x, D)$ the principal part of $B_{qj}(x, D)$ and by $B^0(x, D)$ the matrix $\| B_{qj}^0(x, D) \|$, for every $x \in \Gamma$, every $\xi \in \mathbf{R}^n$ tangent to Γ at x, every $\xi' \in \mathbf{R}^n$ normal to Γ at x, the rows of the matrix

$$B^0(x, \xi + \tau\xi')\, \mathscr{A}(x, \xi + \tau\xi')$$

(the elements of which are polynomials in τ) are linearly independent modulo the polynomial

$$\prod_{k=1}^{r} (\tau - \tau_k^+(x, \xi, \xi')).$$

The Dirichlet system covers every strongly elliptic system (see Agmon-Douglis-Nirenberg [1]) but does not cover every elliptic system (see Bitzadze [1]).

There is not yet a general definition of *normal* boundary operator systems; the problem is tied to the validity of a *Green's formula*, analogous to formula (2.3), for systems. We refer the reader to Geymonat [3], Lipko-Eidelman [1], Roitberg-Sheftel' [1] for results on this subject.

Many results for elliptic equations have been generalized to elliptic systems: in particular, the a priori estimates and the existence of the index, in L^2 as well as in L^p, for boundary value problems $\{A, B\}$, with A properly elliptic in the sense of Douglis-Nirenberg and with B covering A; but many other problems still remain unsolved on this subject (see, for example, problem 11.1).

Consult Agmon-Douglis-Nirenberg [1], Agranovich-Dynin [1], Agranovich-Dynin-Volevic [1], Avantaggiati [1, 2], Campanato [1], Canfora [1], Cattabriga [1], Douglis-Nirenberg [1], de Figuereido [1], Geymonat [2, 3], Gobert [1, 2], Hörmander [6], Lawruk [1], Morrey [1, 2], Necas [1, 2], Pini [2], Roitberg-Sheftel' [1], Sheftel' [1], Solonnikov [1], Vishik [1, 5], Vishik-Eskin [1, 3], Volevic [2], Volpert [1], etc. Pseudo-differential systems are studied in Vishik-Eskin [5].

19) Elliptic systems of the first order which generalize the Cauchy-Riemann system, the theory of pseudo-analytic functions and quasi-conformal representation: see Bers [1], Hörmander [10], Lavrentiev [1], Miranda [5], Morrey [3], Stampacchia [6], Vekua [2].

20) Problems pertaining to the approximation of the solution by finite difference methods: there exists a vast literature on this subject, especially for homogeneous variational problems, but the study of the general case (even from just the point of view of convergence) and the estimate of the error still cause numerous problems. Particular results may be found in Bramble [1], Jamet [1], Lions [29].

Finally, questions concerning analytic regularity, or regularity in Gevrey classes, of solutions and the study of boundary value problems in classes of distributions or ultradistributions will be investigated in Volume 3 of this book.

11. Problems

11.1 Extension of this chapter's results to elliptic systems.

For problems $\{A, B\}$, with A properly elliptic in the sense of Douglis-Nirenberg and B covering A, the a priori bounds are known (see 18) in the Comments); the essential difficulty resides in Green's formula, for the results of Geymonat [3], Lipko-Eidelman [1] and Roitberg-Sheftel' [1] are still incomplete.

11.2 Is it possible to obtain the results of Section 7 directly, *without interpolation*? Among other things, this would no doubt allow a weakening of the regularity hypotheses on the coefficients (see Campanato [2]).

11.3 In the extension of the theory of Sections 6 and 7 to spaces constructed on $L^p(\Omega)$, $p \neq 1, 2, \infty$ (see Lions-Magenes [1], (III–VI)), is it possible to avoid the "exceptional values" of s (as was done here for the case $p = 2$)?

Similarly, it would not be without interest to extend the theory of $\Xi^s(\Omega)$-spaces to "analogous" $\Xi^{s,p}(\Omega)$-spaces constructed on $L^p(\Omega)$, $p \neq 1, 2, \infty$ (see Problem 18.4, Chapter 1).

11.4 Regularity results on Lip_α-spaces are known [see 16) in the Comments]. "Abstractly", the methods presented here fit. But the interpretation of the abstract results leads to new interpolation problems which seem delicate.

11.5 Development of the idea mentioned in Section 9.9: which boundary value problems fit the variational setting, after introduction of a "suitable" operator Λ (see Remark 9.7)?

11.6 Systematic study of boundary value problems for quasi-elliptic operators for which we already know the inequalities "in the interior", in L^2 (see Friberg [1], Hörmander [6], Pini [8], Volevich [1]), in L^p (see Kree [4], Giusti [1]), in $\mathscr{L}^{2,\lambda}$ (see Giusti [1]) and the inequalities in the neighborhood of the subsets of the boundary which are not "singular" with respect to the operator (see Cavallucci [1], Matsuzawa [1, 2]; see also, for parabolic operators, Chapter 4, Volume 2 of this book); see also Pagni [1, 3], Pini [9, 11], Ramazanov [2].

11.7 Use of *pseudo-differential* operators for regular, non-homogeneous elliptic boundary value problems for *all* values of s (see Hörmander [8], Vishik-Eskin [4, 5]). Maybe transposition can be avoided, but, in order to *interpret* the problems and for the choice of f, suitable trace theorems, the study of which does not seem to have been initiated yet, will be required.

11.8 Study of non-homogeneous boundary value problems for pseudo-differential elliptic operators.

11.9 We may consider the open sets with boundary $\Gamma = \bigcup_{i=1}^{v} \Gamma_i$, where the Γ_i's are varieties of dimension *less* than $n - 1$ (see Stermin [1]).

11.10 Non-homogeneous problems of "transmission" type (see Section 9.5, example 4 and 3) in the Comments).

11.11 Questions pertaining to approximation by finite difference methods (see 20) in the Comments): for the non-homogeneous problems considered in this chapter, the question is open in general. See Lions [29].

Chapter 3

Variational Evolution Equations

Sections 3, 5.2, 5.3 and Remark 9.5 rely on Chapter 1. For the rest of this chapter, the knowledge of Chapters 1 and 2 is not required. Section 7 may be skipped on first reading. Sections 8–10, with the exception of 8.3, may be read independently of the rest of the chapter.

1. An Isomorphism Theorem

1.1 Notation

Let $\mathscr{V}$ and $\mathscr{H}$ be two Hilbert spaces, with $\mathscr{V} \subset \mathscr{H}$ and $\mathscr{V}$ dense in $\mathscr{H}$; $\| \ \|_{\mathscr{V}}$, $\| \ \|_{\mathscr{H}}$ denote the norms in $\mathscr{V}$ and $\mathscr{H}$; in order to simplify the writing, the scalar product in $\mathscr{H}$ is denoted by $(\, , \,)$. We *identify* $\mathscr{H}$ with its antidual; then, if $\mathscr{V}'$ denotes the antidual of $\mathscr{V}$, we have

$$\mathscr{V} \subset \mathscr{H} \subset \mathscr{V}'. \quad \square$$

Remark 1.1. Except for the notation, we have already seen examples of this situation in Section 9 of Chapter 2. But in the present chapter, the spaces $\mathscr{V}$ and $\mathscr{H}$ will "*contain the time*" – which was not the case for the stationary problems of Chapter 2. □

If $f \in \mathscr{V}'$ and $v \in \mathscr{V}$, their scalar product (antilinear in v) is denoted by (f, v); it coincides with the scalar product in $\mathscr{H}$ when $f \in \mathscr{H}$. □

As we have already agreed upon in the preceding chapters, $(\cdot, \cdot)$ and sometimes $[\cdot, \cdot]$ denote *sesquilinear* scalar products (linear in f, antilinear in v) and $\langle \cdot, \cdot \rangle$ denotes *bilinear* scalar products. □

The semi-group $G(s)$

We introduce a semi-group

$$s \to G(s) \quad \text{of} \quad \overline{\mathbf{R}}^+ \to \mathscr{L}(\mathscr{V}'; \mathscr{V}'), \tag{1.1}$$

continuous and bounded in $\mathscr{V}'$: $\forall f \in \mathscr{V}'$, $s \to G(s) f$ is a continuous function of $s \geqq 0 \to \mathscr{V}'$; we have

$$\begin{cases} G(s)\, G(t)\, f = G(s+t)\, f, & \forall s, t \geqq 0 \\ G(0)\, f = f \quad \text{and} & \| G(s) \|_{\mathscr{L}(\mathscr{V}'; \mathscr{V}')} \leqq \text{constant}. \end{cases} \tag{1.2}$$

Remark 1.2. Consult Hille-Phillips [1] or Yosida [2] for the theory of semi-groups. We shall use only the simplest results of the theory (the same applies to Volumes 2 and 3). □

We make the hypothesis:

$$(1.3)\quad \begin{cases} G(s) \textit{ forms a bounded, continuous semi-group in } \mathscr{V}; \\ \text{therefore } \forall s \geqq 0,\ G(s) \in \mathscr{L}(\mathscr{V};\mathscr{V}) \text{ and } \forall v \in \mathscr{V}, \\ s \to G(s)\,v \text{ is a continuous function of } s \geqq 0 \to \mathscr{V} \text{ with} \\ \text{properties analogous to (1.2).} \end{cases}$$

Remark 1.3. From (1.1), (1.3) and the fact that $[\mathscr{V}, \mathscr{V}']_{1/2} = \mathscr{H}$ (Chapter 1, Section 2.4), we obtain that $G(s)$ *is a bounded, continuous semi-group in* $\mathscr{H}$ (Section 10.4 of Chapter 1, is also necessary to obtain this result). □

Furthermore, we shall assume that $G(s)$ is a *contraction* semi-group in $\mathscr{H}$:

$$(1.4)\qquad \| G(s) \|_{\mathscr{L}(\mathscr{H};\mathscr{H})} \leqq 1 .$$

We *summarize the hypotheses* by:

$$(1.5)\quad \begin{cases} G(s) \textit{ is a bounded, continuous semi-group in } \mathscr{V}' \\ \textit{and } \mathscr{V} \text{ (therefore in } \mathscr{H}) \text{ and } \| G(s) \|_{\mathscr{L}(\mathscr{H};\mathscr{H})} \leqq 1,\ s \geqq 0. \end{cases} \quad □$$

We denote by $-\Lambda$ the *infinitesimal generator* of $G(s)$ and by $D(\Lambda;\mathscr{V}')$, $D(\Lambda;\mathscr{H})$, $D(\Lambda;\mathscr{V})$ its *domain* in the spaces $\mathscr{V}'$, $\mathscr{H}$, $\mathscr{V}$.

For example, the space $D(\Lambda;\mathscr{V}')$ is provided with the norm

$$(\| v \|^2_{\mathscr{V}'} + \| \Lambda v \|^2_{\mathscr{V}'})^{1/2},$$

which makes it a Hilbert space (since Λ is *closed*); similarly for $D(\Lambda;\mathscr{H})$ and $D(\Lambda;\mathscr{V})$.

Therefore, we have

$$(1.6)\qquad \mathscr{V} \cap D(\Lambda;\mathscr{V}') \subset \mathscr{V} \subset \mathscr{H} \subset \mathscr{V}'$$

and

Lemma 1.1. *In* (1.6), *each space is dense in the following one.*

Proof. We only need to verify that $\mathscr{V} \cap D(\Lambda;\mathscr{V}')$ is dense in $\mathscr{V}$; but

$$\mathscr{V} \cap D(\Lambda;\mathscr{V}') \supset D(\Lambda;\mathscr{V})$$

and

$$D(\Lambda;\mathscr{V}) \quad \text{is dense in } \mathscr{V}. \quad □$$

Since $G(s)$ is a contraction semi-group in $\mathscr{H}$, we have (consider the derivative at the origin of the function $s \to \| G(s)\, v \|^2_{\mathscr{H}}$):

$$(1.7)\qquad R(\Lambda v, v) \geqq 0 \qquad \forall v \in D(\Lambda;\mathscr{H}).$$

We also have

Lemma 1.2. *Under hypothesis* (1.5):

$$\operatorname{Re}(\Lambda v, v) \geqq 0 \qquad \forall v \in \mathscr{V} \cap D(\Lambda; \mathscr{V}'). \tag{1.8}$$

Remark 1.4. If $v \in D(\Lambda; \mathscr{V}')$, then $\Lambda v \in \mathscr{V}'$ and, in (1.8), $(\Lambda v, v)$ denotes the scalar product between $\mathscr{V}'$ and $\mathscr{V}$. □

Proof of Lemma 1.2. 1) Let φ be a continuous (scalar) function with compact support in $t \geqq 0$. We denote by $G(\varphi)$ the operator in $\mathscr{V}'$ (resp. $\mathscr{V}$, resp. $\mathscr{H}$), given for $f \in \mathscr{V}'$ (resp. $\mathscr{V}$, resp. $\mathscr{H}$) by

$$G(\varphi) f = \int_0^\infty G(s) f \cdot \varphi(s)\, ds,$$

the integral being taken in $\mathscr{V}'$ (resp. $\mathscr{V}$, resp. $\mathscr{H}$). If, furthermore, φ is once (resp. *infinitely*) differentiable, $G(\varphi) f$ is in $D(\Lambda)$ (resp. $D(\Lambda^\infty)$) and

$$\Lambda G(\varphi) f = G(\varphi') f.$$

2) Let ϱ_n be a *regularizing sequence* of functions of $\mathscr{D}(R)$ with support in $t \geqq 0$ (see Schwartz [1]).

We easily verify that

$$G(\varrho_n) v \to v \quad \text{in} \quad \mathscr{V}, \qquad \forall v \in \mathscr{V},$$

$$\Lambda G(\varrho_n) v \to \Lambda v \quad \text{in} \quad \mathscr{V}', \qquad \forall v \in D(\Lambda; \mathscr{V}')$$

and therefore

$$(\Lambda v, v) = \lim_{n \to \infty} (\Lambda G(\varrho_n) v, G(\varrho_n) v), \qquad \forall v \in \mathscr{V} \cap D(\Lambda; \mathscr{V}').$$

Now, $G(\varrho_n) v \in D(\Lambda; \mathscr{V}) \subset D(\Lambda; \mathscr{H})$ and, according to (1.7), we therefore have

$$\operatorname{Re}(\Lambda G(\varrho_n) v, G(\varrho_n) v) \geqq 0; \qquad \text{whence (1.8).} \ \square$$

The operator M

We introduce an operator M, satisfying:

$$\begin{cases} M \in \mathscr{L}(\mathscr{V}; \mathscr{V}'), \\ \operatorname{Re}(M v, v) \geqq \alpha \|v\|^2_{\mathscr{V}}, \quad \alpha > 0, \quad \forall v \in \mathscr{V}. \end{cases} \tag{1.9}$$

Then M is an isomorphism of $\mathscr{V}$ onto $\mathscr{V}'$ (indeed, it is easily verified that the image of $\mathscr{V}$ under M is dense and closed in $\mathscr{V}'$). □

The Problem

We consider the (operator) equation:

$$\Lambda u + M u = f, \tag{1.10}$$

where $f \in \mathscr{V}'$ is given, and we seek a solution $u \in \mathscr{V} \cap D(\Lambda; \mathscr{V}')$.

Note that

$$\Lambda + M \in \mathscr{L}(\mathscr{V} \cap D(\Lambda; \mathscr{V}'); \mathscr{V}').$$

1.2 Isomorphism Theorem

Theorem 1.1. *Under hypotheses* (1.5) *and* (1.9), *the operator* $\Lambda + M$ *is an isomorphism of* $\mathscr{V} \cap D(\Lambda; \mathscr{V}')$ *onto* $\mathscr{V}'$.

Remark 1.5. Examples are given in Sections 4 and 5. In the applications, Λ is a differential operator in t (time-variable) and M is an "elliptic" operator in the space variables. □

Remark 1.6. Abstract operator equations of the form (1.10), but in a more general setting and with techniques different from the one we shall use, are studied by P. Grisvard [5–7, 9] and, with still different techniques, by Da Prato [4, 5]. □

The proof of Theorem 1.1 is given in Section 1.4. First, we make some remarks concerning the adjoint Λ^* of Λ.

1.3 The Adjoint Λ^*

Let $G^*(s)$ be the adjoint of $G(s)$ in the sense:

$$(1.11) \qquad (G^*(s) f, g) = (f, G(s) g), \quad \forall f \in \mathscr{V}', \quad g \in \mathscr{V}.$$

Then, since $G(s) \in \mathscr{L}(\mathscr{V}'; \mathscr{V}') \cap \mathscr{L}(\mathscr{V}; \mathscr{V}) \cap \mathscr{L}(\mathscr{H}; \mathscr{H})$, $G^*(s)$ has the same properties and we have

$$\| G^*(s) \|_{\mathscr{L}(\mathscr{H};\mathscr{H})} \leqq 1.$$

Let $-\Lambda^*$ be the infinitesimal generator of $G^*(s)$ in $\mathscr{V}'$ (and in $\mathscr{H}$ and $\mathscr{V}$) and let $D(\Lambda^*; \mathscr{V}')$, $D(\Lambda^*; \mathscr{H})$, $D(\Lambda^*; \mathscr{V})$ be its domain in $\mathscr{V}'$, $\mathscr{H}$, $\mathscr{V}$.

The operator Λ^* is the adjoint – *in the sense of unbounded operators* – in $\mathscr{V}'$ (resp. $\mathscr{H}$, resp. $\mathscr{V}$) of Λ in $\mathscr{V}$ (resp. $\mathscr{H}$, resp. $\mathscr{V}'$) and vice-versa (see Hille-Phillips [1]). Therefore we have

Lemma 1.3. *The necessary and sufficient condition for given* $u \in \mathscr{V}$ *to be in* $D(\Lambda^*; \mathscr{V})$ *is that* $v \to (\Lambda v, u)$ *be continuous on* $D(\Lambda; \mathscr{V}')$ *in the topology induced by* $\mathscr{V}'$.

Analogous results hold on interchanging $\mathscr{V}$ *and* $\mathscr{V}'$ *and also* Λ *and* Λ^*.

1.4 Proof of Theorem 1.1

It is sufficient to show that equation (1.10) has a unique solution.

1) Uniqueness

Assume $f = 0$ in (1.10). Then

$$0 = \operatorname{Re}(\Lambda u + M u, u) \geqq \operatorname{Re}(M u, u) \geqq \alpha \| u \|_{\mathscr{V}}^2,$$

by Lemma 1.2 and (1.9). Therefore $u = 0$.

2) Existence

For $u \in \mathscr{V}$ and $v \in \mathscr{V} \cap D(\Lambda^*; \mathscr{V}')$, we set:

$$E(u, v) = (u, \Lambda^* v) + (M u, v). \tag{1.12}$$

According to Lemma 1.2 (for Λ^*, which is allowed) we have

$$\operatorname{Re}(v, \Lambda^* v) \geqq 0, \quad \forall v \in \mathscr{V} \cap D(\Lambda^*; \mathscr{V}')$$

and therefore

$$\operatorname{Re} E(v, v) \geqq \alpha \, \| v \|_{\mathscr{V}}^2 .$$

By a variant of the projection lemma (see Lions [13], Chapter 3) (see also a different proof in Section 7 below) there exists $u \in \mathscr{V}$, with

$$E(u, v) = (f, v), \quad \forall v \in \mathscr{V} \cap D(\Lambda^*; \mathscr{V}'). \tag{1.13}$$

We show that this *implies that* $u \in D(\Lambda; \mathscr{V}')$; indeed, we may write (1.13) in the form

$$(u, \Lambda^* v) = (f - M u, v)$$

and therefore $v \to (u, \Lambda^* v)$ is continuous on $\mathscr{V} \cap D(\Lambda^*; \mathscr{V}')$ in the topology induced by $\mathscr{V}$, therefore certainly continuous on $D(\Lambda^*; \mathscr{V})$ in the topology induced by $\mathscr{V}$ and therefore, according to Lemma 1.3, $u \in D(\Lambda; \mathscr{V}')$ and

$$(\Lambda u, v) = (f - M u, v), \quad \forall v \in \mathscr{V} \cap D(\Lambda^*; \mathscr{V}').$$

Since $\mathscr{V} \cap D(\Lambda^*; \mathscr{V}')$ is dense in $\mathscr{V}$, we obtain that u satisfies (1.10); hence, the theorem is proved. □

Remark 1.7. At this stage, the reader can consult the examples at the beginning of Section 4. □

2. Transposition

2.1 Generalities

We have already used *transposition* for elliptic problems in Chapter 2, Section 6. Here, we reconsider the general idea in a *heuristic* way.

Let E, F be two topological vector spaces imbedded in the same topological vector space $\mathscr{A}$ and let U be an operator which is known to be an isomorphism of E onto F.

Let U^* be the "adjoint" of U (in $\mathscr{A}$) which is known to be an isomorphism of E_1 onto F_1 (E_1, F_1 are spaces imbedded in $\mathscr{A}$). Then by *transposition*, U^{**} is an isomorphism of F_1' onto E_1'.

But under reasonable hypotheses, U and U^{**} *coincide* on a "dense" subspace; we may extend U from F_1' to E_1' and this extension coincides with U^{**}.

Thus U is an isomorphism of $E \to F$ and of $F_1' \to E_1'$.

Then we can interpolate.

We shall apply this (in a more rigorous fashion!!) to the situation of Section 1.

2.2 Adjoint Isomorphism Theorem

Since $M \in \mathscr{L}(\mathscr{V}; \mathscr{V}')$, its adjoint $M^* \in \mathscr{L}(\mathscr{V}; \mathscr{V}')$ and

(2.1) $$\operatorname{Re}(M^* v, v) \geqq \alpha \| v \|_{\mathscr{V}}^2, \quad \forall v \in \mathscr{V}.$$

Therefore, according to Theorem 1.1, we have:

(2.2) $\Lambda^* + M^*$ is an *isomorphism* of $\mathscr{V} \cap D(\Lambda^*; \mathscr{V}')$ onto $\mathscr{V}'$.

2.3 Transposition

With the symbol ′ still denoting passage to the antidual, we deduce from (2.2) that

(2.3) $(\Lambda^* + M^*)^*$ is an isomorphism of $\mathscr{V}$ onto $(\mathscr{V} \cap D(\Lambda^*; \mathscr{V}))'$. □

Since $\mathscr{V} \cap D(\Lambda^*; \mathscr{V}') \subset \mathscr{V} \subset \mathscr{H} \subset \mathscr{V}'$, each space being dense in the following one, we have:

(2.4) $$\mathscr{V} \cap D(\Lambda^*; \mathscr{V}') \subset \mathscr{V} \subset \mathscr{H} \subset \mathscr{V}' \subset (\mathscr{V} \cap D(\Lambda^*; \mathscr{V}'))'.$$

If $f \in (\mathscr{V} \cap D(\Lambda^*; \mathscr{V}'))'$ and $v \in \mathscr{V} \cap D(\Lambda^*; \mathscr{V}')$, their scalar product is denoted by (f, v), since it coincides with the scalar product of the antiduality between $\mathscr{V}$ and $\mathscr{V}'$ if $f \in \mathscr{V}'$. □

We may define Λ by extension on $\mathscr{V}$ (M is already *given* on $\mathscr{V}$), for

$$\Lambda^* \in \mathscr{L}(\mathscr{V} \cap D(\Lambda^*; \mathscr{V}'); \mathscr{V}'),$$

therefore

$$(\Lambda^*)^* \in \mathscr{L}(\mathscr{V}; (\mathscr{V} \cap D(\Lambda^*; \mathscr{V}'))')$$

and obviously $(\Lambda^*)^* v = \Lambda v$ if $v \in \mathscr{V} \cap D(\Lambda; \mathscr{V}')$. Therefore

(2.5) $$\Lambda \in \mathscr{L}(\mathscr{V}; (\mathscr{V} \cap D(\Lambda^*; \mathscr{V}'))')$$

and we immediately verify that

$$(\Lambda^* + M^*)^* = \Lambda + M.$$

Therefore, we have

Theorem 2.1. *Under hypotheses* (1.5) *and* (1.9), *the operator $\Lambda + M$ is an isomorphism of $\mathscr{V}$ onto $(\mathscr{V} \cap D(\Lambda^*; \mathscr{V}'))'$.*

We shall now interpolate "between" Theorems 1.1 and 2.1.

Remark 2.1. As before, we interpolate only by the method of traces of Chapter 1.

3. Interpolation

3.1 General Application

From Theorems 1.1. and 2.1, we deduce

Theorem 3.1. *Under hypotheses* (1.5) *and* (1.9), *the operator* $\Lambda + M$ *is an isomorphism of*

$$[\mathscr{V} \cap D(\Lambda; \mathscr{V}'), \mathscr{V}]_\theta \to [\mathscr{V}', (\mathscr{V} \cap D(\Lambda^*; \mathscr{V}'))']_\theta$$

for every θ *with* $0 \leqq \theta \leqq 1$.

There remains to characterize the spaces which appear in this theorem.

3.2 Characterization of the Interpolation Spaces

First, we have

Proposition 3.1. *Under hypotheses* (1.5) *and* (1.9), *we have*

$$[\mathscr{V} \cap D(\Lambda; \mathscr{V}'), \mathscr{V}]_\theta = \mathscr{V} \cap [D(\Lambda; \mathscr{V}'), \mathscr{V}]_\theta. \tag{3.1}$$

Proof. We denote by E and F the spaces of the first and second member of (3.1) respectively.

Since obviously $E \subset F$, we need only show that $F \subset E$.

Now

$$\Lambda + M \in \mathscr{L}(F; [\mathscr{V}', (\mathscr{V} \cap D(\Lambda^*; \mathscr{V}'))']_\theta). \tag{3.2}$$

Indeed

$$\Lambda \in \mathscr{L}(D(\Lambda; \mathscr{V}'); \mathscr{V}') \cap \mathscr{L}(\mathscr{V}; (\mathscr{V} \cap D(\Lambda^*; \mathscr{V}'))'),$$

therefore, by interpolation[1]:

$$\Lambda \in \mathscr{L}([D(\Lambda; \mathscr{V}'), \mathscr{V}]_\theta; [\mathscr{V}', (\mathscr{V} \cap D(\Lambda^*; \mathscr{V}'))']_\theta)$$

and therefore a fortiori:

$$\Lambda \in \mathscr{L}(F; [\mathscr{V}', (\mathscr{V} \cap D(\Lambda^*; \mathscr{V}'))']_\theta).$$

And $M \in \mathscr{L}(\mathscr{V}; \mathscr{V}')$, therefore $M \in \mathscr{L}(F; [\mathscr{V}', (\mathscr{V} \cap D(\Lambda^*; \mathscr{V}'))']_\theta)$, whence (3.2).

Then, let $f \in F$; according to Theorem 3.1, there exists a unique $w \in E$ such that

$$(\Lambda + M)\, w = (\Lambda + M)\, f, \quad \text{therefore} \quad (\Lambda + M)\,(w - f) = 0.$$

Since $w - f \in \mathscr{V}$ and according to Theorem 2.1, it follows that

$$w - f = 0.$$

Therefore $f \in E$, i.e. $F \subset E$, whence the proposition. □

---

[1] The definition of the space $[X, Y]_\theta$ is valid without the inclusion of X in Y (or vice-versa).

Proposition 3.2. *Under hypotheses* (1.5) *and* (1.9), *we have*

$$[\mathscr{V}', (\mathscr{V} \cap D(\Lambda^*; \mathscr{V}'))']_\theta = (\mathscr{V} \cap [D(\Lambda^*; \mathscr{V}'), \mathscr{V}]_{1-\theta})'. \tag{3.3}$$

Proof. According to the duality theorem (Chapter 1, Theorem 6.2), we have:

$$[\mathscr{V}', (\mathscr{V} \cap D(\Lambda^*; \mathscr{V}'))']_\theta = [\mathscr{V}, \mathscr{V} \cap D(\Lambda^*; \mathscr{V}')]_\theta'.$$

Applying Proposition 3.1 (with Λ^* replacing Λ) this is equivalent to

$$(\mathscr{V} \cap [\mathscr{V}, D(\Lambda^*; \mathscr{V}')]_\theta)'$$

and since

$$[\mathscr{V}, D(\Lambda^*; \mathscr{V}')]_\theta = [D(\Lambda^*; \mathscr{V}'), \mathscr{V}]_{1-\theta},$$

(3.3) follows. □

With the help of Propositions 3.1 and 3.2, Theorem 3.1 may be stated in the following form:

Theorem 3.1a. *Under hypotheses* (1.5) *and* (1.9), *the operator* $\Lambda + M$ *is an isomorphism of* $\Phi^\theta = \mathscr{V} \cap [D(\Lambda; \mathscr{V}'), \mathscr{V}]_\theta$ *onto* $(\Phi_*^{1-\theta})'$, *where*

$$\Phi_*^\theta = \mathscr{V} \cap [D(\Lambda^*; \mathscr{V}'), \mathscr{V}]_\theta$$

(for $0 \leqq \theta \leqq 1$).

3.3 The Case "$\theta = 1/2$"

We could ask whether it is possible to find a space Ψ such that $\Lambda + M$ is an isomorphism of Ψ *onto its antidual* Ψ'.

The following is a *partial* result:

Theorem 3.2. *Assume that* (1.5) *and* (1.9) *hold. Also assume that*

$$[D(\Lambda; \mathscr{V}'), \mathscr{V}]_{1/2} = [D(\Lambda^*; \mathscr{V}'), \mathscr{V}]_{1/2}. \tag{3.4}$$

Then $\Lambda + M$ *is an isomorphism of* $\Phi^{1/2}$ *onto* $(\Phi^{1/2})'$.

Proof. Apply Theorem 3.1a with $\theta = \frac{1}{2}$. But $\Phi_*^{1/2} = \Phi^{1/2}$, according to (3.4), whence the result. □

4. Example: Abstract Parabolic Equations, Initial Condition Problems (I)

4.1 Notation

We consider the setting of Chapter 2, Section 9.1.

Let V and H be two Hilbert spaces, $V \subset H$, V dense in H; let $\| \ \|$ and $| \ |$ denote the norms in V and H and $[\ ,\ ]$ the (*sesquilinear*) scalar product in H. We identify H with its antidual; then

$$V \subset H \subset V'.$$

If $f \in V'$ and $v \in V$, $[f, v]$ denotes, *in this chapter* (there is a slight difference with the notation of the other chapters), their scalar product in the antiduality. □

Let t be *the time variable.* We shall assume that

$$t \in [0, T], \quad T \text{ finite or } +\infty.$$

With the notation of Section 1, we take

$$\mathscr{V} = L^2(0, T; V), \quad \mathscr{H} = L^2(0, T; H), \tag{4.1}$$

where, if X is a Hilbert space, $L^2(0, T; X)$ denotes the space of (classes of) *measurable* functions f of $(0, T) \to X$ for the Lebesgue measure dt, and such that

$$\left(\int_0^T \| f(t) \|_X^2 \, dt \right)^{1/2} < \infty \quad \text{(see Chapter 1, Section 1).}$$

Then we have

$$\mathscr{V}' = L^2(0, T; V'). \tag{4.2}$$

4.2 The Operator M

For $t \in [0, T]$, let $a(t; u, v)$ be a continuous sesquilinear form on V.

In order to slightly simplify the measurability considerations, we assume V to be *separable*.

The conditions on the family $a(t; u, v)$ are as follows:

$$\left\{ \begin{array}{l} \forall u, v \in V, \quad \text{the function} \quad t \to a(t; u, v) \quad \text{is measurable} \\ \text{and} \\ |a(t; v, u)| \leqq c \, \|u\| \, \|v\|, \quad c = \text{constant}, \quad \forall t \in [0, T], \end{array} \right. \tag{4.3}$$

$$\left\{ \begin{array}{l} \text{(uniform coerciveness on } V\text{) there exists a } \lambda \text{ such that} \\ \operatorname{Re} a(t; v, v) + \lambda |v|^2 \geqq \alpha \|v\|^2, \quad \alpha > 0, \quad \forall v \in V, \quad \forall t \in [0, T]. \end{array} \right. \tag{4.4$_1$}$$

Since, for fixed t, the antilinear form $v \to a(t; u, v)$ is continuous on V, we have:

$$a(t; u, v) = [A(t) u, v], \quad A(t) u \in V', \tag{4.5}$$

which defines

$$A(t) \in \mathscr{L}(V; V'). \tag{4.6}$$

In this section, we want to solve the equations

$$A(t) u + \frac{du}{dt} = f, \quad 0 < t < T$$

$$u(0) = 0,$$

under suitable hypotheses on u and f.

We may always make the change of functions

$$u = e^{-\lambda t} w$$

and obtain the *equivalent* equations

$$(A(t) + \lambda I) w + \frac{dw}{dt} = (e^{-\lambda t} f), \qquad 0 < t < T,$$

$$w(0) = 0.$$

At the risk of changing $A(t)$ to $A(t) + \lambda I$, we may always consider the case where $A(t)$ verifies $(4.4)_1$ with $\lambda = 0$; *therefore, in the sequel, we assume that*

$$\text{(4.4)} \qquad \operatorname{Re} a(t; v, v) \geqq \alpha \|v\|^2, \quad \alpha > 0, \quad \forall v \in V.$$

[We must be careful about the *changing behavior at infinity* if we use the same classical device for $0 < t < \infty$.]

For $v \in \mathscr{V}$, we define:

$$\text{(4.7)} \qquad M v = \text{function "} t \to A(t) v(t) \text{"},$$

or

$$\text{(4.7a)} \qquad M v(t) = A(t) v(t) \quad \text{a.e.}$$

We have:

Lemma 4.1. *Under hypotheses* (4.3)–(4.4), *the operator* M *defined by* (4.7) *satisfies condition* (1.9).

Proof. For $u, v \in \mathscr{V} = L^2(0, T; V)$, set

$$\text{(4.8)} \qquad \mathscr{M}(u, v) = \int_0^T a(t; u(t), v(t)) \, dt.$$

Thanks to (4.3), the function $t \to a(t; u(t), v(t))$ is measurable and bounded in modulus by $c \|u(t)\| \|v(t)\|$, which is integrable, therefore (4.8) has meaning and

$$|\mathscr{M}(u, v)| \leqq c \|u\|_{\mathscr{V}} \|v\|_{\mathscr{V}}.$$

The antilinear form

$$v \to \mathscr{M}(u, v)$$

is continuous on $\mathscr{V}$, therefore in the form

$$\mathscr{M}(u, v) = (\hat{M} u, v), \qquad \hat{M} u \in \mathscr{V}',$$

or

$$\mathscr{M}(u, v) = \int_0^T [\hat{M} u(t), v(t)] \, dt.$$

Therefore

$$\hat{M}\, u(t) = M\, u(t) \quad \text{a.e.,}$$

and

$$\mathcal{M}(u, v) = (M\, u, v), \qquad M \in \mathcal{L}(\mathcal{V}; \mathcal{V}'). \tag{4.9}$$

Finally,

$$\operatorname{Re}(M\, v, v) = \int_0^T \operatorname{Re} a(t; v(t), v(t))\, dt \geqq \alpha\, \|v\|^2_{\mathcal{V}},$$

which proves the lemma. □

4.3 The Operator Λ

For the semi-group $G(s)$ (see Section 1), we take the *semi-group of right translations (in t)*. More precisely, if $f \in \mathcal{V}'$, we set

$$G(s)\, f(t) = \begin{cases} 0, & \text{if } 0 < t < s \\ f(t-s), & \text{if } s < t < T \end{cases} \quad \text{a.e. in } t. \tag{4.10}$$

We define, in this manner, a *contraction semi-group in $\mathcal{V}'$, $\mathcal{V}$* and $\mathcal{H}$.

Then, we have

$$\Lambda\, u = \frac{du}{dt} = u', \tag{4.11}$$

$$D(\Lambda; \mathcal{V}') = \left\{ u \mid u \in \mathcal{V}', \frac{du}{dt} \in \mathcal{V}', u(0) = 0 \right\} \tag{4.12}$$

$\Bigl($where $\dfrac{du}{dt}$ is taken in the sense of distributions taking their values in V'; then u, after possibly a modification on a set of measure zero, is a continuous function of $t \in [0, T] \to V'$, and therefore the condition "$u(0) = 0$" is well-defined$\Bigr)$.

Of course, a description analogous to (4.12) holds for $D(\Lambda; \mathcal{H})$ and $D(\Lambda; \mathcal{V})$. □

The *adjoint semi-group* $G^*(s)$ is defined by:

i) if $T < \infty$,

$$G^*(s)\, f(t) = \begin{cases} f(t+s), & \text{if } 0 < t < T - s \\ 0, & \text{if } T - s < t < T \end{cases} \quad \text{a.e. in } t; \tag{4.13}$$

ii) if $T = +\infty$,

$$G^*(s)\, f(t) = f(t+s), \qquad t > 0. \tag{4.14}$$

Then we have

$$\Lambda^* u = -\frac{du}{dt} \tag{4.15}$$

with

$$D(\Lambda^*; \mathscr{V}') = \left\{u \mid u \in \mathscr{V}', \frac{du}{dt} \in \mathscr{V}', u(T) = 0, \text{ if } T < \infty\right\}, \tag{4.16i}$$

$$D(\Lambda^*; \mathscr{V}') = \left\{u \mid u \in \mathscr{V}', \frac{du}{dt} \in \mathscr{V}', \text{ if } T = +\infty\right\}. \tag{4.16ii}$$

Analogous description for $D(\Lambda^*; \mathscr{H})$ and $D(\Lambda^*; \mathscr{V})$. □

4.4 Application of the Isomorphism Theorems

The results of Sections 1–3 apply to the setting of this section. The application of Theorem 1.1 yields:

Theorem 4.1. *Assume that* (4.3)–(4.4) *hold and that* Λ *is given by* (4.11)–(4.12). *Then, for given* $f \in \mathscr{V}' = L^2(0, T; V')$, *there exists a unique* $u \in \mathscr{V}$ *such that*

$$A(t)\, u + u' = f \tag{4.17}$$

$$u(0) = 0. \tag{4.18}$$

Note that (4.17) implies that $u' \in L^2(0, T; V')$, so that

$$u \in C^0([0, T]; H)$$

and $u(0)$ is well-defined (Chapter 1, Theorem 3.1 and Proposition 2.1).

Proof. Indeed, Theorem 1.1 shows the existence and uniqueness of u in

$$\mathscr{V} \cap D(\Lambda; \mathscr{V}'),$$

solution of (4.17). But then we have (4.18). Conversely, if $u \in \mathscr{V}$ and verifies (4.17), then $u' = f - A(t)\, u \in \mathscr{V}'$, which together with (4.18) shows that $u \in D(\Lambda; \mathscr{V}')$. □

Remark 4.1. In the examples (see Section 4.7), the $A(t)$'s will be *elliptic differential operators*. Then the operators $A(t) + \partial/\partial t$ are *parabolic operators*. □

Remark 4.2. In Volume 3, we shall consider problems of *type* (4.17) to (4.18) with

$$A(t) = A, \quad \text{independent of } t$$

and, *furthermore*, where $-A$ is the infinitesimal generator of a semi-group $H(s)$ (which, in fact, we shall denote by $G(s)$). This semi-group is of course *essentially different* from the semi-group for which $-\Lambda$ is the infinitesimal generator. □

Remark 4.3. Let u be given in H. *Under the hypotheses of Theorem* 4.1, *there exists a unique u satisfying* (4.17) *and*

$$u(0) = u_0. \tag{4.19}$$

Indeed, there exists (trace theorem, Chapter 1, Section 3) $w \in \mathscr{V}$, with $w' \in \mathscr{V}'$ and $w(0) = u_0$. Then $u - w$ must verify

$$A(t)(u - w) + (u - w)' = f - (A(t) w + w') = \tilde{f} \in \mathscr{V}',$$
$$(u - w)(0) = 0$$

and we are led back to Theorem 4.1.

In the following chapters, we shall see how we may consider conditions of type (4.19) with u_0 taken from much more general classes.

The application of Theorem 2.1 yields:

Theorem 4.2. *Assume that* (4.3) – (4.4) *hold and that Λ is given by* (4.11) – (4.12). *Then, for L given in $(\mathscr{V} \cap D(\Lambda^*; \mathscr{V}'))'$, there exists a unique $u \in \mathscr{V} = L^2(0, T; V)$, satisfying*

$$(u, A^*(t) v - v') = (L, v), \quad \forall v \in \mathscr{V} \cap D(\Lambda^*; \mathscr{V}'). \tag{4.20}$$

First of all, let us write out $\mathscr{V} \cap D(\Lambda^*; \mathscr{V}')$:

$$\begin{cases} \mathscr{V} \cap D(\Lambda^*; \mathscr{V}') = \Big\{ v \mid v \in L^2(0, T; V), \\ \qquad v' = \dfrac{dv}{dt} \in L^2(0, T; V'), \text{ with } v(T) = 0 \text{ if } T < \infty \Big\}; \end{cases}$$

so that the subspace $\mathscr{D}(]0, T[; V)$ of C^∞-functions with compact support in $]0, T[$ and with values in V *is not dense* in $\mathscr{V} \cap D(\Lambda^*; \mathscr{V}')$ (its closure is $\mathscr{V} \cap D(\Lambda^*; \mathscr{V}') \cap D(\Lambda; \mathscr{V}')$). Therefore, in (4.20), $L \in (\mathscr{V} \cap D(\Lambda^*; \mathscr{V}'))'$, *is not necessarily a distribution taking its values in* V'.

Here, we find again a problem of a type which we have already met in Chapter 2 (and which we shall often meet again in the sequel): choose L "*in the best way*" so as to "*interpret* (4.20) *suitably*".

4.5 Choice of L in (4.20)

Formally, at first, we take

$$(L, v) = \int_0^T [f(t), v(t)]\, dt + [u_0, v(0)], \tag{4.21}$$

where:

f is a "function" given so that the integral "is well-defined",
u_0 is given in a suitable space.

Then, still formally, u satisfies

$$(4.22)\qquad \begin{cases} A(t)\,u + u' = f \\ u(0) = u_0 . \end{cases}$$

Now, we have to make this rigorous. □

First of all, as v describes $\mathscr{V} \cap D(\Lambda^*; \mathscr{V}')$, $v(0)$ *describes* H and therefore we *must* take

$$u_0 \in H .$$

The choice of f is more delicate.

We introduce the space Ξ (compare with Chapter 2, Section 6): let ϱ be defined by

$$(4.23)\qquad \varrho(t) = \begin{cases} t/t_0, & \text{if } 0 \leqq t \leqq t_0, \\ 1, & \text{if } t \geqq t_0, \end{cases} \qquad t_0 > 0, \quad \text{fixed.}$$

Then

$$(4.24)\qquad \Xi = \{v \mid v \in L^2(0,T;V') = \mathscr{V}', \varrho v' \in \mathscr{V}', v(T) = 0 \text{ (if } T < \infty)\}$$

(provided with the "usual" norm

$$(\|v\|^2_{\mathscr{V}'} + \|\varrho\, v'\|^2_{\mathscr{V}'})^{1/2},$$

which makes Ξ a Hilbert space).

The space $\mathscr{D}(]0, T[; V')$ is *dense* in Ξ; indeed let v be given in Ξ and θ_n be a sequence of functions belonging to $C^0[0, T]$,

$$\theta_n(t) = 1 \text{ if } t \geqq 2\xi_n, \quad \theta_n(t) = 0 \text{ if } t \leqq \xi_n \text{ and } |t\,\theta_n'(t)| \leqq C,$$

with analogous conditions in the neighborhood of T if $T < \infty$, where ξ_n is a decreasing sequence which tends towards zero.

Then, $\theta_n v \to v$ in Ξ (immediate verification: $\varrho\theta_n' v \to 0$ in $L^2(0,T;V')$); we then regularize $\theta_n v$ in t, whence the result. Therefore Ξ' is a space of distributions on $]0, T[$ taking their values in V; more precisely:

Proposition 4.1. *Every* $f_* \in \Xi'$ *may be written – non-uniquely – as*

$$(4.25)\qquad \begin{cases} f_* = f_0 + \dfrac{d}{dt}(\varrho\, f_1), \\ f_i \in L^2(0, T; V). \end{cases}$$

Proof. The proposition is a consequence of the Hahn-Banach Theorem and of the density of $\mathscr{D}(]0, T[; V')$ in Ξ. □

Since $\mathscr{V} \cap D(\Lambda^*; \mathscr{V}') \subset \Xi$, we have:

Proposition 4.2. *In* (4.20), (L, v) *may be chosen in the form*

$$(4.26)\qquad (L, v) = (f_*, v) + (f_{**}, v) + [u_0, v(0)],$$

where:

f_* *is given in* Ξ' (see Proposition 4.1),

f_{**} *is given in* $\mathscr{V}' = L^2(0, T; V')$,

u_0 *is given in* H.

(In (4.26), (f_*, v) – resp. (f_{**}, v) – represents the scalar product between Ξ' and Ξ – resp. between $\mathscr{V}'$ and $\mathscr{V}$). □

4.6 Interpretation of the Problem

Now, we need to make (4.22) more precise. If in (4.26) we take

$$v(t) = \varphi(t)\, k,$$

$$\varphi \in \mathscr{D}(]0, T[), \qquad k \in V,$$

then we deduce from (4.20) that

(4.27) $A(t)\, u + u' = f$ in the sense of $\mathscr{D}'(]0, T[; V')$, $f = f_* + f_{**}$.

Thus, if we define Y by:

$$Y = \{v \mid v \in \mathscr{V}, v' \in \Xi' + \mathscr{V}'\}, \tag{4.28}$$

we see that $u \in Y$ (since $u' = f - A(t)\, u$ and $A(t)\, u \in \mathscr{V}'$).

We have the following *trace theorem* in Y:

Theorem 4.3. *The space* $\mathscr{D}([0, T]; V)$ *is dense in* Y. *The mapping* $u \to u(0)$ *of* $\mathscr{D}([0, T]; V) \to V$ *extends by continuity to a continuous linear mapping, still denoted by* $u \to u(0)$, *of* $Y \to H$.

Proof. 1) *Density*. Let $u \to M(u)$ be a continuous antilinear form on Y; it may be written

$$M(u) - (g_0, u) + (g_1, u'), \qquad g_0 \subset \mathscr{V}', \qquad g_1 \in \mathscr{V} \cap \Xi, \tag{4.29}$$

where (g_0, u) (resp. (g_1, u')) denotes the antiduality between $\mathscr{V}'$ and $\mathscr{V}$ (resp. between $\mathscr{V} \cap \Xi$ and $\mathscr{V}' + \Xi'$).

We assume that

$$M(\varphi) = 0 \qquad \forall \varphi \in \mathscr{D}([0, T]; V)$$

and, from this, we shall deduce that $M(u) = 0$, $\forall u \in Y$.

Let $\tilde{g}_0$, $\tilde{g}_1$ be the extensions of g_0 and g_1 to $\mathbf{R}_t$ by 0 outside of $]0, T[$.

Then $\forall \Phi \in \mathscr{D}(\mathbf{R}_t; V)$, of restriction φ to $]0, T[$, we have:

$$(\tilde{g}_0, \Phi) + (\tilde{g}_1, \Phi') = M(\varphi) = 0$$

and therefore

$$\tilde{g}_0 - \frac{d}{dt}\tilde{g}_1 = 0 \quad \text{in} \quad \mathscr{D}'(\mathbf{R}_t; V'). \tag{4.30}$$

But $\tilde{g}_0 \in L^2(\mathbf{R}_t; V')$, therefore necessarily:

$$(4.31) \quad \begin{cases} g_1 \in L^2(0, T; V), \quad \dfrac{dg_1}{dt} \in L^2(0, T; V'), \quad g_1(0) = 0, \\ \qquad\qquad \text{and} \quad g_1(T) = 0 \quad \text{if} \quad T < \infty. \end{cases}$$

Thus, we can find $g_{1n} \in \mathscr{D}(]0, T[; V)$ with

$$(4.32) \quad g_{1n} \to g_1 \quad \text{in} \quad L^2(0, T; V), \quad \frac{dg_{1n}}{dt} \to \frac{dg_1}{dt} \quad \text{in} \quad L^2(0, T; V').$$

Therefore, *in particular*, $g_{1n} \to g_1$ in $\Xi \cap \mathscr{V}$ and, if $u \in Y$,

$$(g_1, u') = \lim_{n\to\infty} (g_{1n}, u') = \lim_{n\to\infty} -(g'_{1n}, u)$$

and

$$-(g'_{1n}, u) \to -(g'_1, u) \quad (\text{duality } \mathscr{V}', \mathscr{V}).$$

So that

$$(g_1, u') = -(g'_1, u)$$

and

$$M(u) = (g_0 - g'_1, u) = 0, \quad \text{according to (4.30)},$$

whence the density.

2) We shall now define $u(0)$ by Green's formula (note that the idea of this proof is entirely analogous to the proof of Theorem 6.5 of Chapter 2).

According to Section 3.2 of Chapter 1, there exists a continuous linear mapping:

$$h \to v_h$$

of $H \to \{v \mid v \in \mathscr{V}, v' \in \mathscr{V}', v(T) = 0\}$ such that

$$(4.33) \qquad v_h(0) = h.$$

For u given in Y, we set

$$(4.34) \qquad \mathscr{L}(h) = -(u', v_h) - (u, v'_h),$$

where (u', v_n) (resp. (u, v'_h)) denotes the antiduality between $\Xi' + \mathscr{V}'$ and $\Xi \cap \mathscr{V}$ (resp. $\mathscr{V}$ and $\mathscr{V}'$).

Formula (4.34) *does not depend* on the choice of the "right-inverse" v_h, as long as it satisfies (4.33). Indeed, if $\bar{v}_h$ is a second "right-inverse", $w = v_h - \bar{v}_h$ vanishes at 0 and T and then we see, as in 1), that

$$-(u', w) - (u, w') = 0.$$

Furthermore, since the form $h \to \mathscr{L}(h)$ is continuous antilinear on H, it may be written:

$$(4.35) \qquad \mathscr{L}(h) = [\tau u, h], \quad \tau u \in H,$$

which defines a continuous linear mapping $u \to \tau u$ of $Y \to H$.

3) There remains to see that $\tau u = u(0)$ if $u \in \mathscr{D}([0, T]; V)$, which follows immediately by integration by parts on the expression (4.34) for $\mathscr{L}(h)$. □

We are now in a position to prove.

Theorem 4.4. *Assume that* (4.3), (4.4), (4.11) *and* (4.12) *hold. Let* u_0 *be given in* H *and* f *be given by*

$$f = f_* + f_{**}, \quad f_* \in \Xi', \quad f_{**} \in L^2(0, T; V'). \tag{4.36}$$

There exists a unique $u \in \mathscr{V}$ *such that*

$$A(t)\, u + u' = f, \tag{4.37}$$

$$u(0) = u_0. \tag{4.38}$$

(Note that (4.37) implies that $u \in Y$ and $u(0)$ is taken in the sense of Theorem 4.3.)

Proof. Choose L as in Proposition 4.2. Then $u \in Y$ and, by *definition* of $u(0)$ (see Theorem 4.3), we have

$$[u(0), v(0)] = -(u', v) - (u, v') \quad \text{if} \quad v \in \mathscr{V} \cap D(\Lambda^*; \mathscr{V}'). \tag{4.39}$$

But according to (4.37):

$$(A(t)\, u, v) + (u', v) = (f, v) = L, v) - [u_0, v(0)]$$

and with (4.39) (and since $(A(t)\, u, v) = (u, A^*(t)\, v)$) we obtain

$$(u, A^*(t)\, v - v') - [u(0), v(0)] = (L, v) - [u_0, v(0)],$$

from which we obtain

$$[u(0), v(0)] = [u_0, v(0)] \qquad \forall v \in \mathscr{V} \cap D(\Lambda^*; \mathscr{V}');$$

whence (4.38) (since $v(0)$ describes H as v describes $\mathscr{V} \cap D(\Lambda^*; \mathscr{V}')$). For uniqueness, go back over the calculations. □

4.7 Examples

We give some examples of applications of the preceding theorems.

4.7.1 Example 1

Let Ω be an *arbitrary* bounded open set in $\mathbf{R}^n$, Γ its boundary (which may be "arbitrarily irregular"); let Q be the cylinder

$$Q = \Omega \times]0, T[, \quad T < \infty, \quad \text{and} \quad \Sigma = \Gamma \times]0, T[.$$

Take

$$a(t; u, v) = \sum_{i,j=1}^{n} \int_{\Omega} \left(a_{ij}(x, t) \frac{\partial u}{\partial x_i} \frac{\partial \bar{v}}{\partial x_j} dx + c(x, t)\, u\, \bar{v} \right) dx, \tag{4.40}$$

where

(4.41) $$a_{ij} \in L^\infty(Q),$$

(4.42) $$\begin{cases} \text{there exists } \alpha > 0 \text{ such that} \\ \sum_{i,j=1}^{n} a_{ij}(x,t)\,\lambda_i\,\bar{\lambda}_j \geqq \alpha \sum_{i=1}^{n} |\lambda_i|^2, \quad \forall \lambda_i \in \mathbf{C}, \ x \in \Omega, \ t \in]0,T[, \end{cases}$$

(4.43) $$c \in L^\infty(Q).$$

Eventually changing u into $e^{\lambda t} u$, with sufficiently large $\lambda > 0$, we may always assume that

(4.44) $$c(x,t) \geqq 0, \quad \forall x \in \Omega, \quad t \in]0,T[.$$

Next, take

$$H = L^2(\Omega), \quad V = H_0^1(\Omega).$$

Theorems 4.1 and 4.4 apply to the boundary value problem (of Cauchy-Dirichlet type):

(4.45) $$\frac{\partial u}{\partial t} + A\,u = f \quad \text{in} \quad Q,$$

(4.46) $$u = 0 \quad \text{on} \quad \Sigma,$$

(4.47) $$u(x,0) = u_0 \quad \text{in} \quad \Omega,$$

where

$$A\,u = -\sum_{i,j=1}^{n} \frac{\partial}{\partial x_i}\left(a_{ij}\frac{\partial u}{\partial x_j}\right) + c\,u.$$

Let us specify the results thus obtained. Take

$$f \in L^2(0,T;H^{-1}(\Omega)), \qquad u_0 \in L^2(\Omega).$$

Then, from Theorem 4.1 (and Remark 4.3), we deduce that there exists a unique u in

$$L^2(0,T;H_0^1(\Omega)) \cap C^0(0,T;L^2(\Omega)),$$

satisfying (4.45) *in the sense of distributions in* Q, and (4.47) in $L^2(\Omega)$. The boundary condition (4.46) on Σ is contained in the fact that

$$u \in L^2(0,T;H_0^1(\Omega))$$

(if Ω is regular, then see Chapter 4, Section 2). □

Let us now apply Theorem 4.4; then we may take f in a larger space: in fact (see Proposition 4.1), we may take f in the form

(4.48) $$f = f_0 + \frac{\partial}{\partial t}(\varrho\,f_1) + f_2,$$

ϱ defined by (4.23),

$$f_i \in L^2(0,T;H_0^1(\Omega)), \quad i = 0,1, \quad f_2 \in L^2(0,T;H^{-1}(\Omega)).$$

From Theorem 4.4, we deduce that if $u_0 \in L^2(\Omega)$, then there exists a unique u in

$$L^2(0, T; H_0^1(\Omega)),$$

satisfying (4.45) *in the sense of distributions in* Q and (4.47) *in the sense of Theorem* 4.3, condition (4.46) still being contained in

$$u \in L^2(0, T; H_0^1(\Omega)). \quad \square$$

4.7.2 Example 2

Again, we choose Ω and $a(t; u, v)$ as in 4.7.1; we may also assume, eventually changing u into $e^{\lambda t} u$, $\lambda > 0$ and sufficiently large, that

$$\text{(4.49)} \qquad \begin{cases} \text{there exists } m > 0 \text{ such that} \\ c(x, t) \geq m, \quad \forall x \in \Omega, \quad t \in]0, T[. \end{cases}$$

Let

$$H = L^2(\Omega), \qquad V = H^1(\Omega).$$

"Formally", we obtain the problem (of Cauchy-Neumann type):

$$\text{(4.45)} \qquad \frac{\partial u}{\partial t} + A u = f \quad \text{in} \quad Q,$$

$$\text{(4.50)} \qquad \frac{\partial u}{\partial \nu_A} = 0 \quad \text{on} \quad \Sigma,$$

$$\text{(4.47)} \qquad u(x, 0) = u_0 \quad \text{in} \quad \Omega,$$

where $\dfrac{\partial}{\partial \nu_A}$ is the "co-normal" (or "transversal") derivative with respect to A,

$$\left(\text{i.e. if } \Omega \text{ is regular:} \frac{\partial}{\partial \nu_A} = - \sum_{i,j=1}^{n} a_{ij} \cos(x_j, \nu) \frac{\partial}{\partial x_i}\right).$$

We must call attention to the fact that in this case $V' = (H^1(\Omega))'$ is not a space of distributions on Ω and therefore $L^2(0, T; V')$ *is not a space of distributions on* Q. In order to obtain the usual interpretations we may introduce a space $\Xi^1(\Omega)$ (see Chapter 2, Section 6.3); but since we have made no regularity assumptions on Ω, this requires some further developments.

The space $\Xi^1(\Omega)$.

We introduce

$$\text{(4.51)} \qquad \delta(x) = \inf(d(x, \Gamma), 1),$$

where

$$d(x, \Gamma) = \text{distance from } x \text{ to } \Gamma.$$

Note that $\delta \in L^\infty(\Omega)$.

We define (analogously to Chapter 2, Section 6.3, but now δ is not infinitely differentiable):

$$\Xi^1(\Omega) = \left\{ v \mid v \in L^2(\Omega), \delta \frac{\partial v}{\partial x_i} \in L^2(\Omega), i = 1, \ldots, n \right\}. \tag{4.52}$$

We provide $\Xi^1(\Omega)$ with the norm

$$\left(|v|^2 + \sum_{i=1}^{n} \left| \partial \frac{\delta v}{\partial x_i} \right|^2 \right)^{1/2}.$$

In this manner we obtain a Hilbert space. We now extend Proposition 6.2 of Chapter 2 to

Proposition 4.3. *The space $\mathscr{D}(\Omega)$ is dense in $\Xi^1(\Omega)$.*

Proof. Define:

$$E_\varepsilon = \{x \mid x \in \Omega, d(x, \Gamma) \geqq \varepsilon\},$$

$$\chi_\varepsilon = \text{characteristic function of } E_\varepsilon,$$

$$\varphi_\varepsilon = \text{function belonging to } \mathscr{D}(\mathbf{R}^n), \geqq 0,$$

$$\text{of the form } \varphi_\varepsilon(x) = \varepsilon^{-n} \varphi\left(\frac{x}{\varepsilon}\right),$$

$$\varphi \in \mathscr{D}(\mathbf{R}^n), \quad \varphi \geqq 0, \quad \int_{R^n} \varphi(x)\, dx = 1,$$

φ with support in $|x| \leqq \frac{1}{2}$, and finally let

$$\delta_\varepsilon = \chi_\varepsilon * \varphi_\varepsilon.$$

We have:

$$\delta_\varepsilon(x) = 0, \quad \text{if} \quad d(x, \Gamma) \leqq \frac{\varepsilon}{2}$$

$$\delta_\varepsilon(x) = 1, \quad \text{if} \quad d(x, \Gamma) \geqq \frac{3\varepsilon}{2}.$$

We immediately see that if $u \in \Xi^1(\Omega)$, $\delta_\varepsilon u \to u$ in $L^2(\Omega)$ as $\varepsilon \to 0$.

To show that

$$\delta_\varepsilon u \to u \quad \text{in} \quad \Xi^1(\Omega) \quad \text{as} \quad \varepsilon \to 0, \tag{4.53}$$

we still have to show that

$$\delta \frac{\partial}{\partial x_i}(\delta_\varepsilon u) \to \delta \frac{\partial u}{\partial x_i} \quad \text{in} \quad L^2(\Omega)$$

and for this purpose it is sufficient to show that

$$\delta \left(\frac{\partial \delta_\varepsilon}{\partial x_i} \right) u \to 0 \quad \text{in} \quad L^2(\Omega).$$

But

$$\frac{\partial \delta_\varepsilon}{\partial x_i} = 0 \quad \text{if} \quad d(x, \Gamma) \geqq \frac{3\varepsilon}{2} \quad \text{and} \quad \delta \left| \frac{\partial \delta_\varepsilon}{\partial x_i}(x) \right|$$

is bounded, whence the result and (4.53).

To complete the proof it suffices (as for the case where the boundary of Ω is regular) to regularize $\delta_\varepsilon u$, with fixed ε. □

Thus, we define

$$(4.54) \qquad \Xi^{-1}(\Omega) = (\Xi^1(\Omega))' \quad (\text{dual of } \Xi^1(\Omega));$$

$\Xi^{-1}(\Omega)$ is a space of distributions on Ω.

According to Proposition 4.3 and the Hahn-Banach Theorem, we have

Proposition 4.4. *Every element f of $\Xi^{-1}(\Omega)$ may be represented, non-uniquely, by*

$$f = f_0 + \sum_{i=1}^{n} \frac{\partial}{\partial x_i}(\delta f_i), \quad f_i \in L^2(\Omega), \quad i = 0, \ldots, n. \quad \square$$

Note that we have

$$(4.55) \qquad H^1(\Omega) \subset \Xi^1(\Omega), \quad \Xi^{-1}(\Omega) \subset (H^1(\Omega))'$$

($H^1(\Omega)$ is dense in $\Xi^1(\Omega)$, by Proposition 4.3, and therefore we may identify $\Xi^{-1}(\Omega)$ with a subspace of $(H^1(\Omega))'$).

It follows that

$$(4.56) \qquad L^2(0, T; \Xi^{-1}(\Omega)) \subset L^2(0, T; (H^1(\Omega))')$$

and that $L^2(0, T; \Xi^{-1}(\Omega))$ is a space of distributions on Q.

We come back to the Cauchy-Neumann problem and take

$$f \in L^2(0, T; \Xi^{-1}(\Omega)), \quad u_0 = 0$$

(we could also take $u_0 \in L^2(\Omega)$, but we take $u_0 = 0$ for the sake of simplicity).

We may apply Theorem 4.1 and find a *unique* u in

$$L^2(0, T; H^1(\Omega)) \cap C^0(0, T; L^2(\Omega)),$$

with $u(0) = 0$ *and* $\dfrac{du}{dt} \in L^2(0, T; (H^1(\Omega))')$, *satisfying the equation*

$$\text{(4.57)} \qquad \int_0^T a(t; u(t), v(t))\, dt + \int_0^T \left[\frac{du(t)}{dt}, v(t)\right] dt = \int_0^T \langle f(t), \overline{v(t)}\rangle\, dt, \quad \forall v \in L^2(0, T; H^1(\Omega)),$$

where $[\ ,\ ]$ denotes the antiduality between $(H^1(\Omega))'$ and $H^1(\Omega)$ (and the scalar product in $L^2(\Omega)$) and $\langle\ ,\ \rangle$ denotes the duality between $\Xi^{-1}(\Omega)$ and $\Xi^1(\Omega)$.

It follows from (4.57) that u satisfies (4.45) in the sense of distributions on Q.

Furthermore, we have *the boundary condition* (4.50) *on* Σ which is contained in (4.57), but in a *formal* manner; indeed, *formally* integrating by parts in (4.57), we obtain, thanks to the fact that v is arbitrary, $\dfrac{\partial u}{\partial \nu_A} = 0$ on Σ.

But this is *formal* for two reasons:

(i) no regularity conditions have been imposed on Σ (and therefore one may not speak of a "co-normal" to Σ);

(ii) if we assume Σ to be regular, we must *justify*, in a certain sense, the integrations by parts.

The justification of (ii) will be given in Chapter 4 of Volume 2; but without regularity assumptions on Σ, there exists (at the moment) no other interpretation of the condition "$\dfrac{\partial u}{\partial \nu_A} = 0$ on Σ", other than equation (4.57) itself. □

Theorem 4.4 also applies to problem (4.45), (4.50), (4.47); this allows us to take f in a more general space. Indeed, we may take f in the form

$$\text{(4.58)} \qquad f = f_0 + \frac{\partial}{\partial t}(\varrho f_1) + f_2 \quad (\varrho \text{ defined by } (4.23)),$$

with $f_i \in L^2(0, T; H^1(\Omega))$, $i = 0, 1$, $f_2 \in L^2(0, T; \Xi^{-1}(\Omega))$.

Then, if furthermore $u_0 \in L^2(\Omega)$, *there exists a unique u in* $L^2(0, T; H^1(\Omega))$ *such that*

$$(4.59)\quad \begin{cases} \int_0^T a(t; u(t), v(t))\, dt - \int_0^T \left[u(t), \frac{dv(t)}{dt} \right] dt = \int_0^T [f_0(t), v(t)]\, dt - \\ - \int_0^T \left[\varrho(t) f_1(t), \frac{dv}{dt} \right] dt + \int_0^T \langle f_2(t), \overline{v(t)} \rangle\, dt + \\ + [u_0, v(0)], \quad \forall v \in L^2(0, T; H^1(\Omega)), \quad \text{with} \\ \frac{dv}{dt} \in L^2(0, T; (H^1(\Omega))' \quad \text{and} \quad v(T) = 0. \end{cases}$$

It follows that u satisfies (4.45) in the sense of distributions on Q and $u(0) = u_0$ in the sense of Theorem 4.3; furthermore (4.59) contains, in a *formal* way, the boundary condition (4.50) (considerations analogous to those made for equation (4.57) hold). □

4.7.3 Example 3

Consider a *bounded* open set in $\mathbf{R}^n$, *with* (suitably) *regular boundary* Γ and assume that the subset Γ_0 of Γ is an $(n - 1)$-dimensional variety with boundary (therefore $\Gamma_0 \neq \Gamma$); $\Gamma_1 = \Gamma - \Gamma_0$. Again, take $a(t; u, v)$ defined by (4.40) with (4.41), (4.42), (4.43), (4.49) (we could also replace (4.49) with (4.44)). Finally, take

$$(4.60)\quad H = L^2(\Omega), \qquad V = \{v \mid v \in H^1(\Omega), v = 0 \text{ on } \Gamma_0\}^{((1))}.$$

The general theory applies again; "formally", the problem is (of Cauchy type with mixed conditions on Σ):

$$(4.45)\quad \frac{\partial u}{\partial t} + A u = f \quad \text{in } Q$$

$$(4.61)\quad \begin{cases} u = 0 \quad \text{on } \Sigma_0 = \Gamma_0 \times]0, T[, \\ \dfrac{\partial u}{\partial \nu_A} = 0 \quad \text{on } \Sigma_1 = \Gamma_1 \times]0, T[, \end{cases}$$

$$(4.47)\quad u(x, 0) = u_0 \quad \text{in } \Omega.$$

For the *interpretation* of the problem, we meet the same difficulties as in 4.7.2 concerning the space V'. We could use the space $\Xi^{-1}(\Omega)$, but we can do better in the following way.

((1)) Γ being assumed sufficiently regular, we may define $\gamma_0\, v$ if $v \in H^1(\Omega)$ (Chapter 1, Section 8); we impose on $v \in V$ that the restriction of $\gamma_0\, v$ to Γ_0 vanishes.

The space $\Xi^1_{\Gamma_0}(\Omega)$.

We introduce

$$\delta_0(x) = \inf(d(x, \Gamma_1), 1), \tag{4.62}$$

and

$$\hat{\Xi}^1_{\Gamma_0}(\Omega) = \left\{ v \mid v \in L^2(\Omega), \delta_0 \frac{\partial v}{\partial x_i} \in L^2(\Omega), i = 1, \ldots, n \right\}. \tag{4.63}$$

Provided with the norm

$$\left(|v|^2 + \sum_{i=1}^n \left| \delta_0 \frac{\partial v}{\partial x_i} \right|^2 \right)^{1/2},$$

$\hat{\Xi}^1_{\Gamma_0}(\Omega)$ is a Hilbert space.

Let

$$\Gamma_0^\beta = \{x \mid x \in \Gamma_0, d(x, \partial\Gamma_0) \geqq \beta > 0, \partial\Gamma_0 = \text{boundary of } \Gamma_0 \text{ in } \Gamma\}$$

and let $\mathcal{O}_\beta$ be an open set in $\mathbf{R}^n$ contained in Ω, a subset of whose boundary $\partial\mathcal{O}_\beta$ is Γ_0^β, the rest of $\partial\mathcal{O}_\beta$ being contained in Ω; then if $v \in \hat{\Xi}^1_{\Gamma_0}(\Omega)$, we have: $v \in H^1(\mathcal{O}_\beta)$ and we can define

$$v|_{\Gamma_0^\beta}, \quad \text{for all} \quad \forall \beta > 0.$$

Therefore we can define $v|_{\Gamma_0}$ (in particular $\in L^2(\Gamma_0)$).

Now, we can define

$$\Xi^1_{\Gamma_0}(\Omega) = \{v \mid v \in \hat{\Xi}^1_{\Gamma_0}(\Omega), v = 0 \text{ on } \Gamma_0\}, \tag{4.64}$$

which is a Hilbert space for the norm induced by the norm of $\hat{\Xi}^1_{\Gamma_0}(\Omega)$.

Of course, we have

$$\begin{cases} V \subset \Xi^1_{\Gamma_0}(\Omega) \subset \Xi^1(\Omega) \\ V \text{ defined by (4.60), } \Xi^1(\Omega) \text{ defined by (4.52).} \end{cases} \tag{4.65}$$

We verify – by the same type of proof as for Proposition 4.3 – that

$$\mathcal{D}(\Omega) \text{ is dense in } \Xi^1_{\Gamma_0}(\Omega). \tag{4.66}$$

We introduce

$$\Xi^{-1}_{\Gamma_0}(\Omega) = (\Xi^1_{\Gamma_0}(\Omega))'. \tag{4.67}$$

Every element f of $\Xi^{-1}_{\Gamma_0}(\Omega)$ may be represented, non-uniquely, by

$$f = f_0 + \sum_{i=1}^n \frac{\partial}{\partial x_i}(\delta_0 f_i), \quad (f_i \in L^2(\Omega), \quad i = 0, \ldots, n). \quad \square$$

Now we may use the space $\Xi_{\Gamma_0}^{-1}(\Omega)$ for problem (4.45), (4.61), (4.62), (4.47), with $u_0 \in L^2(\Omega)$; if we take

$$f \in L^2(0, T; \Xi_{\Gamma_0}^{-1}(\Omega)),$$

we can apply Theorem 4.1.

If, more generally, we take

$$f = f_0 + \frac{\partial}{\partial t}(\varrho f_1) + f_2$$

with $f_i \in L^2(0, T; V)$, $i = 0, 1$, $f_2 \in L^2(0, T; \Xi_{\Gamma_0}^{-1}(\Omega))$, we can apply Theorem 4.4.

The interpretation of the results is completely analogous to the one given for the Cauchy-Dirichlet and Cauchy-Neumann problems. Note that condition (4.61) will be expressed by the fact that, in both cases, u belongs to $L^2(0, T; V)$, with V given by (4.60); condition (4.62) will be verified in a *formal* way; in the first case it will be contained in an equation like (4.57) and in the second case in an equation like (4.59). □

4.7.4 Example 4

Examples 1, 2 and 3 may be generalized to the case of operators A of order $2m$, $m > 1$.

Consider a form of the type

$$a(t; u, v) = \sum_{|p|,|q| \leq m} \int_\Omega a_{pq}(x, t)\, D^q u\, \overline{D^p v}\, dx \tag{4.68}$$

and take

$$H = L^2(\Omega), \quad H_0^m(\Omega) \subset V \subset H^m(\Omega), \tag{4.69}$$

V closed subspace of $H^m(\Omega)$.

We assume that the form $a(t; u, v)$ satisfies (4.3) and (4.4) so that we can apply Theorems 4.1 and 4.4. Also, applying the results of Sections 9.5, 9.6, 9.7 of Chapter 2, we are able to interpret the results in the same way as for examples 1, 2 and 3.

Formally, we solve boundary value problems of the type

$$\begin{cases} \dfrac{\partial u}{\partial t} + \displaystyle\sum_{|p|,|q| \leq m} (-1)^{|p|} D^p(a_{pq}(x, t)\, D^q u) = f \quad \text{in } Q \\ u(x, 0) = u_0 \\ \text{with certain boundary conditions on } \Sigma, \text{ which depend on } V. \end{cases}$$

But we shall not go into the details of this type of application of the theory here. □

4.7.5 Example 5

The "general" choice of H for the applications is $L^2(\Omega)$. But this must not necessarily be the case; see Chapter 2, Section 9.9. We shall give an example of this kind here. We take

$$H = H_0^1(\Omega), \tag{4.70}$$

with

$$(u, v)_H = \sum_{i=1}^{n} \int_\Omega \frac{\partial u}{\partial x_i} \frac{\partial \bar{v}}{\partial x_i} dx; \tag{4.71}$$

in this way we define a *hilbertian* scalar product on H, assuming Ω to be bounded.

Further, we take

$$V = \left\{ v \mid v \in H, \frac{\partial}{\partial x_i} \Delta v \in L^2(\Omega), i = 1, \ldots, n \right\} \tag{4.72}$$

and

$$\| v \|_V^2 = \| v \|_H^2 + \sum_{i=1}^{n} \left\| \frac{\partial}{\partial x_i} \Delta v \right\|_{L^2(\Omega)}^2 . \tag{4.73}$$

Finally, for all $u, v \in V$, let

$$a(u, v) = \sum_{i=1}^{n} \int_\Omega \left(\frac{\partial}{\partial x_i} \Delta u \right) \left(\frac{\partial}{\partial x_i} \Delta \bar{v} \right) dx. \tag{4.74}$$

Of course, we have

$$a(v, v) + \lambda \| v \|_H^2 \geqq \min(1, \lambda) \| v \|_V^2 \quad (\forall \lambda > 0), \qquad \forall v \in V, \tag{4.75}$$

and consequently, according to Theorem 4.1, we have:

Proposition 4.5. *Let H and V be defined by* (4.70) *and* (4.72). *Let V' be the antidual of V* ***when H is identified with its antidual.*** *Let f be given, with*

$$f \in L^2(0, T; V'). \tag{4.76}$$

Then, if u_0 is given in H, there exists a unique $u \in L^2(0, T; V)$, such that, a.e. in $[0, T]$,

$$a(u(t), v) + [u'(t), v] = [f(t), v], \qquad \forall v \in V, \tag{4.77}$$

$$u' \in L^2(0, T; V'), \tag{4.78}$$

$$u(0) = u_0, \tag{4.79}$$

where the scalar product $[f(t), v]$ denotes the duality between V' and V. □

Remark 4.4. We must be careful with the *interpretation* of V'. □

In particular, let us take

$$(4.80) \qquad f \in L^2(0, T; H) \qquad (\text{i.e. } f \in L^2(0, T; H_0^1(\Omega)))$$

in (4.77).

Then (4.77) may be written:

$$(4.81) \qquad a(u(t), v) + \frac{d}{dt}[u(t), v] = [f(t), v)], \qquad \forall v \in V.$$

But

$$[f(t), v] = (f(t), -\Delta v)_{L^2(\Omega)}, \qquad [u(t), v] = (u(t), -\Delta v)_{L^2(\Omega)}.$$

If, $\forall \varphi, \psi \in H^1(\Omega)$, we set:

$$(4.82) \qquad b(\varphi, \psi) = \sum_{i=1}^{n} \int_\Omega \frac{\partial \varphi}{\partial x_i} \frac{\overline{\partial \psi}}{\partial x_i} dx,$$

then

$$a(u(t), v) = b(-\Delta u(t), -\Delta v)$$

and (4.81) may be written:

(4.83)

$$b(-\Delta u(t), w) + \frac{d}{dt}(u(t), w)_{L^2(\Omega)} = (f(t), w)_{L^2(\Omega)}, \quad \forall w = -\Delta v, \quad v \in V.$$

But, as v describes V, Δv describes a space W containing $H^1(\Omega)$ *and therefore* (4.83) *holds* $\forall w \in H^1(\Omega)$. *But then it follows* (see Chapter 2, Section 9.9) *that* u satisfies

$$(4.84) \qquad \frac{\partial u}{\partial t} + \Delta^2 u = f \quad \text{in } Q, \qquad u(x, 0) = u_0 \quad \text{in } \Omega,$$

$$(4.85) \qquad u = 0, \qquad \frac{\partial \Delta u}{\partial \nu} = 0 \quad \text{on } \Sigma.$$

Remark 4.5. The operator $-A$ is the infinitesimal generator of an *analytic* semi group in H, see Yosida [2]. □

Remark 4.6. Problem (4.84), with the boundary conditions

$$(4.86) \qquad \frac{\partial u}{\partial \nu} = 0, \qquad \frac{\partial \Delta u}{\partial \nu} = 0 \quad \text{on } \Sigma,$$

can be treated in the same way. □

4.7.6 Example 6

Let Ω be an open set in $\mathbf{R}^2$. Take

$$(4.87) \quad H = L^2(\Omega), \quad V = \left\{ v \mid v \in L^2(\Omega), \frac{\partial v}{\partial x_1} \in L^2(\Omega), \frac{\partial^2 v}{\partial x_2^2} \in L^2(\Omega) \right\},$$

$$(4.88) \qquad a(t; u, v) = \int_\Omega \frac{\partial u}{\partial x_1} \frac{\overline{\partial v}}{\partial x_1} dx + \int_\Omega \frac{\partial^2 u}{\partial x_2^2} \frac{\overline{\partial^2 v}}{\partial x_2^2} dx.$$

The general theory applies. For the interpretation of the problem, *one* possibility is to introduce (see Chapter 2, Section 6.3 for analogous considerations) a space $K(\Omega)$ such that

$$(4.89)\quad \begin{cases} V \subset K(\Omega) \subset L^2(\Omega) \\ \mathscr{D}(\Omega) \text{ is dense in } K(\Omega) \\ K(\Omega) \text{ is "the smallest possible".} \end{cases}$$

We shall make the hypothesis that Ω is bounded and has a *regular* boundary. Then every $v \in V$ also verifies $\dfrac{\partial v}{\partial x_2} \in L^2(\Omega)$. We define δ by (4.51) and $K(\Omega)$ by

$$(4.90)\quad K(\Omega) = \left\{ v \mid v \in L^2(\Omega),\ \delta \frac{\partial v}{\partial x_1},\ \delta \frac{\partial v}{\partial x_2},\ \delta^2 \frac{\partial^2 v}{\partial x_2^2} \in L^2(\Omega) \right\}.$$

We provide $K(\Omega)$ with the norm

$$\left(\| v \|^2_{L^2(\Omega)} + \left\| \delta \frac{\partial v}{\partial x_1} \right\|^2_{L^2(\Omega)} + \left\| \delta \frac{\partial v}{\partial x_2} \right\|^2_{L^2(\Omega)} + \left\| \delta^2 \frac{\partial^2 v}{\partial x_2^2} \right\|^2_{L^2(\Omega)} \right)^{1/2},$$

which makes it a Hilbert space.

With the same kind of proof as for Proposition 4.3, we verify that

$$\mathscr{D}(\Omega) \text{ is dense in } K(\Omega).$$

Using $K(\Omega)$ in an analogous way as $\Xi^1(\Omega)$ in 4.7.2, we may apply the general theory; in particular, taking

$$f \in L^2(0, T; K'(\Omega)), \qquad u_0 \in L^2(\Omega),$$

we can apply Theorem 4.1; we obtain $u \in L^2(0, T; V) \cap C^0(0, T; L^2(\Omega))$ such that

$$\begin{cases} \dfrac{\partial u}{\partial t} - \dfrac{\partial^2 u}{\partial x_1^2} + \dfrac{\partial^4 u}{\partial x_2^4} = f \text{ in the sense of distributions on } Q, \\ u(x, 0) = u_0. \end{cases}$$

Thanks to the fact that $u \in L^2(0, T; V)$, the solution also verifies certain boundary conditions on Σ, which, for example when Ω is the rectangle

$$\Omega =]0, a_1[\times]0, a_2[,$$

are "formally" given by

$u = 0$, on the sides of Γ parallel to the x_2-axis, $\quad \forall t \in]0, T[$,

$u = \dfrac{\partial u}{\partial x_2} = 0$ on the sides of Γ parallel to the x_1-axis, $\quad \forall t \in]0, T[$.

Remark 4.7. For non-homogeneous parabolic problems treated from the "variational" point of view, the reader may also consult Lions [13], Chapter 6, Section 9. □

5. Example: Abstract Parabolic Equations, Initial Condition Problems (II)

5.1 Some Interpolation Results

Generally, if F is a Hilbert space and $\mathcal{O}$ an open set in $\mathbf{R}_t$, we define

$$H^1(\mathcal{O}; F) = \{f \mid f \in L^2(\mathcal{O}; F), f' \in L^2(\mathcal{O}; F)\} \tag{5.1}$$

which is a Hilbert space for the norm

$$\left(\int_{\mathcal{O}} (\|f(t)\|_F^2 + \|f'(t)\|_F^2)\, dt \right)^{1/2}.$$

Also, we recall (see Chapter 1, Section 7.1 and Remark 9.5) that

$$H^{1/2}(\mathbf{R}; F) = \{f \mid (1 + |\tau|)^{1/2} \hat{f} \in L^2(\mathbf{R}_\tau; F)\}$$

(where $\hat{f}$ = Fourier transform in t of f).

We have (see Remark 9.4, Chapter 1):

Proposition 5.1.

$$[H^1(\mathbf{R}; V'), L^2(\mathbf{R}; V)]_{1/2} = H^{1/2}(\mathbf{R}; H). \tag{5.2}$$

Proof. We consider V as the domain of a positive, self-adjoint, unbounded operator in H, which we diagonalize (as in Chapter 1, Section 2.1) by a unitary operator $\mathcal{U}$. Then if $\mathcal{U} H = \mathfrak{h}$ and if we apply the Fourier transform in t and the operator $\mathcal{U}$ to $L^2(\mathbf{R}, V)$ (resp. $H^1(\mathbf{R}; V')$), we see that $L^2(\mathbf{R}, V)$ (resp. $H^1(\mathbf{R}, V')$) is isomorphically transformed to the space of f's such that

$$\lambda f \in L^2(\mathbf{R}; \mathfrak{h}) \tag{5.3}$$

(resp.

$$(1 + |\tau|) \frac{1}{\lambda} f \in L^2(\mathbf{R}; \mathfrak{h})). \tag{5.4}$$

Then, by definition of the spaces $[X, Y]$ (Chapter 1, Section 2.1), we see that $[H^1(\mathbf{R}; V'), L^2(\mathbf{R}; V)]_{1/2}$ is isomorphically transformed to the space of f's such that

$$\lambda^{1/2}(1 + |\tau|)^{1/2} \frac{1}{\lambda^{1/2}} f = (1 + |\tau|)^{1/2} f \in L^2(\mathbf{R}; \mathfrak{h}) \tag{5.5}$$

which is the transformed space of $H^{1/2}(\mathbf{R}; H)$. □

Now, we define

(5.6) $\quad {}_0H^1(0, T; F) = \{v \mid v \in H^1(]0, T[; F) \text{ (see (5.1))}, v(0) = 0\}$,

closed vector subspace of $H^1(]0, T[; F)$, provided with the induced norm. Further, we *define*

(5.7) $$H^{1/2}(0, T; F) = [H^1(0, T; F), L^2(0, T; F)]_{1/2},$$

(5.8) $${}_0H^{1/2}(0, T; F) = [{}_0H^1(0, T; F), L^2(0, T; F)]_{1/2}.$$

Then

Proposition 5.2. *We have*

(i) $\quad H^{1/2}(0, T; F) =$ space of restrictions to $]0, T[$ of $H^{1/2}(\mathbf{R}; F)$;

(ii) $\quad {}_0H^{1/2}(0, T; F) =$ space of restrictions to $]0, T[$ of the elements of $H^{1/2}(\mathbf{R}; F)$ *which vanish for all* $t < 0$.

Proof. Result (i) is verified exactly as in Chapter 1, Theorem 9.1.

For (ii), we shall show that

(5.9) $\quad {}_0H^{1/2}(0, T; F) =$ space of restrictions to $]0, T[$ of the elements of $H^{1/2}(-\infty, T; F)$ *which vanish for* $t < 0$;

then we obtain the result by applying (i), in which we can replace 0 by $-\infty$.

To show (5.9), we first consider the mapping $u \to \tilde{u} =$ extension of u by 0 for $t < 0$, which is a continuous linear mapping of

$${}_0H^1(0, T; F) \to H^1(-\infty, T; F) \quad \text{and of} \quad L^2(0, T; F) \to L^2(-\infty, T; F),$$

therefore, by interpolation, of

$${}_0H^{1/2}(0, T; F) \to H^{1/2}(-\infty, T; F).$$

On the other hand

$$v \to r\,v, \quad r\,v(t) = v(t) - v(-t),$$

is a continuous linear mapping of

$$H^1(-\infty, T; F) \to {}_0H^1(0, T; F), \quad L^2(-\infty, T; F) \to L^2(0, T; F),$$

therefore, by interpolation, of $H^{1/2}(-\infty, T; F) \to {}_0H^{1/2}(0, T; F)$.

Since $r\,\tilde{u} = u$, the desired result follows. □

Proposition 5.3. *We have*:

(i) $\quad [H^1(0, T; V'), L^2(0, T; V)]_{1/2} = H^{1/2}(0, T; H)$,

(ii) $\quad [{}_0H^1(0, T; V'), L^2(0, T; V)]_{1/2} = {}_0H^{1/2}(0, T; H)$.

(of course, as usual, with *equivalent* norms).

Proof. The *same* proof as for Proposition 5.2 shows that

$[{}_0H^1(0, T; V'), L^2(0, T; V)]_{1/2}$ = space of restrictions to $]0, T[$ of the elements of

$$[H^1(-\infty, T; V'), L^2(-\infty, T; V)]_{1/2}$$

which vanish for $t < 0$ and that

$[H^1(-\infty, T; V'), L^2(-\infty, T; V)]_{1/2}$ = space of restrictions to $]-\infty, T[$ of the elements of $[H^1(\mathbf{R}, V'), L^2(\mathbf{R}; V)]_{1/2}$.

From which, with Proposition 5.1:

$[{}_0H^1(0, T; V'), L^2(0, T; V)]_{1/2}$ = space of restrictions to $]0, T[$ of the elements of $H^{1/2}(\mathbf{R}; H)$ which vanish for $t < 0$, whence (ii) (and (i) with obvious modifications), thanks to Proposition (5.2) (ii). □

Proposition 5.4. *Every element* u *of* ${}_0H^{1/2}(0, T; F)$ *satisfies*

$$\int_0^T t^{-1} \| u(t) \|_F^2 \, dt < \infty . \tag{5.10}$$

Proof. Same as for Theorem 11.7 and Remark 11.5 of Chapter 1. □

5.2 Interpretation of the Spaces $\Phi^{1/2}$ and $\Phi_*^{1/2}$

We now apply the theory of Sections 3.2 and 3.3 to the setting of Section 4. In particular:

$$\text{(5.11)} \quad \begin{cases} \textit{under the hypotheses of Theorem 4.1, } A(t) + \dfrac{d}{dt} \textit{ is an} \\ \textit{isomorphism of } \Phi^{1/2} \textit{ onto } (\Phi_*^{1/2})', \end{cases}$$

where

$$\Phi^{1/2} = \mathscr{V} \cap [D(\Lambda; \mathscr{V}'), \mathscr{V}]_{1/2},$$
$$\Phi_*^{1/2} = \mathscr{V} \cap [D(\Lambda^*; \mathscr{V}'), \mathscr{V}]_{1/2}.$$

Therefore, in the notation of Section 5.1, we have

$$\Phi^{1/2} = L^2(0, T; V) \cap [{}_0H^1(0, T; V'), L^2(0, T; V)]_{1/2}.$$

Applying Proposition 5.3 (ii), we obtain:

$$\text{(5.12)} \quad \begin{cases} \Phi^{1/2} = L^2(0, T; V) \cap {}_0H^{1/2}(0, T; H). \\ {}_0H^{1/2}(0, T; H) \text{ defined by (5.8) or Proposition 5.2 (ii).} \end{cases} \quad \square$$

Now, we set:

$$\text{(5.13)} \quad \begin{cases} {}_TH^1(0, T; F) = \{v \mid v \in H^1(0, T; F), v(T) = 0\}; \\ \text{if } T < \infty. \end{cases}$$

Then (interchanging the roles of 0 and T):

$$\Phi_*^{1/2} = L^2(0, T; V) \cap {}_T H^{1/2}(0, T; H) \tag{5.14}$$

(eliminating the *index* T if $T = +\infty$). □

Remark 5.1. It follows from Proposition 5.4 that the spaces $\Phi^{1/2}$ and $\Phi_*^{1/2}$ are *distinct*, since $u \in \Phi^{1/2}$ satisfies (5.10), where $F = H$, whereas $u \in \Phi_*^{1/2}$ satisfies

$$\int_0^T (T - t)^{-1} |u(t)|_H^2 \, dt < \infty. \quad □$$

Remark 5.2. For $u \in \Phi^{1/2}$, we have (see Chapter 1, Theorem 4.1)

$$D_t^{1/4} u \in L^2(0, T; H),$$

$D_t^{1/4}$ being defined by Fourier transform after extension to $\mathbf{R}_t$. □

Remark 5.3. If, instead of $]0, T[$, we consider $\mathbf{R}_t$, taking for $G(s)$ the semi-group (in fact the group!) of *translations* on $\mathbf{R}_0$, we satisfy the hypotheses of Theorem 3.2; in this case $\Phi^{1/2} = \Phi_*^{1/2}$. □

Remark 5.4. The results of this section apply to the examples given in Section 4.7. □

6. Example: Abstract Parabolic Equations, Periodic Solutions

6.1 Notation. The Operator Λ

We consider the setting of Section 4.

The spaces $\mathscr{V}$, $\mathscr{H}$, $\mathscr{V}'$ are the same as in Section 4, but this time we *always* have $T < \infty$. The operator M is defined as in Section 4 (4.7).

On the other hand, the semi-group $G(s)$ is different than the one in Section 4.3.

Here, we define:

$$G(s) f(t) = \begin{cases} f(t - s + T), & \text{if } \ 0 < t < s \\ f(t - s), & \text{if } \ s < t < T. \end{cases} \tag{6.1}$$

Then, we have

$$\Lambda v = \frac{dv}{dt} = v' \tag{6.2}$$

with

$$D(\Lambda; \mathscr{V}') = \left\{ v \mid v \in \mathscr{V}', \frac{dv}{dt} = v' \in \mathscr{V}', v(0) = v(T) \right\}, \tag{6.3}$$

with analogous descriptions for $D(\Lambda; \mathscr{H})$ and $D(\Lambda; \mathscr{V})$.

The semi-group $G(s)$ (in fact the group) is a contraction semi-group (in fact unitary) and we may apply the general theory. □

6.2 Application of the Isomorphism Theorems

Theorem 1.1 yields:

Theorem 6.1. *Assume that* (4.3)–(4.4) *hold. Then, for* f *given in* $\mathcal{V}' = L^2(0, T; V')$, *there exists a unique* $u \in \mathcal{V}$ *such that*

$$A(t)\, u + u' = f \tag{6.4}$$

$$u(0) = u(T). \tag{6.5}$$

(As in Theorem 4.1, (6.4) implies that $u' \in L^2(0, T; V')$ and then (6.5) has meaning according to Theorem 3.2 of Chapter 1.)

Proof. Choose Λ as in (6.2), (6.3) and note that if $u \in \mathcal{V} \cap D(\Lambda; \mathcal{V}')$, we have (6.5). □

Remark 6.1. Condition (6.5) is a *periodicity* condition: if f is given on $\mathbf{R}$, takes its values in V' and has *period* T and if $A(t)$ is defined on $\mathbf{R}$, with period T, we deduce the existence of *a* (*unique*) *solution with period* T from Theorem 6.1. □

Theorem 2.1 yields:

Theorem 6.2. *Assume that* (4.3)–(4.4) *hold. For* L *given in* $(\mathcal{V} \cap D(\Lambda; \mathcal{V}'))'$, *there exists a unique* $u \in \mathcal{V}$ *such that*

$$(u, A^*(t)\, v - v') = (L, v), \quad \forall v \in \mathcal{V} \cap D(\Lambda^*; \mathcal{V}'). \tag{6.6}$$

Proof. Application of Theorem 2.1 and of the fact that, Λ being defined by (6.2), (6.3), we have

$$\Lambda^* v = -v', \tag{6.7}$$

$$D(\Lambda^*; \mathcal{V}') = D(\Lambda; \mathcal{V}'). \quad □ \tag{6.8}$$

6.3 Choice of L

Let ϱ_1 (compare with (4.23)) be defined by:

$$\varrho_1(t) = \begin{cases} t/t_0, & \text{if } 0 \leqq t \leqq t_0, \quad \text{fixed } t_0 < \dfrac{T}{2}, \\ 1, & \text{if } t_0 \leqq t \leqq T - t_0, \\ \dfrac{T - t}{T - t_0}, & \text{if } T - t_0 \leqq t \leqq T. \end{cases} \tag{6.9}$$

Next, we define Ξ_1 (compare with (4.24)) by:

$$\Xi_1 = \{v \mid v \in L^2(0, T; V') = \mathscr{V}', \varrho_1 v' \in \mathscr{V}'\}, \tag{6.10}$$

a Hilbert space for the norm

$$(\|v\|^2_{\mathscr{V}'} + \|\varrho_1 v'\|^2_{\mathscr{V}'})^{1/2}. \tag{6.11}$$

We have (as in Section 4):

Proposition 6.1. *The space $\mathscr{D}(]0, T[; V')$ is dense in Ξ_1.*
Every $f_* \in \Xi_1'$ may be written — non-uniquely — as

$$f_* = f_0 + \frac{d}{dt}(\varrho_1 f_1), \quad f_i \in L^2(0, T; V). \quad \square \tag{6.12}$$

Then we choose L in (6.6) by (compare with (4.26)):

$$\begin{cases} (L, v) = f_*, v) + (f_{**}, v) + [u_0, v(0)] \\ f_* \in \Xi_1', \quad f_{**} \in \mathscr{V}', \quad u_0 \in H. \quad \square \end{cases} \tag{6.13}$$

6.4 Interpretation of the Problem

Theorem 6.3. *Hypothesis of Theorem* 6.1. *Let f and u_0 be given with*

$$f = f_* + f_{**}, \quad f_* \in \Xi_1', \quad f_{**} \in \mathscr{V}', \quad u_0 \in H. \tag{6.14}$$

There exists a unique $u \in \mathscr{V}$ such that

$$A(t) u + u' = f, \tag{6.15}$$

$$-u(0) + u(T) = u_0, \tag{6.16]}$$

where $u(T) - u(0)$ is taken (as in Theorem 4.3) *by extension by continuity.*

We only give the

Outline of the proof.

First, we verify (as in Theorem 4.4) that u, solution of (6.6) with the choice (6.13) for L, satisfies (6.15).

Then $u \in Y_1$, where this time (compare with (4.28)):

$$Y_1 = \{v \mid v \in \mathscr{V}, v' \in \Xi_1' + \mathscr{V}'\}. \tag{6.17}$$

As in Theorem 4.3, we show that $\mathscr{D}([0, T]; V)$ is dense in Y_1 and that the mapping

$$u \to \{-u(0) + u(T)\},$$

of

$$\mathscr{D}([0, T]; V) \to V,$$

extends by continuity to a mapping, still denoted $u \to \{-u(0) + u(T)\}$, of $Y_1 \to H$.

Then (6.16) follows by integrations by parts, as in Theorem 4.4. $\square$

6.5 The Isomorphism of $\Phi^{1/2}$ onto its Dual

According to (6.8), we may apply Theorem 3.2, therefore:

Theorem 6.4. *Hypothesis of Theorem* 6.1. *Then* $A(t) + \dfrac{d}{dt}$ *is an isomorphism of* $\Phi^{1/2}$ *onto its dual* $(\Phi^{1/2})'$. □

There remains to *interpret* $\Phi^{1/2}$.

Proposition 6.2. *The space* $\Phi^{1/2} = \mathscr{V} \cap [D(\Lambda; \mathscr{V}'), \mathscr{V}]_{1/2}$ *is the space of functions* u:

$$u = \sum_{-\infty}^{+\infty} u_n \exp\left(2\pi \frac{i\,n\,t}{T}\right), \quad u_n \in V, \tag{6.18}$$

such that

$$|||u||| = \left(\sum_{-\infty}^{+\infty} [\|u_n\|_V^2 + (1 + |n|)\, \|u_n\|_H^2]\right)^{1/2} < \infty, \tag{6.19}$$

($|||u|||$ being a norm equivalent to the initial norm).

This result is a consequence of

Proposition 6.3. *We have*

$$[D(\Lambda; \mathscr{V}'), \mathscr{V}]_{1/2} = \text{space of } u\text{'s of the form (6.18)}, \tag{6.20}$$

with

$$\left(\sum_{-\infty}^{+\infty} (1 + |n|)\, \|u_n\|_H^2\right)^{1/2} < \infty .$$

Proof. The idea is the same as for the proof of Proposition 5.1: we consider $D(\Lambda; \mathscr{V}')$ (resp. $\mathscr{V}$) as domain of J_1 (resp. J_2), a positive self-adjoint operator in $\mathscr{V}'$. The operator J_1 is given by (here we replace the Fourier transform by the Fourier series):

(6.21)
$$J_1 f = \sum_n (1 + |n|)\, f_n \exp\left(\frac{2\pi\, i\, n\, t}{T}\right), \quad \text{if} \quad f = \sum_n f_n \exp\left(\frac{2\pi\, i\, n\, t}{T}\right)$$

and the operator J_2 is given by the diagonalization operator.

The operators J_1 and J_2 are commutative and we use Section 13 of Chapter 1. □

7. Elliptic Regularization

7.1 The Elliptic Problem

In this section, we shall give a new proof of Theorem 1.1 by the so-called "elliptic regularization method", which consists in "approaching" the operator $\Lambda + M$ with a family of operators of a more "elliptic" nature.

Theorem 7.1. *Let* $\varepsilon > 0$ *be fixed. Under the hypotheses of Theorem* 1.1, *there exists a unique* $u_\varepsilon \in \mathscr{V} \cap D(\Lambda; \mathscr{H})$ *such that*

$$(7.1)\quad (\Lambda u_\varepsilon, v) + (M u_\varepsilon, v) + \varepsilon(\Lambda u_\varepsilon, \Lambda v) = (f, v), \quad \forall v \in \mathscr{V} \cap D(\Lambda; \mathscr{H}).$$

Proof. For $u, v \in \mathscr{V} \cap D(\Lambda; \mathscr{H})$, set

$$(7.2)\qquad \Pi_\varepsilon(u, v) = (\Lambda u, v) + (M u, v) + \varepsilon(\Lambda u, \Lambda v).$$

We have

$$\operatorname{Re} \Pi_\varepsilon(v, v) = \operatorname{Re}(\Lambda v, v) + \operatorname{Re}(M v, v) + \varepsilon \|\Lambda v\|_{\mathscr{H}}^2$$

and, according to (1.5)–(1.9), we therefore have

$$(7.3)\qquad \operatorname{Re} \Pi_\varepsilon(v, v) \geqq \alpha \|v\|_{\mathscr{V}}^2 + \varepsilon \|\Lambda v\|_{\mathscr{H}}^2,$$

whence the result, by Theorem 9.1 of Chapter 2. □

Remark 7.1. (7.1) is called the "regularized elliptic problem" for problem (1.10). □

Let us now verify the following estimates:

Proposition 7.1. *For* $\varepsilon > 0$, *we have*

$$(7.4)\qquad \|u_\varepsilon\|_{\mathscr{V}} \leqq c,$$

$$(7.5)\qquad \sqrt{\varepsilon}\, \|\Lambda u_\varepsilon\|_{\mathscr{H}} \leqq c,$$

$$(7.6)\qquad \Lambda u_\varepsilon \in D(\Lambda^*; \mathscr{V}'),$$

$$(7.7)\qquad \|\Lambda u_\varepsilon\|_{\mathscr{V}'} \leqq c.$$

Proof. (7.4) and (7.5) follow from (7.3).

(7.1) may be written in the form:

$$\varepsilon(\Lambda u_\varepsilon, \Lambda v) = (f - \Lambda u_\varepsilon - M u_\varepsilon, v), \qquad v \in \mathscr{V} \cap D(\Lambda; \mathscr{H}).$$

and we see that $v \to (\Lambda u_\varepsilon, \Lambda v)$ is continuous on the space

$$D(\Lambda; \mathscr{V}) \ (\subset \mathscr{V} \cap D(\Lambda; \mathscr{H}))$$

in the topology induced by $\mathscr{V}$ and therefore (Lemma 1.3) we have (7.6) and

$$\varepsilon(\Lambda u_\varepsilon, \Lambda v) = \varepsilon(\Lambda^* \Lambda u_\varepsilon, v),$$

from which we obtain

$$(7.8)\qquad \varepsilon \Lambda^* \Lambda u_\varepsilon + \Lambda u_\varepsilon + M u_\varepsilon = f.$$

Therefore, since, for $\varepsilon > 0$, the operator $\varepsilon \Lambda^* + I$ is invertible in $\mathscr{L}(\mathscr{V}', \mathscr{V}')$:

$$\Lambda u_\varepsilon = (\varepsilon \Lambda^* + I)^{-1} (f - M u_\varepsilon)$$

and therefore

$$\|\Lambda u_\varepsilon\|_{\mathscr{V}'} \leqq \|(\varepsilon \Lambda^* + I)^{-1}\|_{\mathscr{L}(\mathscr{V}';\mathscr{V}')} \|f - M u_\varepsilon\|_{\mathscr{V}'}. \tag{7.9}$$

But, according to (7.4), $\|f - M u_\varepsilon\|_{\mathscr{V}'}$ is bounded and since $-\Lambda^*$ is the infinitesimal generator of a bounded semi-group in $\mathscr{V}'$,

$$\|(\varepsilon \Lambda^* + I)^{-1}\|_{\mathscr{L}(\mathscr{V}';\mathscr{V}')} \leqq \text{constant},$$

we deduce (7.7) from (7.9) and (7.4). □

7.2 Passage to the Limit

Theorem 7.2. *Assume that* (1.5) *and* (1.9) *hold. Let* u_ε *be the solution of the "regularized elliptic problem"* (7.1). *Then, as* $\varepsilon \to 0$, *we have*:

$$u_\varepsilon \to u \quad \textit{in} \quad \mathscr{V}, \tag{7.10}$$

$$\Lambda u_\varepsilon \to \Lambda u \quad \textit{in} \quad \mathscr{V}', \tag{7.11}$$

where u *is the solution of* (1.10).

Remark 7.2. The proof which follows *again yields the existence of* u satisfying (1.10); the uniqueness must, at any rate, be shown separately as in 1) of Section 1.4. □

Proof of Theorem 7.2. From the estimates (7.4) and (7.7) it follows that we can extract u_η, $\eta \to 0$, so that

$$\begin{cases} u_\eta \to w & \text{weakly in } \mathscr{V}, \\ \Lambda u_\eta \to \chi & \text{weakly in } \mathscr{V}'. \end{cases} \tag{7.12}$$

But since Λ is closed in $\mathscr{V}'$, we see that

$$w \in D(\Lambda; \mathscr{V}'), \quad \chi = \Lambda w. \tag{7.13}$$

Now, $\Lambda \in \mathscr{L}(D(\Lambda; \mathscr{V}); \mathscr{V})$ yields, by transposition,

$$\Lambda^* \in \mathscr{L}(\mathscr{V}'; (D(\Lambda; \mathscr{V}))') \tag{7.14}$$

and therefore

$$\Lambda^* \Lambda u_\eta \to \Lambda^* \Lambda w \text{ weakly in } (D(\Lambda; \mathscr{V}'))',$$

and consequently

$$\eta \Lambda^* \Lambda u_\eta \to 0 \quad \text{in} \quad D(\Lambda; \mathscr{V}'))'$$

and (7.8) (for $\varepsilon = \eta$) yields

$$\Lambda w + M w = f, \quad w \in \mathscr{V} \cap D(\Lambda; \mathscr{V}')$$

in the limit.

Therefore $w = u$ and we have

$$u_\varepsilon \to u \text{ weakly in } \mathscr{V}, \quad \Lambda u_\varepsilon \to \Lambda u \text{ weakly in } \mathscr{V}'. \tag{7.15}$$

There remains to show the *strong* convergence. For this purpose, let

$$p_\varepsilon = (M(u_\varepsilon - u), u_\varepsilon - u) + (\Lambda(u_\varepsilon - u), u_\varepsilon - u) + \varepsilon(\Lambda u_\varepsilon, \Lambda u_\varepsilon).$$

According to (7.1) and since $\Lambda u + M u = f$:

$$p_\varepsilon = (f, u_\varepsilon) - (f, u_\varepsilon - u) - (M u_\varepsilon + \Lambda u_\varepsilon, u) \to 0, \quad \text{as} \quad \varepsilon \to 0.$$

But

$$\operatorname{Re} p_\varepsilon \geqq \alpha \, \|u_\varepsilon - u\|_{\mathscr{V}}^2,$$

whence (7.10).

Finally

$$\Lambda u_\varepsilon = (\varepsilon \Lambda^* + I)^{-1} (f - M u_\varepsilon) \to f - M u$$

strongly in $\mathscr{V}'$, whence the theorem. □

Remark 7.3. In the setting of Section 4, if for example $A(t) = -\Delta_x$ (Laplacian), the "initial" (non-regularized) problem being for example:

$$(7.16) \quad \begin{cases} -\Delta_x u + \dfrac{\partial u}{\partial t} = f, & x \in \Omega, \text{ an open set in } \mathbf{R}^n, \quad t \in]0, T[, \\ u(x, 0) = 0 & \\ u = 0, & \text{for } x \in \partial\Omega = \text{boundary of } \Omega, \quad t \in]0, T[, \end{cases}$$

the "regularized elliptic" problem is

$$(7.17) \quad \begin{cases} \Delta_x u_\varepsilon + \dfrac{\partial u_\varepsilon}{\partial t} - \varepsilon \dfrac{\partial^2 u_\varepsilon}{\partial t^2} = f \\ u_\varepsilon(x, 0) = 0 \\ \dfrac{\partial u_\varepsilon}{\partial t}(x, T) = 0 \\ u_\varepsilon = 0 \quad \text{for} \quad x \in \partial\Omega, \quad t \in]0, T[, \end{cases}$$

which is indeed a (mixed) *elliptic* problem.

If A is a differential operator of order $2m$, we can replace

$$A + \frac{\partial}{\partial t} \quad \text{by} \quad A + \frac{\partial}{\partial t} - \varepsilon \frac{\partial^2}{\partial t^2}$$

or by

$$A + \frac{\partial}{\partial t} + (-1)^m \varepsilon \frac{\partial^{2m}}{\partial t^{2m}};$$

similarly, in the general theory, the term $\varepsilon(\Lambda u_\varepsilon, \Lambda v)$ in (7.1) may be replaced by $\varepsilon(\Lambda^m u_\varepsilon, \Lambda^m v)$, *for example.* □

8. Equations of the Second Order in t

8.1 Notation

Let V and H be defined as in Section 4.1. Let $a(t; u, v)$ be a family of continuous sesquilinear forms on V, such that

$$\text{(8.1)} \quad \begin{cases} \text{the function } t \to a(t; u, v) \text{ is, } \forall u, v \in V, \text{ once continuously} \\ \text{differentiable in } [0, T], \text{ with } T < \infty, \text{ for example,} \end{cases}$$

$$\text{(8.2)} \quad \begin{cases} a(t; u, v) = \overline{a(t; u, v)}, \quad \forall u, v \in V, \quad \text{and for a suitable } \lambda: \\ a(t; v, v) + \lambda |v|^2 \geqq \alpha \|v\|^2, \quad \alpha > 0, \quad \forall v \in V. \end{cases}$$

Remark 8.1. The *hermitian symmetry* hypothesis: $a(t;u,v) = \overline{a(t;u,v)}$ can be generalized to the case for which only the "principal part" of $a(t; u, v)$ is hermitian. See, for example, Lions [13]. □

Let $A(t) \in \mathscr{L}(V; V')$ be the operator defined by $a(t; u, v)$ as in (4.5), (4.6).

We consider the equations

$$\text{(8.3)} \qquad A(t)\, u(t) + u''(t) = f(t), \qquad u'' = \frac{d^2}{dt^2} u,$$

with the Cauchy data:

$$\text{(8.4)} \qquad \begin{cases} u(0) = u_0, \\ u'(0) = u_1. \quad \square \end{cases}$$

8.2 Existence and Uniqueness Theorem

Theorem 8.1. *Assume that* (8.1) *and* (8.2) *hold. Let* f, u_0, u_1 *be given with*

$$\text{(8.5)} \qquad f \in L^2(0, T; H)$$

$$\text{(8.6)} \qquad u_0 \in V, \quad u_1 \in H.$$

Then there exists a unique function u *satisfying* (8.3), (8.4) *and*

$$\text{(8.7)} \qquad u \in L^2(0, T; V),$$

$$\text{(8.8)} \qquad u' \in L^2(0, T; H).$$

Remark 8.2. If (8.7) holds, then $A(t)\, u \in L^2(0, T; V')$, so that (8.3) implies

$$\text{(8.9)} \qquad u'' \in L^2(0, T; V').$$

Then $u(0)$ and $u'(0)$ are well-defined, so that (8.4) has meaning. □

The proof of Theorem 8.1 proceeds in two stages. In the first part, *energy inequalities* are established (see (8.15), below); in the second part, we show how to use these inequalities to obtain the *existence* of a solution (we give *one* method: Faedo-Galerkin, see Faedo [1] (and also Green [1]); there are others, see Lions [13] and the Comments; and also Section 8.4).

Uniqueness is treated separately; if the solution u is "regular", uniqueness is immediate. □

First part. *A priori estimates.*

Equation (8.3) is *equivalent to*

$$a(t; u(t), v) + [u''(t), v] = [f(t), v], \quad \forall v \in V. \tag{8.10}$$

Formally taking "$v = u'(t)$" in (8.10) and taking twice the real part of the equality thus obtained, we obtain (using the hermitian symmetry of $a(t; u, v)$):

(8.11)
$$a(t; u(t), u'(t)) + a(t; u'(t), u(t)) + \frac{d}{dt}|u'(t)|^2 = 2\,\mathrm{Re}[f(t), u'(t)].$$

Set

$$a'(t; u, v) = \frac{d}{dt} a(t; u, v), \quad \forall u, v \in V;$$

(8.11) is equivalent to

$$\frac{d}{dt}(a(t; u(t), u(t)) + |u'(t)|^2) - a'(t; u(t), u(t)) = 2\,\mathrm{Re}[f(t), u'(t)] \tag{8.12}$$

and integrating from 0 to t, we obtain:

$$a(t; u(t), u(t)) + |u'(t)|^2 = a(0; u_0, u_0) + \\ + |u_1|^2 + \int_0^T a'(\sigma; u(\sigma), u(\sigma))\, d\sigma + 2\,\mathrm{Re} \int_0^T [f(\sigma), u'(\sigma)]\, d\sigma. \tag{8.13}$$

The second term of (8.13) is less than or equal to (the c's denoting various constants):

$$c\, \|u_0\|^2 + |u_1|^2 + c \int_0^T \|u(\sigma)\|^2\, d\sigma + 2 \int_0^T |f(\sigma)|\, |u'(\sigma)|\, d\sigma$$

and the first term is $\geqq \alpha\, \|u(t)\|^2 + |u'(t)|^2$. Therefore:

$$\|u(t)\|^2 + |u'(t)|^2 \leqq c \left(\|u_0\|^2 + |u_1|^2 + \int_0^T |f(\sigma)|^2\, d\sigma \right) + \\ + c \int_0^T (\|u(\sigma)\|^2 + |u'(\sigma)|^2)\, d\sigma. \tag{8.14}$$

Applying Gronwall's lemma, it follows that

$$\| u(t) \|^2 + |u'(t)|^2 \leqq \tag{8.15}$$

$$\leqq c\left(\| u_0 \|^2 + |u_1|^2 + \int_0^T |f(\sigma)|^2 \, d\sigma \right), \quad 0 \leqq t \leqq T. \quad \square$$

Second part. *Proof of the existence.*

In order to slightly simplify matters (but there is nothing essential to this), let us assume V to be separable. Let $w_1, \ldots, w_m, \ldots$ form a "basis" for V in the following sense:

- for all m, $w_1, \ldots, w_m$ are linearly independent;
- the combinations $\sum_{\text{finite}} \xi_i w_i$, $\xi_i \in \mathbf{C}$, are dense in V.

(Such a basis always exists if V is separable.)

We define the "*approximate solution*" $u_m(t)$ *of order* m of the problem in the following way:

$$u_m(t) = \sum_{i=1}^{m} g_{im}(t)\, w_i, \tag{8.16}$$

the $g_{im}(t)$'s being determined so that

$$a(t; u_m(t), w_j) + [u''_m(t), w_j] = [f(t), w_j], \quad 1 \leqq j \leqq m, \tag{8.17}$$

$$g_{im}(0) = \xi_{im}, \quad g'_{im}(0) = \eta_{im} \tag{8.18}$$

with

$$\begin{cases} \sum_{i=1}^{m} \xi_{im} w_i \to u_0 & \text{in } V \text{ as } m \to \infty, \\ \sum_{i=1}^{m} \eta_{im} w_i \to u_1 & \text{in } H \text{ as } m \to \infty. \end{cases} \tag{8.19}$$

Thus, the g_{im}'s are determined by a *linear differential system* which admits a unique solution. The *same calculations* as in the first part show that

$$\| u_m(t) \|^2 + |u'_m(t)|^2 \leqq \textit{constant independent of } m. \tag{8.20}$$

Therefore, in *particular* (and it may be seen that the result cant be improved upon; see Section 8.4):

$$\begin{cases} u_m \text{ (resp. } u'_m\text{) remains in a bounded set of } L^2(0, T; V) \\ \text{(resp. } L^2(0, T; H)) \end{cases} \tag{8.21}$$

and we may therefore extract u_μ from u_m so that

$$u_\mu \to u \quad \text{weakly in} \quad L^2(0, T; V)$$

$$u'_\mu \to \chi \quad \text{weakly in} \quad L^2(0, T; H).$$

But $\chi = u'$.

There remains *to be shown that the function u constructed in this manner is a solution of the problem.*

To this end, we introduce

$$C_T^1[0, T] = \{\varphi \mid \varphi \in C^1([0, T)], \varphi(T) = \varphi'(T) = 0\}$$

and consider the functions

$$\psi = \sum_{j=1}^{\mu_0} \varphi_j \otimes w_j. \tag{8.22}$$

For $m = \mu > \mu_0$, we deduce from (8.17) (by multiplying by $\overline{\varphi_j(t)}$ and summing over j from 1 to μ_0) that

$$\int_0^T [a(t; u_\mu, \psi) - [u'_\mu, \psi']] \, dt = \int_0^T [f, \psi] \, dt + [u'_\mu(0), \psi(0)].$$

It follows from this and from the second condition in (8.19) that

$$\left\{ \begin{array}{l} \int_0^T (a(t; u, \psi) - [u', \psi']) \, dt = \int_0^T [f, \psi] \, dt + [u, \psi(0)], \\ \forall \psi \text{ of the form (8.22).} \end{array} \right. \tag{8.23}$$

But since the w_m's form a basis for V, the set of functions of the form (8.22) is dense in the space of functions $\psi \in L^2(0, T; V)$ such that

$$\psi' \in L^2(0, T; H), \quad \psi(T) = 0.$$

It follows from this that u is a solution of the problem. □

Remark 8.2. The preceding proof yields, without further effort, a more precise result than the one in the theorem: it follows from (8.20) that

$$\left\{ \begin{array}{l} u_m \text{ (resp. } u'_m) \text{ remains in a bounded set of } L^\infty(0, T; V) \\ \text{(resp. } L^\infty(0, T; H)), \end{array} \right. \tag{8.24}$$

from which it follows that

$$\left\{ \begin{array}{l} \textit{the solution } u \textit{ of problem } (8.3), (8.4), (8.7), (8.8) \textit{ satisfies} \\ u \in L^\infty(0, T; V), \quad u' \in L^\infty(0, T; H). \end{array} \right. \tag{8.25}$$

This result will be improved upon in Section 8.4. □

Proof of uniqueness in Theorem 8.1. Let $s \in]0, T[$. Set

$$\psi(t) = \left\{ \begin{array}{ll} -\int_t^s u(\sigma) \, d\sigma, & t < s, \\ 0 \quad , & t \geqq s, \end{array} \right.$$

where u is a solution of (8.3) – (8.4) (with (8.7) – (8.8)) for $f = 0$, $u_0 = 0$, $u_1 = 0$.

Then, obviously,

$$\int_0^T [A(t)\, u + u'', \psi]\, dt = 0$$

and by integration by parts (*permissible*), we obtain

$$\int_0^T [a(t; u, \psi) - [u', \psi']]\, dt = 0$$

or

$$\int_0^s [a(t, \psi', \psi) - [u', u]]\, dt = 0,$$

and thus

$$\int_0^s \frac{d}{dt}(a(t; \psi, \psi) - |u(t)|^2)\, dt - \int_0^s a'(t; \psi, \psi)\, dt = 0,$$

from which

$$a(0; \psi(0), \psi(0)) + |u(s)|^2 = \int_0^s a'(t; \psi; \psi)\, dt.$$

It follows that

$$\|\psi(0)\|^2 + |u(s)|^2 \leqq c_1 \left(\int_0^s \|\psi(t)\|^2\, dt + |\psi(0)|^2 \right).$$

But, if we set

$$w(t) = \int_0^t u(\sigma)\, d\sigma,$$

we may write this last inequality:

$$\|w(s)\|^2 + |u(s)|^2 \leqq c_1 \left(\int_0^s \|w(t) - w(s)\|^2\, dt + |w(s)|^2 \right),$$

whence

$$(1 - 2c_1 s)\, \|w(s)\|^2 + |u(s)|^2 \leqq c_2 \int_0^s (\|w(t)\|^2 + |u(t)|^2)\, dt.$$

Choose s_0 with, for example, $(1 - 2c_1 s_0) = \frac{1}{2}$.

Then, for $s \leqq s_0$, we have

$$\| w(s) \|^2 + |u(s)|^2 \leqq c_3 \int_0^s (\| w(t) \|^2 + |u(t)|^2)\, dt$$

therefore $u = 0$ in $[0, s_0]$.

The length of s_0 being independent of the choice of the origin, it follows that $u = 0$ in $[s_0, 2s_0]$ and so on, whence the uniqueness. □

8.3 Remarks on the Application of the General Theory of Section 1

8.3.1 "Vector" Notation

We shall investigate how the general method of Section 1 may be applied to the problem

$$\text{(8.26)} \qquad \begin{cases} A(t)\, u + u'' = f, \\ u(0) = u'(0) = 0. \end{cases}$$

Setting $u = u^1$, $\dfrac{du}{dt} = u^2$, and introducing

$$\vec{u} = \{u^1, u^2\},$$

we may write (8.26) in the form

$$\text{(8.27)} \qquad \frac{d\vec{u}}{dt} + \begin{pmatrix} 0 & -I \\ A(t) & 0 \end{pmatrix} \vec{u} = \{0, f\}, \qquad \vec{u}(0) = 0.$$

Replacing $\vec{u}$ by $\exp(k\,t)\,\vec{u}$ and setting $g = \exp(-k\,t)\,f$, system (8.27) is equivalent to

$$\text{(8.28)} \qquad \frac{d\vec{u}}{dt} + \begin{pmatrix} k\,I & -I \\ A(t) & k\,I \end{pmatrix} \vec{u} = \{0, g\}.$$

We choose (which is permissible):

$$\text{(8.29)} \qquad \begin{cases} k = \lambda + \mu, \\ a_\lambda(t; v, v) = a(t; v, v) + \lambda [v]^2 \geqq \alpha \| v \|^2 + \tfrac{1}{2} [v]^2 & \forall v \in V, \\ 2\mu\, a_\lambda(t; v, v) \geqq a'_\lambda(t; v, v) = a'(t; v, v) & \forall v \in V. \end{cases}$$

Then, we introduce

$$\text{(8.30)} \qquad \Lambda = \frac{d}{dt} + \mu \begin{pmatrix} I & 0 \\ 0 & I \end{pmatrix} = \frac{d}{dt} + \mu\, I$$

$$\text{(8.31)} \qquad M = \begin{pmatrix} \lambda I & -I \\ A(t) & \lambda I \end{pmatrix}$$

and (8.28) may be written

$$\Lambda \vec{u} + M \vec{u} = \{0, \vec{g}\}. \tag{8.32}$$

As we shall see in Section 8.3.2 below, *the theory of Section 1 does not apply to* (8.32), *but we shall indicate a simple "regularization" which allows the application of Section 1.*

8.3.2 Regularization

For $\varepsilon > 0$, we introduce

$$M_\varepsilon = \begin{pmatrix} \lambda I & -I \\ A(t) & \varepsilon(A(t) + \lambda I) + \lambda I \end{pmatrix} \tag{8.33}$$

and we show that the theory of Section 1 applies to

$$\Lambda \vec{u}_\varepsilon + M_\varepsilon \vec{u}_\varepsilon = \{0, g\}. \tag{8.34}$$

(It will then be easy to let $\varepsilon \to 0$.) □

The space $\mathscr{H}$.

We define:

$$\mathscr{H} = \{\vec{v} \mid v^1 \in L^2(0, T; V), v^2 \in L^2(0, T; H)\} \tag{8.35}$$

provided with the scalar product (thanks to the *symmetry of* $a(t; u, v)$, it is indeed a scalar product):

$$(\vec{u}, \vec{v}) = \int_0^T (a_\lambda(t; u^1, v^1) + [u^2, v^2])\, dt \tag{8.36}$$

(where $a_\lambda(t; u^1, v^1) = a(t; u^1, v^1) + \lambda[u^1, v^1]$).

Thanks to the second hypothesis in (8.29), this scalar product is *equivalent* to the "natural" scalar product:

$$\int_0^T ((u^1, v^1)_v + [u^2, v^2])\, dt.$$

The space $\mathscr{V}$:

$$\mathscr{V} = \{\vec{v} \mid v^1 \in L^2(0, T; V), v^2 \in L^2(0, T; V)\}, \tag{8.37}$$

provided with the scalar product

$$(\vec{u}, \vec{v})_{\mathscr{V}} = \int_0^T (a_\lambda(t; u^1, v^1) + (u^2, v^2)_v)\, dt.$$

The *space* $\mathscr{V}'$ is characterized by

$$\mathscr{V}' = \{\vec{v} \mid v^1 \in L^2(0, T; V), v^2 \in L^2(0, T; V')\}. \;\square \tag{8.38}$$

The operator M_ε.

If $\vec{v} \in \mathscr{V}$, we have

$$M_\varepsilon \vec{v} = \{\lambda v^1 - v^2, A(t) v^1 + \varepsilon(A(t) + \lambda) v^2 + \lambda v^2\} \in \mathscr{V}'$$

and

$$\operatorname{Re}(M_\varepsilon, \vec{v}, \vec{v}) = \operatorname{Re} \int_0^T (a_\lambda(t; \lambda v^1 - v^2, v^1) + [A(t) v^1 + \varepsilon(A(t) + \lambda) v^2 +$$

$$+ \lambda v^2, v^2]) \, dt = \lambda \int_0^T \{a(t; v^1, v^1) + \lambda [v^1]^2 - \operatorname{Re}[v^1, v^2] + [v^2]^2\} \, dt +$$

$$+ \operatorname{Re} \int_0^T \{- a(t; v^2, v^1) + [A(t) v^1, v^2]\} \, dt +$$

$$+ \varepsilon \int_0^T a_\lambda(t; v^2, v^2) \, dt$$

$$= \lambda \int_0^T \{a_\lambda(t; v^1, v^1) - \operatorname{Re}[v^1, v^2] + [v^2]^2\} \, dt +$$

$$+ \varepsilon \int_0^T a_\lambda(t; v^2, v^2) \, dt \geqq \lambda \int_0^T (\alpha \| v^1 \|^2 + \tfrac{1}{2} [v^2]^2) \, dt +$$

$$+ \varepsilon \alpha \int_0^T \| v^2 \|^2 \, dt.$$

Therefore

$$\operatorname{Re}(M, \vec{v}, \vec{v}) \geqq \alpha_1 \| \vec{v} \|_{\mathscr{H}}^2 + \varepsilon \alpha \int_0^T \| v^2 \|^2 \, dt \geqq \min(\alpha_1, \varepsilon \alpha) \| \vec{v} \|_{\mathscr{V}}^2. \tag{8.39}$$

Therefore hypothesis (1.9) is satisfied, thanks to the introduction of the regularizing term. □

The semi-group $G(s)$ and the operator Λ.

For $\vec{v} \in \mathscr{H}$ (for example), we define:

$$G(s) \vec{v} = \begin{cases} \vec{v}(t - s) \, e^{-\mu s}, & t > s \\ 0, & t < s \end{cases} \qquad t \in \,]0, T[. \tag{8.40}$$

The infinitesimal generator (formally) of $G(s)$ is of course given by (8.30), its domain consisting of the functions which vanish at the origin. The *fundamental point* is to see that $G(s)$ is a *contraction semi-group in* $\mathscr{H}$ (the *same* proof will yield the result in $\mathscr{V}$ and in $\mathscr{V}'$), *for the norm corresponding to* (8.36). That is, we have to verify that

$$\|G(s)\,v\|^2_{\mathscr{H}} = \int_s^T a_\lambda(t\,; v^1(t-s), v^1(t-s))\,e^{-2\mu s}\,dt + \\ + \int_s^T |v^2(t-s)|^2\,e^{-2\mu s}\,dt \leqq \\ \leqq \int_0^T a_\lambda(t; v^1(t), v^1(t))\,dt + \int_0^T |v^2(t)|^2\,dt.$$

Of course, it suffices to show that

$$e^{-2\mu s}\int_s^T a_\lambda(t; v^1(t-s), v^1(t-s))\,dt \leqq \int_0^T a_\lambda(t; v^1(t), v^1(t))\,dt$$

or (eliminating the index "1") that

(8.41) $$\Phi(s) = e^{-2\mu s}\int_0^{T-s} a_\lambda(t+s; v(t), v(t))\,dt \leqq \Phi(0).$$

But

$$\Phi'(s) = e^{-2\mu s}\left[-2\mu\int_0^{T-s} a_\lambda(t+s; v(t), v(t))\,dt + \right.\\ \left. + \int_0^{T-s} a'_\lambda(t+s; v(t), v(t))\,dt\right] - e^{-2\mu s}\,a_\lambda(T; v(T-s), v(T-s))$$

and, thanks to the third condition in (8.29), we have

(8.42) $$\Phi'(s) \leqq 0.$$

From which (8.41) and the desired result follow. □

Remark 8.3. Inequality (8.41) holds if

$$a(t+s; v, v) \leqq a(t; v, v), \quad \text{i.e. if}$$

(8.43) *the function* $t \to a(t; v, v)$ *is decreasing*, $\forall v \in V$.

If (8.43) *holds, then, with no differentiability hypothesis on* $a(t; v, v)$, *we have existence and uniqueness of a solution for problem* (8.34). □

We may now apply Section 1; we obtain

Proposition 8.1. *There exists a unique solution* $\vec{u}_\varepsilon \in \mathscr{V} \cap D(\Lambda; \mathscr{V}')$ *of* (8.39) *for* $\{0, g\} \in \mathscr{V}'$. □

Furthermore, if $g \in L^2(0, T; H)$, we deduce from (8.39) that

$$(8.44) \quad \begin{cases} \vec{u}_\varepsilon = \{u_\varepsilon^1, u_\varepsilon^2\} \text{ verifies, as } \varepsilon \to 0: \\ u_\varepsilon^1 \text{ remains in a bounded set of } L^2(0, T; V), \\ u_\varepsilon^2 \text{ remains in a bounded set of } L^2(0, T; H). \end{cases}$$

Therefore, we may extract from $\vec{u}_\varepsilon$ a sequence, still denoted by $\vec{u}_\varepsilon$, such that $\vec{u}_\varepsilon \to \vec{w}$ *weakly in* $\mathcal{H}$. But we can specifiy (8.34):

$$(8.45) \qquad \Lambda \vec{u}_\varepsilon + M \vec{u}_\varepsilon + \varepsilon\{0, (A(t) + \lambda) u_\varepsilon^2\} = \{0, g\}.$$

We introduce

$$L^2(0, T; D(A(t))) = \{v \mid v \in L^2(0, T; V), v(t) \in D(A(t)) = \text{domain of } A(t) \text{ in } H \text{ a.e., } A(t)\, v(t) \in L^2(0, T; H)\}.$$

Then

$$(A(t) + \lambda) u_\varepsilon^2 \to (A(t) + \lambda) w^2 \quad \text{weakly in} \quad (L^2(0, T; D(At)))'$$

and therefore

$$\varepsilon\{0, (A(t) + \lambda) u_\varepsilon^2\} \to 0$$

in, for example, the space

$$L^2(0, T; V) \times (L^2(0, T; D(A(t)))'$$

and (8.45) yields

$$(8.46) \qquad \Lambda \vec{w} + M \vec{w} = \{0, g\}, \qquad \vec{w} \in \mathcal{H}$$

in the limit.

If $\vec{w} = \{w^1, w^2\}$ and if we set $w = w^1$, w is a solution of

$$\frac{d^2 w}{dt^2} + A(t)\, w + 2k \frac{dw}{dt} + k^2 w = g$$

and satisfies $w(0) = 0$ (for $u_\varepsilon'(0) = 0$ and $u_\varepsilon'(0) \to w(0)$ weakly in H).

Finally, we verify that $w'(0) = 0$ by passage to relations "integrated in t", as at the end of the proof of Theorem 8.1. □

Remark 8.4. The preceding regularization could be called "*parabolic regularization*".

Of course, we may then apply *elliptic regularization* to equation (8.34), which yields

$$(8.47) \quad \Lambda \vec{u}_\varepsilon + M_\varepsilon \vec{u}_\varepsilon + \varepsilon_1 \Lambda^* \Lambda \vec{u}_\varepsilon = \{0, g\}, \qquad \varepsilon, \varepsilon_1 > 0. \quad □$$

8.4 Additional Regularity Results

Our aim is to show the following result:

Theorem 8.2. *Assume that* (8.1) *and* (8.2) *hold. Then, after possibly a modification on a set of measure zero, the solution* $\vec{u}$ *of* (8.3), (8.4), (8.7), (8.8) *satisfies*

$$\vec{u} = \left\{u, \frac{du}{dt}\right\} \in C^0([0, T]; V) \times C^0([0, T]; H), \tag{8.48}$$

$\{f, u_0, u_1\} \to \vec{u}$ *being a continuous mapping of*

$$L^2(0, T; H) \times V \times H \to C^0([0, T]; V) \times C^0([0, T]; H).$$

Remark 8.5. Other regularity theorems, for the "concrete" case of differential operators, will be found in Chapters 4 and 5 of Volume 2 (and results of M_k regularity, "abstract" and "concrete", in Volume 3). □

We shall first prove some lemmas.

Lemma 8.1. *Let* X, Y *be two Banach spaces,* $X \subset Y$ *with continuous injection,* X *being reflexive. Set:*

$C_S(0, T; Y)$ = space of functions $f \in L^\infty(0, T; Y)$ which are *scalarly* continuous[1] mappings of $[0, T] \to Y$.

Then

$$L^\infty(0, T; X) \cap C_S(0, T; Y) = C_S(0, T; X). \tag{8.49}$$

(Therefore if $f \in L^\infty(0, T; X) \cap C_S(0, T; Y)$, $f(t)$ – which is defined in Y for all t – is, in fact, in X and f is scalarly continuous and takes its values in X.)

Proof. 1) We show that $f(t) \in X$ and that

$$\| f(t) \|_X \leqq \text{constant}. \tag{8.50}$$

We may always assume that f is defined on $\mathbf{R}_t$, with properties analogous to those of f on $[0, T]$.

Let ϱ_n be a regularizing sequence of even functions of $\mathscr{D}(\mathbf{R}_t)$, with $\int_{\mathbf{R}_t} \varrho_n(t)\, dt = 1$.

Since $f \in L^\infty(\mathbf{R}_t; X)$, $f * \varrho_n$ satisfies

$$\| f * \varrho_n(t) \|_X \leqq \text{constant} = M.$$

Therefore, for arbitrary fixed t we can find a sequence ν such that (X being reflexive):

$$f * \varrho_\nu(t) \to \tilde{f}(t) \quad \text{weakly in } X$$

[1] That is $t \to \langle f(t), y' \rangle$ is continuous on $[0, T]$, $\forall y' \in Y'$, dual of Y.

and

$$\| \tilde{f}(t) \|_X \leqq M. \tag{8.51}$$

But, for arbitrary $y' \in Y'$, $t \to \langle f(t), y' \rangle$ is continuous, therefore

$$\langle f * \varrho_n(t) - f(t), y' \rangle = (\varrho_n * \langle f, y' \rangle)(t) - \langle f, y' \rangle(t) \to 0$$

and so, in particular

$$f * \varrho_\nu(t) \to f(t) \quad \text{weakly in } Y.$$

Therefore $\tilde{f}(t) = f(t)$ and (8.51) yields (8.50) with the constant equal to M.

2) Now we show that f is a *scalarly* continuous function of $[0, T] \to X$. If $t_n \to t$, we can (since $\| f(t_n) \|_X \leqq M$) extract t_ν such that $f(t_\nu) \to \chi$ weakly in X. But $f(t_\nu) \to f(t)$ weakly in Y, therefore $\chi = f(t)$ and the lemma is proved. □

Lemma 8.2. *If u satisfies* (8.3), (8.4), (8.7), (8.8) *we can always assume (after possibly a modification on a set of measure zero) that*

$$u \in C_S(0, T; V), \quad \frac{du}{dt} \in C_S(0, T; H). \tag{8.52}$$

Proof. It is known that $u \in L^\infty(0, T; V)$ and $u' \in L^\infty(0, T; H)$, therefore, in particular, u is (after possibly a modification on a set of measure zero) a continuous mapping of $[0, T] \to H$, therefore $u \in$ $\in C_S(0, T; V)$ by Lemma 8.1 with $X = V$ and $Y = H$.

Then, $u'' = f - A\,u \in L^2(0, T; V')$, therefore u' is a continuous mapping of $[0, T] \to V'$, whence $u' \in C_S(0, T; H)$ by another application of Lemma 8.1. □

Lemma 8.3. (Energy equality). *Let u verify* (8.3), (8.4), (8.7), (8.8) and (8.52). *Then, for all t:*

$$\left\{ \begin{aligned} & a(t; u(t), u(t)) + |u'(t)|^2 = a(0; u_0, u_0) + |u_1|^2 + \\ & \qquad + \int_0^t a'(\sigma; u, u)\, d\sigma + 2 \operatorname{Re} \int_0^t (f, u')\, d\sigma. \end{aligned} \right. \tag{8.53}$$

Proof. — 1) *Notation.* We take $t = t_0$ in (8.53).

We introduce $\mathcal{O} = \mathcal{O}_\delta = \mathcal{O}_\delta(t) = \{1$ in $[\delta, t_0 - \delta]$, 0 outside $[0, t_0]$, linear on $[0, \delta]$ and $[t_0 - \delta, t_0]$, continuous on $\mathbf{R}_t\}$; $\varrho = \varrho_n =$ regularizing sequence of even functions, $\int_{\mathbf{R}_t} \varrho_n(t)\, dt = 1$.

We shall assume that $A(t)$ is defined for all $t \in \mathbf{R}$ — with the same properties as on $[0, T]$.

In the same way, we shall assume that u is defined on $\mathbf{R}_t$, with properties analogous to those of u on $[0, T]$ (which is permissible by "extension by reflexion").

We recall that $[\ ,\ ]$ denotes the scalar product in H or the antiduality V', V; we shall denote $(\ ,\ )$ the antiduality $L^2(\mathbf{R}_t; V')$, $L^2(\mathbf{R}_t; V)$ or the scalar product in $L^2(\mathbf{R}_t; H)$.

2) We have

$$(8.54)\quad \begin{cases} (A'(t)(\varrho * (\mathcal{O}_0 u)), \varrho * (\mathcal{O}_0 u)) + 2\,\mathrm{Re}(\varrho * (\mathcal{O}_0 A u), \varrho * (\mathcal{O}_0 u')) + \\ + 2\,\mathrm{Re}(A(\varrho * (\mathcal{O}_0 u)) - \varrho * (A\, \mathcal{O}_0 u), \varrho' * (\mathcal{O}_0 u)) + \\ + 2\,\mathrm{Re}[(\varrho * \varrho * (\mathcal{O}_0 A u))(0), u(0)] - \\ - 2\,\mathrm{Re}[(\varrho * \varrho (\mathcal{O}_0 A u))(t_0), u(t_0)] = 0. \end{cases}$$

Indeed, starting with

$$\int_{-\infty}^{+\infty} \frac{d}{dt}[A(t)(\varrho * (\mathcal{O} u)), \varrho * (\mathcal{O} u)]\, dt = 0,$$

we obtain

$$2\,\mathrm{Re}(A(\varrho * \mathcal{O} u), (\varrho * \mathcal{O} u)') + (A'(\varrho * \mathcal{O} u), \varrho * \mathcal{O} u) = 0$$

and thus

$$(8.55)\quad \begin{cases} (A'(\varrho * \mathcal{O} u), \varrho * \mathcal{O} u) + 2\,\mathrm{Re}(\varrho * (A\, \mathcal{O} u), \varrho * (\mathcal{O} u')) + \\ + 2\,\mathrm{Re}(\varrho * (A\, \mathcal{O} u), \varrho * (\mathcal{O}' u)) + \\ + 2\,\mathrm{Re}(A(\varrho * (\mathcal{O} u)) - \varrho * (A\, \mathcal{O} u), \varrho' * (\mathcal{O} u)) = 0. \end{cases}$$

Now we let $\delta \to 0$ in (8.55).

In the first, second and fourth terms of (8.55), it suffices to replace $\mathcal{O}$ with $\mathcal{O}_0$. The third term may be written

$$(8.56)\quad 2\,\mathrm{Re}(\varrho * (A(\mathcal{O} - \mathcal{O}_0) u), \varrho * \mathcal{O}' u) + 2\,\mathrm{Re}(\varrho * (A\, \mathcal{O}_0 u), \varrho * \mathcal{O}' u).$$

But $(\mathcal{O} - \mathcal{O}_0) u \to 0$ in $L^2(\mathbf{R}_t; V)$, therefore $\varrho * (A(\mathcal{O} - \mathcal{O}_0) u) \to 0$ in $L^\infty(\mathbf{R}_t; V')$ (the supports are compact) and since $\varrho * (\mathcal{O}' u)$ is bounded in $L'(\mathbf{R}_t; V)$ $\left(\text{since} \int_{\mathbf{R}_t} |\mathcal{O}'|\, dt = 2\right)$, the first term of (8.56) goes to zero.

The second term is equal to

$$2\,\mathrm{Re}(\varrho * \varrho * (A\, \mathcal{O}_0 u), \mathcal{O}' u).$$

But $t \to [\varrho * \varrho * (A\, \mathcal{O}_0 u)(t), u(t)]$ is *continuous*, so that this expression tends towards

$$2\,\mathrm{Re}[(\varrho * \varrho * (\mathcal{O}_0 A u))(0), u(0)] - 2\,\mathrm{Re}[(\varrho * \varrho * (\mathcal{O}_0 A u)(t), u(t))].$$

Whence (8.54).

3) We have

$$(8.57)\quad \begin{cases} 2\operatorname{Re}(\varrho * (\mathcal{O}_0 u''), \varrho * (\mathcal{O}_0 u')) + \\ + 2\operatorname{Re}[(\varrho * \varrho * (\mathcal{O}_0 u'))(0), u'(0)] - \\ - 2\operatorname{Re}[(\varrho * \varrho * (\mathcal{O}_0 u'))(t_0), u'(t_0)] = 0. \end{cases}$$

The idea of the *proof* is the same as for (8.54). We start with

$$\int_{-\infty}^{+\infty} \frac{d}{dt} (\varrho * (\mathcal{O} u'), \varrho * (\mathcal{O} u'))\, dt = 0$$

from which we obtain

$$(8.58)\quad \begin{cases} 2\operatorname{Re}(\varrho * (\mathcal{O} u'), \varrho * (\mathcal{O} u'')) + 2\operatorname{Re}(\varrho * (\mathcal{O} - \mathcal{O}_0 u'), \varrho * \mathcal{O}' u') + \\ + 2\operatorname{Re}(\varrho * (\mathcal{O}_0 u'), \varrho * (\mathcal{O}' u')) = 0. \end{cases}$$

As $\delta \to 0$, the second term of (8.58) $\to 0$, the first tends towards

$$2\operatorname{Re}(\varrho * (\mathcal{O}_0 u'), \varrho * (\mathcal{O}_0 u'')).$$

The last term is equal to $2\operatorname{Re}(\varrho * \varrho\, (\mathcal{O}_0 u'), \mathcal{O}' u')$ and tends towards

$$2\operatorname{Re}[\varrho * \varrho * (\mathcal{O}_0 u')(0), u'(0)] - 2\operatorname{Re}[\varrho * \varrho * \mathcal{O}_0 u'(t_0), u'(t_0)],$$

since $t \to [\varrho * \varrho * \mathcal{O}_0 u'(t), u'(t)]$ is *continuous*. Whence (8.57).

4) We add (8.54) to (8.57); taking account of the fact that $A(t) u + u'' = f$, we obtain:

$$(8.59)\quad \begin{cases} (A'(\varrho * (\mathcal{O}_0 u)), \varrho * (\mathcal{O}_0 u)) + 2\operatorname{Re}(\varrho * (\mathcal{O}_0 f), \varrho * (\mathcal{O}_0 u')) + \\ + 2\operatorname{Re}((A(\varrho * (\mathcal{O}_0 u)) - \varrho * (A\, \mathcal{O}_0 u))', \varrho * \mathcal{O}_0 u) + \\ + 2\operatorname{Re}[\varrho * \varrho * (\mathcal{O}_0 A u)(0), u(0)] - \\ - 2\operatorname{Re}[\varrho * \varrho * (\mathcal{O}_0 A u)(t_0), u(t_0)] + \\ + 2\operatorname{Re}[\varrho * \varrho * (\mathcal{O}_0 u')(0), u'(0)] - \\ - 2\operatorname{Re}[\varrho * \varrho * (\mathcal{O}_0 u')(t_0), u'(t_0)] = 0. \end{cases}$$

According to the "vector" Friedrichs' Lemma (see Lions [13], Lemma 7.2, p. 72, for example), the third term of (8.59) tends towards 0.

The first and second terms tend towards $(A' \mathcal{O}_0, \mathcal{O}_0 u) + 2\operatorname{Re}(\mathcal{O}_0 f, \mathcal{O}_0 u')$ and therefore then remains only to pass to the limit in the four last terms. Since 0 and t_0 play symmetrical roles, we shall have (8.53) if we can verify that (setting $\varrho * \varrho = \sigma$):

$$(8.60)\quad 2\operatorname{Re}[\sigma * (\mathcal{O}_0 A u)(t_0), u(t_0)] \to a(t_0; u(t_0), u(t_0)),$$

$$(8.61)\quad 2\operatorname{Re}[\sigma * (\mathcal{O}_0 u')(t_0), u'(t_0)] \to |u'(t_0)|^2.$$

But $(\sigma = \sigma_n)$:

$$2\,\mathrm{Re}[\sigma_n * (\mathcal{O}_0 A\, u)\,(t_0),\, u(t_0)] = 2\,\mathrm{Re}\int_0^{t_0} \sigma_n(t)\,[A\, u(t_0 - t),\, u(t_0)]\,dt$$

and since σ_n is *even*,

$$\int_0^{t_0} \sigma_n(t)\,dt = \tfrac{1}{2},$$

therefore

$$2\,\mathrm{Re}[\sigma_n * (\mathcal{O}_0 A\, u)\,(t_0),\, u(t_0)] - [A(t_0)\, u(t_0),\, u(t_0)]$$

$$= 2\,\mathrm{Re}\int_0^{t} \sigma_n(t)\,[(A\, u)\,(t_0 - t) - (A\, u)\,(t_0),\, u(t_0)]\,dt,$$

which goes to 0 as $\delta_n \to \delta$, whence (8.60) — and analogously for (8.61). This ends the proof of Lemma 8.3. □

Proof of Theorem 8.2. We already know that (8.52) holds.

On the other hand, according to (8.53), the function

$$t \to a(t;\, u(t),\, u(t)) + |u'(t)|^2$$

is *continuous* on $[0, T]$ and since $t \to u(t)$ is a strongly continuous function of $[0, T] \to H$, we see that

(8.62) $\quad t \to \varphi(t) = a_\lambda(t;\, u(t),\, u(t)) + |u'(t)|^2$ *is continuous on* $[0, T]$.

Let $t_n \to t$ and

$$\xi_n = |u'(t_n) - u'(t)|^2 + a_\lambda(t_n;\, u(t_n) - u(t),\, u(t_n) - u(t)).$$

We have

$$\xi_n = \varphi(t_n) + \varphi(t) + 2\,\mathrm{Re}[a_\lambda(t_n;\, u(t),\, u(t)) - a_\lambda(t;\, u(t),\, u))] -$$
$$- 2\,\mathrm{Re}[u'(t_n),\, u'(t)] - 2\,\mathrm{Re}\, a_\lambda(t;\, u(t_n),\, u(t)) -$$
$$- 2\,\mathrm{Re}\,[a_\lambda(t_n;\, u(t_n),\, u(t)) - a_\lambda(t;\, u(t_n),\, u(t))].$$

But

$$|a_\lambda(t_n;\, u(t),\, u(t)) - a_\lambda(t;\, u(t),\, u(t))| \leqq C\,|t - t_n|,$$
$$|a_\lambda(t_n;\, u(t_n),\, u(t)) - a_\lambda(t;\, u(t_n),\, u(t))| \leqq C\,|t - t_n|,$$

and therefore

$$\xi_n \to 2\varphi(t) - 2\,\mathrm{Re}(|u'(t)|^2 + a_\lambda(t,\, u(t),\, u(t))) = 0.$$

Since $\xi_n \geqq |u'(t_n) - u'(t)|^2 + \alpha\, \|u(t_n) - u(t)\|^2$, we have the theorem. □

8.5 Parabolic Regularization; Direct Method and Application

We take up problem (8.3), (8.4) again and associate to it, as in Section 8.3, the *regularized parabolic problem*

$$A(t)\, u_\varepsilon + u_\varepsilon'' + \varepsilon(A(t) + \lambda)\, u_\varepsilon' = f, \quad \varepsilon > 0, \tag{8.63}$$

$$u_\varepsilon(0) = 0, \quad u_\varepsilon'(0) = u_1. \tag{8.64}$$

Theorem 8.3. *Assume that* (8.1), (8.2), (8.5), (8.6) *hold.*

1) *Then, for every* $\varepsilon > 0$, *problem* (8.63), (8.64) *admits a unique solution* u_ε *which satisfies*

$$u_\varepsilon \in C^0([0, T]; V), \tag{8.65}$$

$$u_\varepsilon' \in L^2(0, T; V) \cap C^0([0, T]; H). \tag{8.66}$$

2) *As* $\varepsilon \to 0$,

$$\begin{cases} u_\varepsilon \to u & \textit{uniformly in} \quad C^0([0, T]; V), \\ u_\varepsilon' \to u' & \textit{uniformly in} \quad C^0([0, T]; H), \end{cases} \tag{8.67}$$

where u *is the solution of* (8.3), (8.4), (8.7), (8.8) (solution which belongs to $C^0([0, T]; V)$ and has derivatives in $C^0([0, T]; H)$, according to Theorem 8.2).

Proof of Theorem 8.3, *first part.* As in Section 8.2, we can use the *method* of Faedo-Galerkin. In the notation of Section 8.2, we introduce $u_{\varepsilon m}(t) = \sum_{i=1}^{m} g_{im}(t)\, w_i$, satisfying

$$a(t; u_{\varepsilon m}, w_j) + [u_{\varepsilon m}'', w_j] + \varepsilon\, a_\lambda(t; u_{\varepsilon m}', w_j) = [f(t), w_j], \quad 1 \leqq j \leqq m; \tag{8.68}$$

$$(a_\lambda(t; u, v) = a(t; u, v) + \lambda[u, v]), \quad \text{with} \quad g_{im}(0), g_{im}'(0)$$

as in (8.18), (8.19).

Multiplying (8.68) by $\overline{g_{im}'(t)}$ and summing over j, we obtain:

$$\begin{cases} \dfrac{d}{dt}\big(a(t; u_{\varepsilon m}(t), u_{\varepsilon m}(t)) + |u_{\varepsilon m}'(t)|^2\big) + 2\varepsilon\, a_\lambda(t; u_{\varepsilon m}', u_{\varepsilon m}') - \\ \quad - a'(t; u_{\varepsilon m}, u_{\varepsilon m}) = 2\,\mathrm{Re}\big[(f(t), u_{\varepsilon m}'(t))\big], \end{cases} \tag{8.69}$$

from which we deduce (as in Section 8.2, first part of the existence proof) that

$$\begin{cases} \text{as } m \to \infty,\ u_{\varepsilon m} \text{ (resp. } u_{\varepsilon m}') \text{ remains in a bounded set of} \\ L^\infty(0, T; V) \text{ (resp. } L^2(0, T; V) \cap L^\infty(0, T; H)). \end{cases} \tag{8.70}$$

We then extract $u_{\varepsilon\mu}$ such that

$$u_{\varepsilon\mu} \to \tilde{u}_\varepsilon \quad \textit{weak star in} \quad L^\infty(0, T; V)$$

$$\frac{du_{\varepsilon\mu}}{dt} \to \frac{d\tilde{u}_\varepsilon}{dt} \quad \textit{weak star in} \quad L^\infty(0, T; H) \quad \text{and} \quad L^2(0, T; V).$$

We verify that $\tilde{u}_\varepsilon$ is a solution of (8.63), (8.64).

We already know that $\tilde{u}_\varepsilon \in L^\infty(0, T; V)$, $\dfrac{d\tilde{u}_\varepsilon}{dt} \in L^2(0, T; V)$, therefore $\tilde{u}_\varepsilon$ is (after possibly a modification on a set of measure zero) a continuous function of $[0, T] \to V$. Now

$$\tilde{u}_\varepsilon'' = f - A(t)\,\tilde{u}_\varepsilon - \varepsilon(A(t) + \lambda)\,\tilde{u}_\varepsilon'$$

and since $\tilde{u}_\varepsilon' \in L^2(0, T; V)$, we have

$$\tilde{u}_\varepsilon'' \in L^2(0, T; V'), \tag{8.71}$$

which, together with $\tilde{u}_\varepsilon' \in L^2(0, T; V)$ and Theorem 3.1 of Chapter 1, shows that (after possibly a modification on a set of measure zero) $\tilde{u}_\varepsilon'$ is a continuous function of $[0, T] \to H$. This shows the *existence* of a solution $u_\varepsilon = \tilde{u}_\varepsilon$ having the stated properties. □

Let us verify the *uniqueness*. In fact, we obtain from (8.63) that

$$\frac{d}{dt}\left(a(t; u_\varepsilon(t), u_\varepsilon(t)) + |u_\varepsilon'(t)|^2\right) - a'(t; u_\varepsilon(t), u_\varepsilon(t)) +$$
$$+ 2\varepsilon\, a_\lambda(t; u_\varepsilon'(t), u_\varepsilon'(t)) = 2\,\mathrm{Re}[f(t), u_\varepsilon'(t)] \quad \text{a.e.,}$$

whence

$$\left\{\begin{aligned} & a(t; u_\varepsilon(t), u_\varepsilon(t)) + |u_\varepsilon'(t)|^2 - \int_0^t a'(\sigma; u_\varepsilon(\varrho), u_\varepsilon(\varrho))\, d\sigma + \\ & + 2\varepsilon \int_0^t a_\lambda(\sigma; u_\varepsilon'(\sigma), u_\varepsilon'(\sigma))\, d\sigma = a(0; u_0, u_0) + |u_1|^2 + \\ & + 2\,\mathrm{Re} \int_0^t [f(\sigma), u_\varepsilon'(\sigma)]\, d\sigma, \end{aligned}\right. \tag{8.72}$$

this last equality in fact holding *everywhere*, thanks to the continuity properties of u_ε and u_ε'.

Uniqueness follows immediately from (8.72): if $u_0 = 0$, $u_1 = 0$, $f = 0$, it follows that $u_\varepsilon = 0$. □

Proof of Theorem 8.3, second part. It follows from (8.72) that

$$\left\{\begin{aligned} & u_\varepsilon \text{ (resp. } u_\varepsilon') \text{ remains in a bounded set of } L^\infty(0, T; V) \\ & (\text{resp. } L^\infty(0, T; H)) \text{ as } \varepsilon \to 0 \end{aligned}\right. \tag{8.73}$$

and that

(8.74) $\sqrt{\varepsilon}\, u'_\varepsilon$ remains in a bounded set of $L^2(0, T; V)$.

Therefore, we can extract a sequence — still denoted u_ε — such that

$$u_\varepsilon \to \tilde{u} \quad \textit{weak star} \text{ in } \quad L^\infty(0, T; V)$$

$$u'_\varepsilon \to (\tilde{u})' \quad \textit{weak star} \text{ in } \quad L^\infty(0, T; H).$$

Then

$$u''_\varepsilon = f - A(t)\, u_\varepsilon - \varepsilon(A(t) + \lambda) u'_\varepsilon \to f - A(t)\, \tilde{u} \quad \text{weakly in} \quad L^2(0, T; V')$$

(for example), therefore $(\tilde{u})'' + A(t)\, \tilde{u} = f$ and

$$u_\varepsilon(0) \to \tilde{u}(0) \quad \text{weakly in } H \text{ (for example)}$$

and

$$u'_\varepsilon(0) \to (\tilde{u})'(0) \quad \text{weakly in } V' \text{ (for example)},$$

so that $\tilde{u}(0) = u_0$, $(\tilde{u})'(0) = u_1$, therefore $\tilde{u} = u$ and finally we have, without extracting a subsequence:

$$\text{(8.75)} \quad \begin{cases} u_\varepsilon \to u \text{ (resp. } u'_\varepsilon \to u') \textit{ weak star} \text{ in } L^\infty(0, T; V) \\ \text{(resp. } \textit{weak star} \text{ in } L^\infty(0, T; H)). \end{cases}$$

There remains to show the uniform convergence.

We introduce

$$\text{(8.76)} \qquad \varphi_\varepsilon = u_\varepsilon - u;$$

we have

$$\text{(8.77)} \quad \begin{cases} \varphi''_\varepsilon + A(t)\, \varphi_\varepsilon = -\varepsilon(A(t) + \lambda)\, u'_\varepsilon, \\ \varphi_\varepsilon \in L^\infty(0, T; V), \quad \varphi'_\varepsilon \in L^\infty(0, T; H), \quad \varphi_\varepsilon(0) = 0, \quad \varphi'_\varepsilon(0) = 0. \end{cases}$$

According to the energy equality (Lemma 8.3):

$$a(t; \varphi_\varepsilon(t), \varphi_\varepsilon(t)) + |\varphi'_\varepsilon(t)|^2 = \int_0^t a'(\sigma; \varphi_\varepsilon, \varphi_\varepsilon)\, d\sigma - \varepsilon \int_0^t a_\lambda(\sigma; u'_\varepsilon, \varphi_\varepsilon)\, d\sigma,$$

whence

$$\|\varphi_\varepsilon(t)\|^2 + |\varphi'_\varepsilon(t)|^2 \leqq C_1 \int_0^t \|\varphi_\varepsilon(\sigma)\|^2\, d\sigma + C_1\, \varepsilon \left| \int_0^t a_\lambda(\sigma; u'_\varepsilon, \varphi_\varepsilon)\, d\sigma \right|.$$

By Gronwall's inequality, it follows that

$$\text{(8.78)} \quad \|\varphi_\varepsilon(t)\|^2 + |\varphi'_\varepsilon(t)|^2 \leqq (\exp(C\, t))\, C_1\, \varepsilon \left| \int_0^T a_\lambda(\sigma; u'_\varepsilon, \varphi_\varepsilon)\, d\sigma \right| \leqq$$

$$\leqq \text{(according to (8.74))}\; C_2\, \varepsilon^{1/2} \left(\int_0^T \|\varphi_\varepsilon(t)\|^2\, dt \right)^{1/2} \leqq C_3\, \varepsilon^{1/2},$$

from which (8.67) follows. □

9. Equations of the Second Order in t; Transposition

9.1 Adjoint Isomorphism

As usual (see Section 2.2), we first consider the adjoint isomorphism. For φ given in $L^2(0, T; H)$, let v be the solution of

$$\begin{cases} A(t)\, v + v'' = \varphi, \\ v(T) = v'(T) = 0. \end{cases} \tag{9.1}$$

We define:

$$\begin{cases} X = \text{space described by the solution } v \text{ of (9.1) as } \varphi \text{ describes} \\ L^2(0, T; H). \end{cases} \tag{9.2}$$

According to Theorems 8.1 and 8.2, the functions of X have the following properties:

$$\begin{cases} v \in C^0([0, T]; V), \quad v' \in C^0([0, T]; H), \\ v'' \in L^2(0, T; V'). \end{cases} \tag{9.3}$$

Providing X with the topology carried over by the mapping $\varphi \to v$, we have (done the necessary to have) the result:

$$A(t) + \frac{d^2}{dt^2} \quad \textit{is an isomorphism of } X \textit{ onto } L^2(0, T; H). \tag{9.4}$$

9.2 Transposition

From (9.4) we deduce

Theorem 9.1. *Hypotheses of Theorem 8.1. Let* L *be a continuous antilinear form on* X. *There exists a unique* $u \in L^2(0, T; H)$ *such that*

$$(u, A(t)\, v + v'') = (L, v) \qquad \forall v \in X, \tag{9.5}$$

(where the first parentheses denote the antiduality between $L^2(0, T; H)$ and itself and the second between X' and X). □

Now we have to *choose* L.

Formally (compare with (4.5); see also (3.11), Chapter 6, Volume 2) we shall take

$$(L, v) = \int_0^T [f(t), v(t)]\, dt + [u_1, v(0)] - [u_0, v'(0)], \tag{9.6}$$

giving suitable meanings to the various scalar products occuring in this formula. Still *formally*, we then see that the solution u of (9.5) satisfies

$$A(t)\, u + u'' = f \tag{9.7}$$

$$u(0) = u_0, \quad u'(0) = u_1. \tag{9.8}$$

Now we have to make this rigorous.

9.3 Choice of L

Compare with Section 4.5 and with the introduction of Ξ in (4.24). Let ϱ be the function defined by (4.23). We define the space Ξ by

$$\text{(9.9)}\quad \begin{cases} \Xi = \{v \mid v \in L^2(0,T;V),\ \varrho\, v' \in L^2(0,T;H), \\ \varrho^2 v'' \in L^2(0,T;V'),\ v(T) = v'(T) = 0\} \end{cases}$$

(provided with norm

$$\left(\int_0^T (\|v(t)\|^2 + \varrho^2\,|v'(t)|^2 + \varrho^4\,\|v''(t)\|^2_{V'})\,dt\right)^{1/2}$$

which makes it a Hilbert space). We have:

Proposition 9.1. *The space $\mathcal{D}(]0,T[;V)$ is dense in Ξ (defined by (9.9)).*

Proof. Let $\mathcal{O}_n$ be a sequence of scalar functions belonging to $C^2([0,T])$, $\mathcal{O}_n$ vanishing in $(0,\varepsilon_n)$,

$\varepsilon_n \to 0$ if $n \to \infty$, $\mathcal{O}_n = 1$ if $t \geqq 2\varepsilon_n$ with $|\varrho\,\mathcal{O}_n'| + |\varrho^2\,\mathcal{O}_n''| \leqq C$.

Then if $v \in \Xi$, $\mathcal{O}_n v \to v$ in Ξ (since, as is easily seen,

$$\varrho\,\mathcal{O}_n' v \to 0 \quad\text{in}\quad L^2(0,T;H),$$
$$\varrho^2\,\mathcal{O}_n'' v + 2\varrho^2\,\mathcal{O}_n' v' \to 0 \quad\text{in}\quad L^2(0,T;V')).$$

Since $v(T) = 0$, $v'(T) = 0$, we may truncate in the neighborhood of T; finally, by regularization in t, we obtain the desired result. □

Corollary 9.1. *The space Ξ' (antidual of Ξ) is a space of distributions on $[0,T]$ taking their values in V'. Every $f \in \Xi'$ may be represented, non-uniquely, by*

$$\text{(9.10)}\qquad f = f_0 + \frac{d}{dt}(\varrho\, f_1) + \frac{d^2}{dt^2}(\varrho^2 f_2),$$

where

$$f_0 \in L^2(0,T;V'),\qquad f_1 \in L^2(0,T;H),\qquad f_2 \in L^2(0,T;V).\ \square$$

Since, still according to (9.3), we have

$$\text{(9.11)}\qquad X \subset \Xi,$$

we see that

$$\text{(9.12)}\quad \begin{cases} \text{for } f \in \Xi',\ v \to (f,v) \text{ (scalar product between } \Xi' \text{ and } \Xi) \\ \text{is a continuous antilinear form on } X.\ \square \end{cases}$$

Also, since $v \in C^0([0, T]; V)$ and $v' \in C^0([0, T]; H)$, we may take

(9.13) $$u_1 \in V', \quad u_0 \in H$$

in (9.6).

Summing up, we have:

Proposition 9.2. *In* (9.6), *we may take* $f \in \Xi'$, $u_0 \in H$ *and* $u_1 \in V'$. *In particular, we may take* $f = f_0 \in L^2(0, T; V')$. □

Now the problem is to *interpret equation* (9.5) for the preceding choice of L, which we shall do (in opposition to what we have done in the parabolic case, see Section 4.6), for a particular choice of f ($f \in L^2(0, T; V')$). The general case presents some difficulties (see Problem 13.11).

9.4 Trace Theorem

We shall make an *additional hypothesis*:

(9.14) $$\begin{cases} \text{the domain } D(A(t)) \text{ of } A(t) \text{ in } H \text{ is independent of } t \text{ and, } \forall v \in H, \\ t \to A(t)\, A(0)^{-1} v \text{ is a continuous function of } [0, T] \to H. \end{cases}$$

Let us be somewhat more precise; $D(A(t))$ in H is the set of u's $\in V$ such that $A(t)\, u \in H$; it is assumed to be provided with the norm of the graph:

$$(|u|^2 + |A(t)\, u|^2)^{1/2} = \|u\|_{D(A(t))};$$

then $D(A(t)) = D(A(0))$ with norm equivalence, and this uniformly in t, therefore

$$C^{-1} \|u\|_{D(A(0))} \leqq \|u\|_{D(A(t))} \leqq C \|u\|_{D(A(0))}$$
$$\forall u \in D(A(0)), \quad t \in [0, T]. \ \square$$

Remark 9.1. In the applications (see, among others, Chapters 4 and 5 of Volume 2), hypothesis (9.14) is satisfied when $A(t)$ is an elliptic operator to which one associates the *Dirichlet boundary conditions.* □

Under hypothesis (9.14), we have

(9.15) $$A(t) \in \mathscr{L}(D(A(0)); H)$$

and by transposition, since $A(t) = A^*(t)$:

(9.16) $$A(t) \in \mathscr{L}(H; D(A(0))').$$

Furthermore, $\|A(t)\|_{\mathscr{L}(D(A(0)); H)} \leqq$ constant and therefore the same is true for $\|A(t)\|_{\mathscr{L}(H; D(A(0))')}$. Also, $D(A(0))$ is dense in V, therefore:

(9.17) $$D(A(0)) \subset V \subset H \subset V' \subset D(A(0))'.$$

Lemma 9.1. *If* u *is given in* $L^2(0, T; H)$, *we have*:

$$A(t)\, u \in L^2(0, T; D(A(0))').$$

Proof. According to the preceding remarks, we only need to show that $t \to A(t)\, u(t)$ is a *measurable* function of $[0, T] \to D(A(0))'$. Since we have assumed (see Section 4.2) that V is separable, H is separable and, $A(0)$ being an isomorphism of $D(A(0))$ onto H, $D(A(0))$ is separable. Therefore, it is sufficient to show that $A(t)\, u(t)$ is scalarly measurable, i.e. that for arbitrary $v \in D(A(0))$, the function

$$t \to (A(t)\, u(t), v)$$

is measurable. But it is equal to $(u(t), A(t)\, v)$ and it suffices to verify that $t \to A(t)\, v$ is measurable with values in H.

But $t \to A(t)\, v = A(t)\, A(0)^{-1}(A(0)\, v)$ is *continuous* in H, according to (9.14), whence the result. □

Remark 9.2. If $D(A(t))$ depends on t, we have $A(t)\, u \in L^2(0, T; D(A(t))')$, if we interpret this last space in a suitable way. □

We then have the following (partial) result:

Theorem 9.2. *Assume that hypotheses* (8.1), (8.2) *are satisfied and that* (9.14) *holds. Then, if f is given with*

$$f = f_0 \in L^2(0, T; V') \tag{9.18}$$

and u_0 and u_1 are given with (9.13), *the solution u of* (9.5) *satisfies*

$$A(t)\, u + u'' = f \quad in \quad \mathscr{D}'(]0, T[; D(A(0))') \tag{9.7a}$$

and

$$\left\{\begin{array}{lll} u(t) \to u(0) = u_0 & in \; [H, V']_{1/2} & as \;\; t \to 0 \\ u'(t) \to u'(0) = u_1 & in \; [V', D(A(0))']_{1/2} & as \;\; t \to 0. \end{array}\right. \tag{9.8a}$$

Proof. 1) In (9.5), we take v given by

$$v(t) = \varphi(t)\, k, \qquad \varphi \in \mathscr{D}(]0, T[), \qquad k \in D(A(0)) \qquad \text{(therefore } v \in X).$$

(9.7a) follows. Then, applying Lemma 9.1, we obtain:

$$u'' = f - A(t)\, u \in L^2(0, T; D(A(0))'), \tag{9.19}$$

from which, by the intermediate derivatives theorem of Chapter 1, Section 2 and since $[H, D(A(0))']_{1/2} = V'$ (Chapter 1, Sections 2.4 and 6), it follows that

$$u' = L^2(0, T; V') \tag{9.20}$$

and therefore (Chapter 1, Section 3) u is a continuous function of $[0, T] \to [H, V']_{1/2}$ and u' is a continuous function of $[0, T] \to [V', D(A(0))']_{1/2}$.

2) Now if we take v defined by

$$v(t) = \varphi(t)\, k, \quad \varphi \in \mathcal{D}([0, T]), \quad \varphi(T) = \varphi'(T) = 0, \quad k \in D(A(0)),$$
$$(\text{then } v \in X),$$

it follows from (9.7a) that

$$\int_0^T \langle A(t)\, u + u'', \overline{v(t)}\rangle\, dt = (f, v) = (u, A(t)\, v + v'') -$$
$$- \langle u'(0), \overline{v(0)}\rangle + \langle u(0), \overline{v'(0)}\rangle .$$

But according to (9.5):

$$(u, A(t)\, v + v'') = (f, v) + \langle u_1, \overline{v(0)}\rangle - \langle u_0, \overline{v'(0)}\rangle$$

and therefore

$$\langle u_1 - u'(0), k\rangle\, \overline{\varphi(0)} - \langle u_0 - u(0), k\rangle\, \overline{\varphi'(0)} = 0,$$
$$\forall \varphi \in \mathcal{D}([0, T]) \;\; (\text{with } \varphi(T) = \varphi'(T) = 0) \;\; \text{and} \;\; \forall k \in D(A(0)),$$

whence (9.8a). □

Remark 9.3. The result is *partial* in the sense that we ignore whether without additional hypotheses on $A(t)$, u is the *unique* element of $L^2(0, T; H)$ to satisfy (9.7a) and (9.8a) (see Section 9.5 on this point).

9.5 Variant; Direct Method

We start with a remark of a general nature:

Remark 9.4. *Transposition* – which we use systematically in this book – is only *one* procedure for carrying out certain *passages to the limit*. It may be of interest to carry out these passages to the limit *directly*. This becomes particularly useful when the "natural spaces" of regularity *are not reflexive* – which is precisely the case for the operators $A(t) + \dfrac{d^2}{dt^2}$ (since in this case we meet the spaces $L^\infty(0, T; V)$ or $C^0([0, T]; V)$). In this section, we shall give an example of *direct passage to the limit*. □

We shall assume that (8.1) and (8.2) *,with* $\lambda = 0$, are satisfied (in order to simplify the notation, but this is in no way essential) (see Sections 8.3 and 8.5).

We want to solve, in a suitable way, the problem (see Theorem 9.2):

$$(9.21) \qquad \begin{cases} A(t)\, u + u'' = f \\ u(0) = u_0, \quad u'(0) = u_1, \end{cases}$$

with

$$(9.22) \qquad f \in L^2(0, T; V'), \quad u_0 \in H, \quad u_1 \in V'.$$

Theorem 9.3. *Assume that* (8.1), (8.2) (*with* $\lambda = 0$) *are satisfied and that* f, u_0, u_1 *are given with* (9.22). *Then, there exists a* u *satisfying* (9.21) *and*

$$u \in L^\infty(0, T; H), \tag{9.23}$$

$$u' \in L^\infty(0, T; V'). \tag{9.24}$$

Furthermore (see Remark 9.10) *we shall see that*

$$u \in C^0([0, T]; H), \quad u' \in C^0([0, T]; V'). \quad \square$$

Remark 9.5. The function $A(t)\,u$ belongs to the *dual* of

$$\left\{\begin{array}{l} L^2(0, T; D(A(t))) = \{v \mid v \in L^2(0, T; V),\, v(t) \in D(A(t)) \text{ a.e.,} \\ A(t)\, v \in L^2(0, T; H)\}; \end{array}\right. \tag{9.25}$$

it is in this sense that we can interpret (9.21). We can also replace (9.21) with

$$\left\{\begin{array}{l} \int\limits_0^T [u, A(t)\, v + v''] \, dt = \int\limits_0^T [f, v] \, dt + [u_1, v(0)] - [u_0, v'(0)] \\ \forall v \in X \text{ (defined in (9.2)) and such that } v \in L^2(0, T; D(A(t))). \end{array}\right. \tag{9.21 a}$$

(In (9.21 a), $[f, v]$ denotes the function $t \to [f(t), v(t)]$, the brackets then denoting the antiduality between V' and V; the second (resp. third) brackets express the antiduality between V' and V (resp. H and itself).) $\square$

Remark 9.6. The problem of the *uniqueness* of the solution of (9.21 a) is solved only under an additional hypothesis on $A(t)$ (see Theorem 9.4). $\square$

Proof of Theorem 9.3. — 1) We consider (which is permissible):

$$f_n \in L^2(0, T; H), \quad u_{0n} \in V, \quad u_{1n} \in H, \tag{9.26}$$

with

$$\{f_n, u_{0n}, u_{1n}\} \to \{f_0, u_0, u_1\} \quad \text{in} \quad L^2(0, T; V') \times H \times V'. \tag{9.27}$$

We then consider the problem

$$\left\{\begin{array}{l} A(t)\, u_n + u_n'' = f_n \\ u_n(0) = u_{0n}, \quad u_n'(0) = u_{1n}. \end{array}\right. \tag{9.28}$$

Thanks to (9.26), we have (Section 8):

$$u_n \in L^\infty(0, T; V), \quad u_n' \in L^\infty(0, T; H). \tag{9.29}$$

2) *A priori estimates.* — We take the scalar product of the two terms of (9.28) with $A(t)^{-1} u_n'$; according to (9.28), $u_n''(t) \in V'$ a.e. and $A(t)^{-1} u_n'(t) \in V$ a.e. (in fact *everywhere* if we apply Theorem 8.2). Then, since

$$[A(t)\, u_n(t), A(t)^{-1} u_n'(t)] = [u_n(t), u_n'(t)],$$

we obtain

$$\frac{1}{2}\frac{d}{dt}|u_n(t)|^2 + \operatorname{Re}[u_n''(t), A(t)^{-1} u_n'(t)] = \operatorname{Re}[f_n(t), A(t)^{-1} u_n'(t)],$$

or, again a.e.

$$\frac{1}{2}\frac{d}{dt}\{|u_n(t)|^2 + [A(t)^{-1} u_n'(t), u_n'(t)]\} -$$

$$-\frac{1}{2}\left[\frac{d}{dt}(A(t)^{-1})\, u_n(t), u_n(t)\right] = \operatorname{Re}[f_n(t), A(t)^{-1} u_n'(t)]$$

whence

$$|u_n(t)|^2 + [A(t)^{-1} u_n'(t), u_n'(t)] = |u_{0n}|^2 + [A(0)^{-1} u_{1n}, u_{1n}] + \tag{9.30}$$

$$+ \int_0^t \left[\left(\frac{d}{d\sigma} A(\sigma)^{-1}\right) u_n(\sigma), u_n(\sigma)\right] d\sigma + 2\operatorname{Re}\int_0^t [f_n(\sigma), A(\sigma)^{-1} u_n'(\sigma)]\, d\sigma.$$

But

$$[A(t)^{-1} v, v] \geqq \varrho\, \|v\|_*^2, \quad \varrho > 0, \quad \text{where} \quad \|v\|_* = \text{norm in } V',$$

and

$$\frac{d}{d\sigma} A(\sigma)^{-1} = -A(\sigma)^{-1} A'(\sigma)\, A(\sigma)^{-1} \in \mathscr{L}(V'; V)$$

and remains in a bounded set of $\mathscr{L}(V'; V)$, therefore, in particular, also in a bounded set of $\mathscr{L}(H; H)$.

Therefore, we deduce from (9.30) that

$$|u_n(t)|^2 + \|u_n'(t)\|_*^2 \leqq C_1\left[|u_{0n}|^2 + \|u_{1n}\|_*^2 + \int_0^T |u_n(\sigma)|^2\, d\sigma\right] + \tag{9.31}$$

$$+ C_1\left|\int_0^T [A(\sigma)^{-1} A'(\sigma)\, A(\sigma)^{-1} f_n(\sigma), u_n'(\sigma)]\, d\sigma\right|.$$

It follows that

$$|u_n(t)|^2 + \|u_n'(t)\|_*^2 \leq C_2\left[|u_{0n}|^2 + \|u_{1n}\|_*^2 + \int_0^T \|f_n(\sigma)\|_*^2\, d\sigma\right]. \tag{9.32}$$

Remark 9.7. In fact, we obtain somewhat more. In (9.32), we may replace

$$\int_0^T \| f_n(\sigma) \|_*^2 \, d\sigma \quad \text{with} \quad \| f_n \|^2_{(L^2(0,T;D(A(t))))'}. \quad \square$$

Remark 9.8. Inequality (9.32) is in a way a "shifted energy inequality", taking place in spaces *in duality* with V and H. This procedure is classical for the Cauchy problem for hyperbolic operators, Leray [1]. □

3) From (9.27) – (9.32) we deduce that u_n (resp. u_n') remains in a bounded set of $L^\infty(0, T; H)$ (resp. $L^\infty(0, T; V')$ as $n \to \infty$. Therefore, we may assume, by extracting a sequence u_μ, that

$$(9.33) \quad \begin{cases} u_\mu \to \tilde{u} & \textit{weak star} \text{ in } L^\infty(0, T; H), \\ u_\mu' \to (\tilde{u})' & \textit{weak star} \text{ in } L^\infty(0, T; V'). \end{cases}$$

Let us take v as in (9.21 a). Then, we deduce from (9.28) that

$$(9.34) \quad \int_0^T [u_n, A(t) v + v''] \, dt = \int_0^T [f_n, v] \, dt + [u_{1n}, v(0)] - [u_{0n}, v'(0)]$$

and passing to the limit in (9.34) (for $n = \mu$), we obtain that $\tilde{u}$ satisfies (9.21 a), (9.23), (9.24).

Therefore we may take $u = \tilde{u}$ and we have the theorem. □

Remark 9.9. Following Remark 9.7, we see that Theorem 9.3 *still holds* (in the interpretation (9.21 a)) *under the hypothesis*

$$(9.35) \quad f \in (L^2(0, T; D(A(t))))' \quad [\text{see } (9.25)]. \quad \square$$

Remark 9.10. Inequality (9.32) gives a little more than (9.23), (9.24): as $n \to \infty$, u_n (resp. u_n') converges *uniformly* to u (resp. u') in H (resp. V') and therefore:

$$(9.36) \quad \begin{cases} \textit{under the hypotheses of Theorem } 9.3, & u \in C^0([0, T]; H) \\ \qquad\qquad\qquad\qquad \textit{and} & u' \in C^0([0, T]; V'). \quad \square \end{cases}$$

Remark 9.11. Of course, $\{f, u_0, u_1\} \to \{u, u'\}$ is a continuous mapping of

$$L^2(0, T; V') \times H \times V' \to L^\infty(0, T; H) \times L^\infty(0, T; V'). \quad \square$$

Theorem 9.4. *Assume that* (8.1), (8.2), (9.22) *hold. Furthermore, assume that*

$$(9.37) \quad \forall u, v \in V, \quad \textit{the function } t \to a(t; u, v) \textit{ is in } C^2[0, T].$$

Then problem (9.21) (*or* (9.21 a)), (9.23), (9.24) *admits a unique solution.*

Proof. It is sufficient to show that (9.21 a) holds for all $v \in X$, that is that the subspace

$$X_0 = \{v \mid v \in X, v \in L^2(0, T; D(A(t)))\}$$

is *dense* in X (for the terms of (9.21 a) are continuous in the topology of X). Now, let $v \in X$; set

$$g = A(t)\, v + v'', \qquad g \in L^2(0, T; H),$$

and consider $g_n \in C^1([0, T]; H)$ (for example), $g_n(T) = 0$, with

$$g_n \to g \quad \text{in} \quad L^2(0, T; H).$$

Consider v solution of

$$(9.38) \qquad \begin{cases} A(t)\, v_n'' + v_n = f \\ v_n(T) = 0, \quad v_n'(T) = 0. \end{cases}$$

Then $v_n \in X$ and $v_n \to v$ in X. But we can *differentiate* (9.38) *in* t; it follows that

$$v_n' \in L^2(0, T; V), \qquad v_n'' \in L^2(0, T; H).$$

Then $A(t)\, v_n \in L^2(0, T; H)$, $v_n(t) \in D(A(t))$ a.e. and therefore $v_n \in X_0$, whence the theorem. □

Remark 9.12. We can regularize the data and the operator *simultaneously* and consider, instead of (9.28), the *regularized parabolic* operator (see Section 8):

$$(9.39) \qquad \begin{cases} A(t)\, u_{\varepsilon n} + u_{\varepsilon n}'' + \varepsilon A(t)\, u_{\varepsilon n}' = f_n, \\ u_{\varepsilon n}(0) = u_{0n}, \quad u_{\varepsilon n}'(0) = u_{1n}. \end{cases}$$

It can be shown that $u_{\varepsilon n}$ (resp. $u_{\varepsilon n}'$) converges uniformly in

$$[0, T] \to H \text{ (resp. } V') \quad \text{as} \quad n \to \infty, \quad \varepsilon \to 0. \ \square$$

Remark 9.13. Of course, the generalization of Theorems 9.1–9.3 is *not the "ultimate generalization"*; for example, we could take the scalar product of the two terms of (9.28) with $A(t)^{-2} u_n'$, $A(t)^{-3} u_n', \ldots$ to obtain inequalities of the type (9.32), but where $u_n(t) \in V'$, $u_n'(t) \in D(A(t))'$, etc., which allows passage to the limit under weaker hypotheses on f, u_0, u_1 (we shall go even much further in the case of *differential operators* – see Volumes 2 and 3). □

Remark 9.14. The preceding method also applies to the equations of the first order of Section 4, under hypothesis (8.2). □

Remark 9.15. We may also *interpolate* between the results of Sections 8 and 9.

According to Theorems 8.2 and 9.3 (with (9.36)) respectively,

$$\{f, u_0, u_1\} \to \{u, u'\}$$

is a continuous mapping of

$$L^2(0, T; H) \times V \times H \to C^0([0, T]; V) \times C^0([0, T]; H)$$

and of

$$L^2(0, T; V') \times H \times V' \to C^0([0, T]; H) \times C^0([0, T]; V').$$

By interpolation and using (see Chapter 1, Section 14.2):

$$[L^2(0, T; H), L^2(0, T; V')]_\theta = L^2(0, T; V^{-\theta})$$

(where

$$[V, H]_\alpha = V^{1-\alpha}, \quad (V^{1-\alpha})' = V^{\alpha-1})$$

and

$$[C^0([0, T]; V), C^0([0, T]; H)]_\theta = C^0([0, T]; V^{1-\theta}),$$

we obtain

Theorem 9.5. *Assume that* (8.1), (8.2), (9.37) *hold.* f, u_0, u_1 *are given with*

(9.40) $$f \in L^2(0, T; V^{-\theta}), \quad 0 < \theta < 1,$$

(9.41) $$u_0 \in V^{1-\theta}, \quad u_1 \in V^{-\theta}.$$

Then, the solution u *of* (9.21 a), (9.23), (9.24) *satisfies*

(9.42) $$u \in C^0([0, T]; V^{1-\theta}),$$

(9.43) $$u' \in C^0([0, T]; V^{-\theta}). \quad \square$$

9.6 Examples

We shall give some examples of applications of the preceding theorems. As we have done for the equations of the first order in Section 4.7, we could apply either the regularity theorems of Section 8, or the more general theorems of Section 9, by transposition (Theorems 9.1 and 9.2) or directly (Theorems 9.3 and 9.4).

We shall limit ourselves to giving the results obtained through application of the theorems of Section 9.

9.6.1 Example 1

Let Ω be a bounded open set in $\mathbf{R}^n$. In this example, Ω is *arbitrary*, therefore the boundary Γ may be "arbitrarily irregular".

We set

(9.44) $$a(t; u, v) = \sum_{i=1}^{n} \int_\Omega \frac{\partial u}{\partial x_i} \frac{\partial \bar{v}}{\partial x_i} dx,$$

for

$$u, v \in V = H_0^1(\Omega).$$

If $H = L^2(\Omega)$, then $V' = H^{-1}(\Omega)$.

Let f, u_0, u_1 be given with

(9.45) $$f \in L^2(0, T; H^{-1}(\Omega)),$$

(9.46) $$u_0 \in L^2(\Omega), \quad u_1 \in H^{-1}(\Omega).$$

Then we know that there exists a unique $u \in L^2(Q)$ satisfying (see Theorem 9.1):

(9.47) $$\int_Q u \left(\frac{\partial^2 \bar{v}}{\partial t^2} - \Delta \bar{v} \right) dx\, dt = \int_0^T [f(t), v(t)]\, dt + $$
$$+ [u_1(0), v(0)] - \left[u_0(0), \frac{\partial v}{\partial t}(0) \right],$$

where $[\ ,\ ]$ denotes the antiduality between $H^{-1}(\Omega)$ and $H_0^1(\Omega)$ (and the scalar product in $L^2(\Omega)$), for every function $v \in L^2(0, T; H_0^1(\Omega))$ such that

(9.48) $$\frac{\partial v}{\partial t} \in L^2(0, T; L^2(\Omega)) = L^2(Q),$$
$$\frac{\partial^2 v}{\partial t^2} - \Delta v \in L^2(Q), \quad v(x, T) = 0, \quad \frac{\partial v}{\partial t}(x, T) = 0.$$

The problem fits the conditions of Section 9.4. Therefore we deduce from (9.48) that

(9.49) $$\frac{\partial^2 u}{\partial t^2} - \Delta u = f$$

in the sense of distributions in Q,

(9.50) $$u(x, 0) = u_0(x) \text{ ((1))},$$

(9.51) $$\frac{\partial u}{\partial t}(x, 0) = u_1(x) \text{ ((2))}.$$

Furthermore, we have *the boundary condition*

$$u = 0 \quad \text{on} \quad \Sigma,$$

((1)) According to (9.8a), $u(., t) \to u_0$ in $[H, H^{-1}(\Omega)]_{1/2}$, space which coincides with $(H_{00}^{1/2}(\Omega))'$ *when Γ is regular* (see Chapter 1, Theorem 11.7 and Remark 12.1).

((2)) According to (9.8a), $u'(., t) = \dfrac{\partial u}{\partial t}(., t) \to u_1$ in $[H^{-1}(\Omega), D(A)']_{1/2}$, where

$$D(A) = \{v \mid v \in H_0^1(\Omega), \Delta v \in L^2(\Omega)\}.$$

which is *contained* in (9.47), but in a *formal* manner: indeed, *formally* integrating by parts in (9.47) and taking account of (9.49), (9.50), (9.51), we obtain

$$\int_{\Sigma} u \frac{\partial v}{\partial \nu} d\sigma = 0, \quad \Sigma = \Gamma \times]0, T[. \tag{9.52}$$

But this is formal on two accounts:

(i) no regularity hypothesis is made on Σ;

(ii) if Σ is assumed regular, the integrations by parts would have to be *justified* in a certain sense.

The justification of (ii) will be made in Chapter 5 of Volume 2; but without regularity conditions on Σ, there exists (at the moment) no other interpretation of the condition "$u = 0$ on Σ" than equation (9.47) itself.

9.6.2 Example 2

Let Ω be defined as in the preceding example. We take $H = L^2(\Omega)$ again, and

$$V = H_0^2(\Omega)$$

and then

$$V' = H^{-2}(\Omega).$$

We take

$$a(t; u, v) = \int_{\Omega} \Delta u \overline{\Delta v} \, dx. \tag{9.53}$$

The general theory applies, for, if $\lambda > 0$,

$$|\Delta v|^2 + \lambda |v|^2 \geqq \alpha \| v \|^2_{H^2(\Omega)}, \quad \alpha > 0, \quad \forall v \in H_0^2(\Omega).$$

So let f, u_0, u_1 be given with

$$\begin{cases} f \in L^2(0, T; H^{-2}(\Omega)) \text{ i.e.} \\ f = f_0 + \sum_{i=1}^{n} \frac{\partial f_i}{\partial x_i} + \sum_{i,j}^{n} \frac{\partial^2 f_{ij}}{\partial x_i \partial x_j}, \\ f_0, f_i, f_{ij} \in L^2(Q), \end{cases} \tag{9.54}$$

$$\begin{cases} u_0 \in L^2(\Omega), \\ u_1 \in H^{-1}(\Omega). \end{cases} \tag{9.55}$$

Then there exists a unique function $u \in L^2(Q)$ such that

$$\int_{Q} u \left(\frac{\partial^2 \bar{v}}{\partial t^2} + \overline{\Delta^2 v} \right) dx \, dt = \int_0^T [f(t), v(t)] \, dt + \\ + [u_1, v(0)] - \left[u_0, \frac{\partial v}{\partial t}(0) \right] \tag{9.56}$$

for every function $v \in L^2(0, T; H_0^2(\Omega))$ such that

$$\frac{\partial v}{\partial t} \in L^2(Q), \quad \frac{\partial^2 v}{\partial t^2} + \Delta^2 v \in L^2(Q), \quad v(x, T) = 0, \quad \frac{\partial v}{\partial t}(x, T) = 0,$$

where $[\ ,\ ]$ denotes the antiduality between $H^{-2}(\Omega)$ and $H_0^2(\Omega)$ (and the scalar product in $L^2(\Omega)$).

Then u satisfies

$$\frac{\partial^2 u}{\partial t^2} + \Delta^2 u = f, \tag{9.57}$$

$$u(x, 0) = u_0(x) \text{ ((1))}, \tag{9.58}$$

and

$$\frac{\partial u}{\partial t}(x, 0) = u_1(x) \text{ ((2))}. \tag{9.59}$$

Remark 9.16. The present example shows that the hyperbolic character of the wave operator does not intervene in Example 1. □

Remark 9.17. We have not been able to give the case in which u is *non-zero on* Σ, in the preceding examples. This will be done, *when* Σ is regular, in Chapter 5, Volume 2 (see, in particular, Section 1.2 of Chapter 5). □

9.6.3 Example 3

Again let Ω be an *arbitrary* bounded open set in $\mathbf{R}^n$ and take

$$H = L^2(\Omega), \quad V = H^1(\Omega).$$

Then V' is not a space of distributions on Ω. In order to obtain the "usual" interpretations, we may use the space $\Xi^1(\Omega)$ as defined in Section 4.7.2.

We take $a(t; u, v)$ as in (9.44).

Let f, u_0, u_1 be given with

$$f \in L^2(0, T; \Xi^{-1}(\Omega)), \tag{9.60}$$

$$u_0 \in L^2(\Omega), \quad u_1 \in \Xi^{-1}(\Omega); \tag{9.61}$$

then, since $V \subset \Xi^1(\Omega)$, we may take

$$(L, v) = \int_0^T \langle f(t), \overline{v(t)} \rangle \, dt + \langle u_1, \overline{v(0)} \rangle - \left[u_0, \frac{\partial v}{\partial t}(0) \right], \tag{9.62}$$

((1)) With $u(t) \to u_0$ in $H^{-1}(\Omega)$ if Γ is sufficiently regular and in $[H, H^{-2}(\Omega)]_{1/2}$ in any case.

((2)) With $u'(t) \to u_1$ in $[H^{-2}(\Omega), D(A)']_{1/2}$, where

$$D(A) = \{ v \mid v \in H_0^2(\Omega), \Delta^2 v \in L^2(\Omega) \}.$$

in (9.5), where the brackets denote the duality between $\Xi^{-1}(\Omega)$ and $\Xi^1(\Omega)$ and $[\ ,\ ]$ denotes the scalar product in $L^2(\Omega)$.

From which we deduce the existence and uniqueness of $u \in L^2(Q)$ satisfying

(9.63)
$$\int_H u \left(\overline{\frac{\partial^2 v}{\partial t^2}} - \overline{\Delta v}\right) dx\, dt = (L, v)$$

for all $v \in L^2(0, T; H^1(\Omega))$ such that

$$\frac{\partial v}{\partial t} \in L^2(Q), \quad \frac{\partial^2 v}{\partial t^2} - \Delta v = \varphi \in L^2(Q), \quad v(x, T) = 0, \quad \frac{\partial v}{\partial t}(x, T) = 0$$

and

$$\frac{d^2}{dt^2}[v(t), w] + a(v(t), w) = [\varphi(t), w] \qquad \forall w \in H^1(\Omega).$$

The function u then satisfies

(9.64)
$$\frac{\partial^2 u}{\partial t^2} - \Delta u = f,$$

(9.65)
$$\begin{cases} u(x, 0) = u_0(x)\ ^{((1))}, \\ \dfrac{\partial u}{\partial t}(x, 0) = u_1(x)\ ^{((2))}. \end{cases}$$

Finally, in a *formal way*

(9.66)
$$\frac{\partial u}{\partial \nu} = 0 \quad \text{on} \quad \Sigma$$

Remark 9.18. See also the example given in Remark 6.1, Chapter 5, Volume 2. □

9.6.4 Example 4

Let Ω be a *bounded* open set with *regular boundary* Γ, and assume that Γ_0 is a subset of Γ.

Let Γ_1 be the complement of Γ_0 in Γ.

Take $H = L^2(\Omega)$ and

(9.67)
$$v = \{v \mid v \in H^1(\Omega),\ v = 0 \text{ on } \Gamma_0\}\ ^{((3))}.$$

((1)) With $u(t) \to u_0$ in $[H^1(\Omega), H]'_{1/2}$.

((2)) With $u'(t) \to u_1$ in $[D(A), H^1(\Omega)]'_{1/2}$, where

$$D(A) = \{v \mid v \in H^1(\Omega), \Delta v \in L^2(\Omega) \text{ and } (-\Delta v, w) = a(v, w),\ \forall w \in H^1(\Omega)\}.$$

((3)) i.e. the restriction of $\gamma_0 v$ to Γ_0 vanishes, $\gamma_0 v$ being the trace on Γ defined according to Chapter 1, Section 8.

Then (except if $\Gamma_0 = \Gamma$, in which case we have the Dirichlet problem) V' is not a space of distributions on Ω.

Again let

$$a(t; u, v) = \sum_{i=1}^{n} \int_{\Omega} \frac{\partial u}{\partial x_i} \frac{\partial \bar{v}}{\partial x_i} dx;$$

then the general theory applies.

For the interpretation of the corresponding problem, we meet the same difficulty as in Section 9.6.3 above. We use the space $\Xi_{\Gamma_0}(\Omega)$ as defined in Section 4.7.3.

In (9.5), we may now take (L, v) as given by (9.62), with this time

$$f \in L^2(0, T; \Xi_{\Gamma_0}^{-1}(\Omega)) \tag{9.68}$$

and

$$u_0 \in L^2(\Omega), \quad u_1 \in \Xi_{\Gamma_0}^{-1}(\Omega). \tag{9.69}$$

Then there exists a unique $u \in L^2(Q)$ such that

$$\int_{\Omega} u \left(\frac{\partial^2 \bar{v}}{\partial t^2} - \overline{\Delta v} \right) dx\, dt = (L, v) \tag{9.70}$$

$$\forall v \in L^2(0, T; H^1(\Omega)), \quad \frac{\partial v}{\partial t} \in L^2(Q), \quad v = 0 \quad \text{on} \quad \Gamma_0 \times]0, T[,$$

$$\frac{\partial^2 v}{\partial t^2} - \Delta v = \varphi \in L^2(Q), \quad v(x, T) = 0, \quad \frac{\partial v}{\partial t}(x, T) = 0,$$

$$\frac{d^2}{dt^2}[v(t), w] + a(v(t), w) = [\varphi(t), w], \quad \forall w \in H^1(\Omega),$$

$$w = 0 \quad \text{on} \quad \Gamma_0.$$

The function u satisfies

$$\frac{\partial^2 u}{\partial t^2} - \Delta u = f \quad \text{in } Q, \tag{9.71}$$

(9.67), and the *mixed conditions*

$$\begin{cases} u = 0 \quad \text{on} \quad \Gamma_0 \times]0, T[, \\ \dfrac{\partial u}{\partial \nu} = 0 \quad \text{on} \quad \Gamma_1 \times]0, T[; \end{cases} \tag{9.72}$$

these conditions in a "weak" sense (tied to (9.70)). □

9.6.5 Example 5

Let Ω be a bounded open set in $\mathbf{R}^2$. Take (as in Example 4.7.6) $H = L^2(\Omega)$,

$$V = \left\{ v \mid v \in L^2(\Omega), \quad \frac{\partial v}{\partial x_1} \in L^2(\Omega), \quad \frac{\partial^2 v}{\partial x_2^2} \in L^2(\Omega) \right\}, \tag{9.73}$$

$$a(t; u, v) = \int_\Omega \frac{\partial u}{\partial x_1} \frac{\partial \bar{v}}{\partial x_1} dx + \int_\Omega \frac{\partial^2 u}{\partial x_2^2} \frac{\partial^2 \bar{v}}{\partial x_2^2} dx. \tag{9.74}$$

The general theory applies. For the interpretation of the problem, one possibility is to use the space $K(\Omega)$ introduced in Section 4.7.6.

We then take (L, v) defined by (9.62) with

$$f \in L^2(0, T; K'(\Omega)), \tag{9.75}$$

$$u_0 \in L^2(\Omega), \quad u_1 \in K'(\Omega). \tag{9.76}$$

The corresponding solution $u \in L^2(Q)$ satisfies

$$\begin{cases} \dfrac{\partial^2 u}{\partial t^2} - \dfrac{\partial^2 u}{\partial x_1^2} + \dfrac{\partial^4 u}{\partial x_2^4} = f \text{ in } Q, \\ u(x, 0) = u_0, \\ \dfrac{\partial u}{\partial t}(x, 0) = u_1, \end{cases} \tag{9.77}$$

and, in a weak sense, boundary conditions which are, when Ω is the rectangle $\Omega =]0, a_1[\times]0, a_2[$:

$u = 0$, on the sides of Γ parallel to the x_2-axis,

$u = \dfrac{\partial u}{\partial x_2} = 0$, on the sides of Γ parallel to the x_1-axis. □

Remark 9.19. In this example, the number of derivatives in V depends upon the direction of differentiation.

Other situations may be encountered; for example, let

$$V = \left\{ v \mid v \in L^2(\Omega), \, \Box v = \frac{\partial^2 v}{\partial x_1^2} - \frac{\partial^2 v}{\partial x_2^2} \in L^2(\Omega) \right\}, \tag{9.78}$$

and

$$a(t; u, v) = \int_\Omega \Box u \, \overline{\Box v} \, dx.$$

Then the corresponding partial derivative operator will be

$$\frac{\partial^2}{\partial t^2} + \Box^2.$$

In this case, the choice of a space $K(\Omega)$ (different from $L^2(\Omega)$!) does not seem immediate (see Problem 13.12). □

10. Schroedinger Type Equations

10.1 Notation

Again the operators $A(t)$ are given by the family $a(t; u, v)$ with the properties (8.1), (8.2). In this section we shall consider the (Schroedinger type) equations

$$\mathrm{i}\, A(t)\, u + u' = f \tag{10.1}$$

with the initial condition

$$u(0) = u_0 . \tag{10.2}$$

10.2 Existence and Uniqueness Theorem

Theorem 10.1. *Hypotheses* (8.1), (8.2) *are assumed to be satisfied. Let f be given with*

$$f \in L^2(0, T; H), \quad \frac{df}{dt} = f' \in L^2(0, T; V') \tag{10.3}$$

and let u_0 be given with

$$u_0 \in V. \tag{10.4}$$

There exists a unique u satisfying

$$u \in L^2(0, T; V) \tag{10.5}$$

and (10.1), (10.2).

Remark 10.1. If (10.5) holds, then, according to (10.1),

$$u' = f - \mathrm{i}\, A(t)\, u \in L^2(0, T; V'), \tag{10.6}$$

so that (10.2) has meaning. □

Proof of Theorem 10.1. As for the proof of Theorem 8.1, we first give the a priori estimates which imply (with the standard procedures) the *existence* of a solution.

1) *A priori estimates.* Multiplying (10.1) by $u(t)$, we obtain

$$\mathrm{i}[A(t)\, u(t), u(t)] + [u'(t), u(t)] = [f(t), u(t)],$$

from which, taking twice the real part of the two members:

$$\frac{d}{dt} |u(t)|^2 = 2\,\mathrm{Re}[f(t), u(t)] \tag{10.7}$$

whence

$$|u(t)|^2 = |u_0|^2 + 2\operatorname{Re}\int_0^t [f(\sigma), u(\sigma)]\, d\sigma \tag{10.8}$$

and therefore

$$|u(t)|^2 = |u_0|^2 + \int_0^t |f(\sigma)|^2\, d\sigma + \int_0^t |u(\sigma)|^2\, d\sigma. \tag{10.9}$$

This, together with Gronwall's Lemma, implies

$$|u(t)|^2 \leqq C\left(|u_0|^2 + \int_0^t |f(\sigma)|^2\, d\sigma\right), \quad 0 \leqq t \leqq T. \tag{10.10}$$

Now, multiply "formally" (10.1) by u', taking twice the *imaginary part* of the result; we obtain

$$a(t; u(t), u'(t)) + a(t; u'(t), u(t)) = 2\operatorname{Im}[f(t), u'(t)],$$

from which we get

$$\frac{d}{dt} a(t; u(t), u(t)) - a'(t; u(t), u(t)) = 2\operatorname{Im}[f(t), u'(t)]. \tag{10.11}$$

Integrate (10.11):

$$\left\{\begin{aligned} a(t; u(t), u(t)) = \; & a(0; u_0, u_0) + \int_0^t a'(\sigma; u(\sigma), u(\sigma))\, d\sigma + \\ & + 2\operatorname{Im}\int_0^t [f(\sigma), u'(\sigma)]\, d\sigma. \end{aligned}\right. \tag{10.12}$$

But

$$\int_0^t [f(\sigma), u'(\sigma)]\, d\sigma = [f(t), u(t)] - [f(0), u_0] - \int_0^t [f'(\sigma), u(\sigma)]\, d\sigma$$

(note that $f(0) \in [H, V']_{1/2} \subset V'$). Therefore

$$\left|\int_0^t [f(\sigma), u'(\sigma)]\, d\sigma\right| \leqq \|f(t)\|_{V'} \|u(t)\| + \|f(0)\|_{V'} \|u_0\| + \\ + \int_0^t \|f'(\sigma)\|_{V'} \|u(\sigma)\|\, d\sigma$$

and (10.12) yields:

$$\alpha \|u(t)\|^2 \leqq C\left(\|u_0\|^2 + \int_0^t \|u(\sigma)\|^2\, d\sigma + \int_0^t \|f'(\sigma)\|_{V'}^2\, d\sigma\right) + \\ + 2\|f(t)\|_{V'} \|u(t)\|.$$

The last term is bounded above by

$$\frac{\alpha}{2}\|u(t)\|^2 + \frac{4}{\alpha}\|f(t)\|^2_{V'} \leqq \frac{\alpha}{2}\|u(t)\|^2 + \frac{4}{\alpha}C\int_0^t (|f(\sigma)|^2 + \|f'(\sigma)\|^2_{V'})\,d\sigma$$

and therefore

(10.13)

$$\|u(t)\|^2 \leqq C\left(\|u_0\|^2 + \int_0^t \|u(\sigma)\|^2\,d\sigma + \int_0^t (|f(\sigma)|^2 + \|f'(\sigma)\|^2_{V'})\,d\sigma\right).$$

This inequality, together with Gronwall's Lemma, implies:

(10.14)

$$\|u(t)\|^2 \leqq C\left(\|u_0\|^2 + \int_0^T (|f(\sigma)|^2 + \|f'(\sigma)\|^2_{V'})\,d\sigma\right), \quad 0 \leqq t \leqq T. \ \square$$

2) *Existence of a solution.* We may use the preceding estimates together with the method of Faedo-Galerkin (for example) to show *existence* (we could also use *parabolic regularization,* in analogy with Section 8, as will be done explicitely in Chapter 5, Section 12, Volume 2).

So let $w_1, \ldots, w_m, \ldots$ be a "basis" of V (see Section 8; V is assumed to be separable) and let

$$u_m(t) = \sum_{i=1}^m g_{im}(t)\, w_i$$

be given by

(10.15) $\quad (u'_m(t), w_j) + \mathrm{i}\, a(t; u_m(t), w_j) = (f(t), w_j), \quad 1 \leqq j \leqq m,$

(10.16) $\quad g_{im}(0) = \xi_{im}, \quad \sum_{i=1}^m \xi_{im} w_i \to u_0 \ \text{ in } \ V.$

Then the *same* bounds as in 1) (see (10.14)) show that

(10.17) $\quad u_m$ remains in a bounded set of $L^\infty(0, T; V)$, as $m \to \infty$.

We extract $u_\mu \to \tilde{u}$ *weak star* in $L^\infty(0, T; V)$ and verify that $\tilde{u}$ is a solution of

(10.18) $\quad \mathrm{i}\, A(t)\,\tilde{u} + \dfrac{d\tilde{u}}{dt} = f, \quad \tilde{u}(0) = u_0.$

Then it follows from (10.18) that

$$\frac{d\tilde{u}}{dt} \in L^\infty(0, T; V').$$

According to the *uniqueness* (which is shown separately), we have $\tilde{u} = u$ and therefore existence. $\square$

Remark 10.2. Thus we have shown more than Theorem 10.1, that is, the existence of a solution satisfying

$$u \in L^\infty(0, T; V), \tag{10.19}$$

$$u' \in L^\infty(0, T; V'). \tag{10.20}$$

In fact, by the method of Section 8.4, we can show that u is a *continuous* function of $[0, T] \to V$ and u' a *continuous* function of $[0, T] \to \to V'$. □

Remark 10.13. We shall see other regularity results – established by a somewhat different method – in Chapter 5, Volume 2, and then M_k-regularity results in Volume 3. □

Now we shall briefly investigate the "transposition" of Theorem 10.1.

11. Schroedinger Type Equations; Transposition

11.1 Adjoint Isomorphism

We consider the *adjoint problem*:

$$-\mathrm{i}\, A(t)\, v - v' = \varphi \tag{11.1}$$

$$v(T) = 0, \tag{11.2}$$

where φ describes the spacc Φ *defined* by

$$\Phi = \{\varphi \mid \varphi \in L^2(0, T; H),\ \varphi' \in L^2(0, T; V'),\ \varphi(0) = \varphi(T) = 0\}. \tag{11.3}$$

Remark 11.1. In (11.3), we have imposed the conditions "$\varphi(0) = \varphi(T) = 0$" so that the space $\mathscr{D}(]0, T[; H)$ be dense in Φ, Φ being provided with the natural (hilbertian) norm:

$$\left(\int_0^T (|\varphi(t)|^2 + \|\varphi'(t)\|^2_{V'})\, dt\right)^{1/2}. \quad □$$

We *define*:

$$\left\{ \begin{array}{l} X = \text{space described by the solution } v \text{ of (11.1), (11.2) as } \varphi \\ \text{describes } \Phi. \end{array} \right. \tag{11.4}$$

Providing X with the topology carried over by the topology of Φ in the mapping $\varphi \to v$, we have (done what was necessary to have)

$$-\mathrm{i}\, A - \frac{d}{dt} \textit{ is an isomorphism of } X \textit{ onto } \Phi. \quad □ \tag{11.5}$$

11.2 Transposition of (11.5)

Theorem 11.1. *Assume* (8.1), (8.2) *to be satisfied. Let L be a continuous antilinear form on X. There exists a unique $u \in \Phi'$ with*

$$(u, -\mathrm{i}\, A(t)\, v - v') = (L, v), \quad \forall v \in X, \tag{11.6}$$

(where the first parenthesis denotes the duality between Φ' and Φ and the second between X' and X). □

From Remark 11.1 and the Hahn-Banach Theorem it follows that

$$\begin{cases} \textit{every } u \in \Phi' \textit{ may be written, non-uniquely, as } u = u_0 + \dfrac{d u_1}{dt}, \\ u_0 \in L^2(0, T; H), \quad u_1 \in L^2(0, T; V). \quad \square \end{cases} \tag{11.7}$$

We still have to *choose* L.

11.3 Choice of L

Analogously to Sections 4.5 and 9.3, we define the space Ξ by (ϱ being the function defined by (4.23)):

$$\Xi = \{v \mid v \in L^2(0, T; V), \varrho v' \in L^2(0, T; V'), v(T) = 0\} \tag{11.8}$$

(provided with the norm

$$\left(\int_0^T (\| v(t) \|^2 + \| \varrho\, v'(t) \|_{V'}^2)\, dt \right)^{1/2},$$

which makes it a Hilbert space).

We see again that $\mathscr{D}(]0, T[; V)$ is dense in Ξ and therefore Ξ' = antidual of Ξ, is a space of distributions on $]0, T[$ taking their values in V'; furthermore, every $f \in \Xi'$ may be written, non-uniquely, as

$$f = f_0 + \frac{d}{dt}(\varrho f_1), \quad \text{with } f_0 \in L^2(0, T; V'), \quad f_1 \in L^2(0, T; V). \tag{11.9}$$

Finally, we may take

$$L(v) = (f, v)_{\Xi', \Xi} + (u_0, v(0)), \quad \text{with} \quad f \in \Xi' \quad \text{and} \quad u_0 \in V', \tag{11.10}$$

in (11.6), $(\, , \,)_{\Xi', \Xi}$ denoting the scalar product between Ξ' and Ξ. Then, *formally*, (11.6) yields:

$$\begin{cases} u \in \Phi' \text{ (the structure of which is given by (11.7))} \\ \mathrm{i}\, A(t)\, u + u' = f, \quad \text{given by (11.9)} \\ u(0) = u_0, \quad u_0 \in V'. \quad \square \end{cases} \tag{11.11}$$

Remark 11.2. It is possible [at least under hypothesis (9.14)] to give an interpretation of the type given in Section 9 to (11.11). But we do not specify the results here; see also Problem 13.12. □

Remark 11.3. Of course, we can interpolate, as, for example, in Remark 9.15. □

Remark 11.4. Examples analogous to those of Sections 4.7 and 9.6 could be given here. □

12. Comments

The introduction of weak solutions of evolution equations is classical, in particular since the works of Leray [2–4] (in connection with nonlinear Navier-Stokes equations) and Sobolev [3]: see Ladyzenskaya [1–3], Ladyzenskaja-Vishik [1], Lions [13], Tréves [1, 2], Vishik [4].

The presentation of Sections 1–3 follows the paper of Lions [27] (which contains the case of variational inequality problems in an analogous framework; for more general results in this direction, see Brézis [1, 2]).

For the examples of Section 4, see also Browder [4, 10, 11], and Lions [13]. All the examples given here correspond to the case for which $D(A(t))$ (suitably defined, see Kato [1–3]) is *independent* of t; when $D(A(t))$ depends on t, and in a Hilbert setting, see Baiocchi [3, 4], Bardos [1], Brézis [3], Carroll [1, 2], G. Cooper (1], Lions [13, 18, 22, 23]. We have not touched upon the study of corresponding non-homogeneous problems, see Problems 13.6, 13.8.

In the Banach space setting and through the use of semi-group theory (see Hille-Phillips [1], Yosida [2]), first-order operator equations have been studied by Yosida [1], Kato [1, 2, 5], Kato-Tanabe [1], Poulsen [1], Sobolevski [1–7], Tanabe [1] and S. Krein [4].

Results of the type given in Section 5, have been obtained by Kaplan [1] via different methods.

The method of elliptic regularization used in Section 7 was introduced in this form by Lions [21]. For the case of elliptic-parabolic equations, an analogous regularization method was introduced by Oleinik [4]. Another method consists in replacing $M + \Lambda$ by $M + \frac{1 - G(h)}{h} = M_h$; then one applies elliptic variational theory (Chapter 2, Section 9) and lets h go to zero.

Of course, there are many possible choices for the space in which we seek a solution, not only in the "space variables" (abstract or not) but also in the *time variable*.

Thus Sobolevski has studied the "abstract Cauchy problem" in spaces satisfying *weighted Hölder conditions in t* (see Sobolevski [3]) and in L^p or Lorentz spaces (still in t) (see Sobolevski [4, 5]). Along these lines, see also Da Prato-Giusti [1, 2].

Equations of the second order in t are studied in Sections 8 and 9. Regularity results of the same type as those in Theorem 8.2 are given by Baoicchi [7], Lions [21], Torelli [1], Strauss [1]; we have followed the presentation of Strauss for Lemmas 8.1, 8.2 and 8.3.

Concerning the existence and uniqueness theorems for the Cauchy problem, we also call attention to the method of Baoicchi, who has obtained the results directly, without duality, under slightly different hypotheses; this direct method is valid for first and second order equations (see Baiocchi [6, 7]) and Schroedinger type equations (see Pozzi [1]).

Other existence and uniqueness results, using esentially different methods (Dunford integrals and interpolation) are gives in Grisvard [6, 7, 9] and (still by different methods) in Da Prato [4, 5].

Operator equations of the second order in t have been studied by semi-group methods in Pogorelenko-Sobolevski [1], and Raskin-Sobolevski [1]; also, see the work of S. Krein [4].

The choice of the particular scalar product $(u, v)_{\mathscr{H}}$ (given in (8.36)) is the analog, when the coefficients depend on t, to what is done in the case of *independence* of t, to show (according to Yosida [1]) that $\begin{pmatrix} 0 & -1 \\ A & 0 \end{pmatrix}$ is the infinitesimal generator of a *group* (see also Lions [7]). It is important to choose a suitable scalar product on $\mathscr{H}$; for example, the property for

$$-\mathcal{A} = -\begin{pmatrix} 0 & -1 \\ A & 0 \end{pmatrix}$$

to be the infinitesimal generator of a semi group in $\mathscr{H}$, is *independent of the scalar product*; but in order to apply the theorem of Hille-Yosida, it is *much* easier to come back (if possible) to the case where the semi-group is a *contraction* semi-group — and this is precisely the object of the "suitable" choice of the scalar product on $\mathscr{H}$; see also J. A. Goldstein [1], Mizohata [1].

For the case of operator equations *with lag*, we refer to Artola [1], Baiocchi [7] and Pozzi [1].

The case of stochastic operator equations would be treated by the same type of methods; in fact, this is true for all the results of this book, but the rather long preparation of the framework would have made the presentation too heavy.

Still in the way of "general evolution equations", we point out the following studies (the subjects of which are not touched upon here):

1) the theory of *semi-group distributions*: see Da Prato-Mosco [1, 2], Foias [1], Fujiwara [1], Lions [12], Peetre [9], Shiraishi-Hiraka [1], Yoshinaga [1, 2] and Chazarain [1, 2];

2) the *Cauchy problem* is *not well-posed* for the operators $A + \frac{d^k}{dt^k}$ when $k > 2$, *except* for the (uninteresting) case where A is a *bounded* operator: see Fattorini [1], Chazarain [1]; for the operators $\sum_{j=0}^{k} A_j \frac{d^j}{dt^j}$, unbounded A_j, see *very* partial results in the following chapters.

But for problems different from the Cauchy problem, the general result of Section 1 applies; for example, for the operator $A(t) + \frac{d^{2m+1}}{dt^{2m+1}}$ on a finite interval $(0, T)$, by taking $\Lambda = \frac{d^{2m+1}}{dt^{2m+1}}$, the domain of Λ being defined by $u^{(j)}(0) = 0$, $u^{(j)}(T) = 0$, $0 \leqq j \leqq m - 1$. This reproduces (using the intermediate derivatives theorem) the results of Kadlec [1], Cattabriga [5], Grisvard [7], who also investigates the *non-Hilbert* cases;

3) for the general study of the solutions of the equations

$$A u + \frac{du}{dt} = 0$$

(behaviour at infinity, asymptotic representation of the solutions, norm convexity properties of solutions ...) according to the spectral properties of the unbounded operator A, see Agmon-Nirenberg [1, 2], Geymonat [5], Lax [1], Lioubich [1], Zaidman [1, 2];

4) for the study of almost periodic solutions of evolution equations, see Amerio-Prouse [1] and the bibliography of this work;

5) for the "operator" generalization of Sturm-Liouville problems, see Kostiouchenko-Levitan [1];

6) for the use of Wiener type integrals for evolution equations, see Daletski [1], Nelson [1] (in these works, a result of Trotter [1] is used, a result which incidently can also be used as a starting point for numerical approximation methods; see generalizations in Bardos [1]);

7) certain properties — in particular of *uniqueness* — remain valid when the *equality* $\frac{du}{dt} + A u = f$ is replaced by an *inequality* satisfied (in a suitable norm) by $\left\| \frac{du}{dt} + A u - f \right\|$; see Friedman [3], Tanabe [3];

8) the problems mentioned in 7) should not be confused with the *variational evolution inequalities*, such as are studied in Lions-Stampacchia [1], Brézis [1–3], Brézis-Lions [1];

9) we call attention to the study of operator equations in locally convex topological vector spaces; see Yosida [3];

10) for transport equations, see Jörgens [1] and S. Krein [4];

11) we have left aside the question of (eventual) analyticity of solutions in t; we shall come back to this point in Volume 3; consult Kato [3], S. Krein [4, 5], Yosida [2];

12) for "scattering" theory, see Kato [5], Lax-Phillips [2];

13) we also call attention to the study, in Baouendi-Grisvard [1], of the operator

$$x \frac{\partial u}{\partial t} + (-1)^m \frac{\partial^{2m} u}{\partial x^{2m}},$$

the type of which changes according to the sign of x; this study has been carried on to analogous *nonlinear* cases by Bardos-Brézis [1].

Also consult the Comments to Chapters 4 and 5 (Volume 2).

13. Problems

13.1 In Theorem 1.1, the semi-group $G(s)$ operates in $\mathscr{H}$, in $\mathscr{V}$ and in $\mathscr{V}'$. What can be said when $G(s)$ is a contraction semi-group in $\mathscr{H}$, but does not operate in $\mathscr{V}$ and $\mathscr{V}'$? (see also 13) in Section 12).

13.2 The study of non-homogeneous problems for operators

$$A(t) + \frac{d}{dt},$$

where $A(t)$ is a *singular* integro-differential operator, may be of interest.

For example for the case (which comes up in applications) where $A(t) = A$ is given by

$$A = \text{p.v.} \int_{-1}^{1} \frac{1}{x - y} \varphi'(y)\, dy,$$

see Cherruault [1] and the bibliography of this work.

13.3 Extension of the results of Sections 4 and 5 to operators $A(t) + \dfrac{d}{dt}$ in Banach spaces by the use of semi-group theory; Yosida [2], Kato [1, 2], Kato-Tanabe [1], Sobolevski [1–3], Tanabe [1].

13.4 In the setting of Section 4, when $A(t)$ verifies

$$\operatorname{Re}(A(t)\, v, v) + \lambda(t)\, |v|^2 \geqq \alpha\, \|v\|^2,$$

with $\lambda \in L^1(0, T)$ (see Lions-Raviart [1], Baiocchi [5]); see also Problem 17.15 in Chapter 4 of Volume 2.

13.5 Case for which the space V of Section 4 is replaced by a measurable family of Hilbert spaces $V(t)$; see the bibliography given in the Comments to this chapter.

13.6 Problem analogous to 13.5, but for the operators $A(t) + \dfrac{d^2}{dt^2}$.

13.7 Problems 13.5 and 13.6 contain, as particular cases on the examples, problems in non-cylindrical open sets; then see the problems of Chapters 4 and 5, Volume 2.

13.8 Obtainment of the results of Sections 4, 5 and 6 by passage to the limit $\varepsilon \to 0$ starting from non-homogeneous "elliptic" problems, by the application of elliptic regularization (Section 7).

13.9 Question analogous to 13.8 for the results of Sections 9 and 11; for "elliptic regularization" corresponding to Section 8, see Strauss [1].

13.10 It would be of interest to systematically study the "smallest possible" spaces $K(\Omega)$, such that $V \subset K(\Omega)$, $\mathcal{D}(\Omega)$ is dense in $K(\Omega)$, when V is defined by

$$V = \{v \mid v \in L^2(\Omega), \quad \Lambda_i\, v \in L^2(\Omega), \quad \text{and boundary conditions}\},$$

where the Λ_i's are differential operators with constant or variable coefficients. The spaces $\Xi^s(\Omega)$, studied in Chapter 2 and the examples of Section 9.6 are particular and partial cases. We ignore whether *there exists* a minimal space with these properties.

13.11 General interpretation of equation (9.5) in case $f \in \Xi'$ (see also Section 11, Chapter 5, Volume 2).

13.12 Problem analogous to 13.11 for the case of Schroedinger equations (Section 11).

Bibliography[1]

Adams, R., Aronszajn, H., Smith, K. T.

1. Theory of Bessel potentials. Part II, Ann. Inst. Fourier **17**, 1—135 (1967).

Agmon, S.

1. Multiple layer potentials and Dirichlet problem for higher order elliptic equations in the plane. Comm. Pure Appl. Math. **10**, 179—239 (1957).
2. The coerciveness problem for integro-differential forms. J. Analyse Math. **6**, 183—223 (1958).
3. The L_p approach to the Dirichlet problem I. Ann. Sc. Norm. Sup. Pisa **13**, 405—448 (1959).
4. Remarks on self-adjoint and semi-bounded elliptic boundary value problems. Proc. Intern. Symp. on linear spaces, Jerusalem: Pergamon Press 1960, 1—13.
5. Problèmes mixtes pour les équations hyperboliques d'ordre supérieur. Colloques sur les équations aux dérivées partielles. C.N.R.S., Paris 1962, 13—18.
6. On the eigenfunctions and on the eigenvalues of general elliptic boundary value problems. Comm. Pure Appl. Math. **15**, 119—147 (1962).
7. Lectures on Elliptic Boundary Value problems, Princeton: Van Nostrand Mathematical Studies 1965.
8. On Kernels, eigenvalues and eigenfunctions of operators related to elliptic problems. Comm. Pure Appl. Math. **18**, 627—663 (1965).
9. Maximum properties for solutions of higher order elliptic equations. Bull. Amer. Math. Soc. **60**, 77—80 (1960).

Agmon, S., Douglis, A., Nirenberg, L.

1. Estimates near the boundary for solutions of elliptic partial differential equations satisfying general boundary conditions I. Comm. Pure Appl. Math. **12**, 623—727 (1959); II. Comm. Pure Appl. Math. **17**, 35—92 (1964).

Agmon, S., Kannai, Y.

1. On the asymptotic behavior of spectral functions and resolvent kernels of elliptic operators. Israel J. of Math. **5**, 1—30 (1967).

Agmon, S., Nirenberg, L.

1. Properties of solutions of ordinary differential equations in Banach spaces. Comm. Pure Appl. Math. **16**, 121—239 (1963).
2. Lower bounds and Uniqueness theorems for solutions of differential equations in a Hilbert space. Comm. Pure Appl. Math. **20**, 207—229 (1967).

[1] Some of the references are not explicitly needed until Volume 2; they are given here because of their close relationship with works which are analysed or cited in the present volume.

Agranovich, M. S.

1. Opérateurs elliptiques singuliers intégro-différentiels. Uspehi Mat. Nauk **20**, No. 5, 3—120, (1965) [Russian Math. Surv. **20**, No. 5—6, 1—121 (1965)].

Agranovich, M. S., Dynin, A. S.

1. Problèmes aux limites généraux pour les systèmes elliptiques en plusieurs dimensions. Dokl. Ak. Nauk **146**, 511—514 (1962) [Soviet Math. **3**, 1323 to 1327 (1962)].

Agranovich, M. S., Vishik, I. M.

1. Problèmes elliptiques avec paramètres et problèmes paraboliques de type général. Uspehi Mat. Nauk **19**, 53—161 (1964) [Russian Math. Surv. **19**, 53—157 (1964)].

Agranovich, M. S., Volevic, L. R., Dynin, A. S.

1. Solvability of general boundary-value problems for elliptic systems in higher-dimensional regions. Proc. Joint Soviet-American Symp. on Part. Diff. Equat., Novosibirsk, 1963, 1. 11.

Alexandrov, A. D.

1. Recherches sur le principe du maximum. Izv. Vyss. Ucebn. Zaned. Matematika, 1958, No. 5; 1959, No. 3 and 5; 1960, No. 3 and 5; 1961, No. 1.
2. La méthode de projection dans l'étude des solutions des équations elliptiques. Dokl. Ak. Nauk **169**, 751—754 (1966) [Soviet Math. **7**, 984—987 (1966)].

Amerio, L., Prouse, G.

1. Abstract almost periodic functions and functional analysis, New York: Van Nostrand 1970.

Arima, R., Mizohata, S.

Cf. Mizohata-Arima.

Arkeryd, L.

1. On the L^p estimates for elliptic boundary problems. Math. Scand. **19**, 59 to 76 (1966).

Aronszajn, N.

1. On coercive integro-differential quadratic forms. Tech. Report, No. 14, Univ. of Kansas (1955), 94—106.
2. Boundary values of functions with finite Dirichlet Integral. Tech. Report, No. 14, Univ. of Kansas (1955), 77—94.
3. Associated spaces, interpolation theorems and the regularity of solutions of differential operators. Partial Diff. Equations Proc. Symp. Pure Math., Vol. IV, A.M.S., 1961, 23—32.
4. Potentiels besséliens. Ann. Inst. Fourier **15**, 43—58 (1965).
5. The Rayleygh-Ritz and the Weinstein methods for approximations of eigenvalues (I. Operators in a Hilbert Space). Oklahoma Agr. Math. Coll., Stillwater Okla., 1949.
6. A unique continuation theorem for solutions of elliptic partial differential equations or inequalities of second order. J. Math. Pures appl. **36**, 235—249 (1957) [C. R. Ac. Sci., Paris **242**, 723—725 (1956)].

Aronszajn, N., Gagliardo, E.

1. Interpolation spaces and interpolation methods. Ann. Mat. pura appl. **4**, LXVIII, 51—118 (1965).

Aronszajn, N., Milgram, A. N.

1. Differential operators on Riemannian manifolds. Rend. Circ. Mat. Palermo **2**, 1—61 (1952).

Aronszajn, N., Mulla, F., Szeptycki, P.

1. On spaces of potentials connected with L^p classes. Ann. Inst. Fourier **13**, 211—306 (1963).

Aronszajn, N., Smith, K. T.

1. Functional spaces and functional completion. Ann. Inst. Fourier **6**, 125 to 185 (1956).
2. Theory of Bessel Potentials, Part I. Ann. Inst. Fourier **11**, 385—476 (1961).

Aronszajn, N., Adams, R., Smith, K. T.

Cf. Adams-Aronszajn-Smith.

Artola, M.

1. Equations paraboliques à retardement. C. R. Ac. Sci., Paris **264** A, 668 to 671 (1967).

Atiyah, M. F.

1. Global aspects of the theory of elliptic differential operators. Lecture given at the International Congress in Moscow (1966). Proceedings, 57—64.

Atiyah, M. F., Bott, R.

1. The index problem for manifolds with boundary. Bombay Coll. Diff. Analysis, Oxford: University Press 1964, 175—186.

Atiyah, M. F., Singer, I. M.

1. The index of elliptic operators on compact manifolds. Bull. Amer. Math. Soc. **69**, 422—433 (1963).

Avantaggiati, A.

1. Problemi al contorno per i sistemi ellittici simmetrici di equazioni lineari alle derivate parziali del primo ordine a coefficienti costanti in m ($\geqq 3$) variabili indipendenti. Ann. Mat. pura appl. **4**, LXI, 193—258 (1963).
2. Nuovi contributi allo studio dei problemi al contorno per i sistemi ellittici del primo ordine. Ann. Mat. pura appl. **4**, LXIX, 107—170 (1965).

Babich, V. M.

1. Sur le problème du prolongement des fonctions. Uspehi Mat. Nauk **8**, **2**, 111—113 (1953).

Bade, W. G., Freeman, R. S.

1. Closed extensions of the Laplace operator determined by a general class of boundary conditions. Pacific J. of Math. **12**, 395—410 (1962).

Baiocchi, C.

1. Su alcuni spazi di distribuzioni e sul problema di Dirichlet per le equazioni lineari ellittiche. Ricerche di Mat. **13**, 3—29 (1964).

2. Sui problemi ai limiti per le equazioni paraboliche del tipo del calore. Boll. U.M.I. **3**, 19, 407—422 (1964).
3. Regolarità e unicità della soluzione di una equazione differenziale astratta. Rend. Sem. Mat. Padova **35**, 380—417 (1965).
4. Sul problema misto per l'equazione parabolica del tipo del calore. Rend. Sem. Mat. Padova **36**, 80—121 (1966).
5. Un teorema di interpolazione; applicazioni ai problemi ai limiti per le equazioni a derivative parziali. Ann. Mat. pura appl. **4**, LXXIII, 235—252 (1966).
6. Soluzioni ordinarie e generalizzate del problema di Cauchy per equazioni differenziali astratte lineari del secondo ordine negli spazi di Hilbert. Ricerche di Mat. **16**, 27—95 (1967).
7. Sulle equazioni differenziali astratte lineari del primo e secondo ordine negli spazi di Hilbert. Ann. Mat. pura appl. **4**, LXXVI, 233—304 (1967).

Baouendi, M. S.

1. Sur une classe d'opérateurs elliptiques dégénérés, Thesis, Paris, 1966 [Bull. Soc. Math. France **95**, 45—87 (1967)].

Baouendi, M. S., Goulaouic, C.

1. Commutation de l'intersection et des foncteurs d'interpolation. C. R. Ac. Sci., Paris **265** A, 313—315 (1967).
2. Étude de la regularité et du spectre d'une classe d'opérateurs elliptiques dégénérés. C. R. Ac. Sci., Paris **266** A, 336—338 (1968).

Baouendi, M. S., Grisvard, P.

1. Sur une équation d'évolution changeant de type. C. R. Ac. Sci., Paris **265** A, 556—558 (1967).

Bardos, C.

1. Approximation semi-discrète de la solution d'une équation variationnelle, astreinte à vérifier des conditions aux limites dépendant du temps. Rend. Sem., Mat. Padova **38**, 41—59 (1967).

Bardos, C., Brezis, H.

1. Sur une classe de problèmes d'évolution non linéaires. J. Diff. Equations **6**, 345—394 (1969).

Barkoski, V. V., Roitberg, Ya. A.

1. Sur les opérateurs minimaux et maximaux relatifs aux problèmes aux limites non homogènes généraux. Ukrainskii Mat. J. **18**, No. 2, 91—96 (1966).

Barozzi, C.

1. Si una generalizzazione degli spazi $L^{q,\lambda}$ di Morrey. Ann. Sc. Norm. Sup. Pisa **19**, 609—626 (1965).

Barros-Neto, J.

1. Inhomogeneous boundary value problems in a half space. Ann. Sc. Norm. Sup. Pisa **19**, 331—365 (1965).

Beals, R.

1. Non local boundary value problems fro elliptic operators. Amer. J. Math. **87**, 315—348 (1965).
2. On eigenvalue distributions for elliptic operators without smooth coefficients. Bull. Amer. Math. Soc. **72**, 701—705 (1966).

Berezanski, Yu. M.

1. Quelques exemples de problèmes aux limites "non classiques" pour équations aux dérivées partielles. Dokl. Ak. Nauk **131**, 478—481 (1960) [Soviet Math. **1**, 259—262 (1960)].
2. Sur le problème de Dirichlet pour l'équation des cordes vibrantes. Uspehi Math. Nauk **15**, 363—372 (1960).
3. Espaces avec normes négatives. Uspehi Math. Nauk **18**, No. 1, 63—96 (1963) [Russian Math. Surv. **18**, No. 1, 63—95 (1963)].
4. Développements en fonctions propres d'opérateurs auto-adjoints, Kiev: Naukova Dumka 1965 [English translation: Vol. **17** Transl. Amer. Math. Soc. (1968)].
5. Sur les solutions généralisées des problèmes aux limites. Dokl Ak. Nauk **126**, 1159—1162 (1959).

Berezanski, Yu. M., Krein, S. G., Roitberg, Ya. A.

1. Un théorème sur les homéomorphismes et la croissance locale de régularité à la frontière pour les solutions des équations elliptiques. Dokl. Ak. Nauk **148**, 745—748 (1963) [Soviet Math. **4**, 152—155 (1963)].

Berezanski, Yu. M., Roitberg, Ya. A.

1. Théorèmes d'homomorphisme ... Journal Math. Ukrainien t. **19**, 3—32 (1967).

Bers, L.

1. An outline of the theory of pseudo analytic functions. Bull. Amer. Math. Soc. **62**, 291—375 (1956).

Bers, L., Nirenberg, L.

1. On linear and non linear elliptic boundary value problems in the plane, Convegno Int. Equaz. lin. der. parz. Trieste **1954**, Roma: Cremonese 1955, 141—167.

Besov, O. V.

1. Sur une famille d'espaces fonctionnels. Théorèmes de restriction et de prolongement. Dokl. Ak. Nauk **126**, 1163—1165 (1959).
2. Sur une famille d'espaces fonctionnels. Trudy Steklov **60**, 42—81 (1961) [Amer. Math. Soc. Transl. (2) **40**, 85—126 (1964)].
3. Théorèmes d'injection et de prolongement pour une classe de fonctions différentiables. Matemat. Zametkin **1**, No. 2, 235—244 (1967).
4. Prolongement des fonctions de $L_p^{\vec{l}}$ et $W_p^{\vec{l}}$. Trudy Steklov **89**, 5—17 (1967).

Besov, O. V., Il'in, V. P., Lizorkin, P. I.

1. L_p-estimations pour une classe d'intégrales singulières non isotropes. Dokl. Ak. Nauk **169**, 1250—1253 (1966) [Soviet Math. **7**, 1065—1069 (1966)].

Besov, O. V., Kadlec, Ja., Kufner, A.

1. Certaines propriétés d'espaces avec poids. Dokl. Ak. Nauk **171**, 514—416 (1966) [Soviet Math. **7**, 1497—1499 (1966)].

Bitzadze, A. V.

1. Sur l'unicité de la solution du problème de Dirichlet pour les équations elliptiques aux dérivées partielles. Uspehi Mat. Nauk **3**, No. **6**, 211—212 (1948).

2. Sur le problème homogène de la dérivée oblique pour les fonctions harmoniques en trois variables. Dokl. Ak. Nauk **148**, 749—752 (1963) [Soviet Math. **4**, 156—159 (1963)].
3. Le problème de la dérivée oblique avec coefficients polynômes. Dokl. Ak. Nauk **157**, 1273—1275 (1964) [Soviet Math. **5**, 1102—1104 (1964)].
4. Problèmes aux limites pour les équations elliptiques du deuxième ordre, Moscow 1966.

Bokobza, J., Unterberger, A.

Cf. Unterberger-Bokobza.

Bony, I. M., Courrège, P., Prioret, P.

1. Sur la forme intégro-différentielle du générateur infinitésimal d'un semigroupe de Feller sur une variété différentiable. C. R. Acad. Sc. Paris **263**, 207—210 (1966).

Bott, R., Atiyah, M. F.

Cf. Atiyah-Bott.

Bourbaki, N.

1. Intégration, Chapters 1/4, Act. Sc. Ind., Paris: Hermann 1966.

Boutet de Monvel, L.

1. Pseudo-noyaux de Poisson et opérateurs pseudo-différentiels sur une variété à bord. C. R. Ac. Sc. Paris **261**, 3927—3930 (1965).
2. Comportement d'un opérateur pseudo-différentiel sur une variété à bord, C. R. Ac. Sc. Paris, **261**, 4587—4589 (1965).

Bramble, J. H.

1. Error estimates for elliptic boundary value problems. Course C.I.M.E., Ispra, July 1967. Roma: Cremonese 1968.

Brelot, M., Choquet, G., Deny, J.

1. Séminaire sur la théorie du potentiel, Paris: Institut Poincaré 1957, 1958, ...

Brezis, H.

1. Inéquation d'évolution abstraites. C. R. Acad. Sc. Paris **264**, 732—735 (1967).
2. Equations et inéquations non linéaires dans les espaces vectoriels en dualité. Ann. Inst. Fourier **18**, 115—175 (1968).
3. Inéquations Variatonnelles. J. Math. Pures et Appl., (1972).

Brézis, H., Bardos, C.

Cf. Bardos-Brézis.

Brézis, H., Lions, J. L.

1. Sur certains problèmes unilatéraux hyperboliques. C. R. Acad. Sc. Paris **264**, 928—931 (1967).

Browder, F. E.

1. Estimates and existence theorems for elliptic boundary value problems. Proc. Nat. Acad. Sc. U.S.A. **45**, 365—372 (1959).

2. A priori estimates for solutions of elliptic boundary value problems. Indagationes Math. **22,** 145—169 (1960).
3. Functional Analysis and Partial Differential Equations, I, II. Math. Ann. **138**, 55—79 (1959); **145**, 81—226 (1962).
4. A priori estimates of elliptic and parabolic equations. Proc. Symp. pure math., Vol. IV, Partial diff. equat., Amer. Math. Soc. 73—83 (1961).
5. On the spectral theory of elliptic differential operators, I. Math. Ann. **142,** 22—130 (1961).
6. A continuity property for adjoints of closed operators in Banach spaces and its applications to elliptic boundary value problems. Duke Math. J. **28,** 157—182 (1961).
7. Non local elliptic boundary value problems. Amer. J. Math. **86**, 735—750 (1964).
8. Asymptotic distributions of eigenvalues and eigenfunctions for non local elliptic boundary value problems, I. Amer. J. Math. **87**, 175—195 (1965).
9. Strongly elliptic systems of differential equations, Contribution to the Theory of Partial Diff. Equat. Ann. Math. Studies, Princeton (1954), 15—51.
10. Linear parabolic equations. Proc. Nat. Ac. Sci. **49**, 185—190 (1963).
11. Parabolic systems of differential equations with time dependent coefficients. Proc. Nat. Ac. Sci. **42**, 914—917 (1956).

Butzer, P. L.

1. On Dirichlet's problem for the half-space and the behavior of its solution on the boundary. J. Math. Anal. Appl. **2**, 86—96 (1961).

Caccioppoli, R.

1. Sui teoremi di esistenza di Riemann. Ann. Sc. Norm. Sup. Pisa **6**, 177—187 (1937).

Calderon, A. P.

1. Lebesgue spaces of differentiable functions and distributions, Partial Diff. Equations. Proc. Symp. Pure Math., Vol. 4, A.M.S., 33—49 (1961).
2. Intermediate spaces and interpolation, Studia Math., Special Issue, t. I. Coll Anal. Fonct. of Warsaw (1963), 31—34.
3. Intermediate spaces and interpolation, the complex method. Studia Math. **24**, 113—190 (1964).
4. Boundary value problems for elliptic equations. Proc. Joint Soviet-American Symp. on Part. Diff. Equat., Novosibirsk (1963), 1—4.
5. Algebras of singular integral operators, Lecture given at the International Congress of Moscow (1966). Proceedings, 393—395.
6. Integrales singulares y sus aplicaciones a ecuaciones differentiales hiperbolicas, Universidad de Buenos Aires, fasc. 3, 1960.
7. Uniqueness in the Cauchy problem for partial differential equations. Amer. J. Math. **80**, 16—36 (1958).
8. The analytic calculation of the index of elliptic equations. Proc. Nat. Ac. Sci. **57**, 1193—1194 (1967).

Calderon, A. P., Zygmund, A.

1. On the existence of singular integrals. Acta Math. **88**, 85—139 (1952).
2. Singular integral operators and differential equations. Amer. J. Math. **79**, 901—921 (1957).

Campanato, S.

1. Sui problemi al contorno per sistemi di equazioni differenziali del tipo de l'elasticità, I e II. Ann. Sc. Norm. Sup. Pisa **12**, 223—258 and 275—302 (1959).
2. Sulla regolarità delle soluzioni di equazioni differenziali di tipo ellittico, Pisa: Ed. Tecnico Scientifiche 1963.
3. Proprietà di una famiglia di spazi funzionali. Ann. Sc. Norm. Sup. Pisa **18**, 137—160 (1964).
4. Equazioni ellittiche del secondo ordine e spazi $\mathscr{L}^{2,\lambda}$. Ann. Mat. pura appl. **4**, LXIX, 321—382 (1965).
5. Equazioni paraboliche del secondo ordine e spazi $\mathscr{L}^{2,\lambda}(\Omega, \delta)$. Ann. Mat. pura appl. **4**, LXXIII, 55—102 (1966).
6. Alcune osservazioni relative alle soluzioni di equazioni ellittiche d'ordine $2m$. Atti Conv. Equaz. a deriv. parz., Bologna, April 1967, 17—25.

Campanato, S., Stampacchia, G.

1. Sulle maggiorazioni in L^p nella teoria delle equazioni ellittiche. Boll. U.M.I. **20**, 393—399 (1965).

Canfora, A.

1. Teorema del massimo modulo e teorema di esistenza per il problema di Dirichlet relativo ai sistemi fortemente ellittici. Ricerche di Mat. **15**, 249—294 (1966).

Carroll, R.

1. Problems in linked operators I—II. Math. Ann. **151**, 272—282 (1963); **160**, 233—256 (1965).
2. On the propagator equation. Illinois J. Math. **11**, 506—522 (1967).

Cattabriga, L.

1. Su un problema al contorno relativo al sistema di equazioni di Stokes. Rend. Sem. Mat. Padova **31**, 1—33 (1961).
2. Su un certo spazio funzionale. Un teorema di tracce. Ann. Sc. Norm. Sup. Pisa **20**, 783—796 (1966).
3. Sulla connessione di un teorema di tracce con un certo poliedro convesso. Boll. U.M.I. (3), **22**, 1—12 (1967).
4. Alcuni teoremi di immersione per spazi funzionali generalizzanti gli spazi di S. L. Sobolev. Ann. Univ. Ferrara, Sc. Mat. **12**, 73—88 (1967).
5. Un problema al contorno per una equazione parabolica di ordine dispari. Ann. Sc. Norm. Sup. Pisa **13**, 163—203 (1959).

Cavallucci, A.

1. Sulle proprietà differenziali delle soluzioni delle equazioni quasi-ellittiche. Ann. Mat. pura e appl. **4**, LXVII, 143—168 (1965).
2. Alcuni teoremi di tracce. Sem. Mat. Fis. di Modena **15**, 137—157 (1966).
3. Costruzione di un rilevamento per la traccia iperpiana dello spazio $H_\mu(\mathbf{R}^n)$. Boll. U.M.I. (3), **22**, 1—6 (1967).

Chazarain, J.

1. Problèmes de Cauchy dans les espaces de distribution. C. R. Acad. Sc. **266**, 10—13 (1968).
2. Problèmes de Cauchy dans les espaces d'ultra distribution. C. R. Acad. Sc., Paris **266**, 564—566 (1968).

Cherruault, Y.

1. Approximation d'opérateurs linéaires et applications, Paris: Dunod 1968.

Choquet, G., Brelot, M., Deny, J.

Cf. Brelot-Choquet-Deny.

Cimmino, G.

1. Nuovo tipo di condizioni al contorno e nuovo metodo di trattazione per il problema generalizzato di Dirichlet. Rend. Circ. Mat. Palermo **61**, 177—224 (1937).
2. Sulle equazioni lineari alle derivate parziali di tipo ellittico. Rend. Sem. Mat. Fis. Milano **23**, 225—286 (1952).

Cooper, G.

1. Un problème relatif aux équations d'evolution abstraites. C. R. Acad. Sc. Paris **265**, 422—424 (1967).

Cordes, H. O.

1. Zero order à priori estimates for solutions of elliptic differential equations. Proc. Symp. Pure Math. **IV**, 157—166 (1961).
2. Über die Bestimmtheit der Lösungen elliptischer Differentialgleichungen durch Anfangsvorgaben. Nachr. Akad. Wiss., Göttingen **11**, 239—258 (1956).

Courant, R., Hilbert, D.

1. Methods of Mathematical Physics, Vol. 1 and 2, New York: Interscience Publishers 1953, 1962.

Courrège, P.

1. Book to be published.

Courrège, P., Bony, I. U., Prioret, P.

Cf. Bony-Courrège-Prioret.

Daletski, Yu. L.

1. Intégrales continues et équations opérationnelles d'évolution. Uspehi Mat. Nauk **17**, 3—115 (1962) [Russian Math. Surv. **17**, 1—105 (1962)].

Da-Prato, G.

1. Semi-gruppi di crescenza n. Ann. Sc. Norm. Sup. Pisa **20**, 753—782 (1966).
2. Semi-gruppi regolarizzabili. Ricerche di Matem. **15**, 223—248 (1966).
3. Spazi $\mathscr{L}^{p,\theta}(\Omega, \delta)$ e loro proprietà. Ann. Mat. pura appl. **4**, LXIX ,383—392 (1965).
4. Somma di generatori infinitesimali di semi gruppi ... Ann. Mat. Pura Appl. **4**, 78, 131—158 (1968).
5. Equations opérationnelles dans les espaces de Banach (cas analytique). C. R. Ac. Sci. Paris **266** A, 277—279 (1968).

Da-Prato, G., Giusti, E.

1. Equazioni di evoluzione in L^p. Ann. Sc. Norm. Sup. Pisa **21**, 487—505 (1967).
2. Equazioni di Schroedinger e delle onde per l'operatore di Laplace iterato in $L^p(\mathbf{R}^n)$. Ann. Mat. pura appl. **4**, LXXVI, 377—398 (1967).

Da-Prato, G., Mosco, U.

1. Semi-gruppi distribuzioni analitici. Ann. Sc. Norm. Sup. Pisa **19**, 367—396 and 563—576 (1965).

De-Giorgi, E.

1. Sulla differenziabilità e l'analiticità delle estremali degli integrali multipli. Mem. Acc. Sc. Torino **III**, 3, 25—43 (1957).
2. Un esempio di estremali discontinue per un problema variazionale di tipo ellittico. Boll. U.M.I. (4), **1**, 135—137 (1968).

Deny, J., Brelot, M., Choquet, G.

Cf. Brelot-Choquet-Deny.

Deny, J., Lions, J. L.

1. Les espaces du type de Beppo Levi. Ann. Inst. Fourier **5**, 305—370 (1953—54).

Deutsch, N.

1. Interpolation dans les espaces vectoriels topologiques localement convexes, Thesis. Memoire **13** Soc. Math. France, 1968.

De Wilde, M., Garnir, M. G., Schmets, J.

Cf. Garnir-De Wilde-Schmets.

Dezin, A. A.

1. Théorèmes d'existence et d'unicité pour la solution des problèmes aux limites pour les équations aux dérivées partielles dans des espaces fonctionnels. Uspehi Mat. Nauk **14**, 21—73 (1959) [Amer. Math. Soc. Transl. (2), **42**, 71—128 (1960)].

Dixmier, J.

1. Les algèbres d'opérateurs dans l'espace hilbertien, Paris: Gauthier-Villars 1957.

Douglis, A., Agmon, S., Nirenberg, L.

Cf. Agmon-Douglis-Nirenberg.

Douglis, A., Nirenberg, L.

1. Interior estimates for elliptic systems of partial differential equations. Comm. Pure Appl. Math. **8**, 503—538 (1955).

Dunford, N., Schwartz, J.

1. Linear Operators, I, II, New York: Interscience Publishers 1958, 1963.

Dynkin, E. B.

1. Markov Processes, Grundlehren, Vol. 108, Berlin/Heidelberg/New York: Springer 1965.

Dynin, A. S.

1. Problèmes aux limites elliptiques en dimension n et en une fonction inconnue. Dokl. Akad. Nauk **141**, 285—287 (1961) [Soviet Math. **2**, 1431—1433 (1961)].
2. Opérateurs singuliers d'ordre arbitraire sur une variété. Dokl. Ak. Nauk **141**, 21—23 (1961) [Soviet Math. **2**, 1375—1377 (1962)].

Dynin, A. S., Agranovich, M. S.

Cf. Agranovich-Dynin.

Dynin, A. S., Agranovich, M. S., Volevic, L. R.
Cf. Agranovich-Volevic-Dynin.

Egorov, Ju. V., Kondratiev, V. A.
1. Problème de dérivée oblique. Dokl. Ak. Nauk **170** (1966) [Soviet Mat. **7**, 1271—1273 (1966)].
2. Sur l'existence de la solution des problèmes aux limites avec dérivée oblique. Uspeh Mat. Nauk **22** (133), 165—167 (1967).

Eidelman, S., Lipko, B.
1. Contribution à la théorie des potentiels paraboliques. Dokl. Ak. Nauk **166**, 1050—1053 (1966) [Soviet Math. **7**, 237—240 (1966)].

Eskin, I. G., Vishik, I. M.
Cf. Vishik-Eskin.

Faedo, S.
1. Un nuovo metodo per l'analisi esistenziale e quantitativa dei problemi di propagazione. Ann. Sc. Norm. Sup. Pisa **1**, 1—40 (1949).

Fattorini, H. O.
1. Ordinary differential equations in linear topological spaces (I). J. Diff. Equations **5**, 72—105 (1969).

Fichera, G.
1. Sulla teoria generale dei problemi al contorno per le equazioni differenziali lineari. Rend. Acc. Lincei **21**, 46—55 e 156—172 (1956).
2. Linear elliptic equations of higher order in two independent variables and singular integral equations, with applications to anisotropic inhomogeneous elasticity. Part. Diff. Equat. and Continuum Mech., Madison: Univ. Wisconsin Press (1961), 55—80.
3. On a unified theory of boundary value problems of elliptic-parabolic equations of second order, Boundary value Problems in Diff. Equat., Madison: Univ. of Wisconsin Press, (1960), 97—120.
4. Linear elliptic differential systems and eigenvalue problems, Lecture Notes No. 8, Berlin/Heidelberg/New York: Springer 1965.

Figueiredo, D. G. de
1. The coerciveness problem for forms over vector valued functions. Comm. Pure Appl. Math. **16**, 63—94 (1963).

Finn, R., Serrin, J.
1. On the Hölder continuity of quasi-conformal and elliptic mappings. Trans. Amer. Math. Soc. **89**, 1—15 (1958).

Fishel, B.
1. Boundary value problems for second order, formally self-adjoint, elliptic differential equations. J. London Math. Soc. **33**, 62—70 (1958).

Foias, C.
1. Remarques sur les semi-groupes distributions d'opérateurs normaux. Portug. Math. **19**, 227—242 (1960).

Foias, C., Lions, J. L.

1. Sur certains théorèmes d'interpolation. Acta Sc. Math. Szeged **22**, 269—282 (1961).

Freeman, R. S.

1. On the spectrum and resolvent of homogeneous elliptic differential operators with constant coefficients. Bull. Amer. Math. Soc. **72**, 538—541 (1966).
2. On closed extensions of second order formally self-adjoint uniformly elliptic differential operators. Pacific J. of Math. **22**, 71—97 (1967).

Freeman, R. S., Bade, W. G.

Cf. Bade-Freeman.

Friberg, F.

1. Estimates for partially hypo-elliptic differential operators. Medd. Lund's Univ. Math. Sem. **17**, 1—97 (1963).

Friedman, A.

1. Generalized functions and partial differential equations, New York: Prentice-Hall 1963.
2. Partial differential equations of parabolic type. New York: Prentice-Hall 1964.
3. Uniqueness of solutions of ordinary differential inequalities in Hilbert space. Arch. Rat. Mech. Anal. **17**, 353—357 (1964).

Friedrichs, K. O.

1. On the differentiability of the solution of linear elliptic differential equations. Comm. Pure Appl. Math. **6**, 299—325 (1953).
2. Asymptotic phenomena in mathematical physics. Bull. Amer. Math. Soc. **61**, 485—504 (1955).
3. Symmetric positive linear differential equations. Comm. Pure Appl. Math. **11**, 333—418 (1958).
4. Spektraltheorie hallbeschränkter Operatoren. Math. Ann. **109**, 465—487; 685—713 (1934).

Fujiwara, D.

1. A Characterisation of exponential distribution semi-groups. J. Math. Soc. Japan **18**, 3, 267—274 (1966).
2. Concrete characterisation of the domains of fractional powers of some elliptic differential operators of the second order. Proc. Jap. Acad. **43**, 82—86 (1967).

Gagliardo, E.

1. Proprietà di alcune classi di funzioni in piu' variabili. Ricerche di Mat. **7**, 102—137 (1958); **8**, 24—51 (1959).
2. Caratterizzazione delle tracce sulla frontiera relative ad alcune classi di funzioni in n variabili. Rend. Sem. Mat. Padova **27**, 284—305 (1957).
3. Interpolation d'espaces de Banach et applications. C. R. Acad. Sc. Paris **248**, 1912—1914, 3388—3390, 3517—3518 (1959).
4. Interpolazione di spazi di Banach e applicazioni, Genova: Ediz. Scientifiche 1959.
5. Interpolazioni di spazi di Banach e applicazioni. Ricerche Mat. **9**, 58—81 (1960).

6. A common structure in various families of functional spaces. Quasi linear interpolation spaces, Part I, II. Ricerche mat. **10**, 244—281 (1961) et **12**, 87—107 (1963).

Gagliardo, E., Aronszajn, N.

Cf. Aronszajn-Gagliardo 1.

Gårding, L.

1. Dirichlet's problem for linear elliptic partial differential equations. Math. Scand. **1**, 55—72 (1953).
2. On the asymptotic distribution of the eigenvalues and eigenfunctions of elliptic differential operators. Math. Scand. **1**, 237—255 (1953).

Gårding, L., Malgrange, B.

1. Opérateurs partiellement hypo-elliptiques et partiellement elliptiques. Math. Scand. **9**, 5—21 (1961).

Garnir, H. G., De Wilde, M., Schmets, J.

1. Analyse Fonctionnelle. Théorie Constructive, Vol. 1. Basel: Birkhäuser 1968.

Gelfand, I. M.

1. Quelques problèmes d'analyse et d'équations différentielles. Uspehi Mat. Nauk **14**, No. 3, 3—19 (1959) [Amer. Math. Soc. Transl. (2), **26**, 201—218 (1963)].
2. Sur les équations elliptiques. Uspehi Mat. Nauk **15**, No. 3, 121—132 (1960) [Russian Math. Surv. **15**, 113—123 (1960)].

Gelfand, I. M., Šhilov, G. E.

1. Fonctions généralisées, Vol. I, II, III, Moscow 1958 (German translation: Berlin: VEB Deutscher Verlag der Wissenschaften; French translation: Paris: Dunod; English translation: New York: Academic Press).

Gelfand, I. M., Vilenkin, N. Ya.

1. Fonctions généralisées, Vol. IV, Moscow 1961 (English translation: New York: Academic Press; French translation: Paris: Dunod 1967).

Geymonat, G.

1. Sul problema di Dirichlet per le equazioni lineari ellittiche. Ann. Sc. Norm. Sup. Pisa **16**, 225—284 (1962).
2. Sui problema ai limiti per i sistemi lineari ellittici. Ann. Mat. Pura appl. **4**, LXIX, 207—284 (1965).
3. Su alcuni problemi ai limiti per i sistemi lineari ellittici secondo Petrowski. Le Matematiche **20**, 211—253 (1965).
4. Osservazioni su un teorema di prolungamento di R. T. Seeley. Boll. U.M.I. **22**, 1—8 (1967).
5. Osservazioni sulla convessità per alcune equazioni differenziali del secondo ordine. Rend. Ist. Lombardo (A) **99**, 921—932 (1965).

Geymonat, G., Grisvard, P.

1. Problèmes aux limites elliptiques dans L^p. Orsay, Jan.-March 1964.
2. Problemi ai limiti lineari ellittici negli spazi di Sobolev con peso. Le Matematiche **22**, 1—38 (1967).
3. Alcuni risultati di teoria spettrale per i problemi ai limiti lineari ellittici. Rend. Sem. Mat. Padova **38**, 121—173 (1967).

Gilbarg, D., Serrin, J.

1. On isolated singularities of solutions of second order elliptic differential equations. J. Analyse Math. **4**, 309—340 (1955—1956).

Girardeau, J. P.

1. Sur l'interpolation entre un espace localement convexe et son dual. Rev. Fac. Ciencias de Lisboa **11**, 165—186 (1965).

Giusti, G.

1. Equazioni quasi-ellittiche e spazi $\mathscr{L}^{p,\lambda}(\Omega, \delta)$, I et II. Ann. Mat. pura appl. **4**, LXXV, 313—354 (1967); Ann. Sc. Norm. Sup. Pisa **21**, 353—372 (1967).

Giusti, G., Da-Prato, G.

Cf. Da Prato-Giusti.

Gobert, J.

1. Opérateurs matriciels de dérivations elliptiques et problèmes aux limites. Mém. Soc. Roy. Sc. Liège, coll. in-4^0, **6** (1961).
2. Une inégalité fondamentale de la théorie de l'élasticité. Bull. Soc. Roy. Sc. Liège **31**, No. 3—4, 182—191 (1962).

Gohberg, I. C., Krein, M. G.

1. Les propositions fondamentales sur le défaut, le nombre des racines et l'indice des opérateurs linéaires. Uspehi Math. Nauk **12**, No. 2, 43—119 (1957) [Amer. Math. Soc. Transl. (2) **13**, 185—264 (1960)].

Goldstein, J. A.

1. Abstract evolution equations. Trans. Amer. Math. Soc. **141**, 159—185 (1969).

Golovkin, K. K.

1. Trudi Mat. Inst. Steklov (1970).

Goulaouic, C.

1. Prolongements de foncteurs d'interpolation et applications. Thesis, Paris, 1967; Ann. Inst. Fourier **18**, 1—98 (1968).

Goulaouic, C., Baouendi, M. S.

Cf. Baouendi-Goulaouic.

Green, J. W.

1. An expansion method for parabolic partial differential operators. J. Res. Nat. Bur. Stand. **51**, 127—132 (1953).

Grisvard, P.

1. Espaces intermédiaires entre espaces de Sobolev avec poids. Ann. Sc. Norm. Sup. Pisa **17**, 255—296 (1963).
2. Identités entre espaces de traces. Math. Scand. **13**, 70—74 (1963).
3. Problèmes aux limites elliptiques dans L^2. Séminaire Lions-Schwartz, Paris, 1964—1965.
4. Commutativité de deux foncteurs d'interpolation et applications. Journal de Liouville **45**, 143—290 (1966).
5. Espaces d'interpolation et équations opérationnelles. C. R. Acad. Sc. Paris **260**, 1536—1538 (1965).

6. Problèmes aux limites résolus par le calcul opérationnel. C. R. Acad. Sc. Paris **262**, 1306—1308 (1966).
7. Equations opérationnelles abstraites dans les espaces de Banach et problèmes aux limites dans des ouverts cylindriques. Ann. Sc. Norm. Pisa **21**, 307—347 (1967).
8. Caractérisation de quelques espaces d'interpolation. Ark. Rat. Mech. Anal. **25**, 40—63 (1967).
9. Equations différentielles abstraites. Cours Peccot, Collège de France, Paris 1968, Ann. Sc. Ec. Norm. Sup. (4), **2**, 311—395 (1969).

Grisvard, P., Baouendi, M. S.

Cf. Baouendi-Grisvard.

Grisvard, P., Geymonat, G.

Cf. Geymonat-Grisvard.

Grothendieck, A.

1. Sur certains espaces de fonctions holomorphes. J. reine angew. Math., Ed. 192, 35—64, 77—95 (1953).
2. Produits tensoriels topologiques et espaces nucléaires. Memoirs of the Amer. Mat. Society, No. 16, 1955.

Grubb, G.

1. A characterisation of the non local boundary value problems associated with an elliptic operator, July 1966, Departement of Math., Stanford University. Ann. Sc. Norm. Sup. Pisa **22**, 425—513 (1968).

Grusin, V. V., Vainberg, B. R.

Cf. Vainberg-Grusin.

Guzeva, O.

1. Sur les problèmes aux limites pour les systèmes fortement elliptiques. Dokl. Akad. Nauk **102**, 1069—1072 (1955).

Hanna, M. S., Smith, K. T.

1. The Dirichlet problem in polyhedra. Comm. Pure Appl. Math. **20**, 575 593 (1967).

Hardy, G. G., Littlewood, D. E., Polya, G.

1. Inequalities, Cambridge 1952.

Heinz, E.

1. Über die Eindeutigkeit beim Cauchyschen Anfangswertproblem einer elliptischen Differentialgleichung zweiter Ordnung. Nachr. Akad. Wiss. Göttingen **1**, 1—12 (1955).

Hervé, R. M.

1. Recherches axiomatiques sur la théorie des fonctions surharmoniques et du potentiel. Ann. Inst. Fourier, Grenoble **12**, 415—571 (1962).

Hestenes, M. R.

1. Applications of the theory of quadratic forms in Hilbert space to the Calculus of variations. Pacific J. Math. **1**, 525—581 (1951).

2. Extension of the range of a differentiable function. Duke Math. J. **8**, 183 to 192 (1941).

Hilbert, D., Courant, R.

Cf. Courant, Hilbert.

Hille, E., Phillips, R. S.

1. Functional Analysis and Semi-groups, A.M.S. Coll. Pub., XXXI, 1957.

Hiraka, Y., Shiraishi, R.

Cf. Shiraishi-Hiraka.

Hörmander, L.

1. On the theory of general partial differential operators. Acta Math. **94**, 161 to 248 (1955).
2. On the interior regularity of the solutions of partial differential equations. Comm. pure appl. Math. **11**, 197—218 (1958).
3. On the regularity of the solutions of boundary problems. Acta Math. **99**, 225—264 (1958).
4. Definition of maximal differential operators. Ark. Mat. **3**, 501—504 (1958).
5. Weak and stong extensions of differential operators. Comm. Pure appl. Math. **14**, 371—379 (1961).
6. Linear Partial Differential Operators, Grundlehren, Vol. 116, Berlin/Göttingen/Heidelberg: Springer 1963.
7. Pseudo-differential operators. Comm. pure appl. math. **18**, 501—517 (1965).
8. Pseudo-differential operators and non elliptic boundary problems. Ann. of Math. **83**, 129—209 (1966).
9. Pseudo-differential operators and hypoelliptic equations. Proc. Symp. on singular integrals, Chicago, April 1966. Amer. Math. Soc. **10**, 138—183 (1967).
10. An introduction to complex analysis in several variables, Princeton: Van Nostrand 1966. Amer. Math. Soc., Vol. 10, 138—183.
11. Hypoelliptic second order differential equations. Acta Math. **119**, 147—171 (1967).

Hörmander, L. Lions, J. L.

1. Sur la complétion par rapport à une intégrale de Dirichlet. Math. Scand. **4**, 259—270 (1956).

Horvath, J.

1. Topological vector spaces and distributions, Reading, Mass.: Addison-Wesley 1966.

Huet, D.

1. Phénomènes de perturbations singulières dans les problèmes aux limites. Ann. Inst. Fourier **10**, 1—96 (1960).
2. Phénomènes de perturbations singulières relatives au problème de Dirichlet dans un demi-espace. Ann. Sc. Norm. Sup. Pisa **18**, 425—448 (1964).
3. Sur quelques problèmes de perturbation singulière dans les espaces L_p. Revista Fac. Ciencias de Lisboa XI, 137—164 (1965).
4. Remarque sur un théorème d'Agmon et applications à quelques problèmes de perturbation singulière. Boll. U.M.I. **3**, 21, 219—227 (1966).
5. Perturbations singulières et régularité. C. R. Acad. Sc. Paris **256** A, 316 to 318 (1967).

Il'in, V. P.

1. Sur un théorème de Hardy et Littlewood. Trudy Mat. Inst. Steklova **53**, 128—144 (1959).

Il'in, V. P., Besov, D. V., Lizorkin, P. I.

Cf. Besov-Il'in-Lizorkin.

Itano, M.

1. On a trace theorem for the space $H^{\mu}(\mathbf{R}^N)$. J. Sci. Hiroshima Univ. **30**, 11—29 (1966).

Jamet, P.

1. (To be published).

John, F.

1. Plane waves and spherical means applied to partial differential equations, New York: Intersc. Publ. 1955.

Jones, B. F.

1. A class of singular integrals. Amer. J. Math. **86**, 441—462 (1964).

Jorgens, K.

1. An asymptotic expansion in the theory of neutron transport. Comm. Pure Appl. Math. **11**, 219—242 (1958).

Kadlec, J.

1. Solution du premier problème aux limites pour une généralisation de l'équation de la chaleur dans des classes de fonctions avec dérivée fractionnaire par rapport au temps. Cecoslovack Mat. J. **16**, 91—113 (1966).

Kadlec, J., Kufner, A.

1. Characterization of functions with zero traces by integrals with weight functions. Casopis pro jestov. mat. **91**, 463—471 (1966).

Kadlec, J., Kufner, A., Besov, O. V.

Cf. Besov-Kadlec-Kufner.

Kadlec, J., Necas, J.

1. Sulla regolarità delle soluzioni di equazioni ellittiche negli spaci $H^{k,\lambda}$. Ann. Sc. Norm. Sup. Pisa **21**, 527—546 (1967).

Kannai, Y., Agmon, S.

Cf. Agmon-Kannai.

Kaniel, S., Schechter, M.

1. Spectral theory for Fredholm operators. Comm. Pure Appl. Math. **16**, 423 to 448 (1963).

Kaplan, S.

1. Abstract boundary value problems for linear parabolic equations. Ann. Sc. Norm. Sup. Pisa **20**, 395—420 (1966).

Kato, T.

1. Integration of the equation of evolution in a Banach space. J. Mat. Soc. Japan **5**, 208—234 (1953).

2. Abstract evolution equations of parabolic type in Banach and Hilbert spaces. Nagoya Math. J. **19**, 93—125 (1961).
3. Fractional powers of dissipative operators. J. Math. Soc. Japan **13**, 246—274 (1961).
4. A Generalization of the Heinz inequality. Proc. Japan Acad. **6**, 304—308 (1961).
5. Perturbation theory for linear operators, Grundlehren Vol. 132, Berlin/Heidelberg/New York: Springer 1966.

Kato, T., Tanabe, H.

1. On the evolution equation. Osaka Math. J. **14**, 107—133 (1962).

Kohn, J. J., Nirenberg, L.

1. Non coercive boundary value problems. Comm. Pure Appl. Math. **18**, 443 to 492 (1965).
2. On algebra of pseudo-differential operators. Comm. Pure Appl. Math. **18**, 269—305 (1965).
3. Degenerate elliptic-parabolic equations of second order. Comm. Pure Appl. Math. **20**, 797—872 (1967).

Komatsu, H.

1. Fractional powers of operators. Pac. J. Math. **19**, 285—346 (1966).
2. Id. II. Pac. J. Math. **21**, 89—111 (1967).

Kondratiev, V. A.

1. Problèmes aux limites pour équations elliptiques dans des domaines coniques. Dokl. Ak. Nauk **153**, 27—29 (1963) [Soviet Math. **4**, 1600—1602 (1963)].

Kondratiev, V. A., Egorov, Ju. V.

Cf. Egorov-Kondratiev.

Koselev, A. S.

1. A priori L_p estimates and generalized solutions of elliptic equations and systems. Uspehi Mat. Nauk **13**, 29—89 (1958) [Amer. Math. Soc. Transl. (2) **20**, 105—171 (1962)].

Kostiouchenko, A. G., Levitan, K. M.

1. Sur le comportement asymptotique des valeurs propres des problèmes opérationnels du type Sturm-Liouville. Analyse Fonct. et Applic. **1**, 1, 86—96 (1967).

Kotake, T., Narasimhan, M. S.

1. Fractional power of a linear elliptic operator. Bull. Soc. Math. France **90**, 449—471 (1962).

Krée, P.

1. Sur les multiplicateurs dans $\mathscr{F} L^p$. Ann. Inst. Fourier **16**, 2, 31—89 (1966).
2. Sur les multiplicateurs dans $\mathscr{F} L^p$ avec poids. Ann. Inst. Fourier **16**, 2, 91—121 (1966).
3. Propriétés de continuité dans L^p de certains noyaux. Boll. U.M.I. (3), **22**, 330—344 (1967).
4. Distributions quasi homogènes. Généralisation des intégrales singulières et du calcul symbolique de Calderon-Zygmund. C. R. Ac. Sc. Paris **261**, 2560 to 2563 (1965).

Krein, M. G., Gohberg, I. C.
Cf. Gohberg-Krein.

Krein, S. G.
1. Sur quelques classes de problèmes aux limites bien posés. Dokl. Ak. Nauk **114**, 1162—1165 (1957).
2. Sur un théorème d'interpolation dans la théorie des opérateurs. Dokl. Ak. Nauk **130**, 491—494 (1960) [Soviet Math. **1**, 61—64 (1960)].
3. Sur la notion d'échelle normale d'espace. Dokl. Ak. Nauk **132** 510—513 (1960) [Soviet Math. **1**, 586—589 (1960)].
4. Linear differential equations in Banach spaces, Moscow 1967.
5. Problème de Cauchy bien posé et analyticité de la solution de l'équation d'évolution. Dokl. Ak. Nauk **171** (1966) [Soviet Math. **7**, 1594—1598 (1966)].

Krein, S. G., Laptiev, G.
1. Boundary value problems for second order differential equations in Banach spaces, I, II. Diff. Equations, II **3**, 382—390 (1966); **7**, 919—926 (1966).

Krein, S. G., Petunin, Yu. I.
1. Echelles d'espaces de Banach. Uspehi Mat. Nauk **21**, No. 2, 89—168 (1966) [Russian Math. Surv. **21**, No. 2, 85—159 (1966)].

Krein, S. G., Petunin, Yu. I., Semenov, E. M.
1. Hyperechelles de structures de Banach. Dokl. Ak. Nauk **170**, 265—267 (1966) [Soviet Math. **7**, 1185—1188 (1966)].

Krein, S. G., Roitberg, Ya. I., Berezanski, Yu. M.
Cf. Berezanski-Krein-Roitberg.

Kudryavcev, L. D.
1. Fonctions généralisées et injection pour des classes de fonctions. Applications à la résolution par la méthode variationnelle des équations elliptiques. Trudi Steklov **55**, 1—181 (1959).
2. La méthode variationnelle pour les domaines non bornés. Dokl. Ak. Nauk **157**, 45—48 (1964) [Soviet Math. **5**, 887—890 (1966)].

Kufner, A., Kadlec, J.
Cf. Kadlec-Kufner.

Kufner, A., Kadlec, J., Besov, O. V.
Cf. Besov-Kadlec-Kufner.

Ladyženskaya, O. A.
1. Problèmes mixtes pour les équations hyperboliques, Moscow 1953.
2. Sur la solution d'équations opérationnelles non stationnaires. Mat. Sbornik **39**, 491—524 (1956).
3. Sur les équations opérationnelles non stationnaires et leurs applications aux problèmes linéaires de la Physique Mathématique. Mat. Sbornik **45**, 123 to 158 (1958).

Ladyženskaya, O. A., Uraltzeva, N. N.
1. Equations elliptiques linéaires et quasi linéaires, Moscow 1964.

Ladyženskaya, O. A., Vishik, I. M.

Cf. Vishik-Ladyženskaya.

Landis, E. M.

1. Sur certaines propriétés qualitives dans la théorie des équations elliptiques et paraboliques. Uspehi Mat. Nauk **14**, 21—85 (1959) [Amer. Math. Soc. Transl. (2), **20**, 173—238 (1962)].

Laptiev, G., Krein, S. G.

Cf. Krein-Laptiev.

Lax, P. D.

1. On Cauchy's problem for hyperbolic equations and the differentiability of the solutions of elliptic equations. Comm. Pure Appl. Math. **8**, 615—653 (1955).
2. A Phragmen-Lindelöf theorem in harmonic analysis and its applications to some questions in the theory of elliptic equations. Comm. Pure Appl. Math. **10**, 361—389 (1957).

Lax, P. D., Milgram, N.

1. Parabolc equations, Contributions to the theory of partial differential equations. Annals of Math. Studies, No. 33, Princeton (1954), 167—190.

Lax, P. D., Phillips, R. S.

1. Local boundary conditions for dissipative symmetric linear differential operators. Comm. Pure Appl. Math. **13**, 427—455 (1960).
2. Scattering Theory. New York: Academic Press 1967.

Lavrentiev, M. A.

1. Un problème général de la théorie de la représentation quasi conforme des domaines plans. Mat. Sbornik **21**, 285—320 (1947).

Lawruk, B.

1. Problèmes aux limites paramétriques pour systèmes elliptiques d'équations différentielles linéaires. Bull. Akad. Polon. Sci. Ser. Math. II, 257—267 et 269—278 (1963).

Leray, J.

1. Hyperbolic differential equations, Princeton, 1952.
2. Etude de diverses équations intégrales non linéaires et de quelques problèmes que pose l'hydrodynamique. J. Math. Pures Appl. **12**, 1—82 (1933).
3. Essai sur les mouvements plans d'un liquide visqueux que limitent les parois. J. Math. Pures Appl. **13**, 331—418 (1934).
4. Sur les mouvements d'un liquide visqueux remplissant l'espace. Acta Math. **63**, 193—248 (1934).

Levi, E. E.

1. Sulle equazioni lineari totalmente ellittiche alle derivate parziali. Rend. Circ. Mat. Palermo **24**, 275—317 (1907).

Levitan, V. M.

1. Book to be published.

Levitan, V. M., Kostiouchenko, A. G.

Cf. Kostiouchenko-Levitan.

Lichtenstein, L.

1. Eine elementare Bemerkung zur reellen Analysis. Math. Zeit. **30**, 794—795 (1929).

Lions, J. L.

1. Supports dans la transformation de Laplace. J. Anal. Math. Israel **2**, 369 to 380 (1952—1953).
2. Problèmes aux limites en théorie des distributions. Acta Math. **94**, 13—153 (1955).
3. Sur quelques problèmes aux limites relatifs à des opérateurs différentiels elliptiques. Bull. Soc. Math. France **83**, 225—250 (1955).
4. Some questions on elliptic equations. Tata Inst. of Fundamental Research, Bombay, 1957.
5. Sur les problèmes mixtes pour certains systèmes paraboliques dans des ouverts non cylindriques. Ann. Inst. Fourier **7**, 143—182 (1957).
6. Boundary value problems. Technical report No. 1—3. Lawrence: University of Kansas 1957.
7. Une remarque sur les applications du théoreme de Hille Yosida. J. Math. Soc. Japan **9**, 62—70 (1957).
8. Problèmes mixtes abstraits. Proc. Int. Congress of Math., Edinburgh, 1958, 389—397.
9. Espaces intermédiaires entre espaces hilbertiens et applications. Bull. Math. Soc. Sc. Math. Phys. R. P. Roumaine **2**, 50, 419—432 (1958).
10. Sur les problèmes aux limites du type dérivées oblique. Ann. of Math. **64**, 207—239 (1959).
11. Théorèmes de traces et d'interpolation (I) . . . (V); (I) (II) Ann. Sc. Norm. Sup. Pisa **13**, 389—403 (1959); **14**, 317—331 (1960); (III) Journal de Liouville **42**, 196—203 (1963); (IV) Math. Annalen **151**, 42—56 (1963); (V) Anals de Acad. Brasileira de Ciencias **35**, 1—110 (1963).
12. Les semi-groupes distributions. Port. Math. **19**, 141—164 (1960).
13. Equations différentielles opérationnelles et problèmes aux limites, Grundlehren Vol. **111**, Berlin/Göttingen/Heidelberg: Springer 1961.
14. Sur les espaces d'interpolation; dualité. Math. Scand. **9**, 147—177 (1961).
15. Une construction d'espace d'interpolation. C. R. Acad. Sc. Paris **251**, 1853 to 1855 (1961).
16. On a trace problem. Rend. Sem. Mat. Padova **31**, 232—242 (1961).
17. Espaces d'interpolation et domaines de puissances fractionnaires d'opérateurs. J. Math. Soc. Japan **14**, 233—241 (1962).
18. Remarques sur les espaces d'interpolation et les problèmes aux limites. Colloque C.N.R.S., No. 117. "Les équations aux dérivées partielles", Paris, June 1962, 75—86.
19. Properties of some interpolation spaces. J. Math. and Mech. **11**, 969—978 (1962).
20. Dérivées intermédiaires et espaces intermédiaires. C. R. Acad. Sc. Paris **256**, 4343—4345 (1963).
21. Equations différentielles opérationnelles dans les espaces de Hilbert. Centro Int. Mat. Estivo, Varenna, 1963 (Equazioni differenziali astratte. Roma: Cremonese 1963).
22. Remarques sur les équations différentielles opérationnelles. Osaka Math. Journal **15**, 131—142 (1963).
23. Quelques remarques sur les équations différentielles opérationnelles du premier ordre. Rend. Sem. Mat. Padova **33**, 215—225 (1963).

24. On some optimization problems for linear parabolic equations. Functional Analysis and Optimization. New York: Academic Press 1966.
25. Optimisation pour certaines classes d'équations d'évolution non linéaires. Ann. Mat. Pura appl. **4**, LXXII, 275—294 (1966).
26. Sur le contrôle optimal de systèmes décrits par des équations aux dérivées partielles linéaires, (I) Remarques générales, (II) Equations elliptiques, (III) Equations d'évolution. C. R. Acad. Sc. Paris **263**, 661—663; 713—715; 776—779 (1966).
27. Remarks on evolution inequalities. J. Math. Soc. Japan **18**, 331—342 (1966).
28. Remarks on control problems associated to Partial Diff. Equations. Conf. on the Math. theory of Control, Los Angeles, February 1967.
29. Problèmes aux limites non homogènes à données irrégulières; une méthode d'approximation, Cours C.I.M.E., Ispra, July 1967; Roma: Cremonese 1968.
30. Conditions aux limites de Vishik-Sobolev et problèmes mixtes. C. R. Acad. Sc. Paris **244**, 1126—1128 (1957).
31. Contribution à un problème de M. Picone. Ann. Mat. Pura Appl. **4**, XLI, 201—219 (1955).

Lions, J. L., Brézis, H.
Cf. Brézis-Lions.

Lions, J. L., Deny, J.
Cf. Deny-Lions.

Lions, J. L., Foias, C.
Cf. Foias-Lions.

Lions, J. L., Hörmander, L.
Cf. Hörmander-Lions.

Lions, J. L., Magenes, E.
1. Problèmes aux limites non homogènes (I) . . . (VI); (I) (III) (IV) (V), Ann. Sc. Norm. Sup. Pisa **14**, 259—308 (1960); **15**, 39—101 (1961); 311—326; **16**, 1—44 (1962); (II), Ann. Inst. Fourier **11**, 137—178 (1961); (VI), Journal d'Analyse Math. **11**, 165—188 (1963); (VII), Ann. Mat. Pura Appl. **4**, LXIII, 201—224 (1963).
2. Remarques sur les problèmes aux limites pour opérateurs paraboliques. C. R. Acad. Sc. Paris **251**, 2118—2120 (1960).
3. Remarques sur les problèmes aux limites linéaires elliptiques. Rend. Acc. Naz. Lincei **8**, 32, 873—883 (1962).

Lions, J. L., Nikolski, S. M., Lizorkin, P. I.
Cf. Nikolski-Lions-Lizorkin.

Lions, J. L., Peetre, J.
1. Sur une classe d'espaces d'interpolation, Institut des Hautes Études Scientifiques. Publ. Math. **19**, 5—68 (1964).

Lions, J. L., Raviart, P. A.
1. Remarques sur la résolution et l'approximation d'équations d'évolutions couplées. I.C.C. Bulletin **5**, 1—20 (1966).
2. Remarques sur la résolution, exacte et approchée, d'équations d'évolution paraboliques à coefficients opérateurs non bornés. Calcolo 221—234 (1967)

Lions, J. L., Stampacchia, G.

1. Variational inequalities. Comm. Pure Applied Math. **20**, 3, 493—519 (1967) [Development of the Note C. R. Acad. Sc. Paris **261**, 25—27 (1965)].

Lioubich, Yu. I.

1. Transformation de Laplace, classique et locale, et problème de Cauchy abstrait. Uspehi Mat. Nauk **21**, 3—51 (1966).

Lipko, B., Eidelman, S.

Cf. Eidelman-Lipko.

Littlewood, D. E., Hardy, B. G., Polya, G.

Cf. Hardy-Littlewood-Polya.

Littman, W.

1. Polar sets and removable singularities of partial differential equations. Arkiv für Mat. **7**, 1—9 (1967).

Littman, W., Stampacchia, G., Weinberger, H.

1. Regular points for elliptic equations with discontinuous coefficients. Ann. Sc. Norm. Sup. Pisa **17**, 45—79 (1963).

Ljusternik, L. A., Vishik, I. M.

Cf. Vishik-Ljusternik.

Lizorkin, P. S.

1. Potentiels de Bessel non isotropes. Théorèmes d'inclusion pour espaces de Sobolev $L_p^{(r_1, \ldots, r_n)}$ avec dérivées non entières. Dokl. Ak. Nauk **170**,508—511 (1966) [Soviet Math. **7**, 1222—1225 (1966)].

Lizorkin, P. S., Nikolski, S. M.

1. Sur certaines inégalités pour des espaces avec poids et les problèmes aux limites fortement dégénérés à la frontière. Dokl. Akad. Nauk **159**, 512—515 (1964) [Soviet Math. **5**, 1535—1539 (1964)].

Lizorkin, P. S., Lions, J. L., Nikolski, S. M.

Cf. Nikolskii-Lions-Lizorkin.

Lizorkin, P. S., Besov, O. V., Il'in, V. P.

Cf. Besov-Il'in-Lizorkin.

Lopatinski, Ya. B.

1. Une méthode de réduction des problèmes aux limites pour les systèmes d'équations différentielles elliptiques à une équation intégrale régulière. Ukrain. Mat. Zh. **5**, 123—151 (1953).

Magenes, E.

1. Sul problema di Dirichlet per le equazioni lineari ellittiche in due variabili. Ann. Mat. Pura Appl. **4**, XXXXVIII, 257—279 (1959).
2. Sur les problèmes aux limites pour les équations linéaires elliptiques, Colloques No. 117 du C.N.R.S. sur Les équations aux dérivées partielles, Paris, June 1962, 95—111.
3. Spazi di interpolazione ed equazioni a derivate parziali, Atti VII Congresso U.M.I., Genova 1963. Roma: Cremonese 1964, 134—197.

Magenes, E., Lions, J. L.

Cf. Lions-Magenes.

Magenes, E., Stampacchia, G.

1. I problemi al contorno per le equazioni differenziali di tipo ellittico. Ann. Sc. Norm. Sup. Pisa **12**, 247—357 (1958).

Malgrange, B.

1. Sur une classe d'opérateurs différentiels hypoelliptiques. Bull. Soc. Math. France **85**, 283—306 (1957).

Malgrange, B., Gårding, L.

Cf. Gårding-Malgrange.

Martyniuk, A. E.

1. Certaines nouvelles applications des méthodes du type Galerkin. Math. Sbornik **49**, 85—108 (1959).

Matsuzawa, T.

1. Regularity at the boundary for solutions of hypo-elliptic equations. Osaka J. Math. **3**, 313—334 (1966).
2. On quasi elliptic boundary problems. Trans. Amer. Math. Soc. **133**, 241—265 (1968).

Mikhlin, S. G.

1. Intégrales singulières en plusieurs dimensions et équations intégrales. Moscow: Fizmatgiz 1962 (English translation: New York: Pergamon Press).

Milgram, A. N., Aronszajn, M.

Cf. Aronszajn-Milgram.

Milgram, N., Lax, P. D.

Cf. Lax-Milgram.

Minakshisundaran, S., Plejel, A.

1. Eigenfunctions of the Laplace operator on Riemannian manifolds. Canada J. Math., 242—256, 1949.

Miranda, C.

1. Partial Differential Equations of Elliptic Type, Second revised Edition, Ergebnisse Math. Vol. 2, Berlin/Heidelberg/New York: Springer 1970 (First edition 1955).
2. Teorema del massimo modulo e teorema di esistenza e di unicità per il problema di Dirichlet relativo alle equazioni ellittiche in due variabili. Ann. Mat. Pura Appl. **4**, XLVI, 265—311 (1958).
3. Sulle equazioni ellittiche del secondo ordine di tipo non variazionale a coefficienti discontinui. Ann. Mat. Pura Appl. **4**, LIII, 353—386 (1963).
4. Teoremi di unicità in domini non limitati e teoremi di Liouville per le soluzioni dei problemi al contorno relativi alle equazioni ellittiche. Ann. Mat. Pura Appl. **4**, LIX, 189—212 (1962).
5. Sui sistemi di tipo ellittico di equazioni lineari a derivate parziali del primo ordine, in n variabili indipendenti. Mem. Acc. Lincei **3**, sez. I, 83—121 (1952).

Mizohata, S.

1. Quelques problèmes au bord, de type mixte, pour des équations hyperboliques. Séminaire Leray, Collège de France, Paris, December **1966**.
2. Hypoellipticité des équations paraboliques. Bull. Soc. Math. France **85**, 15—50 (1957).

Mizohata, S., Arima, R.

1. Propriétés asymptotiques des valeurs propres des opérateurs elliptiques auto-adjoints. J. Math. of Kyoto Univ. **4**, 245—254 (1964).

Morel, H.

1. Introduction de poids dans l'étude de problèmes aux limites. Ann. Inst. Fourier **12**, 299—414 (1962).

Morrey, C. B.

1. Second order elliptic systems of differential equations. Contributions to the theory of partial differential equations. Ann. of Math. Studies, No. 33, 101—159. Princeton: University Press 1954.
2. Multiple integrals in the calculus of variations. Grundlehren Vol. 130, Berlin/Heidelberg/New York: Springer 1966.
3. On the solutions of quasi-linear elliptic partial differential equations. Trans. Amer. Math. Soc. **43**, 126—166 (1938).
4. Multiple integral problems in the calculus of variations and related topics. Univ. of California Publ. in Math. **1**, 1—130 (1943).
5. Partial regularity results for non linear elliptic systems. J. Math. and Mech. **17**, 649—670 (1968).

Mosco, M., Da-Prato, G.

Cf. Da Prato-Mosco.

Müller, C.

1. On the behavior of the solutions of the differential equation $\Delta u = F(x, u)$ in the neighborhood of a point. Comm. Pure Appl. Math. **7**, 505—515 (1954).

Murthy, V., Stampacchia, G.

1. Equazioni ellittiche che degenerano. Conv. Eq. deriv. parziali, February 1965. Roma: Cremonese 90—96, and Ann. Mat. Pura Appl. 80 (1968).

Muskhelisvili, N. I.

1. Equations intégrales singulières, Moscow 1946 (English translation: Groningen: Noordhoff).

Mulla, F., Aronszajn, N., Szeptycki, P.

Cf. Aronszajn-Mulla-Szeptycki.

Nagy, B. Sr., Riesz, F.

Cf. Riesz-Nagy.

Narasimhan, M. S., Kotake, T.

Cf. Kotake-Narasimhan.

Nash, J.

1. Continuity of the solution of parabolic and elliptic equations. Amer. J. Math. **80**, 931—954 (1958).

Necas, J.

1. Sur les normes équivalentes dans $W_p^{(k)}(\Omega)$ et sur la coercivité des formes formellement positives. Sémin. de Math. Super., été 1965, Montréal.
2. Les méthodes directes dans la théroie des équations elliptiques, Prague: Editions de l'Académie Tchécoslovaque de Sciences 1967.
3. Sur la coercivité des formes sesquilinéaires elliptiques. Rev. Math. Pures Appl. **9**, 47—69 (1964).
4. L'application de l'égalité de Rellich pour résoudre les problèmes aux limites, Séminaire Leray, Collège de France, Paris, 1963.

Necas, J., Kadlec, J.

Cf. Kadlec-Necas.

Nelson, E.

1. L'équation de Schroedinger et les intégrales de Feyman, Colloque C.N.R.S., No. 117, "Les équations aux derivées partielles", June 1962, 151—158.

Nikolski, S. M.

1. Théorèmes d'inclusion, de prolongement et d'approximation pour les fonctions différentiables de plusieurs variables. Uspehi Mat. Nauk **16**, No. 5, 63—114 (1961) [Russian Math. Surv. **16**, No. 5, 55—104 (1961)].
2. Le premier problème aux limites pour une équation linéaire générale. Dokl. Ak. Nauk **146**, 767—769 (1962) [Soviet Math. **3**, 1388—1390 (1962)].
3. Sur les propriétés à la frontière des fonctions de plusieurs variables. Dokl. Ak. Nauk **146**, 542—546 (1962) [Soviet Math. **3**, 1357—1360 (1962)].
4. Une représentation contructive de la classe zéro des fonctions différentiables de plusieurs variables. Dokl. Ak. Nauk **170**, 512—515 (1966) [Soviet Math. **7**, 1227—1230 (1966)].
5. Théorie de l'approximation des fonctions différentiables de plusieurs variables et théorèmes de plongement, Moscow, 1968.

Nikolski, S. M., Lizorkin, P. I.

Cf. Lizorkin-Nikolski.

Nikolski, S. M., Lions, J. L., Lizorkin, P. I.

1. Integral representation and isomorphism properties of some classes of functions. Ann. Sc. Norm. Sup. Pisa **19**, 127—178 (1965).

Nirenberg, L.

1. On non linear elliptic partial differential equations and Hölder continuity. Comm. Pure Appl. Math. **6**, 103—155 (1953).
2. Remarks on strongly elliptic partial differential equations. Comm. Pure Appl. Math. **8**, 648—674 (1955).
3. On elliptic partial differential equations. Ann. Sc. Norm. Sup. Pisa **13**, 115—162 (1959).

Nirenberg, L., Agmon, S.

Cf. Agmon-Nirenberg.

Nirenberg, L., Agmon, S., Douglis, A.

Cf. Agmon-Douglis-Nirenberg.

Nirenberg, L., Bers, L.

Cf. Bers-Nirenberg.

Nirenberg, L., Douglis, A.

Cf. Douglis-Nirenberg.

Nirenberg, L., Kohn, J.

Cf. Kohn-Nirenberg.

Odhnoff, J.

1. Un exemple de non-unicité d'une équation différentielle opérationnelle. C. R. Acad. Sc. Paris **258**, 1689—1691 (1964).

Oleinik, O. A.

1. Sur un problème de G. Fichera. Dokl. Ak. Nauk **157**, 1297—1300 (1964) [Soviet Math. **5**, 1129—1133 (1964)].
2. Alcuni risultati sulle equazioni lineari e quasi lineari ellittico-paraboliche a derivate parziali del secondo ordine. Bend. Accad. Naz. Lincei (8), XL, 775—784 (1966).
3. Sur les équations linéaires du deuxième ordre à forme caracteristique non négative. Mat. Sbornik **69**, 111—140 (1966).
4. Problèmes de Cauchy et aux limites pour équations hyperboliques du deuxième ordre dégénérant dans le domaine et sur la frontière. Dokl. Ak. Nauk **169**, 525—528 (1966) [Soviet Math. **7**, 969—973 (1966)].
5. Sur les équations de type elliptique qui dégénèrent à la frontière. Dokl. Ak. Nauk **87**, 885—888 (1952).
6. Sur les équations de type elliptique avec un petit paramètre dans les dérivées d'ordre maximum. Mat. Sbornik **31**, 104—117 (1952).

Pagni, M.

1. Problemi al contorno per una certa classe di equazioni lineari alle derivate parziali. Atti Sem. Mat. Fis. Modena **13**, 119—164 (1964).
2. Un teorema di trace. Rend. Acc. Naz. Lincei, VII, **38**, 627—631 (1965).
3. Problemi con condizioni al contorno omogenee per una certa classe di operatori formalmente ipoellittici. Ann. Mat. Pura Appl. **4**, LXXII, 201—212 (1966).
4. Su un problema di tracce. Atti Sem. Mat. Fis. Modena **15**, 175—181 (1966).

Palais, R., and others

1. Seminar on the Atiyah-Singer index Theorem. Annals of Math. Studies, No. 57, Princeton, 1965.

Panejach, B. P., Volevic, L. R.

Cf. Volevich-Panejach.

Pederson, R. N.

1. On the unique continuation theorem for certain second and fourth order elliptic equations. Comm. Pure Appl. Math. **11**, 67—80 (1958).

Peetre, J.

1. Théorèmes de régularité pour quelques classes d'opérateurs différentiels. Med. Lund Univ. Mat. Sem. **16**, 1—122 (1959).
2. Another approach to elliptic boundary problems. Comm. pure appl. Math. **14**, 711—731 (1961).
3. Mixed problems for higher order elliptic equations in two variables, I. Ann. Sc. Norm. Sup. Pisa **15**, 337—353 (1961).

4. Elliptic partial Differential equations of higher order, Lec. series No. **40**, Inst. for Fluid Dynamics and app. Math., University of Maryland, 1962.
5. Nouvelles propriétés d'espaces d'interpolation. C. R. Acad. Sc. Paris **256**, 54—55 (1963).
6. A theory of interpolation of normed spaces. Lectures Brasilia, 1963 (Notas de matematica, No. 39, 1968).
7. On the theory of interpolation spaces. Rev. Un. Mat. Argentina **23**, 49—66 (1967).
8. Espaces d'interpolation, généralisations, applications. Rend. Sem. Mat. Fis. Milano **34**, 133—161 (1964).
9. Sur la théorie des semi-groupes distributions. Sem. Leray, Collège de France, Nov. 1963—May 1964, 79—98.
10. On an interpolation theorem of Foias and Lions. Acta Szeged **25**, 255—261 (1964).
11. Some remarks on continuous orthogonal expansions, and eigenfunction expansions for positive self-adjoint elliptic operators with variable coefficients. Math. Scand. **17**, 56—64 (1965).
12. Espaces d'interpolation et théorème de Sobolev. Ann. Inst. Fourier **16**, 279 to 317 (1966).
13. A formula for the solution of Dirichlet's problem (to be published in Summa Brasil Math.).

Peetre, J., Lions, J. L.

Cf. Lions-Peetre.

Petrowski, J. G.

1. Sur l'analyticité des solutions des systèmes d'équations différentielles. Mat. Sbornik **5**, 47, 3—70 (1939).

Petryshyn, W. V.

1. On a class of *K-p-d* and non *K-p-d* operators and Operator equations. J. Math. Anal. and Applic. **10**, 1—24 (1965).

Petunin, Yu. I., Krein, S. G.

Cf. Krein-Petunin.

Petunin, Yu. I., Krein, S. G., Semenov, E. M.

Cf. Krein-Petunin-Semenov.

Phillips, R. S.

1. Dissipative operators and hyperbolic systems of partial differential equations. Trans. Amer. Math. Soc. **90**, 123—254 (1959).
2. Semi groups of contraction operators, Course C.I.M.E., May—June 1963, Roma: Cremonese 1—48.

Phillips, R. S., Hille, E.

Cf. Hille-Phillips.

Phillips, R. S., Lax, P.

Cf. Lax-Phillips.

Pini, B.

1. Una problema di valori al contorno generalizzato per la equazione lineare a derivate parziali lineare parabolica del secondo ordine. Riv. Mat. Univ. Parma **3**, 153—187 (1952).

2. Sui sistemi di equazioni lineari a derivate parziali del second ordine dei tipi ellittico e parabolico. Rend. Sem. Mat. Padova **22**, 265—280 (1953).
3. Sull'unicità della soluzione del problema di Dirichlet per le equazioni lineari ellittiche in due variabili. Rend. Sem. Mat. Padova **26**, 223—231 (1956).
4. Sul problema di Dirichlet per le equazioni a derivate parziali lineari ellittiche in due variabili. Rend. Sem. Mat. Padova **26**, 177—200 (1956).
5. Sul problema fondamentale di valori al contorno per una classe di equazioni paraboliche lineari. Ann. Mat. pura appl. **4**, XLIII, 261—297 (1957).
6. Sulle equazioni lineari pseudoparaboliche, I and II. Rend Sem. Mat. Padova **30**, 255—280 and 361—375 (1960).
7. Sui sistemi lineari pseudoparabolici. Ricerche di Mat. **10**, 33—65 (1961).
8. Proprietà locali delle soluzioni di una classe di equazioni ipoellittiche. Rend. Sem. Mat. Padova **32**, 221—238 (1962).
9. Su un problema al contorno per certe equazioni ipoellittiche. Revue Romaine Math. pures appl. **9**, 7, 643—653 (1964).
10. Sulle tracce di un certo spazio funzionale, I, II. Rend. Acc. Naz. Lincei VIII, **37**, 28—34 (1964); VIII, **38**, 1—6 (1965).
11. Su un problema tipico relativo a una certa classe di equazioni ipoellittiche. Atti Acc. Sc. Bologna **12**, 1, 1—26 (1964).
12. Proprietà al contorno delle funzioni di classe $H^{p_1,\ldots,p_n}$ per regioni dotate di punti angolosi. Ann. Mat. pura appl. **4**, LXXIII, 33—54 (1966).

Plejel, A.

1. Green's function and asymptotic distribution of eigenvalues and eigenfunctions. Proc. Symp. Spectral Theory and Differential Problems, Stillwater, Okl. (1951), 439—454.

Plejel, A., Minakshisundaran, S.

Cf. Minakshisundaran-Plejel.

Pogorelenko, V. A., Sobolevski, P. E.

1. Equations hyperboliques dans des espaces de Banach. Uspehi Mat. Nauk **22**, 133, 170—172 (1967).

Polya, G., Hardy, G. C., Littlewood, D. E.

Cf. Hardy-Littlewood-Polya.

Poulsen, E. T.

1. Evolutionsgleichungen in Banach-Räumen. Math. Zeit. **90**, 286—309 (1965).

Pozzi, G. A.

1. Problemi di Cauchy e problemi ai limiti per equazione di evoluzione del tipo di Schroedinger lineari e non lineari, I, II. Ann. Mat. Pura Appl. **4**, 78, 197—258 (1968); **4**, 81, 205—248 (1969).

Prioret, P., Courrege, P., Bony, I. M.

Cf. Bony-Courrège-Prioret.

Prodi, G.

1. Tracce sulla frontiera delle funzioni di Beppo Levi. Rend. Sem. Mat. Padova **26**, 36—60 (1956).
2. Tracce di funzioni con derivate di ordine l a quadrato integrabile su varietà di dimensione arbitraria. Rend. Sem. Mat. Padova **28**, 402—452 (1958).

Prouse, G., Amerio, L.

Cf. Amerio-Prouse.

Pucci, C.

1. Una problema variazionale per i coefficienti di equazioni differenziali di tipo ellittico. Ann. Sc. Norm. Sup. Pisa **16**, 159—171 (1962).
2. Regolarità alla frontiera di soluzioni di equazioni ellittiche. Ann. Mat. Pura Appl. **4**, 65, 311—328 (1964).
3. Operatori ellittici estremanti. Ann. Mat. Pura Appl. **4**, LXXII, 141—170 (1966).
4. Limitazioni per soluzioni di equazioni ellittiche. Ann. Mat. Pura Appl. **4**, LXXIV, 15—30 (1966).

Pulvirenti, G.

1. Su un classico problema ai limiti per l'operatore di Laplace iterato. Le Matematiche **16**, 80—104 (1961).
2. Su una classe di problemi ai limiti. Le Matematiche **20**, 87—99 (1965).

Ramazanov, M. D.

1. Sur une classe d'espaces fonctionnels. Differenzialine Uravnienia II, No. 1, 65—82 (1966).
2. A priori estimates of solutions of the Dirichlet problem for certain hypoelliptic equations. Dokl. Akad. Nauk **179**, 783—785 (1968).

Raskin, V. G., Sobolevski, P. E.

1. Problème de Cauchy pour les équations différentielles du deuxième ordre dans les espaces de Banach. Sibirsk Mat. J. **8**, 70—90 (1967).

Raviart, P. A., Lions, J. L.

Cf. Lions-Raviart.

Rham, G. de

1. Variétés différentiables. Paris: Hermann 1955.

Riesz, F., Nagy, B. Sz.

1. Leçons d'Analyse fonctionnelle. Budapest: Akad. Kiado 1952.

Roitberg, Ya. A.

1. Problèmes elliptiques avec conditions aux limites non homogènes et croissance locale et régularité à la frontière des solutions généralisées. Dokl. Akad. Nauk **157**, 798—801 (1964) [Soviet Math. **5**, 1034—1038 (1964)].
2. Théorèmes sur les homéomorphismes définis dans L_p par des opérateurs elliptiques et croissance locale et régularité des solutions généralisées. Ukrainskii Mat. J. **5**, 122—129 (1965).
3. Théorèmes d'homomorphisme réalisés par des opérateurs elliptiques. Dokl. Akad. Nauk **180**, 542—545 (1968) (Soviet Math. **9**, 656—660 (1968)].

Roitberg, Ya. A., Barkoskii, V. V.

Cf. Barkoskii-Roitberg.

Roitberg, Ya. A., Berezanski, Yu. M., Krein, S. G.

Cf. Berezanski-Krein-Roitberg.

Roitberg, Ya. A., Sheftel', Z. G.

1. Conditions aux limites normales et formule de Green pour les systèmes d'équations à dérivées partielles, Tezi Depovider, VIII Zvitnoi Naukovoi Konferentzii, Seria Fys. Mat., 1966, Dzogobich, 125—129.
2. Problèmes aux limites généraux pour équations elliptiques à coefficients discontinus. Dokl. Ak. Nauk **148**, 1034—1037 (1963) [Soviet Math. **4**, 231 to 234 (1963)].

Schechter, M.

1. Integral inequalities for partial differential operators and functions satisfying general boundary conditions. Comm. Pure Appl. Math. **12**, 37—66 (1959).
2. General boundary value problems for elliptic equations. Comm. Pure Appl. Math. **12**, 457—486 (1959).
3. Remarks on elliptic boundary value problems. Comm. Pure Appl. Math. **12**, 561—578 (1959).
4. Negative norms boundary problems. Ann. of Math. **72**, 581—593 (1960).
5. Mixed boundary problems for general elliptic equations. Comm. Pure Appl. Math. **13**, 183—201 (1960).
6. A generalization of the problem of transmission. Ann. Sc. Norm. Sup. Pisa **14**, 207—235 (1960).
7. Various types of boundary conditions for elliptic equations. Comm. Pure Appl. Math. **13**, 407—425 (1960).
8. Coerciveness in L^p. Trans. Amer. Math. Soc. **107**, 10—29 (1963).
9. On L^p estimates and regularity, I. Amer. J. of Math. **85**, 1—13 (1963); II, Math. Scand. **13**, 47—69 (1963); III, Ric. di Mat. **13**, 192—206 (1964).
10. Non local elliptic boundary value problems. Ann. Sc. Norm. Sup. Pisa **20**, 421—442 (1966).
11. Observations concerning a paper of Peetre. Comm. Pure Appl. Math. **14**, 733—736 (1961).

Schmets, J. Garnir, H. G., Wilde, W. De

Cf. Garnir, De Wilde, Schmets.

Schechter, M., Kaniel, S.

Cf. Kaniel-Schechter.

Schwartz, J., Dunford, N.

Cf. Dunford-Schwartz.

Schwartz, L.

1. Théorie des distributions I, II. Paris: Hermann 1950—1951 (2nd ed. 1957).
2. Transformation de Laplace des distributions. Sem. Math. Univ. Lund, Vol. dedicated to M. Riesz, 1952, 196—206.
3. Les équations d'évolution liées au produit de convolution. Ann. Inst. Fourier **2**, 19—49 (1952).
4. Les travaux de Gårding sur le problème de Dirichlet. Sem. Bourbaki, Paris, May 1952.
5. Espaces de fonctions différentiables à valeurs vectorielles. J. d'Analyse Math. **4**, 88—148 (1954—1955).
6. Distributions à valeurs vectorielles I, II. Ann. Inst. Fourier **7**, 1—141 (1957); **8**, 1—209 (1958).

7. Su alcuni problemi della teoria delle equazioni differenziali lineari di tipo ellittico. Rend. Sem. Mat. Fis. Milano **27**, 211—249 (1958).

Seeley, R. T.

1. The index of elliptic systems of singular integral operators. J. Math. Anal. Appl. **7**, 289—309 (1963).
2. Non existence of the resolvent for a Dirichlet problem. Notices of the Amer. Math. Soc. **11**, 570 (1964).
3. Extension of C^∞ functions. Proc. Amer. Math. Soc. **15**, 625—626 (1964).
4. Integro-differential operators on vector bundles. Trans. Amer. Math. Soc. **117**, 167—204 (1965).
5. Singular integrals boundary value problems. Amer. J. Math. **88**, 781—809 (1966).
6. The resolvent of an elliptic boundary problem. Atti Convegno Equaz. Der. Parz., Bologna: Ed. Oderisi, Gubbio, 1967, 164—168, and Amer. J. Math. 1970.

Semenov, E. M.

1. Une échelle d'espaces avec la propriété d'interpolation. Dokl. Ak. Nauk **148**, 1038—1041 (1963) [Soviet Math. **4**, 235—239 (1963)].

Semenov, E. M., Krein, S. G., Petunin, Yu. I.

Cf. Krein-Petunin-Semenov.

Serrin, J.

1. Pathological solutions of elliptic differential equations. Ann. Sc. Norm. Sup. Pisa **18**, 385—387 (1964).

Serrin, J., Finn, R.

Cf. Finn-Serrin.

Serrin, J., Gilbarg, D.

Cf. Gilbarg-Serrin.

Shamir, E.

1. Evaluations dans $W^{s,p}$ pour des problèmes aux limites mixtes dans le plan. C. R. Acad. Sc. Paris **254**, 3621—3623 (1962).
2. Une propriété des espaces $H^{s,p}$. C. R. Acad. Sc. Paris **255**, 448—449 (1962).
3. Mixed boundary value problems for elliptic equations in the plane, the L^p theory. Ann. Sc. Norm. Sup. Pisa **17**, 117—139 (1963).
4. Reduced Hilbert transform and singular integral equations. J. d'Analyse Math. **12**, 277—305 (1964).
5. Wiener-Hopf type problems for elliptic systems of singular integral equations. Bull. Amer. Math. Soc. **72**, 501—504 (1966).
6. Elliptic systems of singular integral operators, I. Trans. Amer. Math. Soc. **127**, 107—124 (1967).
7. Regularization of mixed second order elliptic problems. Israel J. of Math. **6**, 150—168 (1968).

Shapiro, Z. Ya.

1. Sur les problèmes aux limites généraux de type elliptique. Isz. Akad. Nauk, ser. Mat. **17**, 539—562 (1953).

Sheftel', Z. G.

1. Résolution dans L_p et résolution classique des problèmes aux limites généraux pour équations elliptiques avec coefficients discontinus. Uspehi Mat. Nauk **19**, No. 3, 230—232 (1964).
2. Inégalité de l'énergie et problèmes aux limites généraux pour équations elliptiques avec coefficients discontinus. Sibirski Mat. J. **6**, No. 3, 636—668 (1965).
3. Théorie générale des problèmes aux limites pour systèmes elliptiques à coefficients discontinus. Ukrainski Math. J. **18**, No. 3, 132—136 (1966).

Sheftel', Z. G., Roitberg, Ya. A.

Cf. Roitberg, Sheftel'.

Shilov, G. E., Gelfand, I. M.

Cf. Gelfand-Shilov.

Shimakura, N.

1. Sur les domaines des puissances fractionnaires d'opérateurs. Bull. Soc. Math. F. **96**, 265—288 (1968).
2. Problèmes aux limites variationnels du type elliptique. Ann. E.N.S. Paris **2**, 255—310 (1969).

Shiraishi, R., Hiraka, Y.

1. Convolution maps and semi-group distribution. J. Sic. Hiroshima Univ. **28**, 1, 71—88 (1964).

Singer, I. M., Atiayh, M. F.

Cf. Atiyah-Singer.

Slobodetski, L. N.

1. Espaces de Sobolev généralisés et applications aux problèmes aux limites pour les équations différentielles aux dérivées partielles. Uchen. Zap. Leningrad Gos. Ped. Inst. **197**, 54—112 (1958).
2. Estimations des solutions des systèmes elliptiques et paraboliques. Dokl. Akad. Nauk **120**, 468—471 (1958).
3. Estimations dans L_2 des solutions des systèmes linéaires elliptiques et paraboliques. Vestnik Lenin. Gos. Univ. **7**, 28—47 (1960).

Smith, K. T.

1. Functional spaces, functional completion and differential problems. Univ. of Kansas, Tech. report, No. 14, 1954.
2. Inequalities for formally positive integro-differential forms. Bull. Amer. Math. Soc. **67**, 368—370 (1961).

Smith, K. T., Aronszajn, N.

Cf. Aronszajn-Smith.

Smith, K. T., Aronszajn, N. A., Adams, R.

Cf. Adams-Aronszajn-Smith.

Smith, K. T., Hanna, M. S.

Cf. Hanna-Smith.

Sobolev, S. L.

1. Sur les problèmes aux limites pour les équations polyharmoniques. Mat. Sbornik **2**, 44, 467—500 (1937).
2. Applications de l'Analyse fonctionnelle aux équations de la Physique Mathématique, Leningrad, 1950 (German translation: Berlin: Akademie-Verlag; English translation: Amer. Math. Soc.).
3. Méthode nouvelle à résoudre le problème de Cauchy pour les équations hyperboliques normales. Mat. Sbornik **43**, 39—71 (1936).

Sobolev, S. L., Vishik, I. M.

1. Nouvelle formulation générale des problèmes aux limites. Dokl. Ak. Nauk **111**, 521—523 (1956).

Sobolevski, P. E.

1. Sur la fonction de Green des puissances fractionnaires (et en particulier entières) des opérateurs elliptiques. Dokl. Akad. Nauk **142**, 804—807 (1962) [Soviet Math. **3**, 183—187 (1962)].
2. Inégalités de coercivité pour les équations paraboliques abstraites. Dokl. Akad. Nauk **157**, 52—55 (1964) [Soviet Math. **5**, 894—897 (1964)].
3. Solutions généralisées des équations différentielles du 1er ordre dans les espaces de Banach. Dokl. Akad. Nauk **165**, 486—489 (1965) [Soviet Math. **6**, 1465—1468 (1965)].
4. Sur la solution d'équations paraboliques dans les espaces de Lorentz. Uspehi Mat. Nauk **21** (127), 169—171 (1966).
5. Sur les puissances fractionnaires d'opérateurs faiblement positifs. Dokl. Akad. Nauk **166**, 1296—1299 (1966) [Soviet Math. **7**, 287—291 (1966)].
6. Sur les équatiions de type parabolique dans les espaces de Banach. Trudy Moskov Mat. Obs. **10**, 297—350 (1961) [Amer. Math. Soc. Transl. (2) **49**, 1—62 (1966)].

Sobolevski, P. E., Pogorelenko, V. A.

Cf. Pogorelenko-Sobolevski.

Sobolevski, P. E., Raskin, V. R.

Cf. Raskin-Sobolevski.

Solonnikov, V. A.

1. Sur les problèmes aux limites généraux au sens de Douglis et Nirenberg. (I) Isvestia Akad. Nauk **28**, 665—706 (1964); (II) Troudi Steklov C II, 233—297 (1966) [Amer. Math. Soc. Transl. (2), **56**, 193—231 (1966)].
2. Inégalités à priori pour les équations du deuxième ordre de type parabolique. Troudi Steklov LXX, 133—212 (1964).

Solonnikov, V. A., Ladyženskaya, O. A., Uraltzeva, N. N.

Cf. Ladyženskaya-Solonnikov-Uraltzeva.

Stampacchia, G.

1. Formes bilinéaires coercitives sur les ensembles convexes. C. R. Acad. Sc. Paris **258**, 4413—4416 (1964).
2. Le problème de Dirichlet pour les équations elliptiques du second ordre à coefficients discontinus. Ann. Inst. Fourier **15**, 189—258 (1965).

3. The $\mathscr{L}^{p,\lambda}$ spaces and applications to the theory of partial differential equations. Proceedings of the Colloquium "Equa. diff. II", Bratislava, September 1966.
4. Su un problema relativo alle equazioni di tipo ellittico del secondo ordine. Ricerche Mat. **5**, 3—24 (1956).
5. Equations elliptiques à données discontinues. Séminaire Schwartz Paris **5**, No. 4 (1960—1961).
6. Sistemi di equazioni di tipo ellittico a derivate parziali del primo ordine e proprietà delle estremali degli integrali multipli. Ricerche Mat. **1**, 200—226. (1952).

Stampacchia, G., Campanato, S.
Cf. Campanato-Stampacchia.

Stampacchia, G., Lions, J. L.
Cf. Lions-Stampacchia.

Stampacchia, G., Magenes, E.
Cf. Magenes-Stampacchia.

Stampacchia, G., Littman, W., Weinberger, H.
Cf. Littman-Stampacchia-Weinberger.

Stampacchia, G., Murthy, V.
Cf. Murthy-Stampacchia.

Stein, E. M.
1. The Characterization of functions arising as potentials. Bull. Amer. Math. Soc. **67**, 102—104 (1961).

Stermin, B. Ju.
1. Les problèmes aux limites généraux pour les équations elliptiques, pour domaines de frontières de dimension générale. Dokl. Akad. Nauk **159**, 992 to 994 (1964) [Soviet Math. **5**, 1658—1660 (1964)].

Stone, M. H.
1. Linear transformations in Hilbert spaces and their applications to analysis. Amer. Math. Soc. Coll. Publ. **15** (1932).

Strauss, W.
1. On continuity of functions with values in various Banach spaces. Pacific J. Math. **19**, 543—551 (1966).

Szeptycki, P., Aronszajn, N., Mulla, F.
Cf. Aronszajn-Mulla-Szeptycki.

Taibleson, M. H.
1. Lipschitz Classes of functions and distributions in E_n. Bull. Amer. Math. Soc. **69**, 487—493 (1963).
2. On the theory of Lipschitz spaces of distributions on Euclidean n-space I, II. J. Math. Mech. **13**, 407—479 (1964); **14**, 821—839 (1965).

Talenti, G.

1. Sopra una classe di equazioni ellittiche a coefficienti misurabili. Ann. Mat. Pura appl. **4**, LXIX, 285—304 (1965).
2. Equazioni lineari ellittiche in due variabili. Le Matematiche **21**, 1—38 (1966).
3. Soluzioni a simmetria assiale di equazioni ellittiche. Ann. Mat. Pura appl. **4**, LXXIII, 127—158 (1966).

Tanabe, H.

1. On the equations of evolution in a Banach space. Osaka Math. J. **11**, 121—145 (1959); **12**, 145—166 and 363—376 (1960).
2. Note on uniqueness of solutions of differential inequalities of parabolic type. Osaka Math. J. **2**, 191—204 (1965).
3. On estimates for derivatives of solutions of weighted elliptic boundary problems. Osaka Math. J. **3**, 163—183 (1966).

Tanabe, H., Kato, T.

Cf. Kato-Tanabe.

Torelli, G.

1. Un complemento ad un teorema di Lions sulle equazioni differenziali astratte del secondo ordine. Rend. Sem. Mat. Padova **34**, 224—241 (1964).

Treves, F.

1. Relations de domination entre opérateurs différentiels. Acta Math. **101**, 1—139 (1959).
2. Problèmes de Cauchy et problèmes mixtes en théorie des distributions. J. Anal. Math. Israel **7**, 105—187 (1959).
3. Opérateurs différentiels hypoelliptiques. Ann. Inst. Fourier **9**, 2—73 (1959).
4. Distributions with values in variable spaces. Anais Acad. Bresileira de Ciencias **38**, 1, 5—9 (1966).
5. Linear partial differential equations with constant coefficients, New York: Gordon & Breach 1966.

Troisi, M.

1. Su un problema di trasmissione. Ricerche di Mat. **11**, 24—50 (1962).
2. Sui problemi di trasmissione per due equazioni ellittiche di ordine diverso. Ricerche di Math. **12**, 216—247 (1963).
3. Sulle regolarizzazioni delle soluzioni di taluni problemi di trasmissione. Ricerche Mat. **13**, 281—316 (1964).

Trotter, H. F.

1. Approximation of semigroups of operators. Pacific J. Math. **8**, 887—919 (1958).

Unterberger, A., Bokobza, J.

1. Les opérateurs pseudo-différentiels d'ordre variable. C. R. Acad. Sci. Paris **261**, 2271—2273 (1965).

Uraltzeva, N. N., Ladyženskaya, O. A.

Cf. Ladyženskaya-Uraltzeva.

Uraltzeva, N. N., Ladyženskaya, O. A., Solonnikov, V. A.
Cf. Ladyženskaya-Solonnikov-Uraltzeva.

Uspenski, V.
1. Propriétés des classes généralisées W_p^r de Sobolev. Sibirski Math. J. **3**, 418 to 445 (1962).
2. Théorèmes d'inclusion et de prolongement pour une classe de fonctions. Sibirski Math. J. **7**, 192—199 (1966) [Sib. Math. J. **7**, 154—161 (1966)].

Vainberg, B. R., Grusin, V. V.
1. Sur les problèmes uniformément non elliptiques pour les équations elliptiques. Dokl. Ak. Nauk **172**, 518—520 (1967) [Soviet Math. **8**, 118—120 (1967)].

Vekua, I. N.
1. Nouvelles méthodes pour résoudre les équations elliptiques, Moscow-Leningrad, 1948.
2. Fonctions analytiques généralisées, Moscow, 1959.

Vilenkin, N. Ya., Gelfand, I. M.
Cf. Gelfand-Vilenkin.

Vishik, I. M.
1. Sur les systèmes fortement elliptiques d'équations différentielles. Mat. Sbornik **29**, 615—676 (1951).
2. Sur les problèmes aux limites généraux pour les équations aux dérivées partielles elliptiques. Trudy Moscow Mat. Obsc. **1**, 187—246 (1952) [Amer. Math. Soc. Transl. (2) **24**, 107—172 (1963)].
3. Problèmes aux limites pour les équations elliptiques dégénérant à la frontière. Mat. Sbornik **35**, 513—568 (1954) [Amer. Math. Soc. Transl. (2) **35**, 15—78 (1964)].
4. Le problème de Cauchy pour les équations avec coefficients opérateurs. Mat. Sbornik **39** (81), 51—148 (1956) [Amer. Math. Soc. Transl. (2) **24**, 173—278 (1963)].
5. Equations elliptiques de convolution dans un domaine borné et applications. Lecture given at the International Congress in Moscow (1966). Proceedings, 409—420.

Vishik, I. M., Agranovich, M. S.
C.f Agranovich-Vishik.

Vishik, I. M., Eskin, I. G.
1. Equations de convolution dans un domaine borné. Uspehi Mat. Nauk **20**, No. 3 (1965) [Russian Math. Surv. **20**, 85—151 (1965)].
2. Equations de convolution dans un domaine borné dans des espaces avec poids. Mat. Sbornik **69** (111), 65—110 (1966).
3. Equations paraboliques de convolution dans des domaines bornés. Mat. Sbornik **71** (113), 162—190 (1966).
4. Equations elliptiques de convolution dans des domaines bornés et applications. Uspehi Mat. Nauk **133**, 15—76 (1967).
5. Normally solvable problems for elliptic systems of equations of convolution. Mat. Sbornik **74** (116), 326—356 (1967) [Math. U.R.R.S. **3** (1967), No. 3].

Vishik, I. M., Ladyženskaya, O. A.

1. Problèmes aux limites pour les équations aux dérivées partielles et certaines classes d'équations opérationnelles. Uspehi Mat. Nauk **11**, No. 6, 41—97 (1956) [Amer. Math. Soc. Transl. (2) **10**, 223—281 (1958)].

Vishik, I. M., Ljusternik, L. A.

1. Dégénérescence régulière pour les équations différentielles linéaires avec un petit paramètre. Uspehi Mat. Nauk **12**, 1—121 (1957) [Amer. Math. Soc. Transl. (2), **20**, 239—364 (1962)].

Vishik, I. M., Sobolev, S. L.

Cf. Sobolev-Vishik.

Volevich, L. R.

1. Propriétés locales des solutions des systèmes quasielliptiques. Mat. Sbornik **59** (101), 3—52 (1962).
2. Problèmes aux limites pour systèmes elliptiques généraux. Mat. Sbornik **68** (110), 373—416 (1965).

Volevich, L. R. Panejach, B. P.

1. Certains espaces de fonctions généralisées et théorèmes d'inclusion. Uspehi Mat. Nauk **20**, No. 1, 3—74 (1965) [Russian Math. Surv. **20**, 1—73 (1965)].

Volevic, L. R., Agranovich, M. S., Dynin, A. S.

Cf. Agranovich-Volevic-Dynin.

Volkov, E. A.

1. Sur les propriétés différentielles des problèmes aux limites pour les équations de Laplace (I), (II). Troudi Stekloff LXXVII, 89—112, 113—142 (1965).

Volpert, A. I.

1. Sur l'indice et la résolution normale des problèmes aux limites pour les systèmes elliptiques d'équations différentielles dans le plan. Trudy Moskov. Mat. Obsc. **10**, 41—87 (1961).
2. Sur l'indice des systèmes d'équations intégrales singulières en plusieurs dimensions. Dokl. Ak. Nauk **152**, 1292—1293 (1963) [Soviet Math. **4**, 1540 to 1542 (1963)].

Walter, W.

1. Über die Euler-Poisson-Darboux-Gleichung. Math. Zeit. **67**, 361—376 (1957).

Weinberger, H., Littman, W., Stampacchia, G.

Cf. Littman-Stampacchia-Weinberger.

Weinstein, A.

1. Etude des spectres des équations aux dérivées partielles. Mémorial Sci. Math. No. 88, Paris: Gauthier-Villars 1937.

Weyl, H.

1. The method of orthogonal projection in potential theory. Duke Math. J. **7**, 411—444 (1940).

Yoshikawa, A.

1. Remarks on the theory of interpolation spaces. J. Fac. Sci. Univ. Tokyo, Sect. I, **15**, 209—251 (1968).

Yoshinaga, K.

1. Ultra distributions and semi-group distributions. Bull. Kyushu Inst. Tech. Math. Nat. Sci. **10**, 1—24 (1963).
2. Values of vector valued distributions and smoothness of semi-group distributions. Bull. Kyushu Inst. Tech. Math. Nat. Sci. **12**, 1—27 (1965).

Yosida, K.

1. An operator theoretical integration of the wave equation. J. Math. Soc. Japan **8**, 79—92 (1956).
2. Functional Analysis. Grundlehren Vol. 123, Berlin/Heidelberg/New York: Springer 1965.
3. Time dependent evolution equations on a locally convex space. Math. Annalen **162**, 83—86 (1965).

Zaidman, S.

1. Un teorema di esistenza globale per alcune equazioni differenziali astratte. Ricerche di Mat. **13**, 56—69 (1964).
2. Convexity properties for weak solutions of some differential equations in Hilbert spaces. Canad. J. Math. **17**, 802—807 (1965).

Zygmund, A., Calderon, A. P.

Cf. Calderon-Zygmund.

Additional Bibliography

Chapter 1.

For the general theory of interpolation of linear operators one can also consult

Butzer, P. L., Berens, H.

1. Semi groups of operators and approximation. Grundlehren, Vol. **145**, Berlin/Heidelberg/New York: Springer 1967.

Gagliardo, E.

1. Caratterizzazione costruttiva di tutti gli spazi di interpolazione tra spazi di Banach. Ist. Naz. Alta Mat., Symposia Mat., Vol. 2, 95–106 (1968).

Peetre, J.

1. Interpolation functors and Banach couples. Proc. Int. Congress of Math., Nice, 1970.

Scherer, K.

1. Dualität bei Interpolations- und Approximationsräumen, Diss., Techn. Hochschule Aachen, 1969.

Wentzell, T. D.

1. On interpolation functions. Vestnik Moskov Univ., Mat. **5**, 57–65 (1969).

Yoshinaga, K.

1. On a generalization of the interpolation method. Bull. Kyushu Inst. Techn. **17**, 1–23 (1970).

One can extend the theory (or, at least, part of it) to *non linear* operators which are "bounded" in one couple of Banach spaces and "Lipschitzian" in a second couple of Banach spaces; cf.

Lions, J. L.

1. Some remarks on variational inequalities. Proc. Int. Conf. on Functional Analysis, Tokyo, 1969, 270–282.
2. Sur les inéquations variationnelles d'évolution pour les opérateurs du 2eme ordre en t, Ist. Naz. di Alta Mat., Roma, Symposia Mat., 1970.
3. Interpolation linéaire et non linéaire et régularité, Ist. Naz. di Alta Mat. Roma, Symposia Mat., 1971.

Peetre, J.

2. Interpolation of Lipschitz operators and metric spaces. Matematika (Cluj) (to appear).

For the case when the non linear operator is Lipschitzian in the *two couples* of Banach spaces, cf.

Browder, F. E.

1. Remarks on non linear interpolation in Banach spaces. J. Funct. Analysis, **1969**.

Applications and extensions are given in

Tartar, L.

1. Thesis, Paris, **1971**; J. Funct. Analysis, **1971**.

New interpolation results for non-Banach spaces are given by

Baouendi, M. S., Goulaouic, C.

1. J. Funct. Analysis (to appear).

Goulaouic, C.

1. Interpolation entre espaces localement convexes définis à l'aide de semi-groupes; cas des espaces de Gevrey. Ann. Inst. Fourier **19**, 269—278 (1969).

For the interpolation between subspaces (cf. Problem 18.5) cf.

Seeley, R.

1. Interpolation in L^p with boundary condition (to appear); cf. also Proc. Int. Congress of Math., Nice, **1970**.

For interpolation between Hilbert spaces and for Bessel potentials on manifolds cf.

Adams, R. D., Aronszajn, N., Hanna, M. S.

1. Theory of Bessel Potentials. Part III. Ann. Inst. Fourier **19**, 279—338 (1969).

Chapter 2.

For an approach to boundary value problems of elliptic type using pseudo-differential operators, we refer, in addition to the papers already quoted in the Bibliography (cf. particularly Vhishik-Eskin [1], [4], [5]), to

Boutet de Monvel, L.

1. Comportement d'un opérateur pseudo-différentiel sur une variété à bord. I, II. J. d'Analyse Math. **17**, 241—253, 255—304 (1966).
2. Opérateurs pseudo-differentiels analytiques et problèmes aux limites elliptiques. Ann. Inst. Fourier **19**, 169—268 (1970).
3. Indice des problèmes aux limites elliptiques, Proc. Int. Congress of Math., Nice, **1970**.
4. Boundary problems for pseudo-differential operators. Acta Math. (**1971**).

Boutet de Monvel, L., Geymonat, G.

1. Solutions irrégulières d'un problème aux limites elliptiques. Ist. Naz. Alta Mat., Symposia Mat., Roma, **1971**.

Krée, P.

1. Introduction à la théorie des opérateurs pseudo-différentiels. Confer. Sem. Mat. Univ. Bari, n. 112—114, 1968.
2. Problèmes aux limites en théorie des distributions. Ann. Mat. Pura Appl. **4**, 83, 113—132 (1969).

Seeley, R.

2. Topics in pseudo-differential operators, C.I.M.E. School on Pseudo-differential Operators, Stresa, September 1968, Roma: Cremonese, 167—305.

Shamir, E.

1. Elliptic systems of singular integral operators. II. Boundary value problems in a half-space, to appear (cf. C.I.M.E. School on Pseudo-differential Operators, Stresa, September 1968. Roma: Cremonese, 309—331).

For papers which are directly related to the content of Chapter 2, let us refer for the theory of "variational" problems to

Fujiwara, D., Shimakura, N.

1. Sur les problèmes aux limites elliptiques stablement variationnels. J. Math. Pures Appl. **49**, 1—28 (1970).

Fujiwara, D.

1. On some homogeneous boundary value problems bounded below. J. Fac. Sci. Univ. Tokyo **XVII**, 123—152 (1970).

Grubb, G.

1. Les problèmes aux limites généraux d'un opérateur elliptique, provenant de la théorie variationnelle. Bull. Soc. Math. France, to appear.
2. Coerciveness of the normal boundary problem for an elliptic operator. Bull. Amer. Math. Soc. **76**, 64—69 (1970).
6. On coerciveness and semiboundedness of general boundary problems (to appear).

Torelli, A.

1. Sulla teoria variazionale dei problemi ai limiti ellittici. Rend. Ist. Lombardo Sci. Lett., A, **103**, 573—617 (1969).

These papers give solutions to the general question of Problem 11.5. For variational methods cf. also

Krée, P.

3. Application des méthodes variationnelles aux équations de convolution. C. R. Acad. Sci. Paris, A, **268**, 1193—1196 (1969).

For Green's formula and adjoint problems cf.

Troisi, M.

1. Sulla nozione di problema aggiunto per i problemi al contorno relativi ad una equazione ellittica in due variabili. I, II. Ricerche di Mat. **17**, 109—143 and 164—215 (1968).

For questions related to Sections 6 and 7 (trace theorems and non-homogeneous problems in the spaces $H^s(\Omega)$ with $-\infty < s < 2m$) cf.:

Baouendi, M. S., Geymonat, G.

1. Résultats de dualité dans les problèmes aux limites linéaires elliptiques (I), J. of Differ. Equat. (to appear).
This paper also considers a problem similar to the "best" choice of the space $K^r(\Omega)$ (cf. Section 6.3).

Boutet de Monvel, L., Geymonat, G.

1. Already quoted in this Additional Bibliography.

Goulaouic, C., Grisvard, P.

1. Existence de traces pour les éléments d'espaces de distributions définis comme domaines d'opérateurs différentiels maximaux. Inventiones Math. **9**, 308—317 (1970).

Krée, P.

2. Already quoted in this Additional Bibliography.

Martsinkovsa, G.

1. Linear functionals over Sobolev spaces and boundary problems generated by theorems on homeomorphisms. Ukrainskii Math. J. **21**, 610—626 (1969).

Roitberg, Yu. A.

1. On boundary values of generalized solutions of elliptic equations. Dokl. Akad. Nauk **188**, 41—44 (1969) [Soviet Math. **10**, 1079—1083 (1969)].
2. Green's formula and the homeomorphism Theorem for general elliptic boundary value problems with boundary conditions that are not normal. Ukrainskii Math. J. **21**, 398—405 (1969).

Rushchitskaya, S. O.

1. A theorem on homeomorphisms for an elliptic differential operator and pseudodifferential boundary conditions. Ukrainskii Math. J. **21**, 268—273 (1969).

For "fractional powers" of operators, cf.

Fujiwara, D.

2. L^p-theory for characterizing the domain of the fractional powers of $-\Delta$ in the half space. J. Fac. Sc. Univ. Tokyo 1, **15**, 169—177 (1968).
3. On the asymptotic behavior of the Green operators for elliptic boundary problems and the pure imaginary powers of some second order operators. J. Math. Soc. Japan **21**, 481—522 (1969).

Komatsu, H.

1. Fractional powers of operators, III. Negative powers. J. Math. Soc. Japan **21**, 205—220 (1969).
2. id. IV, Potential operators. J. Math. Soc. Japan **21**, 221—228 (1969).
3. id. V, Dual operators. J. Fac. Sci., Univ. Tokyo **17**, 1 and 2, 373—396 (1970).

Seeley, R.

3. Norms and domains of the complex powers A_B^z (to appear; cf. also Proc. Int. Congress of Math., Nice, 1970).

Shimakura, N.

1. Les puissances fractionnaires de $I - \Delta$ sous les conditions de certaines derivées obliques. J. Math. Kyoto Univ. **9**, 363—379 (1969).

Yoshikawa, A.

1. Remarques sur la théorie d'espaces d'interpolation. Espaces de moyennes de plusieurs espaces de Banach. J. Fac. Sic. Univ. Tokyo, Sect. 1, **16**, 407 to 468 (1970).
2. An abstract formulation of Sobolev type imbedding theorems and its applications to elliptic boundary value problems. J. Fac. Sic. Univ. Tokyo, Sect. 1, **17**, 543—558 (1970).
3. An operator theoretical remark on the Hardy-Littlewood-Sobolev inequality. J. Fac. Sci. Univ. Tokyo, Sect. 1, **17**, 559—566 (1970).
4. Fractional powers of operators, interpolation theory and imbedding theorems, to appear.

For the regularity of the solution of elliptic problems at "corners" using an "operational" method, cf.

Grisvard, P.

1. Equations opérationnelles et problèmes aux limites dans les domaines non réguliers. Proc. Int. Congress of Math., Nice, 1970.

For Dirichlet's problem in the half-space, cf.

Brezzi, F.

1. Sul problema di Dirichlet nel semispazio. Ann. Mat. Pura Appl. **4**, 86, 261 to 298 (1970).

For the problem of "transmission" cf.

Krée, P.

4. Problèmes de transmission pour des opérateurs pseudo-différentiels elliptiques. C. R. Acad. Sc. Paris, A, **269**, 699—701 (1969).

For the problem of the "oblique" derivative and generalizations cf.

Bitsadze A. V.

1. Linear non-Fredholm elliptic boundary value problems. Proc. Int. Congress of Math., Nice, 1970.

Egorov Ju. V., Kondratiev, V. A.

1. On a problem of oblique derivative. Mat. Sbornik **78**, 148—176 (1969).

Talenti, G.

1. Problemi di derivata obliqua per equazioni ellittiche in due variabili. Boll. U.M.I., 3, **22**, 505—526 (1967).

For elliptic operators which *degenerate* at the boundary, cf.

Baouendi, M. S., Goulaouic, C.

1. Régularité et théorie spectrale pour une classe d'opérateurs elliptiques dégénérées. Arch. Rat. Mech. Anal. **34**, 361—379 (1969).
2. Etude de l'analyticité et de la regularité Gevrey pour une classe d'opérateurs elliptiques dégénérés. Ann. Sc. Ec. Norm. Sup. **4**, 31—46 (1971).

3. Regularité analytique et itérés d'opérateurs elliptiques dégénérés: applications (to appear).

Bolley, P., Camus, J.

1. Etude de la régularité de certains problèmes elliptiques dégénérés dans des ouverts non réguliers, par le méthode de réflexion. C. R. Acad. Sc. Paris, A, **268**, 1462—1464 (1969).
2. Sur certains problèmes aus limites, elliptiques et dégénérés. C. R. Acad. Sc. Paris **271**, 980—983 (1970).

Hanouzet, B.

1. Regularité pour une classe d'opérateurs elliptiques dégénérés du deuxième ordre. Le Mathematiche **24**, 450—491 (1969).

Shimakura, N.

2. Problèmes aux limites généraux du type elliptique dégénéré. J. Math. Kyoto Univ. **9**, 275—335 (1969).

Troisi, M.

1. Problemi ellittici con dati singolari. Ann. Mat. Pura Appl. **4**, 83 (1969).
2. Ulteriori contributi allo studio dei problemi ellittici con dati singolari. Ricerche Mat. **19**, 9—25 (1970).

Vishik, M. I., Grusin, V. V.

1. On a class of degenerate elliptic equations of higher order. Mat. Sbornik **79**, 3—36 (1969).
2. Boundary value problems for elliptic equations that degenerate at the boundary. Mat. Sbornik **80**, 455—491 (1969).
3. Degenerate elliptic differential and pseudo differential operators. Uspehi Mat. Nauk **154**, 29—56 (1970).

For the asymptotic distributions of eigenvalues, cf.

Agmon, S.

1. Asymptotic formula with remainder estimates for eigenvalues of elliptic systems. Arch. Rat. Mech. Anal. **28**, 165—183 (1968).

Hörmander, L.

1. The spectral functions of an elliptic operator. Acta Math. **121**, 193—218 (1968).

For elliptic systems, cf.

Roitberg, Ya. A., Sheftel', Z. G.

1. A theorem on homeomorphisms for elliptic systems. Mat. Sbornik **78**, 446 to 472 (1969).

For "variational" second order elliptic equations with discontinuous coefficients, cf.

Chicco, M.

1. Principio di massimo generalizzato e valutazione del primo autovalore per problemi ellittici del secondo ordine di tipo variazionale. Ann. Mat. Pura Appl. **4**, 87, 1—10 (1970).

Hervé, R. M. and M.

1. Les fonctions surharmoniques associées a un opérateur elliptique du second ordre à coefficients discontinus. Ann. Inst. Fourier **19**, 305—359 (1969).

Marino, A., Spagnolo, S.

1. Un tipo di approssimazione dell'operatore $\sum_{l.i,j}^{n} D_i(a_{ij}(x))\, D_j$ con operatori $\sum_{l,j}^{n} D_j(\beta(x)\, D_j)$. Ann. Sc. Norm. Sup. Pisa **23**, 657—673 (1969).

Spagnolo, S.

1. Una caratterizzazione degli operatori differenziali autoaggiunti del 2^0 ordine a coefficienti misurabili e limitati. Rend. Sem. Mat. Padova **39**, 56—64 (1967).
2. Sulla convergenza di soluzioni di equazioni paraboliche ed ellittiche. Ann. Sc. Norm. Sup. Pisa **32**, 571—597 (1968).

For "non-variational" elliptic equations, cf.

Campanato, S.

1. Equazioni ellittiche non variazionali a coefficienti continui. Ann. Mat. Pura Appl., 1971.
2. Un risultato relativo ad equazioni ellittiche del secondo ordine di tipo non variazionale. Ann. Sc. Norm. Sup. Pisa **21**, 701—707 (1967).

Chicco, M.

2. Equazioni ellittiche del secondo ordine di tipo Cordes con termini di ordine inferiore. Ann. Mat. Pura Appl. **4**, 85, 347—356 (1970).

For the numerical solution of elliptic problems and error estimates, we refer to

Aubin, J. P.

1. Approximation of non homogeneous elliptic boundary value problems, New York: Wiley 1971.

Bramble, J. H., Schatz, A. H.

1. Least square methods for $2m^{\text{th}}$ order elliptic boundary value problems (to appear).
2. Rayleigh-Ritz-Galerkin methods for Dirichlet's problem using subspaces without boundary conditions. Comm. pure appl. Math. **23**, 653—675 (1970).

Ciarlet, Ph., Wagschal, C.

1. (to appear).

Fix, D., Strang, G.

1. Book on Finite elements (to appear).

Guglielmo, F. di

1. Construction d'approximation des espaces de Sobolev sur des reseaux en simplexe. Calcolo **6**, 279—331 (1969).
2. Méthodes des élements finis: une famille d'approximation des espaces de Sobolev par les translatés de p fonctions. Calcolo **7**, 185—234 (1970).

Zlamal, M.

1. On the finite method. Numer. Math. **12**, 394—409 (1968).

2. On some finite element procedures for solving second order boundary value problems. Numer. Math. **14**, 42—48 (1969).

These papers give solutions to Problem 11.11.

Chapter 3.

For the general approach to abstract differential equations, cf.

Baiocchi, C.

1. Teoremi di regolarità per le soluzioni di equazioni differenziali astratte. Ist. Naz. Alta Mat., Symposia Math., Roma, **1970**.

Carrol, R.

1. Abstract Methods in Partial Differential Equations, New York: Harper & Row 1969.

Da-Prato, G.

1. Weak solutions for linear abstract differential equations in Banach spaces. Advances in Math. **5**, 181—245 (1970).

Krée, P.

5. Synthèse et extension de certaines théories variationnelles linéaires. C. R. Acad. Sc. Paris, A, **271**, 457—460 (1970).

For problems with 1st order t-derivative and variable spaces $V(t)$, cf.

Baiocchi, C.

2. Problemi misti per l'equazione del calore. Rend. Sem. Mat. Fis. Milano, **1970**.

Carroll, R. W., Cooper, J. M.

1. Remarks on some variable domain problems in abstract evolution equations. Math. Ann. **188**, 143—164 (1970).

Carrol, R., Mazumdar, T.

1. Solutions of some possibly noncoercive evolution problems with regular data (to appear).

For questions related to Problem 13.4 and its applications cf.

Baiocchi, C.

3. Soluzioni deboli dei problemi ai limiti per le equazioni paraboliche del tipo del calore. Rend. Ist. Lombardo Sc. Let., A, **103**, 704—726 (1969).

Marino, M.

1. Sull'esistenza delle soluzioni deboli dei problemi al contorno per operatori parabolici. Le Matematiche **13**, 387—407 (1968).

For differential equations with "delay" terms

Artola, M.

1. Sur les perturbations des équations d'évolution. Application à des problèmes de retard. Ann. Sc. Ec. Norm. Sup. **4**, 2, 137—253 (1969).

Comincioli, V.

1. Problemi periodici relativi a equazioni d'evoluzione paraboliche con termini di ritardo . . . Rend. Ist. Lombardo Sc. Let. A, **104**, 356—381 (1970).
2. Ulteriori osservazioni sulle soluzioni del problema periodico per equazioni paraboliche lineari con termini di perturbazione. Rend. Ist. Lombardo Sc. Let., A, **104**, 726—735 (1970).

A general regularity theorem has been given in

Bardos, C.

1. A regularity theorem for Parabolic equations. J. Functional Analysis **7**, 311—322 (1971).

For problems with second-order t-derivative cf.

Kato, T.

1. Linear evolution equations of "hyperbolic" type. J. Fac. Sci. Univ. Tokyo **17** (1, 2), 241—258 (1970).

and the papers:

Carroll, R., State, E.

1. Existence theorems for some weak abstract variable domain hyperbolic problems (to appear).

Fattorini, H. O.

1. Uniformly bounded cosine functions in Hilbert space. Indiana Univ. Math. J. **20**, 411—425 (1970).

Giusti, E.

1. Funzioni coseno periodiche. Boll. U.M.I. **22**, 478—485 (1967).

Goldstein, J. A.

1. Abstract evolution equations. Trans. Amer. Math. Soc. **141**, 159—185 (1969).

For various aspects of stochastic evolution equations we refer to

Bensoussan, A.

1. Filtrage optimal des systèmes linéaires, Paris: Dunod 1971.

Bensoussan, A., Teman, R.

1. (to appear).

For the asymptotic behavior of the solutions of abstract differential equations cf.:

Artola, M.

2. Dérivées intermédiaires dans les espaces de Hilbert pondérés. Application au comportement à l'∞ des solutions des équations d'évolution. Rend. Sem. Mat. Padova **43**, 177—202 (1970).

Pazy, A.

1. Asymptotic expansions of solutions of ordinary differential equations in Hilbert space. Arch. Rat. Mech. Anal. **24**, 113—218 (1967).

2. Asymptotic behavior of the solution of an abstract equation and some applications. J. Diff. Equat. **4**, 493–509 (1968).

For somewhat related problems for *variational inequalities* we refer to the work of H. Brezis mentioned in the bibliography and to

Lions, J. L.

4. Quelques méthodes de résolution des problèmes aux limites non linéaires. Paris: Dunod, Gauthier-Villars 1969.

For applications cf.

Duvaut, G., Lions, J. L.

1. Sur les inéquations en Mécanique et en Physique. Paris: Dunod 1971.

721/52/71 — III/18/203

Die Grundlehren der mathematischen Wissenschaften in Einzeldarstellungen mit besonderer Berücksichtigung der Anwendungsgebiete

Eine Auswahl

---

65. Fejes Tóth: Lagerungen in der Ebene, auf der Kugel und im Raum. DM 66,–
67. Byrd/Friedman: Handbook of Elliptic Integrals for Engineers and Scientists, DM 64,–
68. Aumann: Reelle Funktionen. DM 68,–
74. Boerner: Darstellungen von Gruppen. DM 58,–
76. Tricomi: Vorlesungen über Orthogonalreihen. DM 68,–
77. Behnke/Sommer: Theorie der analytischen Funktionen einer komplexen Veränderlichen. DM 79,–
78. Lorenzen: Einführung in die operative Logig und Mathematik. DM 54,–
86. Richter: Wahrscheinlichkeitstheorie. DM 68,–
87. van der Waerden: Mathematische Statistik. DM 68,–
94. Funk: Variationsrechnung und ihre Anwendung in Physik und Technik. DM 120,–
97. Greub: Linear Algebra. DM 39,20
99. Cassels: An Introduction to the Geometry of Numbers. DM 78,–
104. Chung: Markov Chains with Stationary Transition Probabilities. DM 56,–
107. Köthe: Topologische lineare Räume I. DM 78,–
114. MacLane: Homology. DM 62,–
116. Hörmander: Linear Partial Differential Operators. DM 42,–
117. O'Meara: Introduction to Quadratic Forms. DM 68,–
120. Collatz: Funktionalanalysis und numerische Mathematik. DM 58,–
121./122. Dynkin: Markov Processes. DM 96,–
123. Yosida: Functional Analysis. DM 66,–
124. Morgenstern: Einführung in die Wahrscheinlichkeitsrechnung und mathematische Statistik. DM 38,–
125. Itô/McKean jr.: Diffusion Processes and their sample Paths. DM 58,–
126. Lehto/Virtanen: Quasikonforme Abbildungen. DM 38,–
127. Hermes: Enumerability, Decidability, Computability. DM 39,–
128. Braun/Koecher: Jordan-Algebren. DM 48,–
129. Nikodým: The Mathematical Apparatus for Quantum Theories. DM 144,–
130. Morrey jr.: Multiple Integrals in the Calculus of Variations. DM 78,–
131. Hirzebruch: Topological Methods in Algebraic Geometry. DM 38,–
132. Kato: Perturbation Theory for Linear Operators. DM 79,20
133. Haupt/Künneth: Geometrische Ordnungen. DM 68,–
134. Huppert: Endliche Gruppen I. DM 156,–
135. Handbook for Automatic Computation. Vol. 1/Part a: Rutishauser: Description of ALGOL 60. DM 58,–
136. Greub: Multilinear Algebra. DM 32,–
137. Handbook for Automatic Computation. Vol. 1/Part b: Grau/Hill/Langmaack: Translation of ALGOL 60. DM 64,–
138. Hahn: Stability of Motion. DM 72,–
139. Mathematische Hilfsmittel des Ingenieurs. 1. Teil. DM 88,–
140. Mathematische Hilfsmittel des Ingenieurs. 2. Teil. DM 136,–
141. Mathematische Hilfsmittel des Ingenieurs. 3. Teil. DM 98,–
142. Mathematische Hilfsmittel des Ingenieurs. 4. Teil. DM 124,–
143. Schur/Grunsky: Vorlesungen über Invariantentheorie. DM 32,–
144. Weil: Basic Number Theory. DM 48,–
145. Butzer/Berens: Semi-Groups of Operators and Approximation. DM 56,–

147. Lamotke: Semisimpliziale algebraische Topologie. DM 48,—
148. Chandrasekharan: Introduction to Analytic Number Theory. DM 28,—
149. Sario/Oikawa: Capacity Functions. DM 96,—
150. Iosifescu/Theodorescu: Random Processes and Learning. DM 68,—
151. Mandl: Analytical Treatment of One-dimensional Markov Processes. DM 36,—
152. Hewitt/Ross: Abstract Harmonic Analysis. Vol. 2: Structure and Analysis for Compact Groups. Analysis on Locally Compact Abelian Groups. DM 140,—
153. Federer: Geometric Measure Theory. DM 118,—
154. Singer: Bases in Banach Spaces I. DM 112,—
155. Müller: Foundations of the Mathematical Theory of Electromagnetic Waves. DM 58,—
156. van der Waerden: Mathematical Statistics. DM 68,—
157. Prohorov/Rozanov: Probability Theory. DM 68,—
159. Köthe: Topological Vector Spaces I. DM 78,—
160. Agrest/Maksimov: Theory of Incomplete Cylindrical Funktions and Their Applications. DM 84,60
161. Bhatia/Szegö: Stability Theory of Dynamical Systems. DM 58,—
162. Nevanlinna: Analytic Functions. DM 76,—
163. Stoer/Witzgall: Convexity and Optimization in Finite Dimensions I. DM 54,—
164. Sario/Nakai: Classification Theory of Riemann Surfaces. DM 98,—
165. Mitrinović/Vasić: Analytic Inequalities. DM 88,—
166. Grothendieck/Dieudonné: Eléments de Géométrie Algébrique I. DM 84,—
167. Chandrasekharan: Arithmetical Functions. DM 58,—
168. Palamodov: Linear Differential Operators with Constant Coefficients. DM 98,—
170. Lions: Optimal Control Systems Governed by Partial Differential Equations. DM 78,—
171. Singer: Best Approximation in Normed Linear Spaces by Elements of Linear Subspaces. DM 60,—
172. Bühlmann: Mathematical Methods in Risk Theory. DM 52,—
173. F. Maeda/S. Maeda: Theory of Symmetric Lattices. DM 48,—
174. Stiefel/Scheifele: Linear and Regular Celestial Mechanics. Perturbed Two-body Motion—Numerical Methods—Canonical Theory. DM 68,—
175. Larsen: An Introduction to the Theory of Multipliers. DM 84,—
176. Grauert/Remmert: Analytische Stellenalgebren. DM 64,—
177. Flügge: Practical Quantum Mechanics I. DM 70,—
178. Flügge: Practical Quantum Mechanics II. DM 60,—
179. Giraud: Cohomologie non abélienne. DM 109,—
180. Landkoff: Foundations of Modern Potential Theory. In preparation
181. Lions/Magenes: Non-Homogeneous Boundary Value Problems and Applications I. DM 78,—
182. Lions/Magenes: Non-Homogeneous Boundary Value Problems and Applications II. In preparation
183. Lions/Magenes: Non-Homogeneous Boundary Value Problems and Applications III. In preparation
184. Rosenblatt: Markov Processes. Structure and Asymptotic Behavior. DM 68,—
185. Rubinowicz: Sommerfeldsche Polynommethode. DM 56,—
186. Wilkinson/Reinsch: Handbook for Automatic Computation II, Linear Algebra. DM 72,—
187. Siegel: Lectures on Celestial Mechanics. DM 78,—
188. Warner: Harmonic Analysis on Semi-Simple Lie Groups I. In preparation
189. Warner: Harmonic Analysis on Semi-Simple Lie Groups II. In preparation